Early Papers on Diffraction of X-rays by Crystals

Volume II

Early Papers

on

Diffraction of X-rays by Crystals

Volume II

Edited by

J. M. Bijvoet, W. G. Burgers and G. Hägg

1972

Springer-Science+Business Media, B.V.

by A. Oosthoek's Uitgeversmaatschappij N.V., Utrecht

ISBN 978-1-4615-6880-3   ISBN 978-1-4615-6878-0 (eBook)
DOI 10.1007/978-1-4615-6878-0

The first volume of *Early Papers on Diffraction of X-rays by Crystals* was published in 1969 just in time for its display at the Eighth General Assembly of the International Union of Crystallography at Stony Brook, U.S.A. As stated in the Prefaces, that volume contained only part of the total manuscript: Chapters I–V. The remainder of the manuscript, Chapters VI–XII, covering the development of X-ray crystallography in the 'trial-and-error' period, the (re-)birth of the Fourier method, and the discovery of the Patterson synthesis, is now published in this present Volume.

Although the manuscript for both volumes was completed in 1969, the Editors, Professors J. M. Bijvoet (Netherlands), W. G. Burgers (Netherlands) and G. Hägg (Sweden), have again spent much effort in preparing the last part of the manuscript as copy for the printer. The Union wishes to express its deep gratitude to them, and especially to Professor Bijvoet, who was very actively involved in the technical side of the production.

For a fuller introduction to the background and purpose of these Volumes, reference should be made to the Prefaces of Volume I.

Thanks are again due to authors and publishers who have kindly given the Union permission to reproduce excerpts from the original historic papers. The Union further expresses its appreciation to Unesco for the financial assistance received for the preparation of the manuscript from a contract under the Unesco Pilot Project on the Teaching of Crystallography in relation to the Physics and Chemistry of Solids.

February, 1972   THE INTERNATIONAL UNION OF CRYSTALLOGRAPHY

Preface

In the Preface to *Early Papers on Diffraction of X-rays by Crystals* Volume I (containing Chapters I–V and published in 1969), the history and planning of the complete book were outlined. The publication in two separate and consecutive volumes was merely a matter of management; the compilation of both volumes was done at the same time.

There is a distinct difference in subject-matter between both volumes: Volume I contains the fundamentals of the theory, while Volume II treats the practical development of the 'trial'-method and the genesis and first applications of the Fourier method. In the period covered by *Early Papers* (1912–1935), the trial method leads to the successful conquest of structures with up to a hundred parameters. We conclude the book with Patterson's discovery (1934) of the F^2-series as described in his second, more detailed and extended paper of 1935. With this method the apparatus was completed which led to the present undreamt-of successes of the Fourier method in the field of organic chemistry.

We have considered the inclusion of Robertson's famous synthesis of the structure of phtalocyanine (1936). However, we decided that its proper place would be at the beginning of a book which, no doubt, will appear one day, describing this later period.

Considerations of space caused us to give up the chapter on Texture planned at first.

Volume I concentrated on the discoveries and work of a relatively small number of physicists (von Laue, the Bragg's, Ewald, Darwin, Debije); as a contrast, the development of the practical X-ray analysis of Volume II shows the participation of a large number of physicists, chemists and mineralogists. While in Vol. I the reader feels the tension going with the solution of fundamental questions such as the character of the Laue diffraction, he encounters in Vol. II a more continuous, none the less brilliant, development leading to a penetration into the secrets of crystal structures as complicated as those of the silicates.

A few papers have the nature of a survey: paper **122** on the relation between chemical composition and crystal structure; **134** on photographic techniques; **154** on interstitial solutions; **166** on alloys and **173** on orthosilicates.

It was often difficult to make a choice from the many papers dealing with some interesting aspect of structure determination. Occasionally we had to choose between consecutive versions of a paper. It was not always the earlier one which we preferred, but the more general version instead (papers **134**,

169 and **196**; their texts include references to the respective earlier papers). Sometimes (paper **191** and **193**) it was the connection with papers preceding in our collection which led to our choice (in the category of paper **193** the syntheses of KH_2PO_4 (West) or $CuSO_4.5aq$. (Beevers and Lipson) would have been good alternatives).

As in Volume I we now list a selection of papers for the benefit of the reader who is interested in a shortened version of Volume II. He is suggested to read papers **101, 102, 106, 108, 114, 122, 127, 134, 135, 136, 138, 140**a, **144, 145, 146, 147, 150, 151, 154, 155, 158, 166, 167, 168, 169, 173, –178, 181** and chapters XI and XII.

In the interval between the appearance of Volume I and Volume II crystallography lost Sir Lawrence Bragg, Dame Kathleen Lonsdale and Professor J. D. Bernal; their papers form salient contributions to this book.

The editors, once again, wish to express their sincere thanks to Dr. G. Boom, former General Secretary of the Union, not only for his essential and most careful help in the reproduction of figures and the proof reading, but also for his unfailing interest in the book and very valuable advice.

February 1972

J. M. BIJVOET
W. G. BURGERS
G. HÄGG

Contents*

* The paper-numbers are found in the left-hand top of each page after the number of the
chapter. The papers are numbered starting with 101 in order to distinguish the numbering
from that of Vol. I.

CHAPTER VII

Atomic (Ionic) Radii and Building Principles

A1. IONIC RADII

A2. ATOMIC RADII

D. ALLOYS

CHAPTER X

Structure Determinations of Increasing Complexity

A. CONCLUSIONS FROM SYMMETRY ONLY

CONCLUSIONS FROM CELL-DIMENSIONS ONLY

CHAPTER XII

The Patterson Synthesis

CHAPTER VI

Symmetry

At the end of last century, the geometrical theory of crystallography found its final expression, and the labours of Federow, Schoenflies, *and* Barlow *had succeeded in establishing that there are 230 types of homogeneous structure to which the ultimate structural details of crystals must conform. Twenty more years elapsed before there appeared a method capable of pulling this purely geometrical theory to the practical test. The work of* Fedorov, Schoenflies *and* Barlow *has been justified, and it is satisfactory to know that there is at hand a complete theory, tested and confirmed again and again, which can be used fearlessly.*

W. T. ASTBURY and W. H. BRAGG
Ann. Rep. **20** (1923) p. 230

Jede wissenschaftliche Strukturanalyse muss mit der Raumsystembestimmung beginnen.

P. NIGGLI im Handbuch der
Experimentalphysik Bd XVII Teil 1, Leipzig 1928 p. 198

3

A. THE THEORY OF SPACE GROUPS

101. The Analytical Expression of the Results of the Theory of Space Groups, by R. W. G. WYCKOFF (1922). Introduction

■

102. Die Bestimmung der Kristallstruktur komplizierter Verbindungen, by P. NIGGLI (1918)

■

103. Geometrische Kristallographie des Diskontinuums, by P. NIGGLI (1919) [Only Space Group V_h^{26} reproduced. Ed.]

■

104. *l.c.* paper **101**, by R. W. G. WYCKOFF (1922) [Only Space Group V_h^{26} reproduced]

■

105. Tabulated Data for the Examination of the 230 Space Groups by Homogeneous X-rays, by W. T. ASTBURY and KATHLEEN YARDLEY (1924) General Introduction [Only Space Groups $V_h^{25} - V_h^{28}$ reproduced]

■

106. XVI Zur systematischen Strukturtheorie. I Eine neue Raumgruppen-symbolik, by C. HERMANN (1928)

■

107. Tagung des erweiterten Tabellenkomitees in Zürich, by P. P. EWALD (1930)

■

108. Sur le symbolisme des groupes de répetition ou de symmetrie des assemblages cristallins, by CH. MAUGUIN (1931)

■

109. Kristallbau und Chemische Konstitution, by K. WEISSENBERG (1925)

■

101[1]. In the days when an atomic structure of matter was a crude working hypothesis without any basis in experimentally determined fact, we find

---

[1] Most of the material for this introduction is given by L. Sohncke, Entwickelung einer Theorie der Krystallstruktur (Leipzig, 1879). It is given in English and brought up to date in a report of the Brit. Assoc. (1901) 297–337.

Robert Hooke[1] reproducing the forms of alum by properly piling up 'a company of bullets and some few other very simple bodies', very much as we represent the structure of a crystal on the basis of X-ray measurements.

It was the phenomenon of regular cleavage, however, that supplied the evidence upon which early hypotheses of the regular arrangement of the material of crystals were based. For instance, Westfeld[2] considered calcite as built up of tiny rhombohedrons; and Bergman[3], basing his beliefs partly on the observation of Gahn that a skalenohedron of calcite yields a rhombohedron on cleaving, developed what might be called the first geometrical theory of crystal structure. For just as the crystals of calcite could be considered as an aggregate of minute rhombohedrons placed parallel to one another, so garnet or pyrite or other crystals can be developed similarly from certain fundamental forms. These ideas seem to be essentially the same as those held by Hauy[4]. He, also, considered cleavage as the guiding factor. The cleavage units, his *molecules integrantes*, were either tetrahedra, triangular prisms, or parallelopipeda, and he showed how crystals with variously developed faces could be represented by the aggregation of these units. These ideas of Hauy were built around the law of rational indices, though they were fundamentally independent of it. Many objections to the details of the hypothesis of Hauy arose, as indeed they must arise against any theory based primarily upon cleavage. Not only does the existence of the many crystals which show no cleavage necessitate many supplementary hypotheses, but the observed cleavage of such substances as fluorite (with octahedral cleavage) is not readily accounted for by any kind of close-fitting units.

Simultaneously with the extension of the belief in the atomic nature of substances, and perhaps because of this belief, emphasis came to be shifted from the shape of the crystal units to the relative positions of their centers of gravity as centers of some sort of crystal molecules. Thus there evolved from these different speculations the basis for a suitable geometrical study in the definite conception of a crystal as composed of units of undefined shape repeated in some regular fashion throughout space.

In such a regular pattern for repeating the crystal unit we have a *space lattice*. All of the symmetrical networks of points which can have crystallographic symmetry were found geometrically by Frankenheim[5]. Some years

---

[1] *Micrographia* (London, 1665), p. 85.
[2] *Mineralogische Abhandlungen*, Stück I. (1767).
[3] *Nov. Acta. Reg. Soc. Sc. Upsal.* (1773), i; *Opusc.* (Upsala, 1780), ii.
[4] *Essai de Cristallographie* (Paris, 1772); etc.
[5] Die Lehre von der Cohäsion (Breslau, 1835).

later this was done more accurately and rigidly by Bravais[1]. As a result of his work, Bravais looked upon a crystal as built up by placing units of a suitable symmetry all in the same orientation at the points of one of these symmetrical networks. Thus the unit of a cubic crystal might have cubic or even tetrahedral symmetry, but it could not, for instance, have monoclinic or hexagonal symmetry. As a matter of fact, Bravais thought of his units as groups of atoms forming some sort of a crystal molecule, though such a view is not a necessary part of the geometrical development. In this theory of Bravais, in which a crystal is composed of aggregates of atoms repeated regularly and indefinitely through space, is to be found the beginning of an adequate treatment of the possible groupings of matter in crystalline bodies. The objections to Bravais' theory, however, are many and obvious. In the first place, all of the space lattices have the complete symmetry of some one of the crystal systems, so that, in order to account for the lower degrees of symmetry, it was necessary for him to ascribe the degradation in such cases to the shape of the crystal units, or molecules, without at the same time being able satisfactorily to treat these units. Again this theory implies a distinct restriction, and one which had not been proved necessary, that all of the crystal molecules must have the same orientation throughout the crystal.

In the course of a general study of the theory of groups of movements Jordan[2] gave a perfectly general method for defining all of the possible ways of regularly repeating an identical grouping of points indefinitely throughout space. By combining this treatment of Jordan with the principle (laid down by Wiener) that regularity in the arrangement of identical atoms is attained when 'every atom has the other atoms arranged about it in the same fashion', Sohncke[3] eventually deduced all of the typical ways of regularly repeating identical groupings of atoms throughout space so that the total assemblage will possess crystallographic symmetry[4]. This method of treatment in attacking the problem of the arrangement of the points within what was the crystal unit or molecule of Bravais brings the problem towards its final solution.

---

[1] *Journ. de l'École Polytech.* (Paris) **19** (1850) 127; **20** (1851) 102.

[2] *Annali di matematica pura ed applicata* (2) **2** (1869) 167, 215, 322.

[3] L. Sohncke, *op. cit.*

[4] At first Sohncke seems to have been inclined to view all of the points of a point system as regular and all of one kind. When the insufficiency of this theory was emphasized he postulated the presence of a few different kinds of points (which can be made to correspond with different kinds of atoms). The partial grouping composed of the points of any one kind is homogeneous; at the same time the different groupings all have the axes and the other elements of symmetry in common.

None of the systems of Sohncke can be made to account in an entirely satisfactory manner for the enantiomorphic (mirror-image) characteristics of many crystals. Schoenflies[1] was led to consider that every point of an assemblage must have all of the other points ranged about it in a 'like fashion', where 'likeness' may refer either to an identical arrangement or to a mirror-image similarity. Starting from this basis, he obtained the 230 space-groups which represent all of the possible typical ways of arranging (symmetry-less) points in space so that the grouping will possess the symmetry of one of the thirty-two crystal classes. The same derivation of the space groups was accomplished independently by Federov[2] and by Barlow, but at present the work of Schoenflies is the most useful because it is presented in a form that is of immediate application. With the aid of this final theory of space groups the different degrees of symmetry exhibited by crystals can at last be traced back definitely and precisely to the arrangement of the atoms in the crystals (without postulating any characteristics of symmetry for them).

Besides indicating the elements of symmetry which are characteristic of each of the 230 typical ways of arranging points in space, Schoenflies gives, in general terms, the coordinates of the points in each of these groupings which are equivalent to one another.

The discovery of the diffraction of X-rays and the consequent development of the physical methods for studying the structure of crystals have made this analytical expression of the results of the theory of space groups of the utmost importance.

102. Die allgemein geometrischen Eigenschaften der regelmässigen Punkt-systeme lassen sich nach den von Schoenflies gemachten Angaben vollkommen bestimmen. Gewöhnlich wird bei der Erforschung der Kristallstruk-turen als Hilfsmittel darauf zu wenig Rücksicht genommen. Es liegt das an der fehlenden Übersicht über die Eigenschaften der 230 Raumgrup-pen. In einem unter der Presse befindlichem Buch soll diesem Mangel abgeholfen werden. An Hand von Bestimmungstabellen wird die Mög-lichkeit geschaffen, das Raumsystem (die Raumgruppe) irgendeines kom-plizierten Kristalles sofort eindeutig zu bestimmen*.

Sobald nämlich Massenteilchen in allgemeinster Lage (mit mehr als

[1] A. Schoenflies, *Krystallsysteme u. Krystallstruktur* (Leipzig, 1891).
[2] E. Federov, *Z. Kryst.* **24** (1895) 209; W. Barlow, *Z. Kryst.* **23** (1894) 1. Federov's work appeared, in Russian, before that of either of the other two.
[3] A. Schoenflies, *Kristallsysteme und Kristallstruktur*, Leipzig 1891.
* [for definition of 'Raumgruppe' (space group), *see* this Vol. p. 8. Ed.].

zwei Freiheitsgraden) auftreten, ist sowohl nach dem Braggschen Verfahren als auch aus Laue-Photogrammen das Raumsystem einer gegebenen Symmetrieklasse* ohne weiteres eruierbar. Sind einmal das Raumsystem und die Zahl der Moleküle im Elementarparallelepiped bekannt, so lassen sich die Gleichungen der Ebenenserien und ihrer Belastungen aufstellen und aus den Intensitätsmessungen die Koordinaten der einzelnen Punkte direkt bestimmen.

1. *Das Braggsche Verfahren*

Die den Spektren 1. Ordnung entsprechenden Abstände mögen als Röntgenperioden bezeichnet werden. Die Bestimmung des Raumsystems eines Kristalles, dessen Symmetrieklasse man kennt, wird nun dadurch ermöglicht, daß jeder Raumgruppe individuelle Verhältnisse der Röntgenperioden zukommen. Ein Beispiel soll dies veranschaulichen:

Die Raumsysteme $\mathfrak{V}_h{}^{25}\mathfrak{V}_h{}^{26}\mathfrak{V}_h{}^{27}\mathfrak{V}_h{}^{28}$ [see the self-explanatory nomenclature of these space groups in paper **108** (and **106**) of this Vol. Ed.] besitzen die Symmetrie der rhombisch-holoedrischen Klasse und unterscheiden sich dadurch von allen übrigen Raumsystemen dieser Klasse, daß sie einem innenzentrierten Elementarparallelepiped zugeordnet sind. Die primitiven Translationen in Richtung der kristallographischen Achsen bezeichnen wir mit a, b, c und nennen das aus a, b, c gebildete Parallelepiped elementar.

Den Raumsystemen $\mathfrak{V}_h{}^{25}\mathfrak{V}_h{}^{26}\mathfrak{V}_h{}^{27}\mathfrak{V}_h{}^{28}$ ist nun gemeinsam, daß die Röntgenperioden senkrecht zu (100), (010), (001) bzw. $a/2$, $b/2$, $c/2$ sind und daß alle Röntgenperioden senkrecht zu Pyramidenflächen mit der Summe $(h + k + l) =$ eine ungerade Zahl ebenfalls der Hälfte der aus den Elementarabständen berechneten Größe entsprechen. Für Pyramidenflächen mit $(h + k + l) =$ eine gerade Zahl sind die Röntgenperioden den Elementarperioden gleich. Keine anderen Raumsysteme der rhombisch-holoedrischen Klasse besitzen derartige Beziehungen. Man bestimme somit von einem rhombisch-holoedrischen Kristall zunächst die Röntgenperioden (Abstände, die der 1. Ordnung der Spektren entsprechen) für (100) (010) (001) und zwei Pyramidenflächen wie (111) und (211) mit ungerader und gerader Indizessumme. Führt die Berechnung der Abstände

---

* [The 'Symmetrie Klassen' (crystal classes) denote the macroscopically distinguishable types of crystal symmetry; the 230 symmetry combinations for the arrangement of the atoms in the crystal (space groups) are combined into 32 crystal classes as macroscopically the distinctions vanish between: rotations axes and screw axes; mirror plane and glide planes; different centerings of the Bravais lattice. Ed.]

1. Ordnung für (211) auf ein Parallelepiped mit doppelt so großen Perioden als aus den von (100) (010) (001) und (111) berechnet wurde, so muß ein Raumsystem $\mathfrak{B}_h{}^{25}$ – $\mathfrak{B}_h{}^{28}$ vorliegen. Die Raumsysteme $\mathfrak{B}_h{}^{25}\mathfrak{B}_h{}^{26}\mathfrak{B}_h{}^{27}\mathfrak{B}_h{}^{28}$ unterscheiden sich nun weiterhin hinsichtlich der bei den Prismen auftretenden Verhältnisse der Röntgenperioden voneinander. (Einfluß von Gleitspiegelebenen.)

Es sind drei Prismenzonen ($hk0$) ($h0l$) ($0kl$) vorhanden. Der aus a, b, c (den Elementarperioden) berechnete Abstand zweier identischer Prismenflächen sei d. (d_{hk0}, d_{h0l}, d_{0kl}).

In $\mathfrak{B}_h{}^{25}$ zeigen die Prismen aller drei Zonen mit der Summe der Indizes = eine ungerade Zahl $d/2$ als Röntgenperiode (wobei d natürlich von den Indizes der Flächen abhängig ist).

In $\mathfrak{B}_h{}^{28}$ gilt dies für zwei Prismenzonen, während in der dritten Zone überhaupt alle Flächen $d/2$ als Röntgenperiode besitzen.

In $\mathfrak{B}_h{}^{26}$ haben alle Flächen zweier Prismenzonen $d/2$ als Röntgenperiode, hingegen nur diejenigen mit ungerader Indizessumme in der dritten Zone.

In $\mathfrak{B}_h{}^{27}$ schließlich besitzen die Prismenflächen aller drei Zonen, ohne Rücksicht auf die Indizes, $d/2$ als Röntgenperiode.

Durch die Untersuchung der Spektralabstände 1. Ordnung von sechs Prismenflächen ist somit die Unterscheidung zwischen $\mathfrak{B}_h{}^{25}$ – $\mathfrak{B}_h{}^{28}$ möglich.

Was hier an einem Beispiel gezeigt wurde, gilt ähnlich für alle Raumsysteme; die nötigen Bestimmungstabellen werden in Kürze der Benutzung dargeboten werden[1].

Kennt man nun das Raumsystem und die absoluten Größen der Elementarperioden, so läßt sich die Zahl der Atome pro Elementarparallelepiped berechnen. Anderseits besitzen die Raumsysteme Punktlagen verschiedener Zähligkeit pro Elementarparallelepiped. Die niedrigzähligen sind mit entsprechenden Symmetrieeigenschaften behaftet, ihnen ist auch nur eine beschränkte Zahl von Freiheitsgraden eigen. Wenden wir für die Symmetriebedingungen die Schönfliessche Nomenklatur an, so erhalten wir beispielsweise für $\mathfrak{B}_h{}^{25}$ – $\mathfrak{B}_h{}^{28}$ die kleine Tabelle I.

Das Raumsystem $\mathfrak{B}_h{}^{28}$ besitzt somit beispielsweise viererlei vierzählige Punktarten in genau bestimmter Lage, da wo die Symmetrieelemente der Kombination C_{2h} zusammenstoßen, ferner eine vierzählige Lage mit einem Freiheitsgrad auf der Schnittlinie zweier Spiegelebenen (C_{2v}). Außerdem zwei achtzählige Lagen, die an Drehungsachsen oder Spiegelebenen gebunden sind, währenddem die allgemeinste Lage im Elementarparallelepiped 16 zählig ist.

---

[1] In einem Buche: 'Geometrische Kristallographie des Diskontinuums', das im Verlage von Gebrüder Bornträger, Berlin erscheinen wird. [*See* this Vol. paper **103**.]

Tabelle I.

Raumgruppe	Zähligkeiten und Symmetriebedingungen von Punktlagen ohne Freiheitsgrad (Die Buchstabenindizes sind durchweg unten rechts hingeschrieben worden.)	Zähligkeiten und Symmetriebedingungen von Punktlagen mit einem Freiheitsgrad	Zähligkeit und Symmetriebedingung von Punktlagen mit zwei Freiheitsgraden	Zähligkeit der allgemeinsten Lage ohne Symmetriebedingung
$\mathfrak{B}_h{}^{25}$	$2 = V_h; 2 = V_h;$ $2 = V_h; \ 2 = V_h;$ $8 = C_i$	$4 = C_{2v}$	$8 = C_s$	16
$\mathfrak{B}_h{}^{26}$	$4 = V; \ \ 4 = V;$ $4 = C_{2h}; 4 = C_{2h};$ $8 = C_i$	$8 = C_2$	$8 = C_s$	16
$\mathfrak{B}_h{}^{28}$	$4 = C_{2h}; 4 = C_{2h};$ $4 = C_{2h}; 4 = C_{2h};$	$4 = C_{2v}; 8 = C_2$	$8 = C_s$	16
$\mathfrak{B}_h{}^{27}$	$8 = C_i; \ \ 8 = C_i;$	$8 = C_2$	—	16

[explanation of symbols: $C_i \bar{1}$ (symmetry center); C_2 2 (two fold axis); C_s m (mirror plane); C_{2h} $2/m$; C_{2v} mm; V 222; V_h mmm. Ed.]

Kennen wir nun, nachdem das Raumsystem bestimmt wurde, die Zahl der Atome pro Elementarparallelepiped und sind gewisse Schlußfolgerungen über innerkonstitutionelle chemische Zusammensetzung zulässig, so wird es im allgemeinen nicht schwer halten die Atomschwerpunktslagen auf die einzelnen Punktarten zu verteilen.

Es sei beispielsweise das Raumsystem $\mathfrak{B}_h{}^{25}$ gefunden worden. Auf das Elementarparallelepiped kommen zwei Moleküle einer Verbindung AB_8, also 2 A-Atome und 16 B-Atome. Es sei kein Grund vorhanden anzunehmen, daß es zweierlei Sorten von B-Atomen gebe, dann müssen A einer zweizähligen Punktlage, B einer 16 zähligen Punktlage angehören.

Den A-Atomen kommen die Koordinaten 000 und $\frac{1}{2}\frac{1}{2}\frac{1}{2}$ zu. Von B sind nun die drei Koordinaten x, y, z eines Punktes zu bestimmen. x, y, z beziehen sich auf die Koordinatenachsen a, b, c mit den Größen a, b, c als Einzelmaßstäbe und dem Nullpunkt in einer zweizähligen Punktlage. Sie müssen also echte Brüche sein. Kennen wir die Koordinaten eines der 16 zusammengehörigen Punkte, so sind auch die Koordinaten der übrigen 15 gegeben, da jedes Raumsystem ein individuelles und bekanntes Koordinatensystem zusammengehöriger Punkte besitzt.

Daraus sind nun sofort die Gleichungen für die Ebenenserien und ihre Belastungen parallel irgendeiner Fläche bestimmbar.

103.
*In his 'Geometrische Kristallographie des Diskontinuums',
P.* Niggli *has given a considerable number of the special
cases of the space-groups, and has specified the positions
within the unit cell of each space-group of all its symmetry
elements.*

A. E. H. TUTTON,
Ann. Rep. **19** (1922) p. 252

Raumgruppe $\mathfrak{B}_h{}^{26}$ (Siehe Fig. 104).

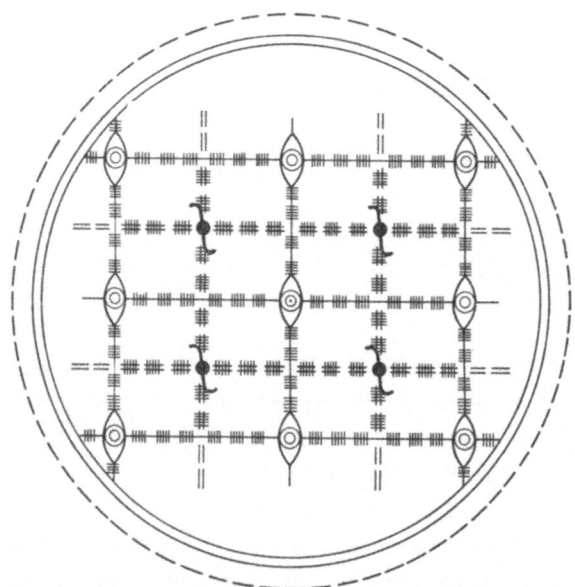

Fig. 104. $V_h{}^{26}$ Ebene $(001)_0$. Die zwei zusammengehörigen Vollständigen Großkreise
entsprechen Spiegelebenen $(001)_{\frac{1}{4}}$ $(001)_{\frac{3}{4}}$. Der gestrichelte Großkreis markiert die Gleitspiegelebenen $(001)_0$ und $(001)_{\frac{1}{2}}$.

|||| |||| |||| Gleitspiegelebenen
$()$: digonale Drehungsachsen $\perp$ Grundebene
$\{$: digonale Schraubenachsen $\perp$ Grundebene
ausgezogene Linien: digonale Drehungsachsen in der Grundebene $(100)_0$ und $(100)_{\frac{1}{2}}$.
gestrichelte Doppellinien: digonale Schraubenachsen, die in $\frac{1}{4}$ und $\frac{3}{4}$ Höhe über die
Grundebene gehen.
ausgefüllte kleine schwarze Kreisen: Symmetriezentren in der Ebene $(001)_0$ und $(001)_{\frac{1}{2}}$;
die kleinen Doppelkreise: in den Abständen $\frac{1}{4}$ und $\frac{3}{4}$ befindliche Symmetriezentren.

Symmetriezentren:

I. $00\frac{1}{4}$ $00\frac{3}{4}$ $\frac{1}{2}\frac{1}{2}\frac{3}{4}$ $\frac{1}{2}\frac{1}{2}\frac{1}{4}$. II. $\frac{1}{2}0\frac{1}{4}$ $\frac{1}{2}0\frac{3}{4}$ $0\frac{1}{2}\frac{3}{4}$ $0\frac{1}{2}\frac{1}{4}$.
III. $\frac{1}{4}\frac{1}{4}\frac{1}{2}$ $\frac{3}{4}\frac{3}{4}0$ $\frac{3}{4}\frac{3}{4}\frac{1}{2}$ $\frac{1}{4}\frac{1}{4}0$ $\frac{3}{4}\frac{1}{4}\frac{1}{2}$ $\frac{1}{4}\frac{3}{4}0$ $\frac{1}{4}\frac{3}{4}\frac{1}{2}$ $\frac{3}{4}\frac{1}{4}0$.

Digonale Drehungsachsen:

$[0001]_{00}$ $[001]_{\frac{1}{2}\frac{1}{2}}$; $[001]_{\frac{1}{2}0}$ $[001]_{0\frac{1}{2}}$. $[100]_{00}$ $[100]_{\frac{1}{2}\frac{1}{2}}$ $[100]_{\frac{1}{2}0}$ $[100]_{0\frac{1}{2}}$.
$[010]_{00}$ $[010]_{\frac{1}{2}\frac{1}{2}}$ $[010]_{\frac{1}{2}0}$ $[010]_{0\frac{1}{2}}$.

Digonale Schraubenachsen:

$[001]_{\frac{1}{4}\frac{1}{4}}$ $[001]_{\frac{3}{4}\frac{3}{4}}$ $[001]_{\frac{3}{4}\frac{1}{4}}$ $[001]_{\frac{1}{4}\frac{3}{4}}$. $[100]_{\frac{1}{4}\frac{1}{4}}$ $(100)_{\frac{3}{4}\frac{3}{4}}$ $[100]_{\frac{3}{4}\frac{1}{4}}$ $[100]_{\frac{1}{4}\frac{3}{4}}$.
$[010]_{\frac{1}{4}\frac{1}{4}}$ $[010]_{\frac{3}{4}\frac{3}{4}}$ $[010]_{\frac{3}{4}\frac{1}{4}}$ $[010]_{\frac{1}{4}\frac{3}{4}}$.

Spiegelebenen: $(001)_{\frac{1}{4}}$ $(001)_{\frac{3}{4}}$.

Gleitspiegelebenen:

$(001)_0$ $(001)_{\frac{1}{2}}$ Gleitkomp. $\dfrac{a}{2}+\dfrac{b}{2}$; $\left.\begin{array}{l}(100)_0\,(100)_{\frac{1}{2}}\\(010)_0\,(010)_{\frac{1}{2}}\end{array}\right\}$ Gleitkomp. $c/2$;

$(100)_{\frac{1}{4}}$ $(100)_{\frac{3}{4}}$ Gleitkomp. $b/2$; $(010)_{\frac{1}{4}}$ $(010)_{\frac{3}{4}}$ Gleitkomp. $a/2$.

Zusammengehörige Koordinatenwerte:

$$m, n, p \quad \bar{m}, \bar{n}, p \quad \bar{m}, n, \bar{p} \quad m, \bar{n}, \bar{p}$$
$$\bar{m}, \bar{n}, \bar{p}+\tfrac{1}{2} \quad m, n, \bar{p}+\tfrac{1}{2} \quad m, \bar{n}, p+\tfrac{1}{2} \quad \bar{m}, n, p+\tfrac{1}{2}$$
$$m+\tfrac{1}{2}, n+\tfrac{1}{2}, p+\tfrac{1}{2} \quad \bar{m}+\tfrac{1}{2}, \bar{n}+\tfrac{1}{2}, p+\tfrac{1}{2} \quad \bar{m}+\tfrac{1}{2}, n+\tfrac{1}{2}, \bar{p}+\tfrac{1}{2}$$
$$m+\tfrac{1}{2}, \bar{n}+\tfrac{1}{2}, \bar{p}+\tfrac{1}{2}$$
$$\bar{m}+\tfrac{1}{2}, \bar{n}+\tfrac{1}{2}, \bar{p} \quad m+\tfrac{1}{2}, n+\tfrac{1}{2}, \bar{p} \quad m+\tfrac{1}{2}, \bar{n}+\tfrac{1}{2}, p \quad \bar{m}+\tfrac{1}{2}, n+\tfrac{1}{2}, p.$$

Punktlagen ohne Freiheitsgrad.

1. $0\,0\,0$ $0\,0\,\tfrac{1}{2}$ $\tfrac{1}{2}\tfrac{1}{2}\tfrac{1}{2}$ $\tfrac{1}{2}\tfrac{1}{2}0$ Zähligkeit $= 4$. Symmetriebedingung $= V$.
2. $\tfrac{1}{2}00$ $\tfrac{1}{2}0\tfrac{1}{2}$ $0\tfrac{1}{2}\tfrac{1}{2}$ $0\tfrac{1}{2}0$ Zähligkeit $= 4$. Symmetriebedingung $= V$.
   Diese beiden Punktlagen bilden jeweilen Gitter von der Form des Elementarparallelepipeds mit zentrierter Basisfläche, zentrierten Kanten [001] und Innenzentrierung.
3. $= \text{I von Symmetriezentren}\big\}$ Zähligkeit jeweilen $= 4$. Symmetrie-
4. $= \text{II von Symmetriezentren}\big\{$ bedingung $= C_{2h}$.
   Auch sie bilden Gitter von der gleichen Form wie 1 und 2.
5. $= \text{III von Symmetriezentren}$. Zähligkeit $= 8$. Symmetriebeding. $= C_i$.
   Das von ihnen gebildete Gitter hat die Form eines Elementarparallelepipeds mit vollständig zentrierten Flächen und Kanten, sowie mit Innenzentrierung.

Punktlagen mit einem Freiheitsgrad. Sie stellen sich bei beliebiger übrigbleibender Lage auf einer der digonalen Drehungsachsen ein. Zähligkeit $= 8$. Symmetriebedingung $= C_2$.

Punktlagen mit zwei Freiheitsgraden. Sie besitzen entweder ein Symbol $m, n, \tfrac{1}{4}$ oder $m, n, \tfrac{3}{4}$. Zähligkeit $= 8$. Die Symmetriebedingung lautet $= C_s$.

Übrigbleibende Punktlagen sind mit drei Freiheitsgraden (ohne Symmetriebedingung) 16 zählig.

Bipyramiden mit der Indizessumme $= 2i$ besitzen normales R und doppelte Belastung, die übrigen ein R von $d/2$ mit einfacher Belastung. Alle Pinakoide besitzen halbelementare Röntgenperiode und vierfache Belastung.

1. In allen drei Prismenzonen treten normale R auf, wenn die Indizessumme gerade ist. $\mathfrak{B}_h{}^{25}$.

2. Das gilt nur für zwei Prismenzonen, während in der anderen Zone alle Flächen bei bloß doppelter Belastung $R = d/2$ haben. $\mathfrak{B}_h{}^{28}$.

3. Nur eine Prismenzone besitzt bei einer Indizessumme von $2i$ normale R, alle Flächen der beiden anderen Zonen haben halbelementare Röntgenperioden. $\mathfrak{B}_h{}^{26}$.

4. Alle drei Prismenzonen besitzen nur Flächen mit doppelter Belastung und $R = d/2$. $\mathfrak{B}_h{}^{27}$.

104.

A memoir of 180 pages in book form by R. W. G. Wyckoff was published by the Carnegie Institute of Washington, in which all the special cases of the space-groups have been worked out. The matter contained is largely tabular, giving the coordinates of the most generally placed equivalent points and all the special cases of these equivalent points, contained within the unit of structure of each of the 230 space-groups. In this form the information will be immediately available to X-ray analysts. This book is, therefore, one of very considerable value.

A. E. H. TUTTON,
Ann. Rep. **19** (1922) p. 252

SPACE-GROUPS AND CRYSTALS

Every crystal, considered as a regular arrangement of atoms in space, must possess the symmetry of some one of the 230 space-groups. The theory of space-groups, then, supplies a method with the aid of which it should be possible to represent all of the ways in which the atoms of a crystal can be arranged in space. If an atom of a crystal occupies such a position that it corresponds with the coordinate position xyz of an equivalent point of the space-group having the symmetry of the crystal, then symmetry demands that exactly similar atoms shall be found at positions corresponding to those of each of the other equivalent points of the space-group. If a compound were of the type AB, where A is one kind of atom and B another, and if the atoms of A occupy the most general equivalent positions one of which is xyz, then there will be as many chemical molecules of AB associated with the unit prism as there are equivalent points in the unit. This number may under certain conditions be relatively great.

If, however, the values of x, y and z which express the positions of the atoms of A and B are such that the atoms lie upon some element of symmetry, two or more of the equivalent positions coincide and this number of molecules to be placed within the unit cell will be reduced. For instance if a point were to lie upon a plane of symmetry, it would of course be identical with its mirror image; or if it stood in a three-fold axis of symmetry, three or four of the equivalent points would occupy the same position.

The results of all of the X-ray experimentation which has thus far been carried out seem to point to the fact that this number of chemical molecules to be contained within a unit cell is in all probability very much less than the number of most generally placed equivalent positions. As a consequence the determination of these special cases of the space-groups becomes of the utmost importance to the person interested in the structure of crystals. It is the purpose of the present work to give these results of the theory of space-groups a detailed expression, thereby putting them into a form in which they will be immediately useful as an aid to the study of the arrangement of the atoms in crystals. [We limit our reproduction to the space-group V_h^{26}. Ed.]

Space-Group V_h^{26}.
Four equivalent positions:

(a) 000; $00\frac{1}{2}$; $\frac{1}{2}\frac{1}{2}\frac{1}{2}$; $\frac{1}{2}\frac{1}{2}0$.
(b) $\frac{1}{2}00$; $\frac{1}{2}0\frac{1}{2}$; $0\frac{1}{2}\frac{1}{2}$; $0\frac{1}{2}0$.
(c) $00\frac{1}{4}$; $00\frac{3}{4}$; $\frac{1}{2}\frac{1}{2}\frac{3}{4}$; $\frac{1}{2}\frac{1}{2}\frac{1}{4}$.
(d) $0\frac{1}{2}\frac{1}{4}$; $0\frac{1}{2}\frac{3}{4}$; $\frac{1}{2}0\frac{3}{4}$; $\frac{1}{2}0\frac{1}{4}$.

Eight equivalent positions:

(e) $\frac{1}{4}\frac{1}{4}0$; $\frac{1}{4}\frac{3}{4}0$; $\frac{3}{4}\frac{1}{4}0$; $\frac{3}{4}\frac{3}{4}0$;
     $\frac{3}{4}\frac{3}{4}\frac{1}{2}$; $\frac{3}{4}\frac{1}{4}\frac{1}{2}$; $\frac{1}{4}\frac{3}{4}\frac{1}{2}$; $\frac{1}{4}\frac{1}{4}\frac{1}{2}$.
(f) $u\,0\,0$; $u\,0\,\frac{1}{2}$; $u+\frac{1}{2},\frac{1}{2},0$; $\quad u+\frac{1}{2},\frac{1}{2},\frac{1}{2}$;
     $\bar{u}\,0\,0$; $\bar{u}\,0\,\frac{1}{2}$; $\frac{1}{2}-u,\frac{1}{2},0$; $\quad \frac{1}{2}-u,\frac{1}{2},\frac{1}{2}$.
(g) $0\,u\,0$; $0\,u\,\frac{1}{2}$; $\frac{1}{2},u+\frac{1}{2},0$; $\quad \frac{1}{2},u+\frac{1}{2},\frac{1}{2}$;
     $0\,\bar{u}\,0$; $0\,\bar{u}\,\frac{1}{2}$; $\frac{1}{2},\frac{1}{2}-u,0$; $\quad \frac{1}{2},\frac{1}{2}-u,\frac{1}{2}$.
(h) $0\,0\,u$; $\frac{1}{2}\frac{1}{2}u$; $\frac{1}{2},\frac{1}{2},u+\frac{1}{2}$; $\quad 0,0,u+\frac{1}{2}$;
     $0\,0\,\bar{u}$; $\frac{1}{2}\frac{1}{2}\bar{u}$; $\frac{1}{2},\frac{1}{2},\frac{1}{2}-u$; $\quad 0,0,\frac{1}{2}-u$.
(i) $0\frac{1}{2}u$; $\frac{1}{2}0\,u$; $\frac{1}{2},0,u+\frac{1}{2}$; $\quad 0,\frac{1}{2},u+\frac{1}{2}$;
     $0\frac{1}{2}\bar{u}$; $\frac{1}{2}0\,\bar{u}$; $\frac{1}{2},0,\frac{1}{2}-u$; $\quad 0,\frac{1}{2},\frac{1}{2}-u$.
(j) $u\,v\,\frac{1}{4}$; $u\,\bar{v}\,\frac{3}{4}$; $u+\frac{1}{2},\frac{1}{2}-v,\frac{1}{4}$; $u+\frac{1}{2},v+\frac{1}{2},\frac{3}{4}$;
     $\bar{u}\,\bar{v}\,\frac{1}{4}$; $\bar{u}\,v\,\frac{3}{4}$; $\frac{1}{2}-u,v+\frac{1}{2},\frac{1}{4}$; $\frac{1}{2}-u,\frac{1}{2}-v,\frac{3}{4}$.

Sixteen equivalent positions:

(k) xyz; $x\bar{y}\bar{z}$; $\bar{x}y\bar{z}$; $\bar{x}\bar{y}z$;

$\bar{x}, \bar{y}, \frac{1}{2}-z$; $\bar{x}, y, z+\frac{1}{2}$; $x, \bar{y}, z+\frac{1}{2}$; $x, y, \frac{1}{2}-z$;

$x+\frac{1}{2}, y+\frac{1}{2}, z+\frac{1}{2}$; $x+\frac{1}{2}, \frac{1}{2}-y, \frac{1}{2}-z$; $\frac{1}{2}-x, y+\frac{1}{2}, \frac{1}{2}-z$;

$$\frac{1}{2}-x, \frac{1}{2}-y, z+\frac{1}{2};$$

$\frac{1}{2}-x, \frac{1}{2}-y, \bar{z}$; $\frac{1}{2}-x, y+\frac{1}{2}, z$; $x+\frac{1}{2}, \frac{1}{2}-y, z$; $x+\frac{1}{2}, y+\frac{1}{2}, \bar{z}$.

105.

Last year, R. W. G. Wyckoff gave us a complete analytical expression of the theory of space-groups—an invaluable work. This year, the subject has been approached from a different point of view, one which we venture to hope will prove intelligible and immediately useful to physicists and chemists even of negligible training in mathematical crystallography. A simple principle has been used to deduce the abnormal spacings to be expected from each of the 230 space-groups when examined by X-rays, thus providing a complete list to which reference can be made with great facility. Together with this table are given all the possible molecular (and therefore ionic and atomic) symmetries for all possible cases, a list which, we trust, will appeal strongly to chemists. Finally, diagrams are given showing in the simplest possible manner all the 230 ways in which molecules can unite to form the various types of crystal structure. With the aid of these the crystal-analyst can readily form a mental picture of what sort of combination he is dealing with. To many—this seems to be characteristic of the science of our own country—such an opportunity is always desirable.

W. T. ASTBURY and W. H. BRAGG,
Ann. Rep. **21** (1924) p. 220

General Introduction

The object of the present paper is to express the conclusions of mathematical crystallography in a form which shall be immediately useful to workers using homogeneous X-rays for the analysis of crystal structures. The results are directly applicable to such methods as the Bragg ionisation method, the powder method, the rotating crystal method, etc., and summarise in as compact a form as possible what inferences may be made from the experimental observations, whichever one of the 230 possible space-groups may happen to be under examination.

It is only in certain cases that the spacings of crystal planes as determined by the aid of homogeneous X-rays agree with the values of those spacings which would be expected from ordinary crystallographic calculations. In the majority of cases the relative arrangement of the molecules in the unit cell leads to apparent anomalies in the experimental results, the observed spacings of certain planes or sets of planes being submultiples of the calculated spacings. The simplest case of such an apparent anomaly is found in the space-group $C_2{}^2$ of the monoclinic system, where the presence of a two-fold screw-axis, because it interleaves halfway the (010) planes by molecules which are exactly like those lying in the (010) planes, except that they have been rotated through 180°, leads to an observed periodicity which is half the periodicity to be inferred from the dimensions of the unit cell, that is, leads to an observed spacing for (010) which is half the calculated. All screw-axes produce similar results, and, in general, a p-fold screw-axis leads to an observed spacing for the plane perpendicular to it which is $1/p$th that to be inferred from the dimensions of the cell. Besides those produced by the screw-axes, other abnormalities arise out of the presence of glide-planes. [For glide-plane extinctions see this Vol. p. 10 Ed.]

A more obvious way by which the periodicity of crystal planes may be reduced is by the interleaving of *completely* identical molecules such as occurs in those Bravais lattices which have similar and similarly orientated molecules at the face-centres or the body-centre of the unit cell. In these cases the interleaving molecules are identical from all points of view and the periodicity of certain sets of general planes is reduced. In a lattice in which there is an identical molecule at the centre of the (001) face, the spacings of all planes $\{hkl\}$ for which $(h + k)$ is odd will be reduced by half, while for a body-centred lattice the same is true for all planes $\{hkl\}$ for which $(h + k + l)$ is odd, and for a face-centred lattice all planes $\{hkl\}$ for which $(h + k)$ or $(k + l)$ or $(l + h)$ is odd. Such abnormalities are well known and must be allowed for in addition to those arising out of the screw-axes and glide-planes.

The conclusions to be drawn from such arguments as are given above hold for the space-groups whatever the material from which they are built up, whether it be ions, molecules or polymers of molecules. They are the conclusions arising out of the theory of mathematical crystallography which in the general case concerns itself, not with the nature of the ultimate asymmetric units of crystalline structures, but with the groups of operations by which infinite systems of such units may be brought into coincidence.

Given a crystal belonging to a particular system and class, the X-ray worker can, with two exceptions, determine to which space-group the

17

crystal belongs, by means of the number of molecules per unit cell and the abnormal spacings found.

In order to facilitate this work, diagrams are given, one for each of the 230 space-groups, showing in as condensed a form as possible the distribution of symmetry elements and the relative positions and orientations of the molecules in the unit cell; tables are accompanying the diagrams [only cases 71–74 reproduced. Ed.]. The diagrams show the ways in which Nature, from completely asymmetric material, builds up crystals of highly symmetrical form and properties. The small arrows marked by the letters u and d represent asymmetric molecules, but it is clear that the theory is unaffected by the substitution of a molecule of p-fold symmetry for p asymmetric molecules of suitable relative orientations.

Description of Tables and Diagrams

The numbers in the first column of the tables [*see* p. 20] correspond to Figs. 1–230. The second column gives the name of the space-group adopted in Hilton's 'Mathematical Crystallography', the third the underlying Bravais lattice, the fourth the number of asymmetric molecules per cell, the fifth the abnormal spacings to be expected, and the sixth and seventh the possible molecular symmetry. If n per cell is the number of asymmetric molecules required to produce the symmetry of the structure, there will be required only n/p molecules of symmetry p. In the majority of crystals the ultimate units undoubtedly correspond in substance to single chemical molecules, and it is chiefly for the elucidation of such structures that the columns described 'Possible Molecular Symmetry' will be found useful.

Figs. 1–230 are in all cases projections on the basal plane (001), (111) or (0001). The explanation of the notation used in the seventh column is as follows: C_1, centre; P, plane; 2-A, dyad axis; (...). The development follows that given in Hilton's 'Mathematical Crystallography'.

Molecules marked with the subscript $_2$ are enantiomorphous to those marked with the subscript $_1$, that is, either of the two can be obtained from the other by an operation involving a reflection. Molecules marked u_1 can be obtained from molecules marked d_1 by rotation about a dyad axis which is parallel to the plane of the paper. This nomenclature simply means that if we think of the molecule u_1 as pointing up, the molecule d_1 is obtained from it by a rotation through 180° (plus a possible translation parallel to the axis) and is therefore pointing down to a corresponding extent. Thick lines denote planes of symmetry perpendicular to the plane of the paper, dotted lines denote glide-planes of symmetry. Axes of symmetry perpendicular to the plane of the paper are represented as follows: dyad rotation-axis ● ; dyad screw-axis, ⟜ ; (...). Centres of symmetry are indicated by crosses.

As mentioned above, the unit cell in these diagrams must be considered as defined by the elements of symmetry. Axes of symmetry actually lying in the top plane of the cell are then marked by thick arrows (e.g., ⟩—→ for a dyad rotation-axis, ⟩⟩—→ for a dyad screw-axis, etc.), while those lying in a plane lower down are drawn with thin lines. If we now commence with a standard molecule, say u_1, near the top left-hand corner of the cell, it must be remembered that its position is arbitrary and it will not in general lie in the top plane of the cell as defined by the elements of symmetry. This standard molecule is drawn with full line, as are also all molecules derived from it by the symmetry elements lying in the top plane of the cell and those planes of pure reflection and axes of pure rotation which lie perpendicular to this plane. In the majority of Figures 1–230 only two kinds

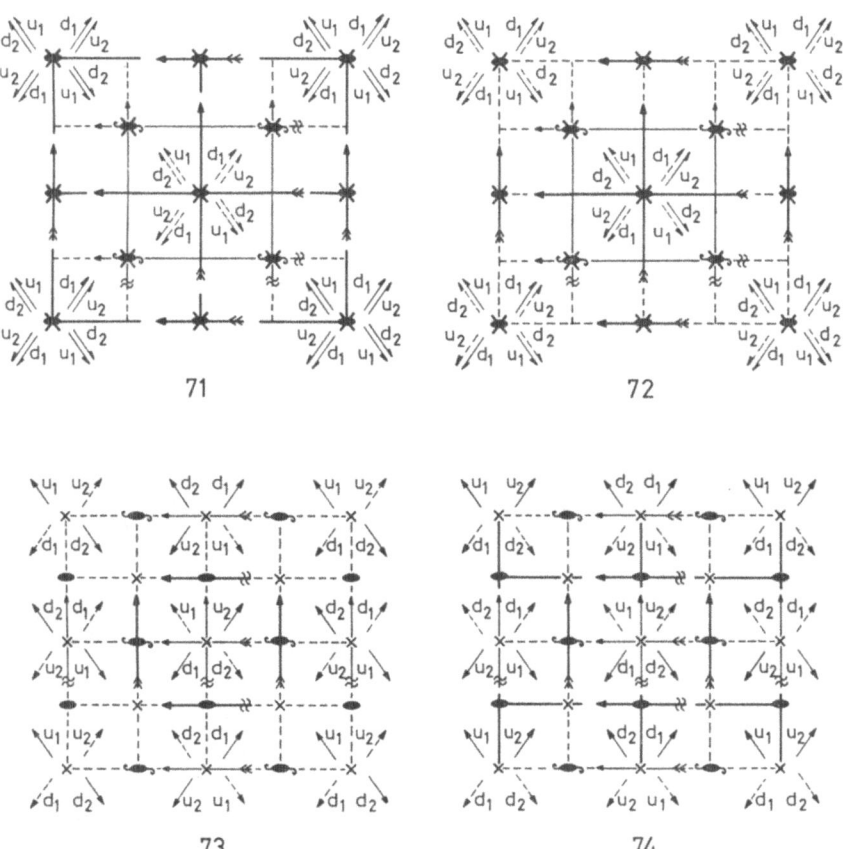

71   72

73   74

No.	Space Group	Bravais Lattice	n.	Abnormal Spacings.	p.	Possible Molecular Symmetry.
71	Q_h^{25}	Γ_0''' [body centred]	16	$\{hkl\}$ halved if $(h+k+l)$ is odd	2	$P\|^l\{100\}$, $\{010\}$ or $\{001\}$; C.
					4	2 P's intersecting in 2–A.
					8	3 P's intersecting in 3 2–A's.
72	Q_h^{26}	Γ_0'''	16	Same as Q_h^{25}; also all $\{0kl\}$, $\{h0l\}$ $\{h0l\}$ halved	2	2–A $\perp^r\{100\}$, $\{010\}$ or $\{001\}$; $P\|^l\{001\}$; C.
					4	3 mutually $\perp^r$2–A's; 2–A $\perp^r$ $P\|^l\{001\}$.
73	Q_h^{27}	Γ_0'''	16	Same as Q_h^{25}; also all $\{0kl\}$, $\{hk0\}$ halved	2	2–A $\perp^r\{100\}$, $\{010\}$ or $\{001\}$; C.
74	Q_h^{28}	Γ_0'''	16	Same as Q_h^{25}, also all $\{hk0\}$ halved	2	2–A $\perp^r\{100\}$ or $\{010\}$; $P\|^l\{100\}$ or $\{010\}$.
					4	2–A $\perp^r$ $P\|^l\{100\}$ or $\{010\}$; 2 P's $\|^l\{100\}$ and $\{010\}$ intersecting in 2–A.

of molecules will be observed, full and dotted. The full ones are as just described, while the dotted ones bear a similar relation to the plane half-way down the cell—that is, they can be obtained from the standard molecule by the operation of a symmetry element perpendicular to the plane of the paper and involving a translation of $c/2$.

The authors wish to express their indebtedness to Sir WM. BRAGG for his unfailing encouragement and advice.

106. Die bisher übliche Symbolik der Raumgruppen nach Schoenflies hat den Nachteil, daß die verschiedenen Raumgruppen, die derselben Kristall-symmetrie isomorph sind, einfach laufend durchnumeriert werden, ohne daß man ihre charakteristischen Symmetrieelemente aus dem Symbol er-kennen kann. Man ist daher bei Arbeiten über Kristallstrukturen darauf angewiesen, sich entweder die Bedeutung der 230 Schoenfliesschen Raum-gruppensymbole gedächtnismäßig einzuprägen oder ein umfangreiches Tabellenwerk zu benutzen.

Die folgende Symbolik beruht auf einer neuartigen Ableitung der 230 Raumgruppen und bezweckt jede Raumgruppe durch wenige charakteristische Symmetrieelemente kenntlich zu machen.

Die Symbole der Raumgruppen enthalten:

1. Den Namen der Kristallklasse.

2. Die Translationsgruppe.

Raumgruppen, in denen die Zelle einfach primitiv ist, erhalten das Symbol P, Zentrierung der a-, b- und c-Fläche wird durch die Symbole A, B bzw. C angedeutet, allseitige Flächenzentrierung durch F und Innenzentrierung durch I.

3. Die charakteristischen Symmetrieelemente.

Spiegelebenen können entweder echte oder Gleitspiegelebenen sein. Zu ihrer Angabe wird ein Koordinatenkreuz benutzt, von dem zwei Achsen in die Spiegelebene fallen. Dann sind die folgenden Gleitkomponenten möglich: $0\,0$ (echte Spiegelebene), $\frac{1}{2}\,0$, $0\,\frac{1}{2}$ und $\frac{1}{2}\,\frac{1}{2}$ immer, außerdem, wenn parallel zu der Spiegelebene eine zentrierte Fläche liegt, auch noch $\frac{1}{4}\,\frac{1}{4}$ oder $\frac{1}{4}\,\frac{3}{4}$. Eine echte Spiegelebene erhält das Symbol m; eine Gleitspiegelebene mit den Gleitkomponenten $\frac{1}{2}\,\frac{1}{2}$ wird mit n bezeichnet. Die übrigen Gleitspiegelebenen erhalten die Symbole a, b, c, d je nachdem die Gleitkomponente in die Richtung der a-, b- oder c-Achse fällt (Gleitkomponenten der Form $(0\,\frac{1}{2})$ oder in eine diagonale Richtung $(\frac{1}{4}\,\frac{1}{4})$.

Achsen können Dreh- oder Schraubenachsen sein. Als Schraubungskomponenten treten auf: 0 (Drehachse), $\frac{1}{2}$, $\frac{1}{3}$, $\frac{2}{3}$, $\frac{1}{4}$, $\frac{3}{4}$, $\frac{1}{6}$, $\frac{5}{6}$. Die entsprechenden Symbole sind: 1, 2, 3, $\bar{3}$, 4, $\bar{4}$, 6 und $\bar{6}$.

Auslöschungen. Die charakteristischen Auslöschungen einer Raumgruppe sind aus diesen Raumgruppensymbolen sofort abzulesen.

107.   *Tagung des erweiterten Tabellenkomitees in Zürich,*
28.–31. Juli 1930

Das in London eingesetzte Tabellenkomitee bestehend aus den Herren Astbury, Bernal, Hermann, Mauguin, Niggli und Wyckoff, einberufen durch Herrn Bernal, hielt am 31. Juli eine Sitzung im Mineralogischen Institut in Zürich ab. An den vorhergehenden Tagen hatten eingehende Diskussionen über die Ausgestaltung gemeinsamer internationaler Tabellen zur Strukturbestimmung ebendort stattgefunden, zu denen durch Einladung des Komitees die Herren Brandenberger, Ewald, Kolkmeijer, Parker, Pauling, Schiebold, Schleede und Schneider zugezogen waren.

Der Vorsitz bei den Verhandlungen wurde Herrn Ewald übertragen. Sitzungen fanden am 28., 29., 30. und 31. Juli statt, im ganzen etwa 25

Stunden. Außerdem wurden manche Fragen auch außerhalb der Sitzungen vorbereitet. Die Teilnehmer hatten zudem ein detailliertes Sachprogramm erhalten, das von Bernal, Ewald, Hermann und Mauguin vorbereitet war und dem die Verhandlungen sich eng anschließen konnten.

Der erste Vormittag brachte bereits eine für alles Weitere sehr wichtige Entscheidung, nämlich die Einigung auf eine gemeinsame rationelle Raumgruppennomenklatur für die Zwecke des ins Auge gefaßten Tabellenwerks. Sie folgt bis auf eine kleine Abänderung dem Vorschlag, der auf Grund der Bezeichnungssysteme von Hermann und von Mauguin den Teilnehmern zugestellt worden war. Über diese Nomenklatur wird Herr Mauguin demnächst berichten. Es wird gehofft, daß die rationelle Nomenklatur sich im Laufe der Zeit durchsetzt; in den Tabellen wird daneben die Schoenflies'sche Bezeichnung angegeben.

108. Il est désirable d'avoir un système de notations rationnelles et simples pour les groupes itératifs qui interviennent à tout instant dans la détermination des structures ou dans l'étude de leurs conséquences physicochimiques. Les notations de C. Hermann[1] me paraissent les meilleures qui aient été proposées. On peut pourtant les simplifier encore et les rendre par là plus pratiques. C'est à quoi je vais m'appliquer dans cette courte note.

C'est ainsi qu'ont été obtenus les symboles qui figurent dans le tableau ci-joint [reproduced for the space groups $V_h{}^{25} - V_h{}^{28}$ only, Ed.]. On pourra si on le désire alléger les symboles en supprimant toute indication qui n'est pas strictement indispensable à la clarté. Nous donnons celles de ces notations abrégées qui nous paraissent le plus commodes.

Symbole de Schoen- flies	Symbole nouveau complet	Symbole abrégè
V_h		
25	$I\,2/m\,2/m\,2/m$	$I\,m\,m\,m$
26	$I\,2/b\,2/a\,2/m$	$I\,b\,a\,m$
27	$I\,2/b\,2/c\,2/a$	$I\,b\,c\,a$
28	$I\,2/m\,2/m\,2/a$	$I\,m\,m\,a$

Les symboles des groupes à trois translations indépendantes résultant ainsi sont unis dans le tableau suivant. C'est le système de notation adopté

[1] *Cf. Z. Krist.* **75** (1930) 159 [this Vol. paper **107**].

par la conférence internationale de Zürich, 28–31 août 1930. On y a résolu de donner ensemble le symbole dérivé ici et, celui de Schoenflies, comme le tableau le montre.

Tableau des 230 groupes à 3 translations indépendantes, selon la notation adoptée à Zürich, 28–31 oaût 1930.

$C_1^1 - P1$	$C_{2v}^{14} - Amm$	$D_{2h}^{13} - Pmmn$	$D_3^1 - H32$
	$C_{2v}^{15} - Abm$	$D_{2h}^{14} - Pbcn$	$D_3^2 - C32$
$C_i^1 - P\bar{1}$	$C_{2v}^{16} - Ama$	$D_{2h}^{15} - Pbca$	$D_3^3 - H3_12$
	$C_{2v}^{17} - Aba$	$D_{2h}^{16} - Pnma$	$D_3^4 - C3_12$
$C_s^1 - Pm$	$C_{2v}^{18} - Fmm$	$D_{2h}^{17} - Cmcm$	$D_3^5 - H3_22$
$C_s^2 - Pc$	$C_{2v}^{19} - Fdd$	$D_{2h}^{18} - Cmca$	$D_3^6 - C3_22$
$C_s^3 - Cm$	$C_{2v}^{20} - Imm$	$D_{2h}^{19} - Cmmm$	$D_3^7 - R32$
$C_s^4 - Cc$	$C_{2v}^{21} - Iba$	$D_{2h}^{20} - Cccm$	
	$C_{2v}^{22} - Ima$	$D_{2h}^{21} - Cmma$	$D_{3d}^1 - H\bar{3}m$
$C_2^1 - P2$		$D_{2h}^{22} - Ccca$	$D_{3d}^2 - H\bar{3}c$
$C_2^2 - P2_1$	$D_2^1 - P222$	$D_{2h}^{23} - Fmmm$	$D_{3d}^3 - C\bar{3}m$
$C_2^3 - C2$	$D_2^2 - P222_1$	$D_{2h}^{24} - Fddd$	$D_{3d}^4 - C\bar{3}c$
	$D_2^3 - P2_12$	$D_{2h}^{25} - Immm$	$D_{3d}^5 - R\bar{3}m$
$C_{2h}^1 - P2/m$	$D_2^4 - P2_12_12_1$	$D_{2h}^{26} - Ibam$	$D_{3d}^6 - R\bar{3}c$
$C_{2h}^2 - P2_1/m$	$D_2^5 - C222_1$	$D_{2h}^{27} - Ibca$	
$C_{2h}^3 - C2/m$	$D_2^6 - C222$	$D_{2h}^{28} - Imma$	$S_4^1 - P\bar{4}$
$C_{2h}^4 - P2/c$	$D_2^7 - F222$		$S_4^2 - I\bar{4}$
$C_{2h}^5 - P2_1/c$	$D_2^8 - I222$	$C_3^1 - C3$	
$C_{2h}^6 - C2/c$	$D_2^9 - I2_12_12_1$	$C_3^2 - C3_1$	$C_4^1 - P4$
		$C_3^3 - C3_2$	$C_4^2 - P4_1$
$C_{2v}^1 - Pmm$		$C_3^4 - R3$	$C_4^3 - P4_2$
$C_{2v}^2 - Pmc$	$D_{2h}^1 - Pmmm$		$C_4^4 - P4_3$
$C_{2v}^3 - Pcc$	$D_{2h}^2 - Pnnn$	$C_{3i}^1 - C\bar{3}$	$C_4^5 - I4$
$C_{2v}^4 - Pma$	$D_{2h}^3 - Pccm$	$C_{3i}^2 - R\bar{3}$	$C_4^6 - I4_1$
$C_{2v}^5 - Pca$	$D_{2h}^4 - Pban$		
$C_{2v}^6 - Pnc$	$D_{2h}^5 - Pmma$	$C_{3v}^1 - C3m$	$C_{4h}^1 - P4/m$
$C_{2v}^7 - Pmn$	$D_{2h}^6 - Pnna$	$C_{3v}^2 - H3m$	$C_{4h}^2 - P4_2/m$
$C_{2v}^8 - Pba$	$D_{2h}^7 - Pmna$	$C_{3v}^3 - C3c$	$C_{4h}^3 - P4/n$
$C_{2v}^9 - Pna$	$D_{2h}^8 - Pcca$	$C_{3v}^4 - H3c$	$C_{4h}^4 - P4_2/n$
$C_{2v}^{10} - Pnn$	$D_{2h}^9 - Pbam$	$C_{3v}^5 - R3m$	$C_{4h}^5 - I4/m$
$C_{2v}^{11} - Cmm$	$D_{2h}^{10} - Pccn$	$C_{3v}^6 - R3c$	$C_{4h}^6 - I4_1/a$
$C_{2v}^{12} - Cmc$	$D_{2h}^{11} - Pbcm$		
$C_{2v}^{13} - Ccc$	$D_{2h}^{12} - Pnnm$		

$D_{2d}^1 - P\bar{4}2m$	$D_4^8 - P4_32_1$	$C_{6h}^1 - C6/m$	$T_h^3 - Fm3$
$D_{2d}^2 - P\bar{4}2c$	$D_4^9 - I42$	$C_{6h}^2 - C6_3/m$	$T_h^4 - Fd3$
$D_{2d}^3 - P\bar{4}2_1m$	$D_4^{10} - I4_12$		$T_h^5 - Im3$
$D_{2d}^4 - P\bar{4}2_1c$			$T_h^6 - Pa3$
$D_{2d}^5 - C\bar{4}2m$	$D_{4h}^1 - P4/mmm$	$D_{3h}^1 - C\bar{6}m2$	$T_h^7 - Ia3$
$D_{2d}^6 - C\bar{4}2c$	$D_{4h}^2 - P4/mcc$	$D_{3h}^2 - C\bar{6}c2$	
$D_{2d}^7 - C\bar{4}2b$	$D_{4h}^3 - P4/nbm$	$D_{3h}^3 - H\bar{6}m2$	$T_d^1 - P\bar{4}3m$
$D_{2d}^8 - C\bar{4}2n$	$D_{4h}^4 - P4/nnc$	$D_{3h}^4 - H\bar{6}c2$	$T_d^2 - F\bar{4}3m$
$D_{2d}^9 - F\bar{4}2m$	$D_{4h}^5 - P4/mbm$		$T_d^3 - I\bar{4}3m$
$D_{2d}^{10} - F\bar{4}2c$	$D_{4h}^6 - P4/mnc$	$C_{6v}^1 - C6mm$	$T_d^4 - P\bar{4}3n$
$D_{2d}^{11} - I\bar{4}2m$	$D_{4h}^7 - P4/nmm$	$C_{6v}^2 - C6cc$	$T_d^5 - F\bar{4}3c$
$D_{2d}^{12} - I\bar{4}2d$	$D_{4h}^8 - P4/ncc$	$C_{6v}^3 - C6cm$	$T_d^6 - I\bar{4}3d$
	$D_{4h}^9 - P4/mmc$	$C_{6v}^4 - C6mc$	
$C_{4v}^1 - P4mm$	$D_{4h}^{10} - P4/mcm$	$D_6^1 - C62$	$O^1 - P43$
$C_{4v}^2 - P4bm$	$D_{4h}^{11} - P4/nbc$	$D_6^2 - C6_12$	$O^2 - P4_23$
$C_{4v}^3 - P4cm$	$D_{4h}^{12} - P4/nnm$	$D_6^3 - C6_52$	$O^3 - F43$
$C_{4v}^4 - P4nm$	$D_{4h}^{13} - P4/mbc$	$D_6^4 - C6_42$	$O^4 - F4_13$
$C_{4v}^5 - P4cc$	$D_{4h}^{14} - P4/mnm$	$D_6^5 - C6_22$	$O^5 - I43$
$C_{4v}^6 - P4nc$	$D_{4h}^{15} - P4/nmc$	$D_6^6 - C6_32$	$O^6 - P4_13$
$C_{4v}^7 - P4mc$	$D_{4h}^{16} - P4/ncm$		$O^7 - P4_33$
$C_{4v}^8 - P4bc$	$D_{4h}^{17} - I4/mmm$	$D_{6h}^1 - C6/mmm$	$O^8 - I4_13$
$C_{4v}^9 - I4mm$	$D_{4h}^{18} - I4/mcm$	$D_{6h}^2 - C6/mcc$	
$C_{4v}^{10} - I4cm$	$D_{4h}^{19} - I4/amd$	$D_{6h}^3 - C6/mcm$	$O_h^1 - Pm3m$
$C_{4v}^{11} - I4md$	$D_{4h}^{20} - I4/acd$	$D_{6h}^4 - C6/mmc$	$O_h^2 - Pn3n$
$C_{4v}^{12} - I4cd$	—————	—————	$O_h^3 - Pm3n$
	$C_{3h}^1 - C\bar{6}$	$T^1 - P23$	$O_h^4 - Pn3m$
$D_4^1 - P42$		$T^2 - F23$	$O_h^5 - Fm3m$
$D_4^2 - P42_1$	$C_6^1 - C6$	$T^3 - I23$	$O_h^6 - Fm3c$
$D_4^3 - P4_12$	$C_6^2 - C6_1$	$T^4 - P2_13$	$O_h^7 - Fd3m$
$D_4^4 - P4_12_1$	$C_6^3 - C6_5$	$T^5 - I2_13$	$O_h^8 - Fd3c$
$D_4^5 - P4_22$	$C_6^4 - C6_4$		$O_h^9 - Im3m$
$D_4^6 - P4_22_1$	$C_6^5 - C6_2$	$T_h^1 - Pm3$	$O_h^{10} - Ia3d$
$D_4^7 - P4_32$	$C_6^6 - C6_3$	$T_h^2 - Pn3$	

[Spacings separate the crystal classes, lines the crystal systems: triclinic, monoclinic, orthorhombic, trigonal, tetragonal, hexagonal and cubic, resp. Ed.]]

[For examples of the determination of space group and the allocation of the atoms to their several *n*-fold positions, *see* a.o. pp. 189, 262 *ff*, 321, 345, 425, 427].

In den meisten Fällen ist freilich heute die Bestimmung der Raumgruppe das nächstliegende Ziel der Strukturarbeiten. Aber es gibt auch direkte Wege, die Massenverteilung im Kristall zu bestimmen, die allein die Kenntnis der Zelle voraussetzen und von der Raumeigenschaft keinen Gebrauch machen. Vgl. die bisher freilich methodisch unvollkommen durchgeführte Methode der Fourierentwicklung (Duane, Havighurst, Bragg).

P. P. EWALD,
Z. Krist. **70** (1929) p. 304

109. Das Teilgebiet, über das nachfolgend berichtet wird,[1] behandelt eine Ergänzung der geometrischen Strukturtheorie der Diskontinua auf Grund eines neu eingeführten Begriffs der *Partikelgruppen*. Dieser Begriff bildet den strengen formalen Rahmen für alle Atomgruppen, die im Kräftespiel im Kristall als Einheiten hervortreten können; die letzteren stehen nun in direktem Zusammenhang einerseits mit dem Kristall und seinen Eigenschaften und andererseits mit den chemischen Molekülen und Radikalen der kristallisierenden Substanz, sie vermitteln somit konkret den Zusammenhang zwischen Kristallbau, physikalischer Eigenschaften und chemischer Konstitution.

Die geometrische Strukturtheorie der Kristalle

Als den Begründer der modernen, geometrischen Strukturtheorie können wir A. Schoenflies ansehen, der gezeigt hat, wie das gesamte Erfahrungsmaterial über Kristalle durch eine einzige Hypothese beherrscht und gedeutet werden kann.

Die Hypothese besagt folgendes:

In jedem Kristall bilden die materiellen Partikeln ein homogenes, d.h. nach den drei Raumrichtungen streng periodisches Diskontinuum.

Diese Hypothese hat sich nun nicht nur als äußerst fruchtbar erwiesen, sondern konnte auch durch die von Lauesche Entdeckung der Röntgen-Interferenzen an Kristallen an einem großen Beobachtungsmaterial quantitativ geprüft und verifiziert werden. Gleich- und Ungleichwertigkeit im Kristall wird durch die Symmetriegruppe des Kristalls bestimmt; diese ist eine Kombination von Symmetrie-Elementen (Dreh- und Schrauben-Achsen, Spiegel-, Gleitspiegel- und Drehspiegel-Ebenen), welche die geo-

---

[1] Gekürzte Zusammenfassung der Arbeiten K. Weissenberg, Habilitations Schrift.

metrische Deutung derjenigen Symmetrie-Operationen sind, die den Kristall mit sich zur Deckung bringen.

Der neue Begriff der Partikelgruppen wird nun in die geometrische Strukturtheorie durch die folgende Definition eingeführt:

Eine beliebig im Gitter eines Kristalls herausgegriffene Menge M von Partikeln soll eine Partikelgruppe genannt werden, wenn jede mit ihr strukturell gleichwertige Partikelmenge alle oder keine Partikel mit M gemeinsam hat.

Um die durch obige Definition postulierte Isoliertheit der gleichwertigen Partikelgruppen im Gitter anzudeuten, werden sie auch Inseln genannt. Gemäß dieser Definition zerfällt die Gesamtheit der Symmetrieoperationen eines Kristalls bezüglich jeder Insel in zwei scharf getrennte Kategorien:

die eine führt die Insel nur in sich über und wird geometrisch in der Symmetriegruppe der Insel zusammengefaßt, sie regelt den Inselbau und stellt gleichsam das auf der Insel herrschende Gesetz dar, dem die Anzahl und relativen Lagen strukturell gleichen aller Partikeln der Insel unterworfen sind;

die andere Kategorie enthält sämtliche übrigen Symmetrieoperationen des Kristalls, von denen jede die Insel in eine andere strukturell gleichwertige Insel überführt, die mit der ersteren keine Partikel gemeinsam hat; diese zweite Kategorie regelt also die Beziehungen zwischen den verschiedenen strukturellgleichwertigen Inseln im Kristall, gibt also das Prinzip des Kristallbaues aus strukturell gleichen Inseln an.

Die Gesamt-Symmetriegruppe des Kristalls wird also durch jede Insel in zwei Teile zerlegt, und zwar in die Symmetriegruppe, die den Inselbau beherrscht, und in den Rest, der den Kristallbau aus strukturell gleichwertigen Inseln beherrscht.

Es gibt nun im Diskontinuum vier und nur vier verschiedene Kategorien von Symmetriegruppen $\Sigma_{III}\,\Sigma_{II}\,\Sigma_{I}$ und Σ_0 je nach der Anzahl von 3, 2, 1 oder 0 Translationen, die in der Symmetriegruppe als Symmetrieoperationen vorkommen.

Dementsprechend gibt es auch vier und nur vier Hauptkategorien von Inseln $J_{III}\,J_{II}\,J_I$ und J_0, die entsprechend ihrem Aufbau bezeichnet werden, und zwar möchten wir folgende Namen vorschlagen:

für J_0 ..... Mikro-Insel
J_I ..... Inselkette
J_{II} ..... Inselnetz
J_{III} ..... Inselraumgitter

Jede herrschende Symmetriegruppe einer Insel schreibt für den Inselbau bestimmte Ganzzahligkeits-, Anordnungs- und Symmetrieverhältnisse vor, die in ein Tabellenwerk angegeben sind, derart, daß die allgemeine Formel[1] für die Partikelgruppe der Insel direkt abgelesen werden kann.

Ein zweites Tabellenwerk enthält die erschöpfende Systematik des Aufbaues jedes einzelnen Kristalls aus Inseln.

1. Die Mikro-Inseln J_0.

Nur die Mikro-Inseln J_0 enthalten endlichviele Atome und ein endliches Volumen.

Der Kristall enthält pro Elementarkörper mindestens soviel strukturell gleichwertige Mikro-Inseln als die kleinste Anzahl strukturell gleichwertiger Punktlagen im Elementarkörper angibt.

2. Die Inselketten J_I.

Die Inselketten J_I enthalten stets ∞ viele Atome und erstrecken sich // zu der in der Symmetriegruppe von J_I enthalten Translationsrichtung, also // zu einer möglichen Kristallkante ins ∞, sie lassen sich stets als eine Kette von Mikro-Inseln J_0 auffassen. Diese Eigenschaft soll durch den Namen Inselkette ausgedrückt werden.

3. Die Inselnetze J_{II}.

Die Inselnetze J_{II} enthalten stets ∞^2 Atome und erstrecken sich nach allen Richtungen // zu der Ebene der beiden Translationen, die ihre Symmetriegruppe enthält, also einer möglichen Kristallfläche ins ∞; sie können stets als ein ∞ Netz von Mikro-Inseln J_0 aufgefaßt werden.

4. Die Inselraumgitter.

Die Inselraumgitter J_{III} enthalten stets ∞^3 Atome und erstrecken sich nach allen Raumrichtungen ins ∞; sie können stets als ein Raumgitter von Mikro-Inseln J_0 aufgefaßt werden.

---

[1] Struktur- und Bruttoformel.

B. THE RELATION BETWEEN MOLECULAR
AND CRYSTAL SYMMETRY

110. The Relation between Molecular and Crystal Symmetry as shown by X-ray Analysis, by G. SHEARER (1923)

■■

111. An X-ray Examination of *i*-Erythritol, by W. G. BURGERS (1927)

■■

112. Crystalline Structure of Hexuronic Acid, by E. G. COX (1932)

■■

113a. W. T. ASTBURY and W. H. BRAGG (1925);
   b. J. D. BERNAL and D. M. CROWFOOT (1933)

■■

110. It is possible to construct a table of the 32 classes and tabulate the minimum number of asymmetric [non-parallel] molecules which must necessarily appear in the elementary cell in order to satisfy the symmetry properties of the crystal. [Table not reproduced. Ed.]

This number may conveniently be called the 'Symmetry Number' of the crystal class. If it is found that the number of molecules in the unit cell is less than the symmetry number for the crystal, it follows that the molecule must possess some symmetry and that this symmetry is reproduced in the crystal. As the number of molecules in the elementary cell must be an integer, it follows that the symmetry which the molecule shares with the crystal must be represented by an integer which is obtained by dividing the number given in the table corresponding to this class of crystal by the number of molecules in the unit cell. Thus, if n is the number characteristic of the class and m the number per unit cell, then n/m is an integer and the crystal shares n/m fold symmetry with the molecule. This number may be called the Symmetry Number of the molecule [see paper **111** for an example].

So far it has not been claimed that the number of molecules given by the table described in our first paragraph is the actual number present when the molecule is asymmetric, but only that the number in the cell must be either this number or an integral multiple of it. It naturally suggests itself that this number is the actual number. This means that the number of molecules in the elementary cell is the minimum necessary to satisfy the symmetry conditions. It seems natural to suppose that a crystal will be constructed with the minimum material. If this is true, it implies that each unit cell contains one and one only of each differently orientated molecule. [See, however, paper **112**.]

The question of molecules which do possess symmetry must now be considered. Here again a natural hypothesis suggests itself, viz., that all the symmetry of the molecule is reproduced in the crystal. It seems reasonable to assume that nature in forming a crystal will make use of all the symmetry already existing in the molecule. If this hypothesis is true it follows that the crystal must always show at least as much symmetry as the molecule from which it is formed. [See, however, papers **113a** and **113b**].

In more complex compounds the molecule certainly appears in the crystal, and it seems reasonable to assume that even in the case of NaCl for example, each chlorine atom has its own particular sodium. [!]

111. The present paper deals with the X-ray examination of an organic compound of relatively simple constitution, *i*-erythritol,

$$CH_2.OH - \overset{*}{CH}.OH - \overset{*}{CH}.OH - CH_2.OH.$$

The investigation was undertaken mainly to determine the symmetry oı the crystal-molecule.

The observed quartering and halvings are in this class characteristic of the space-group C_{4h}^6.

The number of molecules per unit-cell is 8 (7·96).

The number of asymmetric bodies necessary for the construction of a body-centred unit-cell in the tetragonal bipyramidal class is 16; this means that sixteen asymmetric molecules can be disposed about the elements of symmetry in such a way that the symmetry of the whole is tetragonal bipyramidal. When, as in the present case, the cell contains only eight molecules, the molecules cannot be placed in the most general positions, but each molecule must occupy a special position in the unit-cell and possess two-fold symmetry of its own. The tables of Astbury and Yardley show that the possible molecular symmetry in this case is either a centre of symmetry or a two-fold axis.

It is not possible to decide between these two possibilities with such a certainty as accompanies the determination of the space-group and the number of molecules in the unit-cell. However, a consideration of the intensities of the reflexions of a few principal planes, and also of the external form of the crystal, gives strong evidence of the presence of a centre of symmetry in the crystal-molecule.

112. From a purified specimen of 'hexuronic acid' (identified by Szent-Gyorgyi with vitamin C) available in this laboratory I have been able to obtain sufficiently good crystals to carry out an X-ray examination by the single-crystal rotation method. The substance is monoclinic sphenoidal, the spacegroup is C_2^2 ($P2_1$), since the only true halving is ($0k0$) absent when k is odd. The density of the crystals was determined by the flotation method to be 1.65 gm/cc, so that there are four molecules of $C_6H_8O_6$ in the unit cell. Since C_2^2 has only twofold symmetry, a pair of molecules must be associated to form the asymmetric crystal unit. (This does not necessarily imply polymerisation.)

113a. The molecule as it exists in the crystal is part of the structure, and, by the theory of space-groups, its symmetry *in the crystal* is part of the symmetry of the structure. This much is perfectly clear, when there is no misunderstanding as to what is meant by the symmetry in the crystal. The symmetry of the molecule in the crystal is a property which is perceived only indirectly, that is, *via* the symmetry of the structure as a whole. It is merely the symmetry of the immediate environment of the molecule, or, in other words, the symmetry of the points of contact with its neighbours. Whether this can be considered as extending right down into the heart of the molecule is another question.

113b. The actual molecular symmetry is, of course, of stereochemical importance in many cases, but here again the crystal analysis only gives a minimum symmetry, as molecules in general will not be able to pack so as to exhibit their highest symmetry.

CHAPTER VII

Atomic (ionic) Radii and Building Principles

A1. IONIC RADII

114. Welche Radien r_+ und r_- man den positiven Alkali- und negativen Halogenionen (genauer ihren Wirkungssphären) zuschreiben muß, ist aus der folgenden Zusammenstellung der empirischen Gitterkonstanten δ (doppelte Entfernung zweier verschieden geladener Nachbar-ionen) abzuleiten, da ja bei Berührung der Ionen $\delta/2$ einfach gleich der Radiensumme $r_+ + r_-$ anzunehmen ist (Tabelle 1). Durch die Gleichungen $2r_+ + 2r_- = \delta$ mit empirisch bekannten rechten Seiten sind die Einzelwerte r_+ und r_- freilich nur bis auf einen unbestimmten konstanten Summanden festgelegt. Letzteren findet man aber mit Hilfe der Gitterkonstante von LiJ. Da nämlich das Li-Ion mit seinen zwei Elektronen keinen Würfel bilden kann, sondern nur eine einquantige Ringsphäre besitzt, ist es jedenfalls so

Tabelle 1 der empirischen Gitterkonstanten $\delta.10^8$.

LiF	4,00	LiCl	5,11	LiBr	5,45	LiJ	5,95
NaF	4,60	NaCl	5,59	NaBr	5,98	NaJ	6,47
KF	5,31	KCl	6,24	KBr	6,59	KJ	7,05
—	—	RbCl	6,57	RbBr	6,88	RbJ	7,33
—	—	CsCl	6,53	CsBr	6,81	CsJ	7,23

viel kleiner als das Jodion, daß beim Gitteraufbau die um $\delta/\sqrt{2}$ entfernten Jodpartikel eher zur Berührung kommen als je ein Jod- mit einem benachbarten Lithiumion. Daher ist im LiJ und eventuell auch in den anderen Li-Salzen die obige Radienbeziehung durch $r.\sqrt{2}=\delta/2$ zu ersetzen, wodurch der unbestimmte Summand festgelegt ist. Dafür bleibt aber der viel kleinere Radius des Li-Ions ganz unbestimmt.

Für die übrigen Ionendurchmesser bzw. -radien ergeben sich dann die als rohe Näherungen zu betrachtenden Werte der 'Wirkungssphären':

Tabelle 2.

	Na$_+$	K$_+$	Rb$_+$	Cs$_+$	F$_-$	Cl$_-$	Br$_-$	J$_-$
$2r.10^8$	2,2	2,9	3,3	3,2	2,4	3,3	3,6	4,1

Daß sich die empirischen δ-Werte überhaupt annähernd durch additive Zusammensetzung der verschiedenen Kombinationen $2r_+ + 2r_-$ bilden lassen, weist neben den anderen Argumenten ebenfalls auf eine fast lückenlose Berührung der Ionen im Gitter hin.

27 Januar 1920

115. It will be shown that, within certain limits, it is possible to assign to the sphere representing an atom of any element a constant diameter characteristic of that element. The distance between the centres of two neighbouring atoms may be expressed as the sum of two constants, represented by the radii of the corresponding spheres.

This additive law is only intended to be regarded as a working approximation, an aid to the analysis of complex structures. In analysing such a structure, various arrangements of the atoms have to be tried to explain the intensities of the reflected spectra. It will be shown that, when marshalling the atoms together, each atom must be given a certain space in the structure, so that two atoms may not be placed closer together than a distance equal to the sum of the radii of the spheres representing them. This greatly facilitates the determination of the parameters, which are confined to a much more limited range.

3. In the Iron Pyrites structure[1], the iron atoms are situated on a face-centred cubic lattice. If the unit cube of this lattice is divided into eight smaller cubes, each of these latter will have an iron atom situated at four

---

[1] W. L. Bragg, *Proc. Roy. Soc.* (Nov. 1913). [Vol. I. paper **17**]

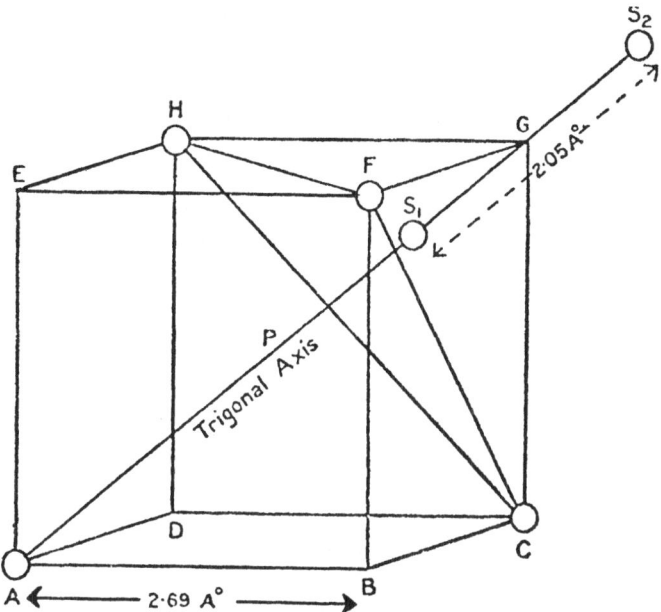

Fig. 1a. Unit of Iron Pyrites Structure.

of its eight corners. Figure 1a represents such a unit of the structure of iron pyrites, the iron atoms being at the corners A, C, H, and F. One diagonal of the cube, the diagonal AG in the figure, is an axis of threefold symmetry, and the sulphur atom lies packed between the three atoms of iron at H, C, and F and the corresponding sulphur atom at S_2.

The structure of metallic iron has been determined by Hull[1]. The iron atoms are situated on a cube-centred lattice, the side of the cube having a length of 2.86 Å, and the distance between the centres of neighbouring iron atoms is 2.47 Å. If this length is taken as a first approximation to the diameter of the sphere representing the iron atom* and spheres of corresponding radius are described with their centres at H, C, and F, the centre of the sulphur atom is thereby fixed. The atom must touch the three iron atoms and also the other sulphur atom at the point G, see fig. 1a. The ratio S_1G/AG which determines the position of S_1 can be calculated to be 0.22/1. This ratio was originally determined by the author as 0.20/1. A more exact determination by Ewald[2], based on the Laue-photograph of Pyrites, gave the value of the parameter as 0.226/1.

[1] Hull, *Phys. Rev.* (10 Dec. 1917).
* [Note the weak point in the following deduction, no distinction being made as to the type of bonding of the particle, *see* this Vol. p. 43 and paper 117. Ed.]
[2] Ewald, *Phys. Zs.* (April 15, 1914). [Vol. I. paper 18]

The exact correspondence of the position of the sulphur atom calculated in this way with that found by experiment is not to be expected, the diameters of the spheres representing the atoms cannot be regarded as absolutely fixed. It will be seen, however, that the conception of the atoms as spheres packed together does lead in this case to an approximate value for the parameter not far from the true one. The diameter of the sulphur atom is given by the distance S_1S_2, which is equal to 2.05 Å. Each sulphur atom is surrounded by three iron atoms and a sulphur atom, each iron atom by six sulphur atoms.

5. The substitution of oxygen for sulphur decreases the distance between atomic centres by 0.38 Å, as will be seen by the following comparison.

		Difference.
MgO	2.11	0.43 Å.
MgS	2.54	
CaO	2.40	0.37 Å.
CaS	2.77	
SrO	2.63	0.36 Å.
SrS	2.99	
BaO	2.81	0.39 Å.
BaS	3.20	
ZnO	1.97	0.38 Å.
ZnS	2.35	

The oxygen atom appears to occupy a smaller space than the sulphur atom, and the diameter 1.30 Å must be assigned to the sphere representing it.

6. Similar relationships are shown by the alkaline halides the molecular volumes of which form a regular series. All these salts crystallize in the same form as NaCl. The following table gives the distances between atomic centres expressed in Ångström Units.

NaF	2.39	KF	2.73				
Diff	.41		.40				
NaCl	2.80	KCl	3.13	RbCl	3.28	CsCl	3.26
Diff	.17		.15		.16		.14
NaBr	2.97	KBr	3.28	RbBr	3.44	CsBr	3.40
Diff	.26		.24		.22		.21
NaI	3.23	KI	3.52	RbI	3.66	CsI	3.61

The replacement of Fluoride by Chlorine, Chlorine by Bromine, and Bromine by Iodine, increases the dimensions of the structure by an approximately constant amount.

8. Two other examples will be taken as affording a cross-check on these measurements. We have the relations

Na–O .... 2.33 Å. (NaNO$_3$)   Ca–O .... 2.30 Å. (CaCO$_3$)
Na–F .... 2.39 Å. (NaF)   Ca–F .... 2.34 Å. (CaF$_2$)

The space occupied by calcium in a crystal is much the same as that occupied by sodium; that occupied by oxygen is much the same as that occupied by fluorine.

9. These examples will indicate the manner in which the results shown in Fig. 3 have been calculated. For instance, the diameter of oxygen has been taken to be 1.30 Å, and that of fluorine to be slightly greater, 1.35 Å. From a comparison of the alkaline halides, the diameters of the halogens are found to be:

Fluorine .... 1.35 Å.
Chlorine .... 2.10 Å.
Bromine .... 2.38 Å.
Iodine .... 2.80 Å.

The diameters of the spheres representing the monovalent alkali metals are calculated from the dimensions of the alkaline halides, diameters having been already fixed for fluorine, chlorine, bromine, and iodine. These data have been used in calculating the diameters of Fig. 3.

10. In Fig. 3 the elements are arranged in the order of their Atomic Numbers. The ordinates represent the diameters of the 'Atomic Domain' measured in Ångström Units. The figure summarizes the empirical relation which has been found to hold, namely, that the distance between neighbouring atomic centres in a crystal is the sum of two constants characteristic of the atoms. The crystal may be imagined as an assemblage of spheres packed together, the constants then representing the radii of the spheres.

It is not intended to assign any physical significance to these 'diameters' other than that discussed below. Sodium, for instance, has been given a diameter much larger than that of chlorine, yet it will be seen that there is every reason for supposing that the group of electrons surrounding the sodium nucleus in sodium chloride has smaller dimensions than that surrounding the chlorine nucleus in the same crystal.

The large diameters assigned to the electropositive elements as compared with the electronegative elements do not imply a corresponding difference in the dimensions of the atomic structure. They are an expression of the fact that the electropositive element does not share electrons with neighbouring atoms, it is always surrounded by a complete stable shell. The repulsion between this outer shell and the shells of neighbouring atoms keeps the

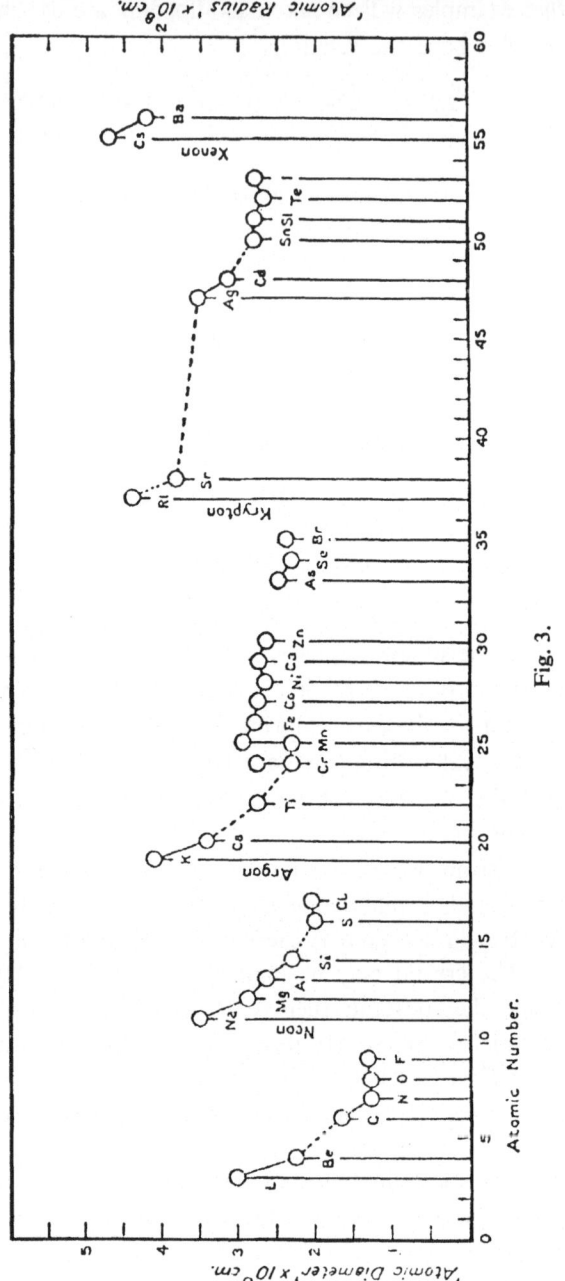

Fig. 3.

atom at a distance from its neighbours, so that it appears to occupy a large space in the crystal structure.

Manchester University, April 1920

116.   § 1. *Refraction equivalents and ionic radii*

In an earlier paper[1] the author has established the fact that the ionic refraction of an element differs materially from the atomic refraction of the same element. In later papers[2] the ionic refractions were calculated for the alkaline metals, the alkaline earth metals, the halogens, and for O^{--} and S^{--}. These calculations are based only on experimental data and on the assumption, that the ionic refraction of the hydrogen ion, even in aqueous solution, equals zero.

The conception, due to Bohr[3] and already pretty generally accepted, respecting the arrangement of the paths of electrons in space, does not at present admit of any theoretical calculation of the difference between the atomic and the ionic refractions of an element. Nevertheless a knowledge of the ionic refractions enables us to draw certain general conclusions with respect to the dimensions of the paths of electrons.

If an ion of the rare gas type is brought into an electrical field it is polarised, whereby the time integral of the vector $-s$ from the nucleus to the centre of gravity of the electronic system assumes a finite value. The average distance from the nucleus to one of the A_p outermost electrons is indicated by ρ_p.

Since the Coulomb force acting on the electrons is inversely proportional to the square of the distance, we can foresee that the quasi-elastic displacement, due to the influence of a given strength of field, is directly proportional to the next highest, that is to say the third power of ρ, multiplied by a function φ characteristic of the ionic structure, namely a function of the effective quantum numbers p^*, n'^* and n''^*. Further we may assume that the value of the vector s is inversely proportional to the effective nuclear charge $Z - \sum_{q<p} A_q$. Since the polarisation has the value $P = N'A_p.e.s.$, where N' signifies the number of ions per cc, the ionic refraction may therefore be expressed by the formula

$$I = \psi(p^*, n'^*, n''^*) \cdot \frac{A_p}{Z - \sum_{q<p} A_q} \cdot N \cdot \tfrac{4}{3}\pi\rho_p{}^3, \tag{4}$$

[1] *Acta Soc. Scient. Fennicae* **50** No 2 (1920).
[2] *Översikt av Finska Vet. Soc. Förh.* **63** A No 4 (1920–1921) and *Comm. Phys.-Math. Soc. Scient. Fennicae* T. **I** No 37.
[3] *Zeitschr. f. Physik* **9** (1922) 1.

39

where Z is equal to the atomic number of the atom and N to the number of ions in a gramme-ion.

That ρ in the expression (4) for the ionic refraction occurs with power 3 can be confirmed also by simple dimensional considerations.

The practical significance of formula (4) is that it admits of a calculation of the *relative* values of ρ, so long as the outermost electron system remains intact with regard of the number of the elctrons A_p and the quantum numbers p, n' and n'', if, as a first approximation, we assume that ψ remains constant in these circumstances.

The ionic refractions of the alkali and the alkaline earth metals, the halogens and elements of the oxygen group, as well as the atomic refractions of the rare gases have, according to the author's previous investigations, the following values

Table I.

$I_O{}^{--} = 4{,}06$	$I_F{}^- = 2{,}20$	$A_{Ne} = 1{,}01$	$I_{Na}{}^+ = 0{,}74$	$I_{Mg}{}^{++} = 0{,}44$
$I_S{}^{--} = 15{,}0$	$I_{Cl}{}^- = 8{,}45$	$A_A = 4{,}23$	$I_K{}^+ = 2{,}85$	$I_{Ca}{}^{++} = 1{,}99$
$I_{Se}{}^{--} = ?$	$I_{Br}{}^- = 11{,}84$	$A_{Kr} = 6{,}42$	$I_{Rb}{}^+ = 4{,}41$	$I_{Sr}{}^{++} = 3{,}22$
$I_{Te}{}^{--} = ?$	$I_I{}^- = 18{,}47$	$A_X = 10{,}56$	$I_{Cs}{}^+ = 7{,}36$	$I_{Ba}{}^{++} = 5{,}24$

With the help of table I and formula (4) we calculate $\rho . \psi^{\frac{1}{3}}$.

Table II.

$\psi^{\frac{1}{3}} \cdot \rho_{--} \cdot 10^8$	$\psi^{\frac{1}{3}} \cdot \rho_{-} \cdot 10^8$	$\psi^{\frac{1}{3}} \cdot \rho \cdot 10^8$	$\psi^{\frac{1}{3}} \cdot \rho_{+} \cdot 10^8$	$\psi^{\frac{1}{3}} \cdot \rho_{++} \cdot 10^8$
$O^{--} = 1{,}06_2$	$F^- = 0{,}91_2$	$Ne = 0{,}73_5$	$Na^+ = 0{,}68_9$	$Mg^{++} = 0{,}60_0$
$S^{--} = 1{,}64_2$	$Cl^- = 1{,}42_8$	$A = 1{,}18_5$	$K^+ = 1{,}08_1$	$Ca^{++} = 0{,}99_3$
$Se^{--} = ?$	$Br^- = 1{,}59_7$	$Kr = 1{,}36_2$	$Rb^+ = 1{,}25_0$	$Sr^{++} = 1{,}16_6$
$Te^{--} = ?$	$I^- = 1{,}85_3$	$X = 1{,}60_8$	$Cs^+ = 1{,}48_3$	$Ba^{++} = 1{,}37_1$

It is important to compare the quotients ρ_-/ρ_+ and ρ_{--}/ρ_{++} for two ions which in the periodic table surround the same rare gas, which lend themselves most readily to a comparison (Table VI).

From the table it appears that scarcely any agreement can be said to exist between the ionic radii calculated according to the different methods. Summarising, it may be said that Fajans-Herzfeld's and especially Grimm's values show a very rapid change and that they are greater than 1. Bragg's values vary rather less and are less than 1, whilst the values calculated

Table VI.

		W. L. Bragg[1]	Fajans & Herzfeld[2]	Wasastjerna
$\dfrac{\rho_-}{\rho_+}$	F⁻, Na⁺	0,38	1,44	1,32
	Cl⁻, K⁺	0,51	1,20	1,32
	Br⁻, Rb⁺	0,53	1,12	1,28
	I⁻, Cs⁺	0,59	1,05	1,25
		W. L. Bragg	Grimm[3]	Wasastjerna
$\dfrac{\rho_{--}}{\rho_{++}}$	O⁻⁻, Mg⁺⁺	0,46	2,28	1,77
	S⁻⁻, Ca⁺⁺	0,60	1,63	1,65
	Se⁻⁻, Sr⁺⁺	0,60	1,32	–
	Te⁻⁻, Ba⁺⁺	0,63	1,16	–

[1] *Phil. Mag.* (6) **40** (1920) 169.
[2] *Zeitschr. für Physik* **2** (1920) 309.
[3] *Zeitschr. für Physikalische Chemie* **98** (1921) 353.

by the help of the ionic refractions are > 1 and change rather slowly. It is of particular interest to compare the above results with Bohr's atomic theory. If as a first approximation we accept the effective quantum number as independent of the nuclear charge, as long as the variations of Z do not exceed ± 2, there follow the equations

$$\frac{\rho_-}{\rho_+} = \frac{Z_+ - \sum_{q<p} A_q}{Z_- - \sum_{q<p} A_q} = \frac{9}{7} = 1{,}29, \tag{5}$$

$$\frac{\rho_{--}}{\rho_{++}} = \frac{Z_{++} - \sum_{q<p} A_q}{Z_{--} - \sum_{q<p} A_q} = \frac{10}{6} = 1{,}67, \tag{6}$$

independently of the horizontal line of the periodic table, in which the quotients in question are formed. It is evident that W. L. Bragg's values are quite irreconcilable with the Bohr-theory. Fajans', Herzfeld's and Grimm's values show such a rapid change that no real agreement is apparent. Such a progression may however possibly be explained on a more precise theory. The values calculated with the help of the ionic refractions agree remarkably well with the theoretical formulae (5) and (6), in as much as ρ_-/ρ_+ is nearly constant and has an average value of 1,29, whilst ρ_{--}/ρ_{++} has an average value of 1,71.

These results being obtained, the following questions arise (...). Can we theoretically explain the empirical law found by W. L. Bragg, according to which the grating intervals are composed, additively, of 'atomic radii'? Can these grating intervals be connected with, and calculated in accordance with the values of the ionic radii as deduced in this paper?

41

§ 2. *The ionic radii in crystals and the atomic volume presented in the kinetic gas theory*

Of far greater importance than all theoretical considerations is the empirical law discovered by W. L. Bragg, according to which, the grating intervals, even in crystals of very complicated structure, can be calculated by assuming that 'atoms' with certain radii lie close-packed in the crystals. This law has, theoretically considered, an extraordinarily wide bearing, and it is by far the most important fact which has as yet come to light with regard to the forces acting between atoms.

The law necessarily suggests the view that the repulsion between ions, though negligible at greater distances, suddenly increases at a certain fixed distance, and that the value of that repulsion does not, to any degree worth mentioning, depend on the relative orientation of the ions. As this particular distance can be additively calculated with the help of constants which are characteristic of the ions in question, it appears indubitable that a close connection exists between the distance and the radii of the electronic systems of the ions. In a crystal consisting of ions the opposite electrical charges tend to compress the material. Then the repulsion which arises at a certain distance, assumes such a magnitude as to bring about a state of equilibrium. This point of view explains Bragg's results. If this view is correct, a close connection should, however, exist between the apparent volumes of the atoms of rare gases in the kinetic gas theory and the grating intervals in crystals. We shall at once subject this connection to a mathematical analysis. (...)

§ 3. *Calculation of the apparent ionic radii in crystals with the help of the refraction equivalents of ions*

The closer investigation of the connection between the grating intervals ε and the ionic radii ρ_i calculated according to ionic refractions, may be effected in different ways. However, in order to abridge the following investigation we will anticipate the final result and establish on that result a theoretical calculation of grating intervals.

It appears that one can divide up the grating distances into additive constants characteristic of the ions, which are proportional to the ionic radii calculated according to the ionic refractions.

Table XI contains the *apparent* ionic radii (in Å) in crystals of the NaCl type, calculated with the help of the ionic refractions, and the apparent atomic volumes of rare gases.

Table XI.

O^{--} = 1,32	F^- = 1,33	Na^+ = 1,01	Mg^{++} = 0,75
S^{--} = 1,69	Cl^- = 1,72	K^+ = 1,30	Ca^{++} = 1,02
Se^{--} = ?	Br^- = 1,92	Rb^+ = 1,50	Sr^{++} = 1,20
Te^{--} = ?	I^- = 2,19	Cs^+ = 1,75	Ba^{++} = 1,40

Table XII [not reproduced] contains a statement of the grating intervals additively calculated with the help of these ionic radii, and the experimentally determined values. Figure 1 facilitates the comparison. The agreement is entirely satisfactory.

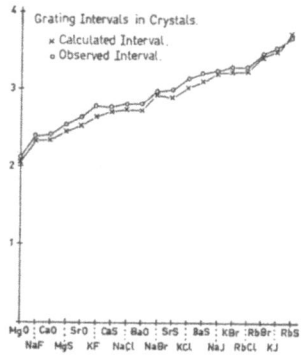

Fig. 1.

Now as the grating intervals in crystals can be calculated both according to the ionic radii stated by W. L. Bragg, see Table IV, and also according

Table IV.

The apparent atomic radii in crystals according to W. L. Bragg. (in Å)

O = 0,65	F = 0,67	(Ne) = 0,65	Na = 1,77	Mg = 1,42
S = 1,02	Cl = 1,05	(A) = 1,03	K = 2,07	Ca = 1,70
Se = 1,17	Br = 1,19	(Kr) = 1,19	Rb = 2,25	Sr = 1,95
Te = 1,33	I = 1,40	(X) = 1,35	Cs = 2,37	Ba = 2,10

to those calculated with the help of ionic refractions, the disagreement between the ionic radii calculated by different methods is due to the assumption which enabled Bragg to divide up the grating intervals into atomic radii. This assumption can therefore not be correct. In reality W. L. Bragg presupposes that the apparent atomic radius in metallic Fe is equal to the 'atomic radius' in Fe-compounds and he further assumes that the interval between the atoms e.g. in metallic sodium is determined

43

by the dimensions of the next outermost electronic system, as the radius for Na^+-ions for instance in NaCl.

Now the atomic radii of alkali- and alkaline earth metals are, as we know, closely connected with the frequencies of the valency electron or valency electrons. In whatever manner we may theoretically explain this connection it, at all events, proves the error of the previously mentioned assumption. All Bragg's radii of positive ions are therefore given too high values, whilst to the negative ions too low values are assigned.

A constant displacement between the radii of the positive and the negative ion is however, from Bragg's point of view, devoid of particular interest; in this connection Bragg does not assign to the radii mentioned by him any further significance than that the grating intervals can be additively calculated by the help of these radii.

117. Limiting consideration in the first place to cases where the bonds are of the polar type, in the generally accepted meaning of the term, there are, broadly speaking, two sets of ionic radii which have been proposed. Both are based on the undoubted fact that interatomic distances are additive in many series of compounds. They differ in that the one set (proposed by the author in 1920, and Niggli[1] in 1921) assigned values of the radii to the cations which were larger than those of the other set (proposed by Wasastjerna[2] in 1932, and foreshadowed by Landé[3] in 1920) by a constant amount of about 0.7 Å. All the anions in the first set are correspondingly smaller by the same amount.

The second set corresponds much more closely with the physical reality, in so far as such an arbitrary idea as that of ionic radius can correspond with reality. The author believes that they are preferable for this reason. Both sets give the same estimate of the point at which the repulsion between cation and anion sets in, but the second set also gives an estimate of the point at which the repulsion between anions sets in. Now in actual crystal structures it is often found that in certain directions anions are nearest neighbours. In these cases (where a non-polar bond between the atoms is absent) the author's figures give far too small an estimate of the distance between the anions, whereas Wasastjerna's figures correspond satisfactorily with the observed value.

[1] Niggli, *Zeitschr. für Kristallogr.* **56** (1921) 167.

[2] Wasastjerna, *Soc. Scient. Fenn. Comm. Phys. Math.* **38** (1923) 1 [this Vol. paper **116**].

[3] Landé, *Zeitschr. für Physik* **1** (1920) 191 [this Vol. paper **114**]

118. Since every atom extends to an unlimited distance, it is evident that no single characteristic size can be assigned to it. Instead, the apparent atomic radius will depend upon the physical property concerned, and will differ for different properties. In this paper we shall derive a set of ionic radii for use in crystals composed of ions which exert only a small deforming force on each other.

The Forces in Crystals

The wave mechanics provides the simple explanation that the repulsive forces arise from the interpenetration of the atoms.

The potential energy of an ionic crystal (ions of valence z) may be written $\Phi = -\alpha(e^2 z^2)/R + \varphi(R)$, the first term representing the Coulomb energy, and the second the potential of the repulsive forces. It has been customary to use the expression

$$\varphi(R) = \beta_n R^{-n} \tag{8}$$

widely applied by Born and his co-workers. The constants β_n and n may be determined from the experimental measurements of the size of the unit of structure in crystals and of their compressibility, which provide values of $d\varphi/dR$ and $d^2\varphi/dR^2$ for $R = R_0$, the equilibrium distance. For many calculations only these two derivatives are significant, so that it is immaterial which form of $\varphi(R)$ is used; examples of such calculations will be given later.

The Sizes of Ions in Crystals

As a first approximation, each electron in a many-electron atom can be considered to have the distribution in space of a hydrogen-like electron under the action of the effective nuclear charge $(Z - S_S)e$, in which S_S represents the screening effect of inner electrons. In the course of a previous investigation[1], values of S_S for a large number of ions were derived.

The size of an atom or ion is determined by the distribution of the outermost electrons. Moreover, for similar ions the electron distribution is similar, but equivalent radii are in the inverse ratio of the corresponding effective nuclear charges. With this principle we shall obtain radii for ions, which we shall define in such a way as to fulfil the following condition: the sum of the radii of two ions equals the inter-ionic distance in a normal

---

[1] Pauling, paper submitted to *Proc. Roy. Soc.*

crystal composed of these ions with the sodium chloride structure[1]. As a starting point we shall use the following experimental inter-ionic distances, for crystals with the sodium chloride structure[2]: $Na^+ - F^-$, 2.31; $K^+ - Cl^-$, 3.14; $Rb^+ - Br^-$, 3.43; $Cs^+ - I^-$, 3.85 Å. The $Cs^+ - I^-$ distance is obtained by subtracting 2.7% from the experimental distance in the crystal with the cesium chloride structure; the justification for this will be given later. From these values all of our radii will be derived. By dividing these distances between the two ions concerned, with the use of the screening constants given in Table I [not reproduced. Ed.] radii for the alkali ions and the halogen ions are obtained. For other ions with similar structures radii are also calculated by assuming them inversely proportional to the effective nuclear charge. These radii are, however, not to be directly applied in the consideration of inter-ionic distances in crystals, for they are the radii which the ions would have *if they were univalent*; they provide, however, a measure of the relative sizes of ions of a given structure, that is, of their extension in space. The actual crystal has the following potential energy.

$$\Phi = -\alpha(z^2 e^2)/R + \beta/R^n \tag{10}$$

For equilibrium, at $R = R_s$, we have the condition

$$(d\Phi/dR)_{R=R_s} = \alpha(z^2 e^2)/R_s^2 - n\beta/R_z^{n+1} = 0 \tag{11}$$

or

$$R_s^{n-1} = n\beta/\alpha e^2 z^2$$

If the ions were univalent, but otherwise unchanged, the potential energy would be $\Phi_1 = -\alpha e^2/R + \beta/R^n$, and for equilibrium, with $R = R_1$

$$R_1^{n-1} = (n\beta)/(\alpha e^2) \tag{12}$$

From Equations 11 and 12 there is obtained

$$R_z = R_1 z^{-2/n-1} \tag{13}$$

From this equation we can calculate the actual crystal radius R_z (for use in normal sodium chloride type crystals) from R_1, the univalent crystal radius. A knowledge of the repulsion exponent n is needed, however.

---

[1] A normal crystal is one in which contact (that is, strong repulsion) occurs only between adjacent anions and cations, and in which there is only so much deformation as that shown by the alkali halides.

[2] The crystal structure data are taken from Wyckoff, 'International Critical Tables,' except where otherwise noted. Inter-atomic distances referred to Goldschmidt are from Goldschmidt, *Skrifter Det. Norske Videnskaps-Akad. Oslo I. Matem.-Naturvid. Klasse* **2** (1926).

This can be derived from the experimental measurement of the compressibility of crystals.

We shall use the following values of n for ions with the structures indicated: He, 5; Ne, 7; Ar, Cu^+, 9; Kr, Ag^+, 10; Xe, Au^+, 12. Averages of these values for the two ions concerned correspond satisfactorily with the experimental results. The radii calculated with these values of n from the univalent radii R_1 by means of Equation 13 are included in Table II[1] [not reproduced] and shown graphically in Fig. 3. The effect of the valence in causing the crystal radius to deviate from the regular dependence on the atomic number shown by the univalent crystal radius is clearly evident.

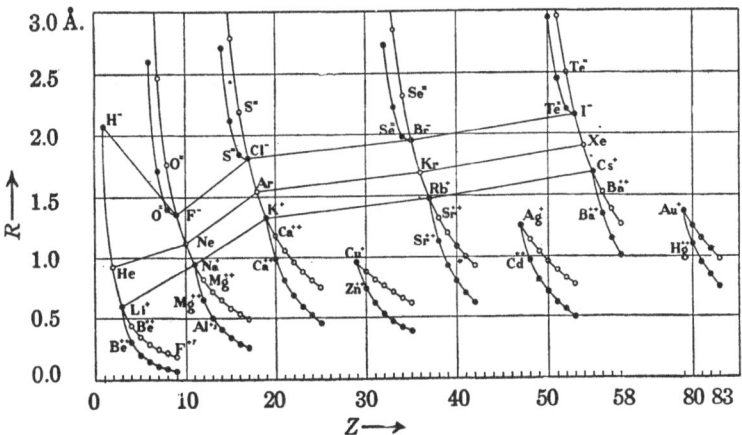

Fig. 3. The crystal radius (solid circles) and the univalent crystal radius (open circles) for a number of ions.

In deriving theoretical values for inter-ionic distances in ionic crystals the sum of the univalent crystal radii for the two ions should be taken, and corrected by means of Equation 13, with z given a value dependent on the ratio of the Coulomb energy of the crystal to that of a univalent sodium chloride type crystal. Thus, for fluorite the sum of the univalent crystal radii of calcium ion and fluoride ion would be used, corrected by Equation 13 with z placed equal to $\sqrt{2}$, for the Coulomb energy of the fluorite crystal (per ion) is just twice that of the univalent sodium chloride structure. This procedure leads to the result 2.34 Å (the experimental distance is 2.36 Å). However, usually it is permissible to use the sodium chloride crystal radius for each ion, that is, to put $z = 2$ for the calcium ion and

---

[1] The same results would be obtained by the use of any other function for $\varphi(R)$, instead of the Born expression; for the calculation depends mainly on the first and second derivatives, which are found experimentally.

$z = 1$ for the fluoride ion, without introducing a large error; the result 2.35 Å is obtained in this way for fluorite.

Comparison with Experiment

The radii in Fig. 3 are in reasonable agreement with Wasastjerna's,[1] and accordingly also account satisfactorily for the empirical data.

In the past, ionic radii have often been compared with observed inter-atomic distances without much regard to the nature of the crystal from which they were derived. Recently several investigators have concluded that in many crystals the bond between atoms does not consist of the electrostatic attraction of only slightly deformed ions. Goldschmidt in particular has divided crystals into two classes, ionic and atomic crystals, and has shown that ionic radii (using Wasastjerna's set) do not account for the observed inter-atomic distances in atomic crystals. In the following pages our crystal radii will be compared with the experimental distances in ionic crystals[2]. It will be shown that the theoretical radii account satisfactorily for the experimental results not only for normal crystals, but also for those deviating from additivity through mutual contact of the anions.

119.   1. *Introduction*

At present V. M. Goldschmidt's set of empirical crystal radii is in wide use[3]. The radii were given the necessary theoretical foundation through L. Paulings valuable investigations[4]. Both Goldschmidt and Pauling were well aware of the approximate character of the set. With the rapid advance in crystal analysis a desire for more accurate predictions of the interatomic distances is making itself felt. In the course of my investigations during the last few years I have repeatedly obtained results concerning interatomic

---

[1] Wasastjerna, *Soc. Sci. Fenn. Comm. Phys. Math.* **38** (1923) 1.

[2] Huggins, who has particularly emphasized the fact that different atomic radii are required for different crystals, has recently [*Phys. Rev.* **28** (1926) 1086] suggested a set of atomic radii based upon his ideas of the location of electrons in crystals. These radii are essentially for use with crystals in which the atoms are bonded by the sharing of electron pairs, such as diamond, sphalerite, etc.; but he also attempts to include the undoubtedly ionic fluorite and cesium chloride structures in this category.

[3] Compare Goldschmidt's various publications, especially Geochemische Verteilungs-gesetze der Elemente VII and VIII.

[4] *J. Am. Chem. Soc.* **49** (1927) 765, *see* also *Z. Krist.* **67** (1928) 377.

distances that were not in agreement with the values predicted from Gold-schmidts radii. E.g. I found in several crystals the Na-O distance in 6 coordination to be 2.46 Å,[1] whereas the sum of the Goldschmidt radii is only 2.30 Å. No other reasonable explanation could be given for the discrepancy than the approximative character of Goldschmidt's radii.

The purpose of the present paper is to show that it has been possible to derive a set of empirical radii for ions with inert gas configuration, by means of which the interionic distances can be calculated with considerably greater accuracy than has been possible with previous sets.

I wish to emphasise that the set of radii given in this paper can be in no way regarded as opposing that of Goldschmidt; it merely represents a closer approximation. With equal emphasis I also want to point out the following fact: The radii obtained have not been derived in a fundamentally new way. Pauling has in his paper actually given a set of theoretical radii corresponding to the empirical ones of this paper. These 'univalent' radii of Pauling were, however, unfortunately put in the background by his final set. The following chapters will show the great advantages obtained by using the radii in the univalent form.

The crystal energy is given by:

$$\Phi = -\frac{Az_1z_2e^2}{R} + \frac{B}{R^n}$$

where A is the Madelung constant and B the Born coefficient.

2. Correction for Coordination Number

At equilibrium $\partial\Phi/\partial R = 0$ and consequently $R = \sqrt[n-1]{nB/Az_1z_2e^2}$. The Madelung constants of the CsCl, NaCl and ZnS lattice types have the following values:

$$A_{CsCl} = 1.762, \quad A_{NaCl} = 1.748, \quad A_{ZnS} = 1.640$$

Not very much is known about the coefficient B. In crystals where no anion-anion contact occurs, the overwhelming contribution to the repulsion potential is given by the cation-anion contacts. With good approximation we may therefore put the coefficient B proportional to the number of these contacts per molecule; i.e. we put the coefficient B proportional to the coordination numbers. We then have:

$$R_{CsCl} = R_{NaCl} \sqrt[n-1]{\frac{8}{6} \cdot \frac{1.748}{1.762}} \qquad R_{ZnS} = R_{NaCl} \sqrt[n-1]{\frac{6}{4} \cdot \frac{1.640}{1.748}}.$$

[1] Z. Krist. **71** (1929) 517, also *Physic. Rev.* **37** (1931) 1295.

We can use the experimentally determined values of the Born exponent n. Following Pauling we will use the following values for the different ionic configurations:

Table I. Characteristic Repulsion Exponents.

Ions with He configuration	$n = 5$
Ne	7
A	9
Kr	10
Xe	12
?	14

For the repulsion exponent operative between ions with different configurations we take the average from the above table.

Using the mean value of $n = 9$ we find 3.3% contraction by transition CsCl-NaCl and 4.3% contraction by transition NaCl- to ZnS-type in excellent agreement with Goldschmidt's empirical values.

Analogously we can calculate the relation between the distances in the CaF_2-lattice (C.N. 8,4), the TiO_2-lattice (C.N. 6,3) and the SiO_2-lattice (C.N. 4,2), by the use of the Madelung constants 5.04, 4.80 and 4.40 respectively for these lattices. We find:

$$R_{CaF_2} = R_{TiO_2} \sqrt[n-1]{\frac{8}{6} \frac{4.80}{5.04}} \qquad R_{SiO_2} = R_{TiO_2} \Big/ \sqrt[n-1]{\frac{6}{4} \frac{4.40}{4.80}}.$$

In the following the coordination number is defined as the number of anions around cations. The C.N. of CaF_2-, TiO_2- and SiO_2-lattices are thus 8, 6, and 4 respectively.

It will be seen that the contraction between the CaF_2-lattice and the TiO_2-lattice is very nearly equal to that between the CsCl- and the NaCl-lattice. The same holds for the pairs TiO_2-SiO_2 and NaCl-ZnS.

We will now make the assumption that the difference in interionic distance for different coordination numbers depends only upon the values of C.N. and not upon the crystal lattice itself. This assumption is not quite correct, but seems to hold with good approximation. By supposing further that the difference in interionic distance between two C.N. is governed by the ratio of the C.N.'s we have the necessary information in order to correct for the effect of C.N. According to the assumptions made above we have:

$$R_{12}/R_8 = R_9/R_6 = R_6/R_4 = R_3/R_2$$
$$R_{12}/R_9 = R_8/R_6 = R_4/R_3.$$

R_8/R_6 and R_6/R_4 we will put equal to $\sqrt[n-1]{1.296}$ and $\sqrt[n-1]{1.391}$ which represent the averages obtained from compounds AX and AX_2.

In agreement with Goldschmidt and Pauling we will use C.N. = 6 as the standard number. The following table shows the variation of the interionic distance with C.N. and n, expressed in terms of the distance unity for C.N. 6. The values for C.N. equal to 10 f.ex. may be obtained by interpolation.

Table II.

Variation of Interionic Distance with Coordination Number.

n	6	7	8	10
C.N. 12	1.126	1.104	1.088	1.068
9	1.069	1.057	1.048	1.038
8	1.053	1.044	1.038	1.029
6	1.00	1.00	1.00	1.00
4	0.936	0.946	0.954	0.963
3	0.889	0.906	0.919	0.936
2	0.834	0.857	0.877	0.902

It will be seen from table II [only partly reproduced] that the correction for the coordination number may be very great. Thus it is absolutely necessary to take this correction into account in order to obtain really accurate values.

For those crystals in which contact between anions occur the correction by means of the above table will not be accurate, since in this case we are not permitted to put the coefficient B proportional to the number of cation-anion contacts. This will, however, be taken care of by the radius ratio correction.

3. Correction for Valency

The second part of the coordination effect is due to the valencies (kernel charges) of the ions in the first sphere. The radius of a sodium ion f.ex. is larger if the surrounding anions are univalent, than if they are bivalent. A second correction is therefore necessary. In reality the correction to be applied is for the coulomb force. Neither Goldschmidt nor Pauling in his final set of radii make any correction for valency. It must be remarked, however, that Pauling made use of the valency correction in deriving his set of radii. As a matter of fact this correction was first given by Pauling in his paper:

$$R_{11} = R_{z_1 z_2}\ ^{n-1}\!\sqrt{z_1 z_2}.$$

By means of this relation we can now correct all observations for the valency effect; i.e. we can reduce all observations to the same standard Coulomb force.

In table III is given the expression $^{n-1}\!\sqrt{z_1 z_2}$ for different values of $z_1 z_2$ and n [only partly reproduced].

Table III.

$^{n-1}\!\sqrt{z_1 z_2}$ as Function of $z_1 z_2$ and n.

n	6	7	8	10
$z_1 z_2 = 2$	1.149	1.122	1.104	1.080
3	1.246	1.201	1.170	1.130
4	1.320	1.259	1.219	1.166
6	1.431	1.348	1.292	1.220
8	1.516	1.414	1.346	1.260

5. An Empirical Set of Univalent Radii for Ions with Inert Gas Configuration

Applying these corrections I have on the basis of the existing material of X-ray data derived a set of ionic radii. Although it is simpler to derive the interionic distances from the data given by Goldschmidt and Pauling, the accuracy attainable by using the set given here more than justifies the additional work. The set of univalent radii given in table IV [reproduced for the greater part only, Ed.] has been derived in the following way. Reliable observations on interionic distances were selected from the existing material. Care was taken not to use any data referring to crystals where anion contacts occur. The distances were corrected by means of the derivations given in sections 2 and 3. From these corrected distances the individual radii were deduced by the use of 1.33 Å for the K^+ ion and 1.81 Å for the Cl^- ion. These two radii were used as starting points for two reasons: firstly because both Goldschmidt and Pauling are in complete agreement as to these radii, secondly because the same values are obtained using Landé's method for deriving the radii from anion contacts.

Conversely the shortest distance between any two ions of the table is calculated in the following manner. Let R_1 denote the sum of the univalent radii. If the valencies of the cation and anion are z_1 and z_2, the interionic

52

Table IV.

A Set of Empirical Univalent Radii.

Valence	He-config.	Ne-config.	A-config.	Kr-config.	Xe-config.
−2		O	S	Se	Te
		1.76	2.20	2.29	2.47
		(1.76)	(2.19)	(2.32)	(2.50)
−1	H	F	Cl	Br	I
	1.36	1.33	1.81	1.96	2.19
	(2.08)	(1.36)	(1.81)	(1.95)	(2.16)
1	Li	Na	K	Rb	Cs
	0.68	0.98	1.33	1.48	1.67
	(0.60)	(0.95)	(1.33)	(1.48)	(1.69)
2	Be	Mg	Ca	Sr	Ba
	0.55	0.89	1.17	1.34	1.49
	(0.44)	(0.82)	(1.18)	(1.32)	(1.53)
3	B	Al	Sc	Y	La
	0.42	0.79	1.03	1.19	1.30
	(0.35)	(0.72)	(1.06)	(1.20)	(1.39)
4	C	Si	Ti	Zr	Ce
	0.38	0.69	0.88	1.07	1.14
	(0.29)	(0.65)	(0.96)	(1.09)	(1.27)

distance in 6-coordination is given by:

$$R_1 /^{n-1}\sqrt{z_1 z_2},$$

where n can be taken from table I. The interionic distance for a C.N. different from 6 is obtained by multiplication with the factor K, K being obtained directly from table II. E.g. we want to find the Na-O distance in 8-coordination. Both ions have Neon-configuration, so that $n = 7$. The sum of the univalent radii is 2.75 Å. Dividing by $\sqrt[6]{2}$ we obtain 2.44, which is the Na-O distance to the expected in 6-coordination. The distance in 8-coordination we obtain by multiplying with the factor 1.044 from table II. I.e. the Na-O distance for C.N. 8 is 2.55 Å.

6. Comparison with Goldschmidt's and Pauling's Radii

Pauling's theoretically derived univalent radii are given in brackets in table IV [partly reproduced]. It will be seen at once that the agreement is excellent for small valencies; but not so good for high valencies. This lack of agreement in the latter case may be due to the fact that the conception of ionic bindings cannot be maintained for such ions. A comparison with Goldschmidt's radii or with Pauling's final set cannot be made directly. From our univalent radii we can calculate the radius for each ion for C.N. 6 in binary compounds AX. Even by this limitation the radius will not be constant, but will depend upon the value used for n. The radii given in table V are obtained by the use of the value of n which is characteristic of the ion. That is, the radii given are the univalent ones divided by $n^{-1}\sqrt{z_2{}^2}$, n having the appropriate values taken from table I.

Table V.

Comparison between Different Sets of Radii.

	H^{-1}	Li^{+1}	Be^{+2}	B^{+3}	C^{+4}
Z.	1.36	0.68	0.39	0.24	0.19
G.	1.27	0.78	0.34	–	0.20
P.	2.08	0.60	0.31	0.20	0.15
	O^{-2}	F^{-1}	Na^{+1}	Mg^{+2}	Al^{+3}	Si^{+4}
Z.	1.40	1.33	0.98	0.71	0.55	0.44
G.	1.32	1.33	0.98	0.78	0.57	0.39
P.	1.40	1.36	0.95	0.65	0.50	0.41
	S^{-2}	Cl^{-1}	K^{+1}	Ca^{+2}	Sc^{+3}	Ti^{+4}
Z.	1.85	1.81	1.33	0.98	0.78	0.62
G.	1.74	1.81	1.33	1.06	0.83	0.64
P.	1.84	1.81	1.33	0.99	0.81	0.68
	Se^{-2}	Br^{-1}	Rb^{+1}	Sr^{+2}	Y^{+3}	Zr^{+4}
Z.	1.96	1.96	1.48	1.15	0.93	0.79
G.	1.91	1.96	1.49	1.27	1.06	0.87
P.	1.98	1.95	1.48	1.13	0.93	0.80
	Te^{-2}	I^{-1}	Cs^{+1}	Ba^{+2}	La^{+3}	Ce^{+4}
Z.	2.18	2.19	1.67	1.31	1.06	0.89
G.	2.11	2.20	1.65	1.43	1.22	1.02
P.	2.21	2.16	1.69	1.35	1.15	1.01

9. Comparison with Observations

In table VIII [reproduced for a few compounds only] I have compared some observed interionic distances with the ones calculated from my set of radii, as well as from Goldschmidt's and Pauling's (the final values) sets.

Table VIII.

Comparison with Observations.

A-X	Obs. Dist.	Compound	C.N.	Calculated Dist.		
				Z.	G.	P.
Li-O	2.15	$LiNO_3$	6	2.13	2.10	2.00
	2.00	Li_2O	4	2.00		
Li-F	2.01	LiF	6	2.01	2.11	1.96
Li-S	2.47	Li_2S	4	2.42	2.38	2.31
Li-Cl	2.57	LiCl	6	2.49	2.59	2.41
Li-Br	2.75	LiBr	6	2.64	2.74	2.55
Li-I	3.03	LiI	6	2.87	2.98	2.76
Be-O	1.65	BeO	4	1.64	1.55	1.60
Be-S	2.10	BeS	4	2.07	1.97	2.00
Be-Se	2.18	BeSe	4	2.18	2.14	2.18
Be-Te	2.43	BeTe	4	2.40	2.34	2.41
B-N	ca. 1.45	BN	3	1.42	–	1.70
B-O	ca. 1.35	$Be_2BO_3 . OH$	3	1.37	–	1.42
Na-N	2.50	$NaNN_2$	6	2.50	–	2.66
Na-O	2.46	$NaClO_3$	6	2.44	2.30	2.35
		Na_2SO_3				

This comparison shows that the set of radii given in the present paper is based on more logical foundations and also allows by far more accurate predictions of interionic distances.

Ryerson Physical Laboratory, University of Chicago

120. Die vorliegende Untersuchung verfolgt den Zweck, zu zeigen, wie vorsichtig man bei der Beurteilung der Atomabstände bei jenen Strukturen sein muß, in welchen nicht allen Atomschwerpunkten durch die Symmetrie fest gebundene Lagen zukommen. Als einfachstes Beispiel in dieser Hinsicht sei der Spinell Al_2MgO_4 gewählt, dessen Atomabstände vielfach als Beispiel gegen die Konstanz der Ionenradien angeführt werden. So stammt der von P. Niggli angeführte, ungewöhnlich kleine Minimalabstand Mg − O = 1,75 Å zweifellos aus der allzu schematisch gedeuteten Spinellstruktur.

Der Strukturtyp des Spinell umfaßt bekanntlich eine große Anzahl chemisch verschiedener Glieder, darunter auch eine beträchtliche Anzahl von Mineralien. Die Struktur ist nach W. H. Bragg (6) und nach S. Nishikawa (7) folgendermaßen charakterisiert:

Die kubische Elementarzelle enthält 8 Moleküle ($Al_2Mg_2O_4$ beim Spinell im engeren Sinne). Sie baut sich aus flächenzentrierten Teilgittern auf. Raumgruppe ist O_h^7. Es liegen

16 Al in 16c: 5/8, 5/8, 5/8 usw.
8 Mg in 8f: 000 usw.
32 O in 32b: uuu usw.

Die Gitterkonstante des gewöhnlichen Spinells beträgt nach W. H. Bragg (l.c.) 8,07 Å.

Der Parameter u, der die Lage der Sauerstoffatome näher fixiert, wird bei allen Spinellstrukturen meistens zu $\frac{3}{8}a$ angegeben. Das bedeutet für den gewöhnlichen Spinell, daß jedes Al von 6O streng oktaedrisch, jedes Mg von 4O streng tetraedrisch umgeben wäre. Die Sauerstoffatome würden in diesem Falle eine ideale (allerdings etwas aufgelockerte) dichteste Kugelpackung bilden. Mit dem Parameterwerte $u = \frac{3}{8}a$ ergeben sich folgende Atomabstände:

$$Mg - O = a/8\sqrt{3} = 1,75 \text{ Å},$$
$$Al - O = a/4 = 2,02 \text{ Å},$$
$$O - O = a/4\sqrt{2} = 2,86 \text{ Å}.$$

Die empirischen Ionenradien von V. M. Goldschmidt (z.B. Lit. 9) ergeben einen Abstand $Mg - O = 2,10$ Å für den Fall, daß Mg von 6O umgeben ist. Wenn dagegen nur 4O das Mg-Ion umgeben, wäre ein etwa 5–8% kleinerer Abstand zu erwarten, also 1,93–1,99 Å, so daß der für Spinell mit $u = \frac{3}{8}a$ gefundene Abstand Mg–O tatsächlich gegen die Konstanz der Ionenradien zu sprechen scheint.

In ähnlicher Weise entspricht der Abstand Al-O nicht dem nach V. M. Goldschmidt zu erwartenden Werte. Danach sollte der Abstand Al–O, wenn Al von 6O umgeben ist, 1,89 Å, vielleicht etwas höher sein.

S. Nishikawa gibt nun aber schon auf Grund von Laue-Aufnahmen am Spinell von Ceylon den Parameter u der Sauerstoffatome nicht zu $\frac{3}{8}a$, sondern zu 0,384a an.

Mit letzterem Parameterwerte findet man folgende Atomabstände:

Mg–O = 1,88 Å;
Al–O = 1,95 Å;
O–O = 2,65 (je 3 Nachbarn), 2,86 (je 6 Nachbarn) und 3,07 Å (je 3 Nachbarn).

Damit ist schon eine beträchtliche Annäherung an die auf Grund der Ionenradien zu erwartenden Werte erreicht.

Ich habe nun versucht, aus Pulveraufnahmen von rotem Spinell von Ceylon unter Verwendung der atomaren F-Kurven, wie sie von W. L. Bragg und J. West angegeben werden (10) einen die beobachteten und geschätzten Intensitäten gut wiedergebenden Wert für u zu ermitteln. Meines Wissens wurde bisher für die Parameterermittlung beim Spinell von den F-Kurven noch nicht Gebrauch gemacht[1].

In Tabelle I [not reproduced] sind die beobachteten und die berechneten Intensitäten, letztere berechnet für $u = \frac{3}{8}a = 0,375a$, für $u = 0,390a$ und z.T. auch berechnet für $u = 0,40a$ einander gegenüber gestellt.

Bei Annahme des Parameters u zu $\frac{3}{8}a$ führt die Berechnung zu merklichen Widersprüchen mit den beobachteten Intensitäten. Auch der Parameter $u = 0,40a$ läßt sich mit ziemlicher Sicherheit ausschließen. Derartige Abweichungen zwischen beobachteter und berechneter Intensität fehlen, wenn man den Parameter u zu 0,39 ansetzt. Man kann keine Feststellung machen, welche gegen diesen Parameter sprechen würde.

Sicherlich sind die F-Kurven nicht so genau bekannt und die Intensitätsschätzungen nicht so zuverlässig, daß man etwa den von Nishikawa (7) angegeben Parameter $u = 0,384a$ mit Sicherheit ausschließen könnte. Die vorliegende Untersuchung will nur zeigen, daß auch ein etwas höherer Parameterwert unter Berücksichtigung der F-Kurven zu einer einwandfreien Übereinstimmung zwischen beobachteten und berechneten Intensitäten führt und auf dieser Grundlage wird man den Parameter u wohl auf den Bereich $0,388 \pm 0,005a$ einschränken können*. Vielleicht wäre es mit Hilfe von absoluten Intensitätsmessungen und unter Anwendung der Fourieranalyse auf dieselben möglich, eine schärfere Parameterbestimmung durchzuführen.

Was bedeutet aber nun der vollkommen im Bereich der Möglichkeit liegende Parameterwert $u = 0,390a$ für die Atomdistanzen im Gitter des gewöhnlichen Spinell?

Die Gitterkonstante des Spinell von Ceylon finde ich zu $8,08 \pm 0,01$ Å. Der Parameter $u = 0,390a$ bedingt folgende kürzeste Abstände:

Mg–O = 1,96 Å;
Al–O = 1,91 Å;
O–O = 2,52 Å (je 3 Nachbarn), 2,87 Å (je 6 Nachbarn) und
    3,20 Å (je 3 Nachbarn).

Die Werte für Al–O und O–O stimmen dann recht gut mit den entsprechenden Abständen im Korund überein. Hier ist der Abstand Al–O

---

[1] Dagegen wohl für den Fall des Magnetites (A. Claasen, *Pr. Physic. Soc. London* **38** (1926) 482).

* [Neutron diffraction (Bacon, 1952) gives $u = 0.327 \pm 0.001$. Ed.]

im Mittel gleich 1,92 Å. Der kürzeste Abstand zweier O-Atome = 2,49 Å. Der Genauigkeitsgrad auch dieser Werte muß natürlich, da es sich nicht um symmetriefixierte Atomlagen handelt, mit Vorsicht beurteilt werden.

Auch im Chrysoberyll liegt der Abstand Al–O nahe bei 1,93 Å.

Mit dem Parameter $u = 0,390a$ ergibt sich für die Struktur des Spinell folgendes Bild: Die Mg-Atome sind tetraedrisch von O-Atomen umgeben. Die Kantenlängen der Tetraeder sind durch den größten O–O-Abstand (3,20 Å) gegeben. Die Al-Atomen werden von den O-Atomen in oktaederähnlicher Gruppierung umgeben. Alle Abstände Al–O sind gleich. Die Kanten der 'Oktaeder' zerfallen aber in zwei Gruppen; 6 Kanten sind durch den O–O-Abstand 2,87 Å gegeben, 6 Kantenlängen aber durch den O–O-Abstand 2,52 Å. Diese Kurzkanten sind jeweils gemeinsam mit je einem benachbarten AlO_6-'Oktaeder'[1], im übrigen stoßen in jedem O-Atom 3 AlO_6-'Oktaeder' und ein MgO_4-Tetraeder zusammen.

Nach dieser Darstellung der Spinellstruktur scharen sich die O-Atome enger um die Al-Atome als um die Mg-Atome.

Besserer Übersichtlichkeit halber seien im folgenden noch einmal die Atomdistanzen im Gitter des Spinell unter Annahme verschiedener Parameterwerte für die Sauerstoffatome zusammengestellt:

	$u=0{,}375a$	$u=0{,}384a$	$u=0{,}390a$	Berechnet[2]
Mg–O	1,75 Å	1,88 Å	1,96 Å	1,93–1,99 Å
				(4-Koordination)
Al–O	2,02 Å	1,95 Å	1,91 Å	mind. 1,89 Å
O–O	2,86 Å	2,65 Å	2,52 Å	2,64 Å

V. M. Goldschmidt (12) sah unter Annahme des Parameters $u = 0,375a$ in der Spinellstruktur einen extremen Kontrapolarisationsfall unter Bildung eines neuen, lockeren Radikals MgO_4 im Kristallgitter. Mit der Feststellung, daß der Parameterwert in der Nähe von 0,39a liegen muß, fällt die Nötigung zu einer solchen Darstellung weg. Die geringere Raumbeanspruchung und höhere Wertigkeit des Ions Al^{+3} gegenüber dem Ion Mg^{+2} macht eine kräftige Kontrapolarisation unter Bildung eines neuen Radikals MgO_4 im Gitter ohnehin unwahrscheinlich.

Die vorliegende Untersuchung zeigt somit, daß die Angabe des Parameterwertes $u = \frac{3}{8}a$ bei Spinellstrukturen im allgemeinen nur die Bedeutung hat, daß dies ein angenehmer Wert für die Intensitätsberechnung ist, der zu keinen derartig großen Widersprüchen zwischen Beobachtung und Berechnung führt, als daß man deshalb die Struktur verwerfen müßte. Irgend-

---

[1] Dies bedeutet ein Abdrängen der Al-Atomen von der gemeinsamen Oktaederkante und damit voneinander, was ebenfalls sonstiger Erfahrung entspricht.

[2] Aus den empirischen Ionenradien von V. M. Goldschmidt.

welche nähere Rückschlüsse über Atomabstände daraus zu ziehen, ist aber nicht angängig. Daher können im allgemeinen schon aus solchen relativ einfachen Strukturen mit nur einem variablen Parameter, wie es beim Spinell der Fall ist, keine Schlüsse gegen die Konstanz der Raumbeanspruchung der Baubestandteile der Kristalle bei sonst vergleichbaren Verhältnissen gezogen werden. Noch viel weniger natürlich, wenn auf die Parameterbestimmung nicht besondere Sorgfalt verwendet wird, aus Strukturen mit einer größeren Anzahl von variablen Parametern.

Literatur

6. W. H. BRAGG, *Nature* **95** (1915) 561; *Phil. Mag.* **30** (1915) 305–315.
7. S. NISHIKAWA, *Proc. Tok. Math. Phys. Soc.* **8** (1915) 199–209.
9. V. M. GOLDSCHMIDT, *Trans. Faraday Soc.* No. 97 **25** (1929) 253–283.
10. W. L. BRAGG und J. WEST, *Z. Krist.* **69** (1928) 118–148.
12. V. M. GOLDSCHMIDT, *Geoch. Vert. Ges.* **7** (1926) 67.

Mineralogisches Institut der Universität, Tübingen

A2. ATOMIC RADII

121. Covalent Radii of Atoms and Interatomic Distances in Crystals containing Electron-Pair Bonds, by L. PAULING and M. L. HUGGINS (1934)

▄▄▄

121. *Abstract*

A simple quantum-mechanical discussion of the orbitals (one-electron orbital eigenfunctions) of an atom is shown to lead to the possibility of formation of various sets of covalent bonds, differing in number and spatial distribution. Among these are tetrahedral sp^3 bonds, octahedral d^2sp^3 bonds, square dsp^2 bonds, etc. After a discussion of the criteria which may be applied to distinguish crystals of essentially covalent rather than ionic or metallic character, observed interatomic distances are used as a basis for formulating sets of radii corresponding to different bond types, including tetrahedral, octahedral, square, trigonal-prism, and normal-

59

valence bonds. [Only discussion of tetrahedral and normal-valence bonds inserted. Ed.] It is found that these radii show various reasonable inter-relations, and that they are in general agreement with the interatomic distances determined experimentally, cognizance being taken when necessary of resonance among several electronic structures. A remarkable and puzzling exception is the octahedral radius of manganese in the pyrite-type crystals MnS_2 and $MnTe_2$, the observed value being over 0.30 Å larger than expected. A somewhat detailed discussion is then given of the structures of several crystals.

Of the three principal classes of crystals, ionic crystals, crystals containing electron-pair bonds (covalent crystals), and metallic crystals, we feel that a good understanding of the first class has resulted from the work done in the last few years. Interionic distances can be reliably predicted with the aid of the tables of ionic radii obtained by Goldschmidt[1] by the analysis of the empirical data and by Pauling[2] by a treatment based on modern theories of atomic structures. The stability, crystal energy, and other properties of simple ionic crystals have been extensively discussed by Born, Goldschmidt, Mayer, and others[3]; and the nature of complex ionic crystals has been elucidated by the recent extensive structure determinations of silicates and related minerals carried out mainly by W. L. Bragg[4] and his school and by the formulation[5] of a set of structural principles for such crystals.

The application of the Lewis theory of the electron-pair bond to crystals was begun early by Huggins[6], who in 1922 ascribed to diamond, sphalerite, wurtzite, carborundum, chalcopyrite, pyrite, marcasite, cobaltite, arsenic, antimony, bismuth, and related crystals the electron-pair-bond structures which are now accepted for them, and which were independently suggested for tetrahedral crystals (diamond, sphalerite, wurtzite, carborundum) by Grimm and Sommerfeld[7] four years later. In 1926 Huggins[8] published a set of atomic radii for use in crystals containing electron-pair bonds. Goldschmidt in the same year published 'atomic radii' obtained from ob-

---

[1] V. M. Goldschmidt, Geochemische Verteilungsgesetze der Elemente. (1926).

[2] L. Pauling, *J. Am. Chem. Soc.* **49** (1927) 765. See also W. H. Zachariasen. *Z. Krist.* **80** (1931) 137 and Huggins and Mayer, *J. chem. Physics* **1** (Nov. 1933).

[3] J. Sherman, *Chem, Rev.* **11** (1932) 93.

[4] W. L. Bragg, *Z. Krist.* **74** (1930) 237.

[5] Linus Pauling, *J. Am. Chem. Soc.* **51** (1929) 1010.

[6] M. L. Huggins, *J. Am. Chem. Soc.* **44** (1922) 1841. A structure of this type was also assigned to calcite, for which an ionic structure (Ca^{++} and CO_3^{--}) is now accepted.

[7] H. G. Grimm and A. Sommerfeld, *Z. Physik* **36** (1926) 36.

[8] M. L. Huggins, *Physic. Rev.* **28** (1926) 1086.

served interatomic distances in metallic crystals as well as covalent crystals; more recently[1] he has collected these and additional radii into a table of radii for use in metals and intermetallic compounds.

During the last two years the application of the quantum mechanics to the problem of the nature of the electron-pair bond and the development of a theory of the paramagnetic susceptibility of molecules and crystals containing electron-pair bonds[2] has clarified the field considerably, and has led us to attack the problem of the formulation of a set of principles governing the structure of crystals containing electron-pair bonds, the first step being the discussion of interatomic distances. In the following sections of this paper there are given a discussion of the covalent bond according to the quantum mechanics, the application of criteria as to the nature of the bonds in various crystals to segregate those containing covalent bonds, the determination of sets of electron-pair-bond or covalent radii, and the discussion of the structures of some covalent crystals.

The Electron-Pair Bond

A semi-quantitative method of treatment of bond eigenfunctions leads to information on the strengths and relative orientation of bonds[3]. The rules are the following.

1. The single electron-pair bond between two atoms involves one orbital[4] for each atom and one pair of electrons.

2. In such a bond the spins of the electrons are opposed, and hence make no contribution to the magnetic moment of the molecule.

3. Two electrons which form a shared pair cannot take part in forming additional pairs.

4. The main resonance terms for a single electron-pair bond are those involving the two orbitals (one from each atom) associated with the bond.

5. Of two orbitals with the same dependence on r, the one with the larger value in the bond direction will give rise to the stronger bond, and for a given orbital the bond will tend to be formed in the direction in which the orbital has its maximum value.

[1] V. M. Goldschmidt, *Trans. Faraday Soc.* **25** (1929) 253.

[2] Linus Pauling, *J. Am. Chem. Soc.* **53** (1931) 1367.

[3] Linus Pauling, *Pr. Nat. Acad.* **14** (1928) 359; *J. Am. Chem. Soc.* **53** (1931) 1367. Hereafter called Ref. I. A detailed treatment of the theory is given by J. C. Slater, *Physic. Rev.* **38** (1931) 1109.

[4] Following Mulliken, we shall use the word 'orbital' to represent a one-electron orbital eigenfunction of an atom.

6. Of two orbitals with the same dependence on ϑ and φ, the one corresponding to the lower energy level for the atom will usually give rise to the stronger bond.

Here r, ϑ, and φ are polar coordinates used in the description of the orbitals of an atom, the nucleus being at the origin of the coordinate system.

We should not expect s orbitals to lead to strong bonds (except in the K-shell), on account of their spherical symmetry. Somewhat stronger bonds, tending to make angles of 90° with each other, are formed by p orbitals. In most cases when bonds are formed, however, the s und p orbitals do not retain their identity, but instead combine linearly to new orbitals especially suited to bond formation. The criterion for the occurrence of this *change in quantization* or *hybridization* of bond orbitals involves the relative bond energies for s and p orbitals and hybrid orbitals, and the difference in energy of s and p orbitals in the atom[1].

It has been found that a simple and powerful approximate method of treatment of bond orbitals can be developed by assuming that the dependence on r of s, p, and d orbitals corresponding to about the same energy is nearly the same, so that the r-portion of the resonance terms can be taken as the same. The consideration of the ϑ, φ portion of the orbitals then leads directly to the determination of the best bond orbitals which can be formed under given circumstances.

Putting

$$\Psi_{n0}(r, \vartheta, \varphi) = R_{n0}(r)s(\vartheta, \varphi) \text{ for } s \text{ orbitals,}$$

$$\left.\begin{array}{l} \Psi_{n1}(r, \vartheta, \varphi) = R_{n1}(r)p_x(\vartheta, \varphi) \\ \qquad\qquad\quad p_y(\vartheta, \varphi) \\ \qquad\qquad\quad p_z(\vartheta, \varphi) \end{array}\right\} \text{for } p \text{ orbitals,}$$

$$\left.\begin{array}{l} \Psi_{n2}(r, \vartheta, \varphi) = R_{n2}(r)d_z(\vartheta, \varphi) \\ \qquad\qquad\quad d_{x+z}(\vartheta, \varphi) \\ \qquad\qquad\quad d_{y+z}(\vartheta, \varphi) \\ \qquad\qquad\quad d_{x+y}(\vartheta, \varphi) \\ \qquad\qquad\quad d_x(\vartheta, \varphi) \end{array}\right\} \text{for } d \text{ orbitals,}$$

the parts s, p_x, etc. of the orbitals depending on ϑ and φ, normalized to 4π, are

---

[1] A detailed discussion of this question and of the assumptions underlying our approximate treatment of bond strengths is given in the interesting papers of J. H. Van Vleck, *J. Chem. Physics* **1** (1933) 177, 219.

$$s = 1 \qquad\qquad d_{x+z} = \sqrt{15}\ \sin\vartheta\cos\vartheta\cos\varphi$$

$$p_x = \sqrt{3}\ \sin\vartheta\cos\varphi \qquad\qquad d_{y+z} = \sqrt{15}\ \sin\vartheta\cos\vartheta\sin\varphi$$

$$p_y = \sqrt{3}\ \sin\vartheta\sin\varphi \qquad\qquad d_{x+y} = \sqrt{15/4}\ \sin^2\vartheta\sin 2\varphi$$

$$p_z = \sqrt{3}\ \cos\vartheta \qquad\qquad d_x = \sqrt{15/4}\ \sin^2\vartheta\cos 2\varphi.$$

$$d_z = \sqrt{(5/4)}(3\cos^2\vartheta - 1)$$

Now by Rule 5 the bond-forming power of an orbital in a given direction increases as the magnitude of the orbital in that direction increases, and it may be conveniently expressed in a qualitative manner by that magnitude, which we shall call the bond-strength of the orbital. Thus the maximum bond-strength of an s orbital is 1, that of a p orbital $\sqrt{3} = 1.732$, and that of a d orbital $\sqrt{5} = 2.237$. In constructing a bond orbital our problem is to find a normalized linear combination of these orbitals which has a large concentration in one direction. When two or more bonds are formed by an atom, a bond orbital must be constructed for each bond with large concentration in the direction of that bond, and these orbitals must be mutually orthogonal. The nature of the bond orbital depends on the orbitals available for linear combination. By Rule 6 orbitals corresponding to unstable states of the atom will not be useful. Thus in carbon or any atom of the first horizontal row of the periodic system the $2s$ and $2p$ levels of the L-shell are stable, and all outer levels, in the M, N, etc. shells, are very unstable, so that the only orbitals to be considered are $2s$, $2p_x$, $2p_y$, and $2p_z$. In general ns and np are to be grouped together, for the difference in term values is usually small. This group of four orbitals ns, np_x, np_y, and np_z gives rise to the octet of Lewis and Langmuir. Even in the M-shell and outer shells where there is a group of five d orbitals they usually correspond to energy values considerably higher than the s and p levels of the same shell (the p–d separation being over twice the s–p separation) and so are not usually brought into use in forming bond orbitals. There are exceptions to this rule, however, especially for very heavy atoms. d orbitals usually become important when the d level of a shell lies close to the s and p levels of the next outer shell, as $3d$, $4s$, $4p$ for the iron group, $4d$, $5s$, $5p$ for the palladium group, and $5d$, $6s$, $6p$ for the platinum group. Then the nature of the bond orbitals which can be formed depends very largely on the number of available d orbitals. The principal types of bond orbitals other than the p orbitals already mentioned are described in the following paragraphs.

The most important case is that of s, p_x, p_y, and p_z. On setting up a general expression for a linear combination of these and varying the coefficients to make the bond-strength a maximum (for details of the calculation see Ref. I), it is found that the best *sp* bond orbital which exists, has a strength of 2.000. The orbital corresponding to a second bond formed by this atom must be orthogonal to this; on evaluating it is found to be equivalent to the first, with a strength of 2.000, and *to make an angle of 109°28′, the tetrahedral angle*, with it. A third and a fourth equivalent orbital, also at tetrahedral angles and orthogonal to the others, can be constructed. These four may be written as

$$\psi_{111} = \tfrac{1}{2}(s + p_x + p_y + p_z)$$
$$\psi_{1\bar{1}\bar{1}} = \tfrac{1}{2}(s + p_x - p_y - p_z)$$
$$\psi_{\bar{1}1\bar{1}} = \tfrac{1}{2}(s - p_x + p_y - p_z)$$
$$\psi_{\bar{1}\bar{1}1} = \tfrac{1}{2}(s - p_x - p_y + p_z).$$

Thus we have shown that when *s* and *p* orbitals are available and $s - p$ quantization is broken an atom can form four (or fewer) equivalent bonds which are directed towards tetrahedron corners. To the approximation involved in these calculations the strength of a bond is independent of the nature of other bonds. This result gives us at once the justification for the tetrahedral carbon atom and other tetrahedral atoms, such as silicon, germanium, and tin in the diamond-type crystals of the elements, and in general, all atoms in tetrahedral structures.

In bivalent nickel, palladium, or platinum there are eight electrons in the outer *d* subshell. Putting them two to an orbital, they occupy a minimum of four of the five *d* orbitals, leaving only one *d* orbital available for bond formation through combination with *s*, p_x, p_y, and p_z of the next outer shell. It is found that only four strong bond orbitals can be formed. These four lie in a plane and are directed towards the four corners of a square so that we may conveniently call them *square bond orbitals*. They have a strength of 2.694. The fifth orbital, p_z, has its maxima normal to the plane. Usually it is not engaged in bond formation, on account of its small strength 1.732. Square bond orbitals are involved in $K_2Ni(CN)_4$, K_2PdCl_4, K_2PtCl_4, $KAuCl_4$, etc.

When six or fewer *d* electrons are present there are two *d* orbitals available for bond formation, which, on combination with *s*, p_x, p_y, and p_z, are found to give rise to six equivalent d^2sp^3 orbitals of strength 2.923, with their maxima directed towards the corners of a regular octahedron. These *octahedral bond orbitals* are involved in a great many complexes, such as $[Co(CN)_6]^{\equiv}$, $[PtCl_6]^{=}$, etc. and are of great importance in crystals containing transition group elements.

When more *d* orbitals are available, six stronger equivalent eigenfunctions with their maxima directed towards the corners of a trigonal prism of unit axial ratio can be formed[1]. The bond-strength of these is 2.985, nearly equal to the maximum 3.000 possible for *dsp* orbitals.

It is probable that eight equivalent bond orbitals can be constructed when four *d* orbitals are available, as in $[Mo(CN)_8]^{=\,=}$ and $[W(CN)_8]^{=\,=}$. The arrangement of these in space has not yet been determined either experimentally or theoretically.

The Determination of the Type of Bond in Crystals. Resonance of Molecules and Crystals Among Several Electronic Structures

Questions such as, for example, whether sphalerite contains Zn^{++} and $S^=$ ions or has a covalent structure similar to that of diamond, and whether ionic or covalent bonds are present in complexes such as $[FeF_6]^\equiv$, $[Fe(CN)_6]^\equiv$, etc., have been extensively discussed; it has, indeed, until recently not been at all clear whether or not they could be definitely answered. The quantum mechanics has thrown much light on this subject[2]. By the methods developed by Slater it is possible to formulate a wave function corresponding to a given Lewis structure for a molecule or crystal, and it has been found that for many molecules and crystals such a wave function provides a good approximation to the normal state. In other cases, however, two or more such functions may be formulated which correspond to nearly the same energy value. If these functions satisfy certain conditions permitting them to combine with one another, in particular if they have the same multiplicity[3], then a more general wave function formed by linear combination of these will provide a better approximation to the normal state than any one of the original functions. The molecule or crystal is then said to *resonate among the various electronic structures* corresponding to the original wave functions. This resonance phenomenon is closely similar to that discovered by Heisenberg and Dirac in treating many-electron atoms and applied by Heitler and London to the hydrogen molecule. It should be pointed out that the resonance obtained depends on the initial choice of wave functions. For example, if hydrogen-atom wave functions are selected originally in discussing the hydrogen molecule-ion H_2^+, as was done by Pauling[4], then the electron may be said to resonate between the two protons; on the other hand, resonance does not appear in Burrau's

---

[1] These were discovered by R. Hultgren, *Physic. Rev.* **40** (1932) 891.

[2] For a detailed discussion see L. Pauling, *J. Am. Chem. Soc.* **54** (1932) 988.

[3] That is, the same number of unpaired electrons (one less than the multiplicity of the state).

[4] L. Pauling, *Chem. Rev.* **5** (1928) 173.

treatment[1] of the molecule-ion by the numerical solution of the wave equation for an electron in the field of two protons. It is consequently the lucidity and convenience of a treatment based on Lewis electronic structures (with some extensions, such as the one-electron bond, the three-electron bond, etc.) which recommends their choice in formulating initial wave functions and discussing their combination.

Arguments based mainly on bond energies[2] and interatomic distances[3] have recently led to the determination of the normal electronic structures of a number of molecules. The hydrogen halides resonate between the extreme ionic structure H^+X^- and the covalent structure $H : \overset{..}{\underset{..}{X}} :$, the contribution of the former being largest for HF and of the latter for HI. Carbon monoxide resonates between the structures $: C :: \overset{..}{O} :$ and $: C ::: O :$, nitrous oxide between the structures $: \overset{..}{N} :: N :: \overset{..}{O} :$ and $: N ::: N : \overset{..}{O} :$, etc.

In such cases as these it is evident that a continuous transition from one extreme structure to another could occur. If, however, the structures have different multiplicities, they cannot be combined with one another (so long as spin-orbit interactions are negligible), so that the transition from one extreme bond type to the other would be effectively discontinuous.

Both the ionic and the covalent structure of sphalerite, for instance, are ainglet structures, with no unpaired electrons, so that either extreme or sny intermediate is possible, and in such a case evidence from various properties of the particular substance must be considered to decide which extreme is more closely approached. On the other hand, in a crystal such as $(NH_4)_3FeF_6$ or $(NH_4)_3Fe(CN)_6$ the lowest ionic state of the $[FeX_6]^=$ complex does not combine with the lowest covalent state, so that the transition from one extreme to the other is discontinuous. The actual state of the complex in the crystal can be determined from the multiplicity. With an ionic state, Fe^{+++} and $: \overset{..}{\underset{..}{F}} :^-$ or $(: C ::: N :)^-$, the F^- or $(CN)^-$ groups contain no unpaired electrons. Fe^{+++} contains five d electrons, which, in accordance with the rules of line spectra, distribute themselves among the five $3d$ orbitals in such a way as to avoid pairing unless required. In this case they remain unpaired, and the ion assumes a sextet state, ^6S, the state of the complex being $^6\Sigma$. If covalent bonds are formed, two of the d orbitals are used in the octahedral d^2sp^3 bonds, leaving only three d orbitals for the five d electrons. Hence two pairs are formed, leaving one

[1] Ø. Burrau, Det. Kgl. Danske Vid. Selsk. Math.-fys. Meddeleker 7 (1927) 14.
[2] L. Pauling and D. M. Yost, Pr. Nat. Acad. 18 (1932) 414; L. Pauling, J. Am. Chem. Soc. 54 (1932) 3570.
[3] L. Pauling, Pr. Nat. Acad. 18 (1932) 293, 498.

unpaired electron, and the complex[1] acquires the doublet state, $^2\Sigma$. The actual multiplicity, r, of the complex can be determined experimentally from the paramagnetic susceptibility, the magnetic moment being given by the equation

$$\mu = \sqrt{(r^2 - 1)} = 2\sqrt{\{S(S + 1)\}}.$$

S is the spin quantum number. The expected magnetic moments for a sextet and a doublet state are 5.91 and 1.73 respectively, measured in Bohr magnetons. The values calculated from the observed paramagnetic susceptibilities of the crystals are 5.88 for $(NH_4)_3FeF_6$ and 2.0 for $K_3Fe(CN)_6$. Accordingly we can say with certainty that the $[FeF_6]^{\equiv}$ ion contains essentially ionic bonds, and the $[Fe(CN)_6]^{\equiv}$ ion covalent bonds.

Further statements can be made with the aid of the available data. In most compounds atoms of the palladium and platinum groups form covalent bonds. The iron-group atoms form covalent bonds with cyanide, sulfur, etc., but usually ionic or ion-dipole bonds with H_2O, NH_3, SO_4, F, Cl, Br, I, etc. The presence of some covalent bonds facilitates the formation of more; thus five CN bonds in $[Fe(CN)_5NH_3]^{\equiv}$ stabilize a covalent bond to NH_3.

When the multiplicity of a complex is the same for ionic or ion-dipole bonds and for covalent bonds, the decision as to which extreme bond type is the more closely approached in any actual case must be made with the aid of less straightforward arguments. Sometimes theoretical energy diagrams can be constructed with sufficient accuracy to decide the question.

Sometimes the atomic arrangement of a crystal is such as not to permit the formulation of a covalent structure. This is the case for the sodium chloride arrangement, as the alkali halides do not contain enough electrons to form bonds between each atom and its six equivalent nearest neighbors. This criterion must be applied with caution, however, for in some cases electron pairs may jump around in the crystal, giving more bonds than there are electron pairs, each bond being of an intermediate type. It must also be mentioned that determinations of the atomic arrangement are sometimes not sufficiently accurate to provide evidence on this point; an atom reported equidistant from six others may be somewhat closer to three, say, than to the other three.

Agreement or non-agreement with the rules governing the structures of ionic crystals also gives evidence regarding the bond type. The complex silicates, which conform excellently to the rules, are probably largely ionic

---

[1] On account of the quenching of the orbital mechanical and magnetic moment of atoms on bound formation, all electron-pair-bond complexes are to be represented as Σ states.

in nature. The tetrahedral arrangement of four oxygen atoms about a silicon atom would be expected whether the bonds are ionic or covalent. An ionic structure would allow the oxygen bonds to assume any angles, large angles being favored, whereas covalent bonds of oxygen would tend to the angles 90° or 109.5°. The observed angles of 180° in β-tridymite suggest that the bonds are largely ionic, and the smaller angles in the α-forms and in quartz suggest that the bonds may also have considerable covalent character. Corundum and hematite contain coordinated octahedra which share faces with one another, and the edges of these octahedra are shortened, as is expected for ionic crystals. The shared octahedral edges in rutile, anatase and brookite are similarly shortened. But the shared edges of octahedra in marcasite are larger than unshared edges, an indication that in this substance the bonds are not ionic. In As_4O_6 three of the six oxygen atoms about each arsenic atom are closer to it than the other three, and an ionic structure provides no explanation of this fact. The observed square arrangement of four chlorine atoms about each platinum atom in K_2PtCl_4 and the trigonal prism of sulfur atoms about molybdenum in molybdenite are not expected for crystals built of ions. These atomic arrangements are, on the other hand, just those expected if covalent bonds are formed. Such agreement or disagreement with the rules governing covalent bonds, especially bond angles, is another valuable criterion of bond type. Light electronegative atoms such as oxygen can form only p-bonds or tetrahedral bonds, at angles of about 90°–110°. The structure of cuprite is such as to surround each oxygen by four copper atoms, so that covalent bonds may well be present in Cu_2O (and Ag_2O).

Ionic bonds may be fully as strong as covalent bonds, so that properties such as hardness, solubility, melting point, ionization in solution, and chemical character are not especially valuable criteria as a rule. Sometimes comparison of properties with those of compounds of known bond type permits reasonably certain conclusions to be drawn. Thus the similarity in physical properties as well as in atomic arrangement of SiC, AlN, and diamond suggests that all three substances contain covalent bonds. PbS is like FeS_2, MoS_2, etc. in properties rather than like CaS, so that it is improbable that PbS is an ionic substance.

Covalent Radii of Atoms

We have constructed a number of sets of atomic radii for use in compounds containing covalent bonds. These radii have been obtained from the study of observed interatomic distances. They are not necessarily applicable only to crystals containing pure covalent bonds (it is indeed probable that very few crystals of this type exist); but also to crystals and molecules in which the bonds approach the covalent type more closely than the ionic or metallic type. The crystals considered to belong to this class are 'tetrahedral' crystals, pyrite and marcasite-type crystals, and others which we have been found on application of the various criteria discussed in the preceding section to contain covalent bonds or bonds which approach this extreme.

Table III.

Standard Tetrahedral Radii.

		Be	B	C	N	O	F
		1.07	0.89	0.77	0.70	0.66	0.64
		Mg	Al	Si	P	S	Cl
		1.40	1.26	1.17	1.10	1.04	0.99
Cu		Zn	Ga	Ge	As	Se	Br
1.35		1.31	1.26	1.22	1.18	1.14	1.11
Ag		Cd	In	Sn	Sb	Te	I
1.53		1.48	1.44	1.40	1.36	1.32	1.28
Au		Hg	Tl	Pb	Bi		
1.50		1.48	1.47	1.46	1.46		

Standard Tetrahedral Radii. Table III, which is closely similar to the table published by Huggins[1] in 1926, contains radii for atoms which form four covalent bonds involving the four tetrahedral sp^3 orbitals. In constructing this table, the radii for C, Si, Ge, and Sn were taken as half the observed interatomic distances in diamond-type crystals of the elements. It was next tentatively assumed that (S) equals half of S–S in pyrite, FeS_2, and hauerite, MnS_2. (The symbol (S) means the radius of sulfur, S–S the distance between two sulfur atoms.) [This common point in Bragg's paper, this Vol. paper **115**, makes many covalent tetrahedral radii of Table III almost equal to the values for these elements in Bragg's paper. Ed.] The radii of Zn, Cd, and Hg were then obtained by subtracting this sulfur radius from the observed interatomic distances in their sulfides, and from these radii and the interatomic distances in the selenides and tellurides radii for Se and Te were calculated. The radius of oxygen was obtained from (Zn) and Zn–O. Smooth curves were then drawn for each row of the table in such a way as to give as good agreement as possible with the observed distances in other tetrahedral crystals involving only elements in these rows.

[1] M. L. Huggins, *Physic. Rev.* **28** (1926) 1086.

In the foregoing treatment the assumption of additivity of interatomic distances in the compounds under discussion has been tacitly made. Examination of Table IV [not inserted. Ed.] shows that this assumption is approximately substantiated by experiment. The agreement between the observed distances and the calculated radius sums is excellent in most cases.

Normal-valence Radii for Non-metallic Atoms

In normal-valence compounds of non-metallic atoms each atom forms covalent bonds to a number given by its valence: one for the halogens, two for oxygen, three for nitrogen, and four for carbon. Examples of such compounds are CH_4, F_2, OF_2, NF_3, diamond, etc. In Table VI[1] and Fig. 3 there are given radii for use in compounds of this type. The sum of the single-bond radii for two atoms gives the expected distance between these two atoms in such a compound when they are connected by a covalent bond. The sum of their double-bond or triple-bond radii similarly gives the expected distance when they are connected by a double or a triple bond.

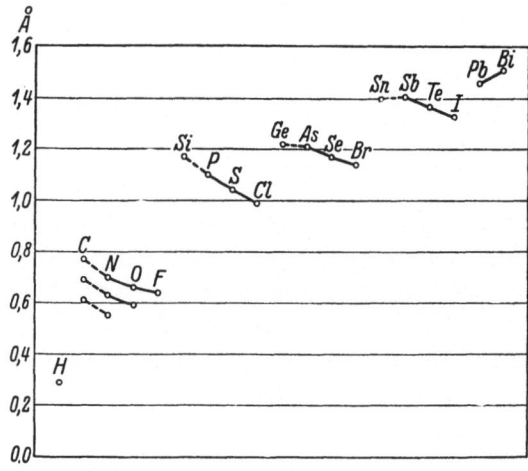

Fig. 3. Normal valence radii of atoms.

The values in Table VI were obtained in the following way. Values for C, Si, Ge, and Sn are the same as in Table III, for the tetrahedral configuration is the normal one for these atoms. Radii for F, Cl, Br, and I were

[1] This table has been published by L. Pauling, *Pr. Nat. Acad.* **18** (1932) 293, in connection with the discussion of the resonance of molecules among several electronic structures.

Table VI.

Normal-valence Radii for Non-metallic Atoms.

Single-bond Radii

H
0.28 Å   (0.375 Å in H_2)

B	C	N	O	F
0.89	0.77	0.70	0.66	0.64
	Si	P	S	Cl
	1.17	1.10	1.04	0.99
	Ge	As	Se	Br
	1.22	1.21	1.17	1.14
	Sn	Sb	Te	I
	1.40	1.41	1.37	1.33
	Pb	Bi		
	1.46	1.51		

Double-bond Radii (factor 0.90)

B	C	N	O
0.80	0.69	0.63	0.59
			S
			0.94

Triple-bond Radii (factor 0.79)

C	N	O
0.61	0.55	0.52

taken as one-half the band-spectral values for het equilibrium separation in the diatomic molecules of these substances. Inasmuch as these radii for F and Cl are numerically the same as the tetrahedral radii for these atoms, the values for N, O, P, and S given in Table III were also accepted as normal-valence radii for these atoms. The differences of 0.03 Å between the normal-valence radius and the tetrahedral radius for Br and of 0.05 Å for I we believe to be due largely to a difference in the nature of the bond orbitals involved. In a tetrahedral atom the bonds are formed by the four tetrahedral bond orbitals. When less than four bonds are formed, the bond orbitals may be pure *p*-orbitals, tetrahedral orbitals, or some intermediate type, depending on the ratio of *s–p* separation to difference in bond energy for different bond orbitals. We believe that in N, O, F, P, S, and Cl the bond orbitals for normal valence compounds lead to about the same radii as tetrahedral orbitals, whereas in atoms below these in the periodic system normal valence bonds involve orbitals which approach *p*-orbitals rather closely, and so lead to weaker bonds, and to radii larger than the tetrahedral radii. This effect should be observed in Br, Se, and As, but not in Ge, and in I, Te, and Sb, but not Sn. For this reason we have added 0.03 Å to the tetrahedral radii for As and Se and 0.05 Å to those for Sb and Te to obtain normal-valence radii for these atoms. The normal-valence radii for Pb and Bi are related to their tetrahedral radii in the same way as for Sn and Sb, respectively.

The effective radius of H in H_2 is 0.375 Å. In other compounds, however, a lower value is operative; from the hydrogen halides and other compounds this is found to be 0,29 Å, as given in Table VI.

It seems probable also that, to within one or two percent, double-bond and triple-bond radii for various atoms should bear constant ratios to single-bond radii. We have chosen 0.79 for the triple-bond factor, which gives agreement with the observed distance in the N_2 molecule, and 0.90 for the double-bond factor. The radii given in Table VI are obtained with these factors.

In Table VII [not reproduced, Ed.] radius sums from Table VI are compared with observed distances from band spectral data. Agreement to within 0.01 or 0.02 Å is usually found. In Table VIII [not reproduced. Ed.] a similar comparison is made with observed distances in crystals. These distances depend in all cases on one or more parameters determined from intensity data, and are accurate to ± 0.05 Å only in exceptional cases. Here too the agreement is satisfactory.

The change of radius with change in number of bonds is strikingly shown by silver, with radius 1.53 Å for four bonds, 1.36 Å for two, and 1.12 Å for one.

The Discussion of the Structure of Certain Crystals

In the preceding sections it has been shown that in a large number of crystals containing covalent bonds, cited as examples, the number and distribution of the bonds are in good agreement with a classification deduced from quantum mechanical considerations, and, moreover, there exist regularities in the observed interatomic distances which may be expressed by assigning covalent radii to the atoms, dependent in a reasonable way on the character of the bonds and the atomic number of the atom. It is evident that these considerations can be used as a basis for judgment regarding the correctness of a reported structure of a covalent crystal, and as a means of predicting a structure or structures for test by comparison with experimental data, in case a rigorous deduction of the structure cannot be made. A few such applications are given below. It must be emphasized, however, that our understanding of the structure of covalent crystals is not complete; occasionally the arrangement of atoms about a given atom is found not to correspond to the bond distribution characteristic of any of the types of bond orbitals described above, and even the observed inter-atomic distances may deviate appreciably from the expected radius sum. The most noteworthy case of such a deviation is the anomalous manganese radius in hauerite.

THE ANOMALOUS MANGANESE RADIUS. In hauerite the Mn–S distance is very close to 2.59 A, corresponding to an octahedral Mn^{II} radius of 1.55 Å. This octahedral radius of manganese is 0.3–0.4 Å larger than expected. From Table X [not reproduced, Ed.] extrapolation of the iso-electronic sequence Ni^{IV} Co^{III} Fe^{II} leads to 1.24 Å for Mn^{I}, and this should be an upper limit for Mn^{II}. Hence the large radius of 1.55 Å found experimentally is in definite and pronounced disagreement with our expectations for octahedral $3d^2 4s 4p^3$ bonds. We are unable to advance a plausible explanation of the anomaly.

It is worthy of mention that the properties of hauerite are in accordance with the conception that the Mn–S bonds are much weaker than the Fe–S bonds in pyrite. The hardness of hauerite is 4, as compared with 6–6.5 for pyrite. All members of the pyrite group have a bright metallic luster except hauerite, which is dull. Hauerite is said to have a much smaller electrical conductivity than the others.

THE STRUCTURE OF PALLADOUS OXIDE. The suggested PbO type structure seems unreasonable to us, inasmuch as we expect bivalent palladium to form four dsp^2 bonds directed towards the corners of a square. More-over, the axial ratio $c/a = 1.75$ of PdO differs greatly from that of PbO (1.26). An entirely reasonable structure, however, has been described by Huggins[1].* In this structure, with 2Pd at $0\frac{1}{2}\frac{1}{4}$, $\frac{1}{2}0\frac{3}{4}$ and 2O at $0\,0\,0$, $0\,0\,\frac{1}{2}$, as shown in Fig. 4, each oxygen atom is bonded to four palladium atoms at the corners of a nearly regular tetrahedron, and each palladium atom forms four coplanar bonds to oxygen atoms at the corners of a rectangle. The geometry of the structure requires some distortion of the bond angles. If the oxygen bonds (sp^3) were directed towards the corners of a regular tetrahedron the axial ratio c/a would have the value 1.414, and if the palladium bonds (dsp^2) were directed towards the corners of a square c/a would be equal to 2.000; the observed value 1.75 shows the distortion to be divided about equally between the two atoms.

[1] M. L. Huggins, *Chem. Rev.* **10** (1932) 427.
* [This structure has been confirmed by X-rays (1941) and neutron diffraction (1953). Ed.]

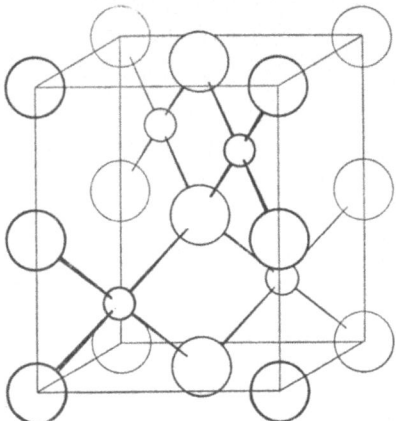

Fig. 4. The structure of PdO (referred to axes 45° from those of the smallest unit). Large circles represent oxygen, small circles palladium atoms.

THE WOLFSBERGITE STRUCTURE. The interesting structure of wolfsbergite, $CuSbS_2$, and emplectite, $CuBiS_2$, determined by Hofmann[1], provides a good example for comparison with our rules and radii. Each copper atom is surrounded tetrahedrally by four sulfur atoms, corresponding to sp^3 bond orbitals, the Cu–S distance being 2.25–2.33 Å, slightly smaller than the radius sum 2.39 Å. We expect trivalent antimony to form three bonds; this is observed, the three nearest sulfur atoms being 2.44, 2.57, and 2.57 Å distant, with others no nearer than 3.11 Å. The radius sum for Sb^{III} and S is 2.45 Å, in good agreement with one of the reported values but somewhat smaller than the other (2.57 Å). It is not unlikely that small changes should be made in the eight parameters, such as to decrease the high Sb–S distance somewhat. It is also noteworthy that the structure is a layer structure on (001), the atoms within the layers being held together by covalent bonds, so that the layers can be separated without any of these bonds being broken. In consequence complete basal cleavage, similar to that in metallic antimony, is observed[2].

EULYTITE. In his investigation of eulytite, $Bi_4Si_3O_{12}$, Menzer[3] used the assumption that each bismuth atom should be equidistant from six oxygen atoms. We, however, believe that trivalent bismuth would form three bonds of considerable covalent character, and would hence have as nearest

---

[1] W. Hofmann, *Z. Krist.* **84** (1933) 177.

[2] It is very improbable that this cleavage should occur in such a way as to rupture Cu—S bonds, as suggested by Hofmann.

[3] G. Menzer, *Z. Krist.* **78** (1931) 136.

neighbours only three oxygen atoms, at a distance of about 2.17 Å.*

AsI_3 AND BiI_3. Heyworth and Braekken[1] in their studies of the hexagonal crystals AsI_3 and BiI_3 assigned to them structures in which each arsenic or bismuth atom is surrounded by six equidistant iodine atoms, the interatomic distances reported being As–I = 2.97 Å and Bi–I = 3.09 Å. As in the case of eulytite, we believe that the trivalent atoms are displaced towards three and away from three of these six atoms, until the smallest interatomic distances become As–I = 2.54 Å and Bi–I = 2.84 Å [2].**

Conclusion. There is no need to multiply examples of the type considered, especially since our knowledge of the principles determining the structures of covalent crystals is still so incomplete.

A number of questions related to those taken up in this paper, such as the distances between atoms not directly connected by bonds, the use of interatomic distances as a criterion for distinguishing between ionic and covalent bonds, etc., have been discussed in a paper by Huggins[3].

Gates Chemical Laboratory, California Institute of Technology,
Chemical Laboratory of the Johns Hopkins University.

---

* [Confirmed by neutron diffraction (1966): each Bi-atom has three oxygen neigbours at a distance of 2.15Å and three more at 2.62Å. Ed.]

[1] D. Heyworth, *Physic. Rev.* **38** (1931) 351, 1792; H. Braekken, *Z. Krist.* **74** (1930) 67; **75** (1930) 574.

[2] The possibility of such a displacement was mentioned by Miss Heyworth.

** [No redetermination performed. Ed.]

[3] M. L. Huggins, *Chem. Rev.* **10** (1932) 427.

B. BUILDING PRINCIPLES

Das Ziel der chemischen Kristallographie des neunzehnten Jahrhunderts, gesetzmäßige Zusammenhänge zwischen Kristalltracht und chemischer Konstitution aufzuklären, hat das Interesse vieler der bedeutendsten Chemiker und Kristallographen dieser Zeitepoche in so hohem Grade auf dieses Grenzgebiet der Chemie und Kristallographie gelenkt, daß um die Jahrhundertwende schon bedeutende Kenntnisse über die Kristallform der verschiedenartigsten chemischen Verbindungen gesammelt waren. Wenn dennoch so tiefgreifende Erkenntnisse nicht gewonnen worden waren, daß zu dieser Zeit von einer Kristallchemie *gesprochen werden konnte, so war dies einfach eine Folge davon, daß man noch nicht die experimentellen Mittel besaß, die uns den inneren Feinbau der Kristalle verraten konnten. Erst die Entdeckung der Röntgeninterferenzen durch* M. v. Laue *und die ersten bahnbrechenden Kristallstrukturbestimmungen* W. H. *und* W. L. Bragg's *eröffneten die Möglichkeit eines Eindringens in die Welt der inneren Architektur der Kristalle.*

Neben dem überwältigenden Material an mehr oder weniger gründlich durchforschten Kristallstrukturen, das an sich schon unmittelbar wichtige Schlüsse erlaubt, haben die ersten zwei Dezennien röntgenkristallographischer Arbeit vor allem dank der Bemühungen V. M. Goldschmidt's *und der* Bragg'schen *Schule, sowie der Arbeiten* L. Pauling's *und des Stockholmer metallographischen Instituts (viele andere Forscher könnten noch erwähnt werden) grundlegende Prinzipien allgemeinerer Natur zu enthüllen vermocht. So kann man heute mit vollem Recht von einer* kristallchemischen Wissenschaft *reden, eine Wissenschaft, die in ihrer ersten Phase wie die Chemie selbst stark empirischen Charakter trägt, die jedoch für jeden Tag eine solidere theoretisch-physikalische Basis erhält.*

O. Hassel,
Kristalchemie (Verlag Steinkopf, 1934)

122. Crystal Structure and Chemical Constitution, by V. M. Goldschmidt (1929)

■

123. The Coördination Theory of the Structure of Ionic Crystals, by L. Pauling (1928)

■

124. The Function of Hydrogen in Intermolecular Forces, by J. D. BERNAL and H. D. MEGAW (1935), *with insertion of part of:*

124a. A Theory of Water and Ionic Solution, with Particular Reference to Hydrogen and Hydroxyl Ions, by J. D. BERNAL and R. H. FOWLER (1933)

122.

> *The structure of a crystal is determined by the ratio of numbers, the ratio of sizes and the properties of polarisation of its building stones. As the building stones we visualise atoms (or ions) and groups of atoms.*
>
> W. M. GOLDSCHMIDT,
> this Vol. p. 91

> *A broad treatment along these lines introduces law and order into the mase of information about inorganic structure, and it is striking to see what an immense number of compounds have structures which approximate to certain simple types.*
>
> W. L. BRAGG,
> *Annual Reports* **23** (1926) p. 277

The task of crystal chemistry is to find systematic relationships between chemical composition and physical properties of crystalline substances, especially to find how crystal structure, the arrangement of atoms in crystals, depends on chemical composition:

From old times numerous data have been known to the student of crystallography and to the chemist, which clearly demonstrate that systematic relationships exist between chemical composition and crystal structure. The first step in the study of such relations was the perception by Haüy that to any homogeneous substance, to any chemical individual, there belongs a certain geometrical complex of crystal faces, which must be determined by the inner structure, by the molecular arrangement of that substance. The next step in the progress of crystallochemical science was the discovery by Mitscherlich that substances which are analogous in chemical composition show in many cases, also, a similarity of crystalline form, a phenomenon to which he gave the name *isomorphism*. The next fundamental perception we owe also to Mitscherlich, namely, that crystal structure is not exclusively determined by chemical composition; in other

77

words, he discovered the phenomenon of *polymorphism*. One generation later Pasteur found the geometrical enantiomorphism of dextro- and laevo-tartaric acid; observation, together with the theory of Le Bel and Van 't Hoff, gave a very clear impression of relations between chemical constitution and crystal structure.

Those were the principal steps in the advance of our empirical knowledge concerning relations between crystalline structure and chemical composition. In the later half of the nineteenth century Th. Hiortdahl and especially P. v. Groth endeavoured to find further systematic relationships between crystal structure and chemical composition. Both scientists tried to obtain regular alterations of crystal structure, a *morphotropism*, by means of systematic chemical substitution. The triumphs of organic structural chemistry, however, were not followed by corresponding progress in crystal chemistry. The work of that period did indeed give distinct signs as to the existence of the relationships in question, but in spite of much laborious work no new universal laws could be found, which might be compared in importance with the discoveries of Mitscherlich and Pasteur. The results of Brøgger on isomorphy and morphotropy in the domain of minerals brought us nearly up to our present level of knowledge. But the very extensive wealth of observations after that epoch, which has been treasured by v. Groth, becomes of great importance for the crystal chemistry of to-day, in view of the establishment of systematic relationships between the structures of substances with very simple composition.

It is indeed easy to understand that an exploration of the relationships between chemical composition and crystal structure has to begin with substances of the very simplest composition; these are, besides the chemical elements, the compounds of the composition AX, AX_2, A_2X_3.

It may be appropriate, after the statement of the systematic relationships between such substances of very simple composition, to approach more complicated substances in order to test the range of validity of the relations.

For some years past I have tried to proceed according to that principle in order to find the laws of crystal chemistry by means of the inductive method.

As a *principle of classification* for those different types of crystalline structure we do not choose the macro-crystallographical symmetry; we do not group the crystals for instance into cubical, hexagonal, tetragonal types, as in ordinary crystallography, but we classify the crystals with regard to their *types of co-ordination*, that is according to the arrangement of the atoms. This view, which is in close accordance with the structural chemistry of Werner, has already been applied to crystal structures by several workers, *e.g.* Ewald and Pfeiffer. The principle of classification now is the number and arrangement of neighbours around any single atom in the crystal lattice. The number of nearest neighbours around any atom we call the co-ordination number for that atom.

Table I. Co-ordination Types for AX Substances.

Number of Co-ordination	1	Single Molecules and Molecule Lattices.
,, ,, ,,	2	Double Molecules, Molecule Chains and their Lattices.
,, ,, ,,	3	Lattices of Boron Nitride Type.
,, ,, ,,	4	Lattices of Zincblende-Wurtzite Types and tetragonal Layer Lattices.
,, ,, ,,	6	Lattices of Sodium Chloride Type and of Nickel Arsenide Type.
,, ,, ,,	8	Lattices of Caesium Chloride Type.

Table II. Co-ordination Types for AX_2 Substances.

Numbers of Co-ordination	2 and 1	Single Molecules and Molecule Lattices.
,, ,, ,,	4 and 2	Lattices of α and β Quartz, α Tridymite, α Cristobalite and Cuprite Types.
,, ,, ,,	6 and 3	Lattices of Anatase, Rutile, Cadmium Iodide, Molybdenite Types.
,, ,, ,,	8 and 4	Lattices of Fluorite Type.

In Tables I and II are given the co-ordination numbers for the most important structures which hitherto have been found for substances AX and AX_2. For AX substances the co-ordination number of atoms X around A must be the same as the co-ordination number for atoms A around X, but the geometrical type of arrangement may be different in the two cases. For AX_2 substances the number of co-ordination of atoms X around A is twice as great as the number of co-ordination of atoms A around X.

The fundamental problem of crystal chemistry is the following. What are the causes which determine the type of structure of any given substance; why, for instance, has magnesium fluoride the structure of rutile, strontium fluoride the structure of fluorite?

If we wish to ascertain the reasons for the appearance of any type of crystal structure we must first ascertain by what operations we can *modify* or *alter* the crystal structure. In any search for causal connections the most direct method is to find out by what means we can modify or alter a pheno-

menon in a measurable manner. Now, we can modify the atomic arrangements to some degree by means of temperature, or by mechanical forces
—sometimes even to a considerable extent; but the most efficient, we may
even say nearly ideal, tool for our work is the *method of chemical substitution*, the most powerful instrument in the hand of the crystallographer.

Any crystal structure may be described by defining the type of atomic
arrangement and the numerical data of atomic distances. The effect of
chemical substitution may be shown by alteration of atomic distances and
by alteration of atomic arrangement.

We may, for instance, consider the series of fluorides of divalent metals,
starting from barium fluoride and proceeding, by chemical substitution, to
the fluorides of strontium, calcium, and magnesium. The types of structure
and the lattice constants, as well as the distances between the centres of
particles (we shall hereafter call them interatomic or, shortly, atomic
distances) are given by Table III.

Table III.

	Structure	Lattice Constants		Distances of Particles A-X	Numbers of Co-ordination
		a	c		
BaF_2	Fluorite-Structure	6.19 Å	–	2.68 Å	8 and 4
SrF_2	,,	5.78 Å	–	2.50 Å	8 ,, 4
CaF_2	,,	5.45 Å	–	2.36 Å	8 ,, 4
MgF_2	Rutile-Structure	4.62 Å	3.06	1.99 Å	6 ,, 3

The sudden alteration of atomic arrangement between calcium fluoride
and magnesium fluoride is most conspicuous at one place in the series,
but every step in the series shows an alteration of the structural dimensions
or, in other words, an alteration of the distances between neighbouring
particles of metal and fluorine.

Such an alteration of distances between atomic centres may be formally
interpreted as the effect of different size of the substituted particles. We
can accordingly state a succession of atomic sizes, corresponding to measurements of interatomic distances, in our case the succession is

$$Mg < Ca < Sr < Ba,$$

in other cases for instance

$$Li < Na < K < Rb < Cs \text{ or } O < S < Se < Te.$$

If we measure the distances between the same kinds of particles in different crystal structures we find in most cases a very close numerical agreement; Such constant or nearly constant interatomic distances lead to the conception that for any particle there is a sphere of action which is practically impenetrable and that the interatomic distances may be interpreted as sums of two radii of particles.

We may determine these radii from the interatomic distances in crystals, provided that we carefully avoid the combination of data from crystals in which the particles or atoms are in different states. The effect of ionisation, in particular, must be taken into account, as the same kind of atom may show very different radii in different degrees of ionisation, an effect which may be foreseen by our present views on atomic structure.

We cannot, generally, compare the atomic distances in ionised salts with those in homopolar compounds or with those in metallic substances. We may divide crystals into several groups within which additivity prevails for interatomic distances; crystals belonging to the same group may be called commensurable crystals.

I tried, in 1926, to draw up a table of radii based on the principle of commensurability, deduced to a large extent from measurements from my laboratory. The starting point of these determinations, so far as they concern the radii of ionised particles, is the calculation of the radii of univalent negative fluorine (1.33 Å) and of divalent negative oxygen (1.32 Å), made by J. A. Wasastjerna as early as 1923 from optical data.

We can now make use of the empirical radii to examine the relations of morphotropism in a number of AX_2 compounds. We may tabulate the difluorides of divalent metals and the dioxides of tetravalent metals according to the quotient of their ionic radii, i.e. the quotient $R_A : R_X$.

We find that a sudden alteration of crystal structure occurs at a certain limit of the quotient $R_A : R_X$. In the series of difluorides as well as in the series of dioxides the alteration of structure type, the morphotropism, between the rutile structure and the fluorite structure takes place when the quotient of radii is about 0.7.

We naturally seek the meaning of that limiting quotient of the radii,

Table IV. The structure of difluorides and dioxides as a function of the quotient of radii $R_A : R_X$.

Rutile structure						Fluorite structure					
MgF_2	NiF_2	CoF_2	FeF_2	ZnF_2	MnF_2	CdF_2	CaF_2	HgF_2	SrF_2	PbF_2	BaF_2
$R_A : R_X$ 0.59	0.59	0.62	0.62	0.62	0.68	0.77	0.80	0.84	0.95	0.99	1.08

MnO_2	VO_2	TiO_2	RuO_2	IrO_2	OsO_2	MoO_2	WO_2	NbO_2	SnO_2	PbO_2	TeO_2	ZrO_2	PrO_2	CeO_2	UO_2	ThO_2
R_X 0.39	0.46	0.48	0.49	0.50	0.51	0.52	0.52	0.52	0.56	0.64	0.67	0.66	0.76	0.77	0.80	0.84

81

and for this purpose we have to look for the geometrical laws for arrangs-
ments of spheres in space. As an anlogy with heteropolar binary compounde
we may consider spacial arrangements of spherical particles of two different
sizes, which conform with the condition that any sphere of one sort is in
contact with as many spheres as possible of the other sort in accordance
with the views of Kossel. Of course, we can in the same manner give a
picture of ternary compounds by arrangements of three sizes of spheres,
always considering the condition of multiple contact between spheres which
represent particles of opposite electrical sign.

An analogy of this kind, of course, can only depict the very simplest
cases of ionic arrangement; in more complicated structures we often meet a
struggle between co-ordination numbers around different ions, a contest the
resultant of which is given by the relative electrostatic energy of the different
arrangements[1].

We shall, however, try to analyse the very simplest cases in correlation
to our picture, and to make the case still simpler we may, first, consider
the arrangement only around one single sphere, in this manner reducing
the ratio of numbers in the infinite crystal lattice. An example from plane
geometry will serve to illustrate this. Circles of two sizes can be arranged
in such a manner that one circle with radius B has contact with as many
circles of radius X as possible, without any intersection of circles. As
Fig. 5 shows, an arrangement of 3 circles X around B is only possible, if
$R_B : R_X$ is greater than 0.15, an arrangement of 4X around B demands
a radius ratio, which exceeds the limit 0.22.

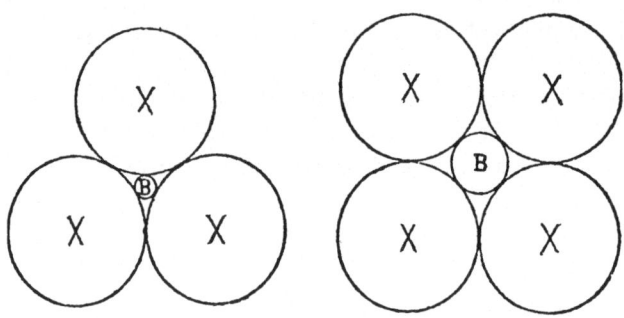

Fig. 5.

[1] The question of the co-ordinative arrangement in such cases has been treated recently
in a very interesting manner by L. Pauling in his contribution to the Sommerfeld Volume,
Probleme der modernen Physik (Leipzig, 1928) 17. [This vol. paper **123**.]

Considerations of this nature were used long ago by A. Magnus[1] in discussing the formation of complex ions and molecules as a function of ionic radii, when such an arrangement has to be formed in space around one single central atom.

For arrangements in space the limits of the radius ratio shown in Table V are valid.

Table V. Table for the arrangement of ions X about an ion A.

Number of Surrounding Ions X	Arrangement of Ions X	Lim $R_A : R_X$
2	Opposite each other	—
3	Equilateral-triangle	0.15
4	Cube-diagonals (tetrahedral)	0.22
4	Square	0.41
6	Cube-edges Octahedron-corners	0.41
8	Cube-diagonals	0.73

We shall, provisionally, set up an hypothesis that the stability of crystal structure of heteropolar compounds requires that anions and cations are in mutual contact. Our table would then give us the limits of radius ratio which could be permitted in any type of crystalline arrangement. For instance we should find as the limiting ratio between rutile structure and fluorite structure, the limit between the co-ordination numbers 6 and 8, that is 0.73. Empirically we find a limiting ratio of about 0.7.

Such a consideration can however, at best, only give us an approximate picture of the factors which determine crystal structure. We have therefore to consider for which kinds of substances our picture may come nearest to reality. Such substances are those the components of which may be considered with sufficient approximation as charged incompressible spheres.

According to our present views on the properties of atoms, the ions of inert gas type, including those of the helium type, correspond to the conception of spherical fields of action, and we can even extend the same consideration to any types of ions which, from spectral data, are known to have the same property. The cations of these types, especially those of high

[1] A. Magnus, 'Die chemischen Komplexverbindungen,' Z. anorgan. Chem. 124 (1922) 288.

ionic charge, may with a sufficient degree of accuracy also fulfil the condition of 'incompressibility'. As to the anions, the smallest anions of inert gas type, with low ionic charge, show the nearest kinship to our very simplified picture, especially fluorine and oxygen.

It is therefore easy to understand that a systematic numerical relationship between crystal structure and radius ratio was first detected through the study of fluorides and oxides of divalent and tetravalent cations, most of which now are understood, from spectral data, to have spherical symmetry. They are ions which either have an S-term as lowest energy level or contain only completed electronic shells.

We have mentioned the transition between rutile structure and fluorite structure as an instance of limiting ratio which accords with the requirements of our conception of charged spheres. We might also mention an example from AX substances, the series of binary compounds between the anions of the oxygen-tellurium series and the divalent cations of inert gas type. The numbers given in Table VI are the ratio $R_A : R_X$ from the empirical radii.

Table VI.

	Mg	Ca	Sr	Ba
O	0.59	0.80	0.96	1.06
S	0.49	0.61	0.73	0.82
Se	0.41	0.56	0.66	0.75
Te	0.37	0.50	0.60	0.68

The only compound which has a radius ratio $R_A : R_X$ less than the geometrical limit, is the telluride of magnesium; it is just this compound which does not show sodium chloride structure, but wurtzite structure.

The relationships between the radius ratio and the type of structure were also found to be valid in a number of cases when, for single ions, we substitute ionised complexes of more than one particle—instead of atom ions, molecule ions, as NH_4, alkyl substituted ammonia, hydrates and ammoniates of metal ions.

The significance of the ratio of radii in the structure of crystals makes it most desirable to determine with a high degree of accuracy the radii of the different types of particles in crystals. This task leads us to ask within what limits the radii of particles remain constant.

If we limit our discussion to such groups of crystals as are built up from particles in a comparable state (I have proposed to call them 'commensurable crystals') we observe that the radii of particles remain constant to a *first approximation*, but that there are measurable, and indeed quite regular, departures from additivity.

Those varations of distances between atomic or ionic centres, which have been found empirically in many series of crystals[1], are partly correlated to the type of co-ordination (that is, to number and arrangement of neighbouring particles) and partly to the peculiar nature of the neighbour particles[2].

Table VIII.

Transition from CsCl type to NaCl type from co-ordination number 8 to 6.	Decrease of distance 3 per cent.
Transition from NaCl type to ZnS types from co-ordination number 6 to 4.	Decrease of distance 5–8 per cent.
Transition from CaF_2 type to rutile type from co-ordination numbers 8 and 4 to 6 and 3	Decrease of distance 3 per cent.

The influence of co-ordination on the atomic distances in ionic lattices is shown by Table VIII which shews that by comparison of different crystal types the departures from constancy of atomic distances are small compared with the distances themselves; from these data we may ascertain within what limits we may expect departures from constancy. It is evident, moreover, that particle distances decrease in a regular manner with decreasing co-ordination number. The fewer neighbours there are around an ion, the shorter is the interionic distance.*

This alteration of interionic distances as a function of co-ordination number is very important in any considerations of the electrostatical lattice energy.

Electrostatic lattice energy[3] for different crystal structures of the same substance has usually been calculated on the assumption of constant ionic

---

[1] One principal aim of the experimental work on crystal structures in our laboratory during the last four years has been to find the laws which govern *variations* of atomic and ionic distances in crystals. It is somewhat disappointing to see that the aims and results of our work have in some cases been so completely misunderstood; as for instance shown by a paper in *Physica* **8** (1928) 129.

[2] The influence of the nature of neighbouring particles has been studied in a number of cases by K. Fajans and H. G. Grimm and has been associated with 'ionic deformation'. Quite recently L. Pauling has published most interesting considerations on this question (*J. Amer. Chem. Soc.* **50** (1928) 1036).

* [Mathematical derivation in Zachariasen's (1931) paper, this Vol. p. 49. Ed.]

[3] Data on lattice energy for a number of structure types have been calculated by Born and his students.

distances. This results in high differences of energy between structures with different co-ordination numbers. These differences are considerably reduced if we take into consideration the influence of co-ordination number on ionic distances. The energy differences between the several crystalline forms of the same substance, then, are diminished very considerably, in accordance with thermochemical experience.

We have, hitherto, considered only substances which, to a first degree of approximation, might be compared with an arrangement of charged spheres of different sizes. If we systematically follow the influence of ionic size on crystal structure, limiting our considerations to such substances, we can empirically get order into whole series of structures from our point of view, and we may parallel those series with arrangements of charged spheres, if we wish to satisfy a demand for a plain picture.

We should not, however, forget that these considerations are limited to those crystals, the particles of which are near the ideal case of incompressible spheres. If we consider crystals for which such an assumption is no longer valid, then the ratio of radii loses its dominating influence.

The departure of ionic properties from the picture of incompressible spheres may be depicted by the idea of ionic *polarisation*, the conception introduced by Haber, viz. that under the influence of an electric field there takes place a mutual displacement of positive nucleus and negative electronic shells. A neutral atom, as well as an ion, may be polarised.

The mutual arrangement of particles is most symmetrical in an arrangement of weakly polarising and weakly polarisable spherical particles, so that any particle of one kind is surrounded at the same distance by as many particles of the other kind as geometrically possible. The increase of polarisation leads to decrease in co-ordination number, to differences in distance between neighbouring particles and, thereby, in many cases, to a lower degree of symmetry of the crystalline arrangement. The extreme case is the formation of one single ionic molecule, for instance formed from two ions A and X; the co-ordination number in this case reaches its minimum, 1, and the distance between particles goes down to much lower numbers than in the crystal lattices described above. For instance, in a gaseous ionic molecule of sodium chloride the distance between sodium and chlorine is much lower than in the sodium chloride lattice. The energetical aspects of polarisation in crystal lattices have recently been investigated by van Arkel[1], who has discussed the contribution of polarisation phenomena to lattice energies.

The phenomenon of polarisation may be described by the following

[1] A. E. van Arkel, *Z. Physik* **50** (1928) 648.

picture. A certain amount of negative electrical charge is drawn for a certain distance towards the cation, and a certain amount of the negative shell of the cation is repelled from the anion. That may happen in one singular direction between only two ions, or it may happen simultaneously in several directions in space, if we have an arrangement of more than two ions.

The main result is in any case a transport of negative charge from anion towards cation. In extreme cases of strong polarisation some proportion of the negative charge may leave the exclusive domain of the anion.

Such extreme cases of polarisation may give rise to very different results, according to the special types of ions involved and according to the state of matter considered. Which result will arise will depend on the energetic properties of the different ions and atoms involved, but may in future, to a large extent be predicted from spectral data.

(a) If the whole of the negative ionic charge is drawn over to the cations and unites with them, there result neutral atoms.

(b) The next possibility is that amounts of negative charge leave the anions which, however, do not completely unite with the cations, but become metallic electrons, common for the crystal as a whole, as probably in the NiAs-type.

(c) The third case is that electrons from anions form electronic bonds, or partake in the formation of such bonds, thereby uniting anions and cations to true molecules by valencies.

There are two sub-cases to be considered, either the formation of valency bonds between a limited number of particles, giving rise to micromolecules, or the uniting of an unlimited number of particles to macromolecules. The macromolecules may be either linear chains, plane nets or three-dimensional spacial arrangements; the last case seems to be represented by AlN.

These different cases and sub-cases of course may originate in a manner which is quite independent of previous ionic polarisation, but we may use the picture of polarised ions as an imaginable step in the formation of these types of bonds. We may here add that the theories of Dr. Sidgwick on electronic pairs and molecular bonds will also, without doubt, become of the very greatest importance in questions of crystal structure.

We shall first consider polarisation phenomena in typical ionic substances. I consider as polarisation phenomena of crystal particles all those alterations which the particles show under the influence of electrical forces. The most simple case is the formation of a dipole under the influence of an electric field. The univalent negative iodine is, under such conditions, a typical polarisable ion. We therefore find in cadmium iodide not a structure of rutile type, as we might expect from analogy with spherical packings, not even a structure of one of the silica types, but another structure which differs in a very important manner from the structures named. It is of the cadmium iodide type, the co-ordinative arrangement of which might be described as such that, though any ion of cadmium is surrounded symmetrically by 6 ions of iodine (in rhombohedrical arrangement), any ion of iodine has contact with 3 ions of cadmium *on one side*; that it is therefore exposed to polarisation. The geometry of the structure is characterised by the peculiarity that an arrangement of *layers* dominates the crystal in the following manner. Packings of three ionic layers build up in the crystal in parallel repetition, those packings of three layers contain in the middle a layer of the weakly polarisable ions, i.e. cadmium, with, on both sides, a layer of strongly polarised ions, i.e. iodine. F. Hund[1], in a very important memoir, introduced the name of 'layer lattice' (Schichtengitter), for structures of this remarkable kind.

Besides cadmium iodide there are known a number of other types of layer lattices. In layer lattices the system of neighbouring layers has become a kind of physical unit, co-ordinated by forces of a second order, as dintinguished from the strong forces inside the system. This peculiarity of layer lattices leads to the prominent cleavage between the layer systems; as H. G. Grimm has said, any of the electrically neutral layer packings may be considered a giant molecule[2] which in one plane has unlimited extensions.

In cadmium iodide the polarisation means the induction of dipoles under the influence of one-sided electrical forces and in the reaction of those dipoles on the field of force. Such polarisation may be shown in ions, as in the polarisation of iodine in cadmium iodide; it may be shown on neutral atoms or molecules without proper dipole moment, as for instance on a molecule I_2, but dipole properties in electrical fields are shown particularly by those molecules, radicals or ions which already possess a dipole moment of their own before the action of an outside field, as for instance H_2O, CN, OH. In cases of this type the outside field first acts on the orientation of the dipoles, then it serves to reinforce the dipole moment.

We should mention here, that from the recent development of the quantum mechanics of molecular systems it follows that such complexes as OH or CN may have states of spherical symmetry, without inherent dipole moment. This is of the utmost importance for a number of applications of space group theory on crystals, where such complexes are one of the constituents.

---

[1] F. Hund, *Z. Physik*, **34**, (1925) 833.
[2] From our point of view a giant *ionic* molecule, *not a valency* molecule.

We shall now look at some instances of the influence of such low symmetrical phenomena of polarisation on the crystal structure.

Let us consider compounds of the formula AX$_2$ and let us again make use of the tool of chemical substitution in order to alter the properties of the building stones of crystals. We may take the fluoride of cadmium, a substance with the structure of fluorite. We replace the fluorine by iodine, and by this substitution we do not obtain the structure of rutile, but as already mentioned, the lattice of cadmium iodide. We now may ask: is it the large size of the iodine ion, compared with fluorine which causes the fundamental alteration of crystal structure or is it another property of the iodine ion? To decide between these possibilities, we may, in the place of the halogen, substitute the radical OH, and then examine the effect on the crystal structure. The hydroxyl ion is, with respect to radius, much more like fluorine than like iodine, as we may find from Table IX.

Table IX.

Radius of univalent negative   F ... 1.33 Å.
     ,,   ,,   ,,   ,,   I ... 2.20 Å.
     ,,   ,,   ,,   ,,   OH ... 1.4–1.5 Å.

We find that a substitution of hydroxyl for fluorine causes the morphotropism from fluorite structure to layer lattice[1]. Therefore it must be the capacity of the iodine to become strongly polarised which is responsible for the special type of morphotropism met in the case CdF$_2$→CdI$_2$.

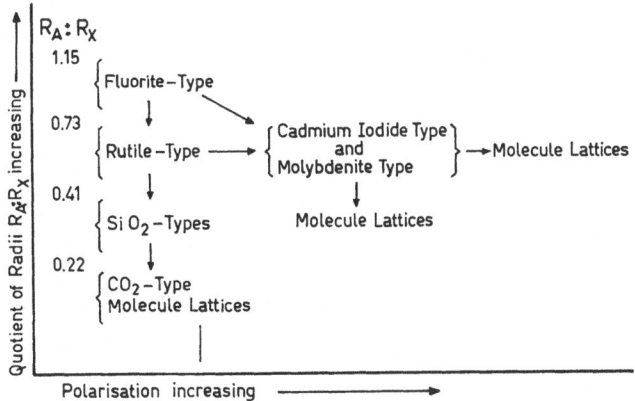

Fig. 6. The influence of radius ratio and polarisation on the crystal structure of compounds AX$_2$.

[1] G. Natta, *Rend. Accad. Naz. Lincei* (1925), has found the structure of cadmium iodide in Cd(OH)$_2$.

In the schematical picture in Fig. 6 those two factors which determine the structure of a substance are shown for some important structure types of substances AX_2.

In his work on substances A_2O_3 and ABO_3 W. H. Zachariasen[1] has recently been able to give such pictures even for such complicated cases of morphotropism as are found in the substances ABO_3.

I have tried in the examples quoted hitherto to treat separately the influence of size relations and of polarisation; we have succeeded in depicting some cases in which the influence of size ratio dominates, and other cases in which properties of polarisation nearly exclusively determine the type of morphotropism. In most morphotropic series which we met in actual cases, however, the two kinds of determinative factors are both of importance, and it is necessary carefully to consider *both* factors if one wishes to predict the crystal structure of an unknown substance.

In principle we can understand or learn to understand the relations between chemical composition and crystal structures for any type of chemical compound, for instance for compounds AX, AX_2, AX_3, A_2X_3, AXY, ABX_3, ABX_4, if we proceed in the following manner. On suitable cases we investigate as far as possible separately the influence of size of particles and the influence of phenomena of polarisation, and afterwards we take into consideration the combined action of these two factors.

We must investigate anew for any type of chemical formula by reason of the purely geometrical fact that the possibilities of arrangement of particles in space are dependent on the relative quantity of the different sorts of particles. Here we find the space group theory, founded by Schoenfliess and Fedorow as a most valuable, even indispensable, tool of structural research, whilst the recent investigations of Weissenberg lead also to most important general results concerning the geometry of arrangements of points in space.

The next task of the space group theory, the necessity of which arises from the results concerning the importance of contact relations, is a study of the packings of spheres of different size, a task which has recently been taken up by Niggli and co-workers. Hitherto the limiting radius ratios have been found empirically and have later on been confirmed by calculation of arrangements of spheres; it would now be convenient to tabulate for all types of structures the limiting conditions of mutual contact, in order to predict possible structures of unknown types.

After having discussed, on the base of the inductive method, what are

[1] W. Zachariasen, Untersuchungen über die Kristallstruktur von Sesquioxyden und Verbindungen ABO_3 *Norska Vidensk. Akad. Oslo I. Mat. Naturv.* (1928) No 4.

the determining factors for the crystal structure of a given substance, we can formulate a general thesis which condenses the experience which we have collected on heteropolar substances.

The structure of a crystal is determined by the ratio of numbers, the ratio of sizes and the properties of polarisation of its building stones. As the building stones of crystals we visualise atoms (or ions) or groups of atoms.

123. In the study of the structure of a crystal with X-rays the effort has been made by many workers, especially in America, to eliminate rigorously all but one of the possible atomic arrangements consistent with the smallest unit of structure permitted by the experimental data, without reference to whether or not the structures were chemically reasonable or were in accord with assumed interatomic distances. The importance of this procedure arises from the certainty with which its results can be accepted. For although structure determinations by less rigorous methods have been found to be false, no important error has yet been detected in any investigation involving the interpretation of photographic data solely with space-group theory and the use of merely qualitative assumptions regarding the factors influencing intensity of reflection. It would accordingly be desirable to conduct all structure determinations by this method; but unfortunately the labor involved in its application to complex crystals, involving more than a very few parameters, is insuperable. Furthermore, if several different kinds of atoms compose the crystal it is often necessary to make quantitative assumptions regarding their relative reflecting powers, so that for this reason too the rigorous method cannot be used.

But complex crystals are of great interest, and it is desirable that structure determinations be carried out for them even at the sacrifice of rigor. The method which has been applied in these cases is this: one atomic arrangement among all the possible ones is chosen, and its agreement with the experimental data is examined. If the agreement is complete or extensive, it is assumed that the structure is the correct one. The principal difficulty underlying this treatment is the selection of the structure to be tested. Striking regularities in the intensities of reflection from simple planes may suggest an approximate structure. This occurred in Dickinson's study of tin tetraiodide[1]; he then considered values of the five parameters involved which differed only slightly from those suggested by the intensities of simple reflections, and found a set giving complete agreement with the Laue photographic data. Because of the large amount of experimental data accounted for by it, this structure can be confidently accepted as

[1] R. G. Dickinson, *J. Amer. Chem. Soc*. **45** (1923) 958; *Ztschr. f. Krist.* **64** (1927) 400.

correct. In other cases the number of possible atomic arrangements has been so greatly decreased by excluding those involving interatomic distances smaller than certain assumed values that a more or less rigorous comparison with X-ray data sufficed to select one of the remaining arrangements as the correct one.

As a result of the recent increase in knowledge of the effective radii of various ions in crystals (Wasastjerna's work in particular), Professor W. L. Bragg has suggested and applied a simple and useful theory leading to the selection of possible structures for certain complex crystals. His fundamental hypothesis is this: if a crystal is composed of large ions and small ions, its structure will be determined essentially by the large ions, and may approximate a close-packed arrangement of the large ions alone, with the small ions tucked away in the interstices. In some cases all of the close-packed positions are not occupied by ions, and an open structure results. To apply this theory one determines the unit of structure in the usual way, and finds by trial some close-packed arrangement of the large ions of known crystal radius (usually oxygen ions with a radius of 1.35 to 1.40 Å) compatible with this unit. The other ions are than introduced into the possible positions in such a way as to give agreement with the observed intensities, the large ions being also shifted somewhat from the close-packed positions if necessary. With the aid of this method Bragg and his co-workers have made a very promising attack on the important problem of the structure of silicates.

During the investigation of the structure of brookite, the orthorhombic form of titanium dioxide, another method for predicting possible structures for ionic compounds was developed, based upon the assumption of the coordination of the large anions in the crystal about the small cations in such a way that each cation designates the center of a polyhedron, the corners of which are occupied by anions. This theory leads for a given crystal to a small number of possible simple structures, for each of which the size of the unit of structure, the space-group symmetry, and the positions of all ions are fixed. In some cases, but not all, these structures correspond to close-packing of the large ions; when they do, the theory further indicates the amount and nature of the distortion from the close-packed arrangement.

The structures of rutile and anatase, the two tetragonal forms of titanium dioxide, have been determined by rigorous methods. They seem at first sight to have little in common beyond the fact that each is a coordination structure, with six oxygen atoms about each titanium atom at octahedron corners. From a certain point of view, however, the structures are closely similar. They are both made up of octahedra sharing edges and corners with each

Fig. 1. The structure of Brookite.

other; in rutile two edges of each octahedron are shared and in anatase four. In both crystals the titanium-oxygen distance is a constant, with the value 1.95 to 1.96 Å. The basic octahedra are only approximately regular; in each case they are deformed in such a way as to cause each shared edge to be shortened from 2.76 Å (the value for regular octahedra) to 2.50 Å, other edges being correspondingly lengthened. As a result of these considerations the following assumptions were made:

1. Brookite is composed of octahedra, each with a titanium atom at its center and oxygen atoms at its corners.

2. The octahedra share edges and corners with each other in such a way as to give the crystal the correct chemical composition.

3. The titanium-oxygen distances throughout are 1.95 to 1.96 Å. Shared edges of octahedra are shortened to about 2.50 Å.

Two possible structures for brookite satisfying these requirements were proposed[1]. The first was not the structure of brookite. The second, however, had the same space-group symmetry as brookite, and the predicted dimensions of the unit of structure agreed within 0.5% with those observed. Structure factors calculated for planes of over fifty forms with the aid of the predicted values of the nine parameters determining the atomic arrangement accounted satisfactorily for the observed intensities of reflections

[1] Linus Pauling and J. H. Sturdivant, *Ztschr. f. Krist.* **68** (1928) 239.

on rotation photographs. This extensive agreement is so striking as to permit the structure proposed for brookite (shown in Fig. 1) to be accepted with great confidence.

The theory has also been used by the author in predicting the structure of topaz, $Al_2SiO_4F_2$. The determination of the unit of structure and space-group symmetry of this orthorhombic crystal has been reported by Leonhardt[1]. In predicting a structure for topaz it was assumed that each aluminium ion is surrounded by four oxygen ions and two fluorine ions at the corners of a regular octahedron, and each silicon ion by four oxygen ions at the corners of a regular tetrahedron. The length of edge of octahedron and tetrahedron was taken as 2.72 Å, corresponding to crystal radii of 1.36 Å for both oxygen and fluorine ions. One structure was built up of these polyhedra. On studying its distribution of microscopic symmetry elements it was found to have the space-group symmetry of V_h^{16}, which is that of topaz. Its unit of structure, containing $4Al_2SiO_4F_2$, has edges $\underline{d}_{100} = 4.72$ Å, $\underline{d}_{010} = 8.88$ Å, and $\underline{d}_{001} = 8.16$ Å, in satisfactory agreement with the observed values $\underline{d}_{100} = 4.64$ Å, $\underline{d}_{010} = 8.78$ Å, and $\underline{d}_{001} = 8.37$ Å. The structure further leads to strong sixth order reflections from {100} and {001}, and strong fourth and eighth order reflections from {010}, in agreement with Leonhardt's observations. This concordance is sufficient to make it highly probable that the correct structure of topaz has been found.

Each aluminium octahedron shares two edges with adjoining octahedra, and four corners with tetrahedra. The octahedron corners occupied by fluorine ions are shared between two octahedra. Each silicon tetrahedron shares its corners with octahedra. There are present in the structure SiO_4 groups; that is, tetrahedra having no elements in common.

The arrangement of the oxygen and fluorine ions is that of double hexagonal close-packing, so that Bragg's methods could be applied to this crystal. But there are very many ways of distributing the cations with this anion arrangement, and the decision among them would have to be made by the more or less laborious comparison of observed and predicted intensities of reflection. Furthermore (as for brookite also) both hexagonal and double hexagonal close-packing are roughly compatible with the observed unit, increasing the number of possibilities to be considered.

The success of the coordination theory with brookite and topaz has led to the attempt to propose a set of principles governing the structures of complex ionic crystals. The crystals considered are to contain small cations, with relatively large electrical charges (that is, usually trivalent and tetra-

[1] J. Leonhardt, *Ztschr. f. Krist.* **59** (1924) 216.

valent cations). The chemical bond between ions need not necessarily be ionic in the sense of the quantum mechanics; it should, however, not be of the extreme non-polar or shared electron pair type; thus compounds of copper and many other eighteen-shell atoms cannot be satisfactorily treated. The principles of major importance for the coordination theory are given in the following paragraphs, with illustrative examples. They have been derived in part empirically, and in part from considerations involving the crystal energy.

1. THE COORDINATED POLYHEDRA Cations and anions are physically differentiated by their differences in size and electric charge; the cations are small and usually trivalent or tetravalent, while the anions are large and univalent or divalent. A coordinated polyhedron of anions is formed about each cation, the cation- anion distance being given by the radius sum, and the coordination number of the cation by the radius ratio. Thus zirconium, with crystal radius 0.80 Å, has a coordination number eight in zircon; aluminium, with radius 0.50 Å, six; and silicon, with radius 0.41 Å, four.

2. THE NUMBER OF POLEHEDRA WITH A COMMON CORNER. The number of polyhedra with a common corner can be determined by the use of an extended conception of electrostatic valence. Let Ze be the electric charge of a cation, and v its coordination number. Then the strength of the electrostatic valence bond going to each corner of the polyhedron is defined as

$$S = \frac{Z}{v}.$$

Let $-\zeta e$ be the charge of the anion located at a corner shared among several polyhedra. Then we postulate the following principle of electrostatic valence:

The state of maximum stability of an ionic crystal is that in which the electric charge of each anion just compensates the strengths of the electrostatic valence bonds reaching to it from the cations at the centers of the polyhedra of which it forms a corner; i.e.,

$$\zeta = \sum_i \frac{Z_i}{v_i} = \sum_i S_i. \tag{1}$$

The strengths of bonds for some common cations are given below:

	Z	v	S
Al^{+3}	3	6	$\frac{1}{2}$
Si^{+4}	4	4	1
Ti^{+4}	4	6	$\frac{2}{3}$
Mg^{+2}, Be^{+2}	2	4	$\frac{1}{2}$
Zr^{+4}	4	8	$\frac{1}{2}$

Equation 1 is satisfied necessarily by all crystals the anions of which are crystallographically equivalent, such as corundum, Al_2O_3, rutile, anatase, spinel, $MgAl_2O_4$, garnet, $Ca_3Al_2Si_3O_{12}$, etc. It is also satisfied by topaz. Each oxygen ion, common to one silicon and two aluminium ions, has $\sum S_i = 2$; while each fluorine ion, attached to two aluminium ions only, has $\sum S_i = 1$. Similarly in beryl some oxygen ions are shared between two silicon ions, and some between one silicon, one beryllium, and one aluminium; in each case $\sum S_i = 2$.

3. SHARING OF EDGES AND FACES. The stability of an ionic arrangement is decreased by an increase in the number of shared edges, and still more by sharing faces. This effect is large for cations with large valence and small coordination number. Thus silicon tetrahedra tend to share only corners with other polyhedra if this is possible (as in topaz), titanium octahedra share only corners and edges, while aluminium octahedra will in some cases share faces (as in corundum).

4. THE NATURE OF CONTIGUOUS POLYHEDRA. Cations with large valence and small coordination number tend to be as far from one another as possible, and hence not to share polyhedron elements. Thus in silicates the silicon tetrahedra will not share any elements with each other if the oxygen-silicon ratio is greater than four (topaz, orthosilicates in general). If necessary, corners will be shared, but not edges or faces. In silicon dioxide all four corners are shared. In $Si_2O_7^{-6}$ there will be groups of two tetrahedra with one corner in common. The metasilicates will not contain groups of two tetrahedra with a common edge, but rather chains or rings, each tetrahedron sharing two corners (as in beryl, with a ring of six tetrahedra, stable because of the approximation of the tetrahedral angle to 120°). The silicon tetrahedron may, however, share edges with polyhedra with a large coordination numbers (zircon, $v_{Zr} = 8$).

5. THE RULE OF PARSIMONY. The number of essentially different kinds of constituents in a crystal tends to be small. First, the electrostatic bonds satisfied by all chemically similar anions should be the same if possible (topaz, all oxygen ions with $2Al + 1Si$, all fluorine ions with $2Al$). This does not require the anions to be crystallographically equivalent (brookite,

two kinds of oxygen ions). Second, the polyhedra circumscribed about all chemically identical cations should be chemically similar ($4O + 2F$ about each aluminium ion in topaz), and similar in their contiguous environment; that is, in the nature of the sharing of corners, edges, and faces with other polyhedra (for example, the titanium octahedron shares two edges in rutile, three in brookite, and four in anatase; but structures are not known in which these different types occur together). The polyhedra which are similar in these respects may or may not be crystallographically equivalent; for they may differ in their remote environment. Thus the contiguously similar tetrahedra in carborundum are crystallographically of several kinds (five in carborundum I).

As an illustration of these principles some predictions may be made regarding the structures of cyanite, andalusite, and sillimanite, the three forms of Al_2SiO_5. From the rule of parsimony we expect all aluminum octahedra to be similar and all silicon tetrahedra to be similar. Let the number of octahedra attached to the ith oxygen ion be α_i; then the stoichiometrical formula of the crystals requires that

$$\Sigma^6_{i=1} \frac{1}{\alpha_i} = \frac{5}{2}, \tag{2}$$

in which the sum is over the six oxygen ions forming one octahedron. Four out of five oxygen ions will be differentiated through being attached to silicon ions (Rule 4); hence

$$\Sigma' \frac{1}{\alpha_i} = 2, \tag{3}$$

in which the prime signifies that the sum is to be taken over these oxygen ions only. (In case the aluminum octahedra were of different kinds average values of the sums would be used in Equations 2 and 3.) Let us now assume that the oxygen ions are of only two kinds, those attached to silicons and those not so attached. Then the equations become

$$\left. \begin{aligned} \frac{n_1}{\alpha_1} + \frac{n_2}{\alpha_2} &= \frac{5}{2}, \\ \frac{n_1}{\alpha_1} &= 2, \\ n_1 + n_2 &= 6. \end{aligned} \right\} \tag{4}$$

The only solution involving small integers is $n_1 = 4$, $\alpha_1 = 2$, $n_2 = 2$, $\alpha_2 = 4$. Thus about each aluminum ion there will be four oxygen ions shared between two aluminum ions and one silicon ion, and two oxygen ions shared between four aluminum ions. In both cases $\sum S_i = 2$, so that the principle of electrostatic valence is satisfied.

This result is not incompatible with Professor Bragg's assignment of a cubic close-packed arrangement of oxygen ions to cyanite. It is, however, in disagreement with the complete atomic arrangement proposed by Taylor and Jackson[1]. Indeed, their suggested structure conflicts with most of the above principles. Each silicon tetrahedron shares a face with an octahedron, contrary to Rules 3 and 4. The structure is far from parsimonious, with four essentially different kinds of octahedra and two of tetrahedra. One oxygen ion, attached to four aluminum ions and one silicon ion, has $\sum S_i = 3$, while another, attached to only two aluminum ions, has $\sum S_i = 1$. For these reasons the suggested atomic arrangement seems highly improbable. An attempt is now being made to derive structures for these three crystals and for a number of others with the aid of the coordination theory.

California Institute of Technology, Pasadena

This rule—'The number of essentially different kinds of constituents in a crystal tends to be small'—is of a much more doubtful character than the others. It is certain that in a number of purely geometrical possibilities the actual one in nature will probably not be the most complicated, but it would seem unsafe to go further.

J. D. BERNAL and W. A. WOOSTER,
Ann. Rep. **26** (1929) 296

124. The evidence for the existence of a hydrogen bond is based largely on crystal structure data which indicates that in compounds containing acid hydrogen certain oxygen, or fluorine, atoms belonging to different complex anions are found to be at the extremely short distance of 2.55 A apart, considerably shorter than the distance of 2.70 A that represents double the radius of the ion O^{--} and very much shorter than that for neutral oxygen. The existence of such a close distance of approach is evidence of considerable mutual energy of oxygen atoms, which can only be due to the presence of hydrogen between them. More direct proof is

---

[1] W. H. Taylor and W. W. Jackson, *Proc. Royal Soc. London* **119** (1928) 132.

difficult as the X-rays cannot show hydrogen directly, but it is hard to imagine where else to put the hydrogen, and the explanation of the hydrogen bond covers so many facts that it is convenient to accept it provisionally until it can be proved or disproved by spectroscopic or other means.

It is not, however, only in acids that short oxygen-oxygen distances are found. The crystal structures of the leading types of basic and amphoteric hydroxides have been determined (except those of the alkalies). Here the distance between oxygens attached to different cations is not constant but gets progressively less with increasing cation charge, that is, with the decrease in alkaline character. Here also the shorter distances 2.9–2.7 Å indicate considerable interionic attraction between hydroxyl groups. This hypothesis is supported, as will be seen below, by much evidence from the detailed structure of hydroxides. It is consequently necessary to postulate another type of binding between hydroxyl groups in basic and amphoteric hydroxides. The chief difference between this and the hydrogen bond is that here both the atoms which it joins possess hydrogen; it was consequently previously called by one of us the double hydrogen bond[1]. It now seems preferable for reasons that will appear later to call it the *hydroxyl bond*. The object of this paper is to bring out the nature of this hydroxyl bond by a discussion of the crystal structures of certain compounds where it occurs, and to attempt to explain it by considering the probable configurations of the hydroxyl ion in different ionic environments.

The clue to the structure of the hydroxyl bond was provided in part by the study of ice[2] and in part by the detailed analysis of the structure of aluminiumhydroxide, hydrargillite[3].

124a. *The structure of ice.* In the normal ice structure the positions of the H nuclei and consequently the orientations of the molecules are fixed. This follows from the x-ray determination of the positions of the O atoms combined with the knowledge of the molecular structure derived from band spectra. Every molecule is surrounded by four others in a tetrahedron and this arrangement will have least energy if the H nuclei, which subtent an angle of about 109° at the centre of the molecule, lie opposite two of the neighbours, while one of the H-nuclei of each of the remaining neighbours lies opposite a negative corner of the original molecule. (See Fig. 4) Such an arrangement has, of course, no trigonal symmetry and is impossible at

[1] Bernal, *Ann. Rep. Chem. Soc.* (1933) 401.
[2] Bernal and Fowler, *J. Chem. Phys.* **1** (1933) 515 [this Vol. paper **124a**].
[3] Megaw, *Z. Krystallog.* **87** (1934) 185.

first sight to fit with the x-ray structure of ice. But in the derivation of the latter, only the effects of the O atoms can be taken into account and so no structure leaving the positions of these unchanged can be rejected on x-ray grounds.

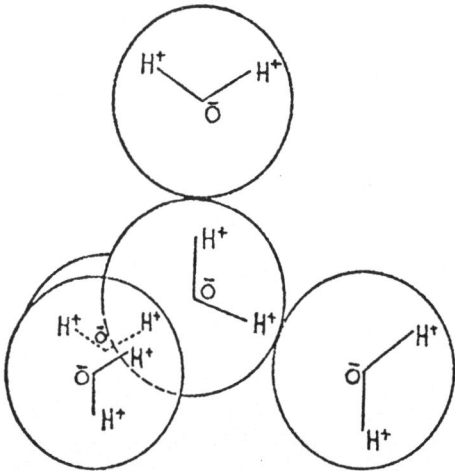

Fig. 4. Tetrahedral coordination of water molecules. The four molecules surrounding one water molecule are shown. Of these, two are in the plane of the paper, one above and one below it.

It is quite conceivable and even likely that at temperatures just below the melting point the molecular arrangement is still partially or even largely irregular, though preserving at every point tetrahedral coordination and balanced dipoles. In that case ice would be crystalline only in the position of its molecules but glass-like in their orientation*.

124. In the structure of hydrargillite the force of attraction between hydroxyl ions is clearly apparent. The structure is built from composite layers, parallel to the c-plane, each consisting of a plane of cations between two planes of hydroxyl ions, as in many other hydroxides. The significant point here, however, is the method of superposition of the layers, which is such that the hydroxyls in one layer come vertically above those in the next, instead of being arranged in close packed positions as in brucite $Mg(OH)_2$. Further, the distance between two such hydroxyls is 2.79 Å, which is not much greater than the minimum distance, 2.70 Å,[1] between

* [An orientation disorder is in accord with the residual entropy of ice. Ed.]
[1] The values given in the literature for the oxygen radius vary from 1.32 (Goldschmidt) to 1.40 (Pauling, Zachariasen). Probably the true value is about 1.35Å.

two oxygen atoms in octahedral coordination round the same cation. Both these facts imply that there must be an attractive force between the two hydroxyls.

This attraction must also be of a directed type more like that of a homopolar bond than of an ionic or residual force. The simplest assumption, that the force is of a dipole character, breaks down completely in this case as the two hydroxy dipoles in each layer would have to be practically head to head, thus $\left(\begin{array}{c} \text{H} \end{array}\right)\!\left(\text{H}\right)$. Only with a tetrahedral model for the OH group, similar to that put forward[1] for the water molecule, does the arrangement in hydrargillite appear at all plausible.

The next stage is to examine whether this tetrahedral character is maintained in all hydroxides and, if not, what are the conditions leading to its appearance. For this a comparative examination of all known hydroxide structures is necessary. As a criterion of the attractive force between two hydroxyls not coordinated round the same cation we may take the distance between their two oxygen atoms, the force being greater the shorter the distance. If this distance is tabulated for a number of hydroxides, it becomes apparent that it is large for small electrostatic valencies and for large cation radius; in other words, that it decreases as the polarization of the hydroxyl increases. Table I [not reproduced] shows this.

The explanation of these changes must be sought inside the hydroxy group itself.

The free (OH)⁻ ion must resemble a neon atom in its electron distribution with either all three p eigenfunctions or only one of them contributing to the binding of the hydrogen atom. In any case the electron distribution must possess cylindrical polar symmetry giving a resulting simple dipole. The hydrogen atom lies inside the p shell partially screened by the binding electrons.

If an increasing and divergent field is applied to such an ion it might be expected to pass through three stages. In the first, where the polarization is small, it retains cylindrical symmetry and merely increases its dipole moment. In layer lattices where this occurs the OH bond must either be perpendicular to the layer or rotate about the perpendicular. The layers are accordingly held together by residual forces or by weak dipole forces between rotating dipoles.

As the polarization increases, the hydrogen atom is pushed further out. At the same time the wave function of the oxygen electrons is subjected by the intenser fields of the cation to increasing perturbation, and the degenerate wave function of cylindrical symmetry becomes under their

[1] Bernal and Fowler, *loc. cit*.

101

influence a tetrahedral wave function similar to that occurring in water. The bonds to the oxygen atom are now definitely *directed*. (It is not implied that the tetrahedron is absolutely regular: the direction of the bonds may be distorted by surrounding atoms.) For example, in hydrargillite, two of the tetrahedral bonds of an oxygen atom go to the aluminium ions, at a third the hydrogen is found, and the fourth represents a concentration of negative electricity. It is evident that the negative charge on one oxygen will attract the hydrogen belonging to another, forming what we propose to call the *hydroxyl bond*. There is one bond of this nature formed per hydroxyl ion.

With still larger polarization the energy required to remove the hydrogen ion from the oxygen altogether becomes very small. In these circumstances the hydrogen ion can migrate into a neighbouring oxygen, if there is one present, as shown in the diagram

$$X\left(\!OH\!\right)\!\left(\!O\!\right)X \rightleftharpoons X\left(\!O\!\right)\!\left(\!HO\!\right)X$$

This will only happen if we are dealing with hydroxy oxides of the type $XO_n(OH)_m$; that is, only in oxy-acids or oxy-acid salts, *e.g.*, H_2SO_4, KH_2PO_4, but not in hydroxy-acids $X(OH)_n$. The initial and final states of this transfer are indistinguishable, so that the position of the H atom on one side or other of the bond is indeterminate, and it may be thought of for many purposes as being a free H^+ ion in 2-coordination, equidistant from the two oxygen atoms. This was Pauling's first picture of the hydrogen bond[1]. The hydrogen bond can be contrasted with the hydroxyl bond by this possibility of interchange, which does not exist in the latter, as such an interchange would in general lead to the formation of a water molecule $\left(\!H\!\right)\!\overset{H}{\left(\!\right)} \rightleftharpoons \overset{}{\left(\!\right)}\!\left(\!H\!\right)\!\overset{H}{}$. It is obvious that the absence of the second hydrogen in the case of the hydrogen bond must lead here to closer linkage. In fact we find that while the shortest hydroxyl bond observed is 2.71 Å, the hydrogen bond has never been observed greater than 2.55 Å.

Considered from the point of view of electrostatic valence, the hydroxyl bond may be thought of roughly as follows. Owing to polarization by the cation, the oxygen atom breaks up into four concentrations of negative charge, each of value $\frac{1}{4}$, in a tetrahedral formation; one of these is occupied by the hydrogen ion, giving a net charge of $+\frac{1}{2}$ there. This is approximately the charge $0.49e$ found for the H atoms in water by a consideration

---

[1] Pauling, *J. Amer. Chem. Soc.* **53** (1931) 1367.

of dipole moments (Bernal and Fowler, *loc. cit.*). Each of these charges is potentially a bond. Where the electrostatic valence from the cation is $\frac{1}{2}$, each surrounding cation forms a bond to one of these negative charges; where the electrostatic valence is 1, two of the negative charges are required to bind it to the cation. The remaining unattached negative charges form hydroxyl bonds with the positive charges. If the electrostatic valence is less than $\frac{1}{2}$, the oxygen wave function does not split tetrahedrally, and hydroxyl bonds are not formed.

The cylindrical OH group is characterized by an OH–OH radius not less than about 1.50 Å, while in the tetrahedral OH group the OH–OH radius varies from 1.41 to 1.35 Å, and the oxygen taking part in a hydrogen bond has an O–O radius of 1.27 Å.

[For metal bonds *see* paper **166** Zur Chemie der Legierungen, by A. WESTGREN (1932) and paper **167** The Electron Theory of Metals, by J. D. BERNAL (1935).]

CHAPTER VIII

Laue - Powder - Rotation - Weissenberg Methods

Long Chain Compounds and Fibre Structures

A. DIFFRACTION METHODS

Continuous progress has been made in the development of more powerful methods. During each quinquennial period it has become possible to determine the structures of crystals which had eluded the efforts of investigators during the previous period. The progress has resulted from the discovery both of new experimental techniques and of new and powerful methods of interpretation of experimental data. The Bragg spectrometric methods and methods based on the use of Laue photographs were used alone in a masterly way in the determination of the first structures studied. In a few years these were supplemented by other experimental methods—the powder method, the rotating crystal (layer line) method and the moving film method.

L. PAULING,
Current Science (1937) p. 20.

LAUE DIAGRAM
see Vol. I in particular paper **1**

125. From: The Crystal Structure of some Carbonates of the Calcite Group, by R. W. G. WYCKOFF (1920): The Method of Projection; The Effect of the Voltage applied to the Tube upon the Resulting Photographs

BRAGG SPECTROMETER
see Vol. I in particular papers **15** and **60**

POWDER DIAGRAM
126. Eine neue Interferenzerscheinung, by W. FRIEDRICH (1913)

127. Interferenz an regellos orientierten Teilchen im Röntgenlicht I, by P. DEBIJE and P. SCHERRER (1916)

128. The Crystal Structure of Iron, by A. W. HULL (1917)

129. Eine einfache Methode zur Erhöhung der Genauigkeit bei Debije-Scherrer Aufnahmen, by A. E. VAN ARKEL (1928)

ROTATION DIAGRAM

130. Enregistrement photographique continu des spectres des rayons de Röntgen, by M. DE BROGLIE (1913)

■

131. Faserstruktur im Röntgenlichte, by M. POLANYI (1921)

■

132. Bemerkung zur Arbeit: Das Röntgenfaserdiagramm von M. Polanyi, by E. SCHIEBOLD (1922)

■

133. Über die Entwicklung des Drehkristallverfahrens, by M. POLANYI, E. SCHIEBOLD and K. WEISSENBERG (1924)

■

134. A Universal X-ray Photogoniometer, combining: Apparatus for single Rotation Photographs—Laue Photographs—X-ray Spectrometry—Powder Photographs—Photographs of Crystal Aggregates, Metals, Materials, Etc., by J. D. BERNAL (1927)

WEISSENBERG DIAGRAM

135. Ein neues Röntgengoniometer, by K. WEISSENBERG (1924)

■

136. Das Weissenbergsche Röntgengoniometer, by J. BÖHM (1926)

■

137. Über die graphische Auswertung von Aufnahmen mit dem Weissenbergschen Röntgengoniometer, by W. SCHNEIDER (1929)

■

LAUE DIAGRAM

125. The indices of the planes producing the spots were determined by plotting the photographs in gnomonic projection.

This form of projection has two more or less obvious advantages. The indices of the reflecting planes are much more easily obtained and the method is capable of a general routine application not only to all systems of crystals but equally to various orientations of a crystal. This second fact is especially important in the present instance because patterns which are somewhat unsymmetrical are of greatest value. A decided saving in time is thus effected by being able to treat such photographs by routine methods.

The positions of the normals to the planes which are reflecting the X-rays are plotted in the gnomonic projection. Both the crystal and the eye are considered to be at the center of the sphere of projection. The plane of the projection is a tangent-plane normal to the direction of the incident X-ray beam (Fig. 4). Thus B is the projection corresponding to A. In the gnomonic projection[1], zones of planes lie along straight lines and if the rays pass parallel (or nearly parallel) to an axis of the crystal, the indices of any plane can be obtained directly from the co-ordinates of its position on the projection.

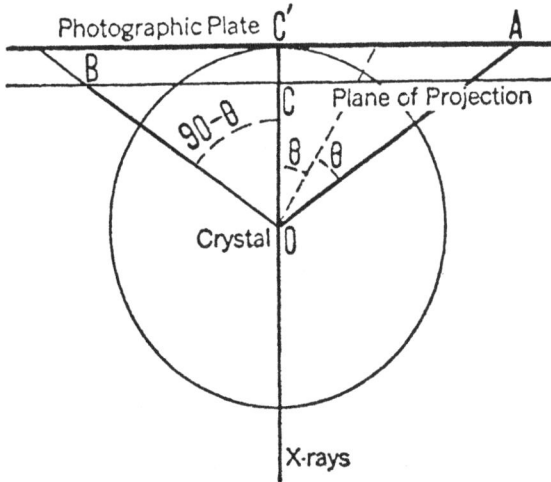

Fig. 4. The method of representing the reflection of a plane (in a Laue photograph) in gnomonic projection.

The following procedure has been found to reduce the time required for reading a Laue photograph to at least one-third or one-fourth of that needed if a stereographic projection is prepared. The accuracy under ordinary conditions of working is also appreciably greater.

An exact reproduction of the positions of the spots in the photograph is prepared in the center of the paper upon which the projection is to be made. A spot and the projection corresponding to it lie on a straight line

[1] The gnomonic projection, its properties and uses to the crystallographer, will be found in V. Goldschmidt, 'Ueber Projektion u. Krystallberechnung' (1887), or in H. E. Boeke, 'Die Gnomonische Projektion' (1913). A simple discussion is given by C. Palache in *Am. Mineralogist* **5** (1920) 67.

which passes through the central spot $(C, C'$, Fig. 4). The distance from a spot to the center C' is the product of the distance from the crystal to the plate, $C'O$, into tan 2θ. BC, the distance of the projection of the plane producing spot A from the center C, is equal to the distance CO, the radius of the sphere of projection, multiplied by cot θ. Then by the use of a straight edge and a table or special scale connecting tan 2θ and cot θ, it is possible to record the distance of a spot from the center and to locate its position on the projection with a single setting. Such a ruler is shown in Fig. 5.

Fig. 5. A ruler used in preparing the gnomonie projection of a Laue photograph. On the right hand is the special projection scale; the left hand scale is an ordinary millimeter ruler.

The Data from the Laue Photographs

Since the distance of any reflection from the central spot on the photographic plate has been measured and since the distance from the crystal to the plate is known, sin θ, where θ is the angle of reflection, can be obtained. Consequently $n\lambda$, where n is the order of the reflection and $\lambda =$ the wave length of the reflected X-rays, can be determined for each spot by using the customary expression, $n\lambda = 2d \sin \theta$. When the values of $n\lambda$, so obtained, were plotted against the estimated intensities of the spots, and those points which coorespond with planes of the same form were connected together, a series of curves, one above another, was obtained. It was thus possible to compare many planes of different forms in the same wave length. These curves are portions of curves which represent in shape the effect of the X-ray beam upon the photographic plate.

The Effect of the Voltage applied to the Tube upon the Resulting Photograph

It is a matter of considerable importance to know the general shape of the curve obtained after plotting $n\lambda$ against intensity for the various spots. The following considerations were made in order to determine the effect upon the curve of impressing different voltages upon the X-ray tube and to find out if filtering screens would prove useful. Two factors are known to influence the photographic effect of a beam of X-rays: the distribution of

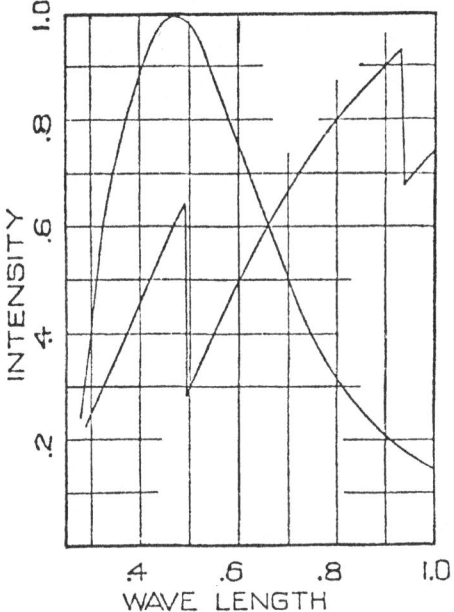

Fig. 10. The smooth curve represents the relative intensity of X-rays of different wave lengths for the 'white' radiation from tungsten when the voltage across the tube is 50,000 volts. The other gives the calculated amount of absorption of a certain thickness of silver bromide.

energy in the beam from the tube, and the selective absorption by the silver 'emulsion' of the plate.

The relative intensity of the X-rays of different wave lengths in the region of the 'white radiation' from a tungsten target for a certain voltage is shown in Fig. 10.[1] The higher the voltages, the greater becomes the intensity of the X-rays of all wave lengths and the shorter is the wave length of maximum intensity. The X-ray spectrum of tungsten has been mapped for various voltages[2] and careful measurements have been made in the region of the white radiation for voltages up to 50 kilovolts[3].

It has also been shown that the effect of X-rays upon the photographic plate, at any rate in this region of the X-ray spectrum, is quite closely proportional to the absorption by the silver bromide of the plate.[4]

The amount of X-rays absorbed by a convenient thickness of AgBr can be readily obtained from this mass absorption curve and is shown in Fig. 10.

[1] C. T. Ulrey, *Phys. Rev.* (2) **11** (1918) 401.
[2] A. W. Hull, *Proc. Nat. Acad. Sci.* **2** (1916) 265.
[3] C. T. Ulrey, *op. cit.*
[4] C. G. Barkla and G. H. Martyn, *Phil. Mag.* **25** (1913) 296.

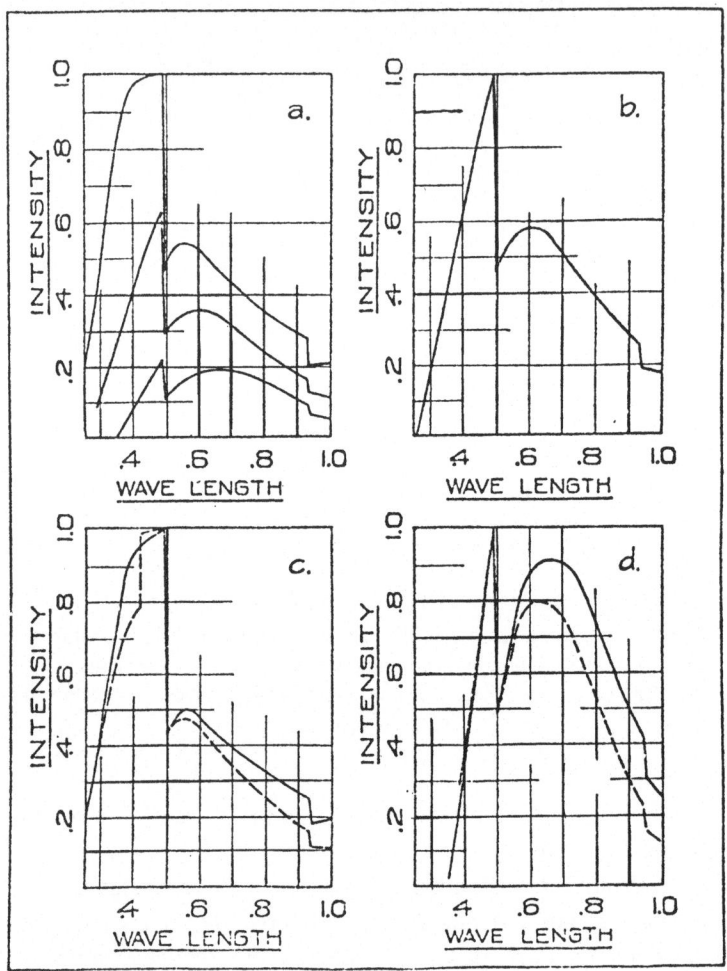

Fig. 12.

Since there is an uneven distribution of energy in the different wave lengths, the effect on the photographic plate is to be obtained by combining this curve (Fig. 10) with wave length-intensity curves, like those of Ulrey[1] for different voltages. The smooth curve of Fig. 10 is the curve for 50,000 volts. The graphs of Fig. 12a show the effect on the plate when the voltage across the tube is respectively (1) 40 K.V., (2) 50 K.V., and (3) 60 K.V. The last, and tallest, (3), is calculated using what seem to be the most probable values at hand. The full curves of Figs. 12 b, c, d are the same ones as those of Fig. 12a, but now each is calculated on the basis

---

[1] C. T. Ulrey, *op. cit.*

of unit effect at the point of maximum absorption. These are more useful because it is the *shape* of these curves that is of importantce; greater intensity can always be obtained by a longer exposure. All agree in showing the greatest effect upon the plate at the position of maximum absorption for silver. In the case of the lowest voltage (40,000) the intensity is seen to rise rapidly to the maximum (Fig. 12d) and then drop, a state of affairs parcularly desirable since the radiation here is very nearly monochromatic. The intensity then rises again between the first and second reflections to another maximum nearly as important as the first, after which it falls off till the critical absorption of bromine is reached, when it takes a further fall. At still lower voltages the continued shift of the maximum in the direction of longer wave lengths would shortly dwarf the silver absorption peak into insignificance compared with the second maximum. This is highly undesirable because only the short wave lengths are useful, for the reason that a reflection in the neighborhood of $n\lambda$ equal to the longer wave length maximum might be either a first order for this region, the second order of a much shorter wave length, or both together.

At 50 K.V. the intensity distribution curve has its maximum near the critical absorption value for silver, and as a result the relative effect of this region of wave lengths is great. Fig. 12b shows that the second maximum has become insignificant. As the voltage is made still higher (60 K.V.) the further shift of the intensity curve (Fig. 12c) towards short wave lengths results in there being a considerable spectrum range, just shorter than the critical value of silver, over which there is a marked effect upon the photographic plate. At the same time the longer wave length maximum has disappeared. Raising the voltage to a higher value would cause this process of increasing the effect of the shorter wave lengths to continue so that the point would soon be reached where the maximum effect would no longer remain at the critical absorption value for silver but would shift towards the shorter wave lengths, more as the voltage is raised.

It is of interest to inquire how closely these relative absorption curves for various voltages approach to a curve for the effect of an X-ray beam upon a photograph plate which would be ideal for Laue photographs. Such a curve would start from zero, rise quite rapidly to a decided maximum (in actual practice this would be the critical silver value ($\lambda = 0.49$) and immediately fall off to a negligible value. Monochromatic radiation, on the one hand, is not desired because so few planes would be in a position to reflect the particular wave length for one setting of the crystal; on the other hand, if the curve rises but slowly over a considerable range of wave lengths, many planes will produce weak spots because they are reflecting X-rays which are not particularly intense. If, again, there is a considerable

113

effect upon the plate for wave lengths *longer* than $n\lambda = 0.49$, there will be produced a considerable number of useless reflections. The result in either of these last two cases is simply that more work is required to interpret the Laue patterns because of the presence of a large number of not particularly valuable reflections. There should be no radiation shorter than half the critical value for silver

$$\frac{0.49 \text{ A.U.}}{2} = 0.24^5 \text{ A.U.,}$$

for otherwise it would be impossible to use the values in the neighborhood of $n\lambda = 0.49$ as assuredly of the first order. This state of affairs is most nearly met by the curve for 50,000 volts. At appreciably higher voltages there is radiation in the wave lengths shorter than 0.25 A.U.; if, again, lower voltages are used, the energy in the wave lengths around 0.49 A.U. becomes small so that a very long exposure is necessary and the second useless maximum becomes relatively of great importance. Except at voltages too low to be useful, the relative amount of energy in the wave lengths about the critical absorption of bromine is so small that the break in photographic effect at this point is negligible.

It was thought that the use of *filters* might furnish a more ideal curve. Such screening might be employed either with a higher voltage to cut down the extremely short wave lengths, or it might serve to reduce the importance of the long wave lengths, especially for low voltages across the X-ray tube. The dotted curve of Fig. 12*c* shows the relative effect of a screen of tin thick enough to absorb one-tenth of the radiation at $\lambda = 0.49$ A.U. for a voltage of 60 K.V. The change of the shape of the curve is not very remarkable and the new one does not show enough advantages to repay the loss in useful wave lengths by absorption. Under the ordinary conditions of experimentation it would not be feasible to use higher voltages and cut down the very short wave lengths, less than $\lambda = 0.25$ A.U., by appropriate screens because of the absorption of these materials in the region of $\lambda = 0.45 - 0.49$ A.U.

The material most effective in reducing the intensity of the longer wave lengths is a substance, like aluminum, which shows no selective absorption in the region of the spectrum employed. This kind of filtering would only be serviceable in improving the lower voltage curves. Fig. 12*d* gives the curve obtained by interposing a thickness of aluminum sufficient to reduce $\lambda = 0.49$ in the 40 K.V. plot to 90 per cent of its unscreened value. Even neglecting the loss of valuable radiation, the result is not as good as the unscreened 50 K.V. curve. Neither does an aluminum filter improve the 50 K.V. curve sufficiently to make its use worth while.

The conclusions to be drawn from this discussion are:

(1) That the most useful Laue photographs are obtained by operating a tungsten tube, if this is used, at 50 K.V.

(2) That the use of various screens is not desirable.

Geophysical Laboratory, Carnegie Institution of Washington

POWDER DIAGRAM

126. In einer jüngst erschienenen Abhandlung[1] hat der Verfasser in Gemeinschaft mit P. Knipping über Interferenzversuche mit Röntgenstrahlen berichtet.

Bei weiteren Versuchen dieser Art, über deren Ergebnisse demnächst an anderer Stelle ausführlich berichtet werden wird, habe ich zu meiner großen Überraschung auch mit nicht kristallinen Körpern Erscheinungen erhalten, die unbedingt als Beugungsbilder und somit als Interferenzen der Röntgenstrahlen angesehen werden müssen. Über sie soll hier eine kurze Mitteilung folgen.

Für die Versuche wurde die gleiche Anordnung benutzt wie bei den Versuchen mit Kristallen (vgl. Fig. 1 daselbst).

Es wurde zunächst gewöhnliches weißes Klebwachs untersucht, das zu einem 3 mm dicken Scheibchen geformt war, und das die Strahlen senkrecht durchsetzten. Nach einer Exposition von 2000 Milliampereminuten mit einer mittelweichen Röhre ergab sich auf der photographischen Platte ein Bild auf dem man deutlich einige Beugungsringe um den Durchstoßungspunkt der Primärstrahlen erkennen kann. Bei anderen Körpern, wie Kanadabalsam, Paraffin, Paraffinöl. Bernstein und Meerschaum, wurden teils ebenfalls Andeutungen von Ringen, stets aber ein viel allmählicherer Abfall der Schwärzung festgestellt, als er ohne eine dieser Substanzen sich zeigte. Diese allgemeine Verbreiterung möchte ich ebensosehr als Beugungserscheinung ansprechen wie die mehr in die Augen springenden Ringe.

Zur Deutung dieser Erscheinungen sei ganz allgemein folgendes vorangeschickt. Im Gegensatz zu den kristallinen Körpern, bei denen die Molekule bzw. Atome eine gesetz- und regelmäßige Anordnung zeigen, ist bei den amorphen Körpern die Anordnung der Teilchen eine vollkommen regellose. Fallen Röntgenstrahlen auf einen amorphen Körper, so werden sie zu einem kleinen Teile nach allen Richtungen hin zerstreut; scharfe Intensitätsmaxima von der Art, wie sie durch Interferenz bei den kristallinen

---

[1] *Münch. Ber.* (1912) 311–322.

Körpern entstehen, können nicht auftreten. Wir haben ein Analogon zu der aus der Optik bekannten Erscheinung, daß Licht durch ein trübes Medium zerstreut wird, und daß, falls die die Trübung verursachenden Partikeln nicht zu klein sind, rund um die Einfallsrichtung Intensitätshäufungen auftreten. Die Beugungsringe, die an behauchten oder mit Lykopodium bestreuten Glasplatten auftreten, sowie die Höfe um Sonne und Mond, die durch Beugung an den Wassertröpfchen dünner Wolkenschichten entstehen, sind Beispiele hierfür. Der Durchmesser der beugenden Teilchen beträgt hierbei das ca. 10 bis 100 fache der Wellenlänge der auffallenden Strahlen.

Nahe das gleiche Größenverhältnis besteht zwischen der Größe der Atome und der Größe der Wellenlange der Röntgenstrahlen. Die erste ist von der Größenordnung 10^{-8} cm, während nach den bisherigen Versuchen der wahrscheinlichste Wert für die Wellenlänge der periodischen Röntgenstrahlung von der Größenordnung 10^{-9} cm ist. Es ist daher offenbar die Möglichkeit gegeben, daß Beugungserscheinungen obengenannter Art auftreten können, wenn Röntgenstrahlen durch einen amorphen Körper hindurchgehen.

Folgender Versuch scheint eine Stütze der Auffassung zu sein, daß wir es mit Beugung am Molekül bzw. Atom zu tun haben.

Paraffin, bei dem sich ein charakteristisches Beugungsbild ergab, wurde während der Exposition in einem dünnen Gläschen, das leer keinerlei Beugung zeigte, geschmolzen gehalten. Als Heizvorrichtung diente eine vom elektrischen Strom durchflossene Drahtspirale, die in das Gläschen eingefügt war, ohne jedoch von den Primärstrahlen getroffen zu werden. Das Resultat war zwar nicht ein deutlich wahrnehmbarer Ring. Jedoch ergab eine Photometrierung des radialen Schwärzungsverlaufes bei beiden Platten den gleichen allgemeinen Abfall nur mit dem Unterschiede, daß beim festen Paraffin an der Stelle, wo das Auge den Ring wahrnahm, ein Inflexionspunkt in der Kurve lag. Die Kurven, die mit dem Kochschen selbstregistrierenden Mikrophotometer aufgenommen wurden, werden später in einer ausführlicheren Mitteilung der Versuche an anderer Stelle veröffentlicht werden.

Von sehr wesentlicher Bedeutung scheint mir folgender Unterschied zwischen den Interferenzerscheinungen an Kristallen und den hier beschriebenen zu sein.

Versuche, die ich im vergangenen Sommer an Kristallen angestellt habe, und die unter anderen demnächst zur Veröffentlichung gelangen werden, hatten gezeigt, daß das Material der Antikathode und demnach die spektrale Zusammensetzung der Primärstrahlen ohne Einfluß auf die Lage der Interferenzmaxima sowie auf die Härte der interferierenden Strahlen war.

Bei den Kristallen scheint die Wellenlänge der interferierenden Strahlen durch die Gitterkonstante des Kristalles bestimmt zu sein. Bei den hier untersuchten Körpern ist nun ein großer Einfluß der Zusammensetzung der Primärstrahlen auf die Erscheinung vorhanden. Mit einer Eisenantikathode ergab sich ein völlig anderes Beugungsbild als mit einer Antikathode, die aus einem platinplattierten Nickelblech bestand.

Auch bei ein und derselben Röhre variierte die Schärfe der Ringe sowie ihr gegenseitiges Intensitätsverhältnis je nach dem Alter der Röhre, was bei Kristallen niemals der Fall war.

Ich will an dieser Stelle noch auf eine andere Erklärungsmöglichkeit der hier beschriebenen Beugungserscheinung hinweisen, die ebenfalls in Erwagung zu ziehen ist.

Man kann sich die untersuchten Körper aufgebaut denken aus vielen Kriställchen, die ungeordnet alle möglichen Richtungen zu den Primärstrahlen haben, eventuell den Molekulen selbst die bei den meisten in Betracht kommenden Substanzen wohl sehr groß sind, gitterartige Struktur beilegen. Die dann nach Wahrscheinlichkeitsgesetzen verteilten Interferenzflecken der einzelnen Kristallindividuen könnten solche Häufungsstellen mit Rotationssymmetrie zeigen. Jedoch müßte dann nach den bisherigen Erfahrungen [diffraction patterns of white radiation. Ed.] die Erscheinung unabhängig sein vom Antikathodenmaterial, was nicht der Fall ist [pattern of characteristic radiation].

Eine Erscheinung möchte ich hier nicht unerwähnt lassen, für die ich noch keine genügende Erklärung habe, die aber von großem Interesse ist. Es zeigte sich, daß das Beugungsbild durch rein mechanische Veränderungen des beugenden Körpers beeinflußt wird. Dieser Einfluß wurde zufällig an einem Stück Wachs entdeckt, das, wie sich später zeigte, beim Formen einen einseitigen Druck erfahren hatte. In dem mit diesem erhaltenen Beugungsbilde fehlte die Intensität in zwei gegenüberliegenden Stellen fast vollkommen, von den Ringen waren nur die beiden gegenüberliegenden Quadranten sichtbar. Ich dachte zunächst an eine Polarisation der Primärstrahlen, weil die Lage der Intensitätsmaxima mit der nach der Impulstheorie zu erwartenden übereinstimmte. Jedoch blieb die Lage der Maxima beim Drehen der Röhre und somit der Polarisationsebene um 90° unverändert. Dagegen wanderten die Maxima mit beim Drehen des Wachses um die Richtung der Primärstrahlen. Ein Stück Wachs, das in einem Schraubstock kräftig einseitig zusammengepreßt war, zeigte dann, daß dieser Effekt in der Tat durch die von der Deformation herrührende Inhomogenität hervorgebracht war. Die Auslöschung der Ringe erfolgt senkrecht zur Druckrichtung.

Institut für theoretische Physik der Universität, München

127. Vor einiger Zeit hat der eine von uns auf eine Methode aufmerksam gemacht, die dazu dienen kann sowohl über die Zahl, wie über die gegenseitige Anordnung der Elektronen im Atom auf experimentellem Wege Aufschluß zu erlangen[1]. Die Möglichkeit einer solchen Messung beruht, wie damals hervorgehoben wurde, darauf, daß, wenn eine Regelmäßigkeit der Anordnung der Elektronen im Atom vorhanden ist, dieselbe auch dann noch erkennbar bleibt, wenn viele solche Atome in regelloser Orientierung miteinander gemischt vorkommen.

Im einzelnen konnte nämlich gezeigt werden, daß wenn eine solche Substanz mit der vorausgesetzten inneren Regelmäßigkeit der Elektronenanordnung versehen mit monochromatischen Röntgenstrahlen bestrahlt wird, die dadurch hervorgebrachte Sekundärstrahlung nicht (im wesentlichen) gleichmäßig von der Substanz aus in den Raum hinein ausgestrahlt wird, sondern Maxima und Minima zeigen muß. Dieselben liegen auf Kegeln, deren Achse mit der Richtung der primären Strahlung zusammenfällt und deren Spitze sich im Innern des als klein angenommenen Sekundärstrahlers befindet. Damit die fraglichen Maxima und Minima zustandekommen ist noch obendrein nötig, daß die Wellenlänge der benutzten Primärstrahlung von derselben Größenordnung, wie die gegenseitigen Elektronenabstände ist. Daß diese zweite Forderung experimentell erfüllbar sein dürfte, wurde damals geschlossen aus einem Vergleich der Wellenlänge der Fluoreszenz-Röntgenstrahlung mit den nach den Bohrschen Quantenansätzen zu erwartenden Elektronenabständen.

Versuche, welche inzwischen von uns in dieser Richtung angestellt wurden, zeigten den erwarteten Erfolg. Nebenbei aber fanden sich in einigen Fällen über den erwarteten Effekt übergelagert anders geartete Interferenzen, welche durch die Schärfe der auftretenden Maxima klar erkennen ließen, daß für sie nicht die regelmäßige Anordnung der doch voraussichtlich recht kleinen Zahl von Elektronen im Atom verantwortlich gemacht werden konnte. In dieser vorliegenden ersten Mitteilung wollen wir uns auf die Beschreibung und Erklärung dieser einen Erscheinung allein beschränken; auf die eigentlichen Elektroneninterferenzen und verwandte Erscheinungen beabsichtigen wir in einer späteren Mitteilung näher einzugehen.

Die Interferenzen sind scharf, also muß es sich um eine Erscheinung handeln, bei der eine recht große Zahl von Strahlungszentren zusammenwirkt. Ist dem aber so, dann liegt es nahe in den Fällen, wo die Interferenzen beobachtet wurden, dieselben zurückzuführen auf die kristallinische Struktur der durchstrahlten Substanz, auch wenn letztere, wie es stets der Fall war, als anscheinend amorphes Pulver benutzt wird, oder sogar

---

[1] *Nachr. d. Kgl. Ges. d. Wiss. Göttingen* vom 27. Febr. 1915, *Ann. der Phys.* **46** (1915) 809.

als 'amorph' in der Chemie bezeichnet wird. Dieser von uns tatsächlich angenommene Standpunkt mag befremdlich erscheinen mit Rücksicht auf das von Friedrich, Knipping und v. Laue in ihrer ersten Arbeit angegebene Versuchsergebnis[1], wonach ein fein gepulverter Kristall keine Interferenzen mehr aufkommen ließ. In Wirklichkeit läßt sich aber einerseits diese Behauptung, wie die hier mitgegebenen Aufnahmen zeigen, nicht aufrecht erhalten, andererseits folgt die Erscheinung mit Notwendigkeit aus der von v. Laue entworfenen Theorie der Kristallinterferenzen, wie im folgenden ausgeführt wird.

Erklärt man sich mit den nachfolgenden Überlegungen einverstanden, dann liefert die Beobachtung der fraglichen Interferenzen ein einfaches Mittel, um mit absoluter Sicherheit über den (mikro-)kristallinischen oder amorphen Zustand einer Substanz zu entscheiden. Die in § 3 vorgeführte Diskussion dreier Photogramme soll zeigen, wie man über die einfache Feststellung einer jener Tatsachen hinausgehend die photographische Aufnahme benutzen kann, um den inneren Aufbau des Einzelkristalls zu erforschen. Tatsächlich gelingt es mit Hilfe einer einzigen Photographie die gegenseitige Lage und die Abstände der Atome im Kristall zu bestimmen, ähnlich wie das bekanntlich Bragg durch die elektrometrische Untersuchung der Reflexion an den verschiedenen Netzebenen eines großen Kristalls gelungen ist. Man kann mit gutem Grund sogar behaupten, daß die ganze Frage mit Hilfe eines vollständig amorph aussehenden Pulvers nach unsrer Methode erheblich leichter beantwortet werden kann, als es durch Beobachtungen an einem großen gut ausgebildeten Kristalle möglich ist.

Hat man für irgendeine Substanz die Atomanordnung bestimmt, dann kann man dieselbe in Pulverform auch umgekehrt benutzen als Gitter zur Analysierung der auffallenden Strahlung nach Wellanlängen. Da natürlich von einer Orientierung der 'amorphen' Substanz keine Rede ist und auf dem Diagramm eine einzige Linie des auffallenden Spektrums an vielen (z.B. an 10) getrennten Orten als klarer schmaler Strich vorhanden ist, dürfte die nachher zu beschreibende Anordnung auch zur Bestimmung der Wellenlängen, welche in der auffallenden Strahlung vorkommen, mit Vorteil anwendbar sein. Unsere Anordnung stellt so aufgefaßt wohl das einfachste Spektroskop überhaupt dar.

§ 1. *Die Versuche*

Die zu untersuchende in Pulverform vorliegende Substanz (es wurden u.a. untersucht Graphit, von Kahlbaum bezogenes: amorphes Bor, amorphes

---

[1] *Sitz.-Ber. d. Kgl Bayer. Ak. d. W.* (1912) 315.

Silizium, Borstickstoff, Lithiumfluorid usw.) wurde in Form eines Stäbchens gepreßt mit etwa 2 mm Durchmesser und etwa 10 mm Länge[1]. Das Stäbchen wurde aufgestellt in der Mitte einer zylinderförmigen Kamera von 57 mm Durchmesser, welche durch einen Deckel lichtdicht abgeschlossen werden konnte. Der Eintritt wurde den Röntgenstrahlen in diese Kamera in horizontaler Richtung gestattet durch ein längeres in der Mitte mit einer Bohrung von 2,5 mm versehenes Bleiröhrchen, das in eine Messingröhre eingegossen war. Aus der Kamera trat das scharf begrenzte Strahlenbündel, ohne die Wandung derselben zu berühren, hinaus, verlief dann weiter in ein längeres aus schwarzem Papier angefertigtes, an die Kamera angeschraubtes Rohr und durchsetzte schließlich den ebenfalls aus dünnem schwarzen Papier gebildeten Boden desselben. In dieser Weise war dafür gesorgt, dasz keine merkliche Sekundärstrahlung an der Kamera selbst erzeugt wurde. Tatsächlich zeigte ein mehrmals wiederholter Kontrollversuch ohne zerstreuende Substanz, der sich über dieselbe Zeitdauer (gewöhnlich zwei, gelegentlich vier Stunden) erstreckte wie der Zerstreuungsversuch, keine Spur zerstreuter Strahlung.

Das Strahlenbündel traf das oben erwähnte Stäbchen in der Mitte; die von demselben ausgehende Sekundärstrahlung wurde photographisch aufgenommen auf zwei halbkreisförmig gebogenen an der Wand der Kamera anliegenden Films. Auf denselben konnte die Strahlung aufgefangen werden in einem Winkelbereich von 9° bis 171°, einer Filmlänge von etwa 80 mm entsprechend. Die an beiden Seiten fehlenden Bereiche von der jeweiligen Winkelausdehnung 2 × 9° waren durch das Vorhandensein der oben erwähnten Ein- und Austrittsvorrichtung bedingt. Gelegentlich wurde zur Vervollständigung der Übersicht über die ganze Erscheinung eine besondere Aufnahme mit einem kleineren, die Austrittsöffnung bedeckenden Filmstück gemacht.

Die Primärstrahlung wurde erzeugt mit Hilfe von Röntgenröhren, welche einem von Herrn Rausch v. Traubenberg entworfenen Modell im wesentlichen nachgebildet waren. Sie blieben während des Betriebs dauernd an der Pumpe. Die Strahlung verließ die Röhre durch ein Aluminiumfenster von 0,05 mm Dicke. Der Abstand Antikathode–zerstreuende Substanz betrug etwa 12 cm.

Die Fig. 1 auf der beigegebenen Tafel [see p. 126. Ed.] zeigt eine Aufnahme mit LiF als zerstreuende Substanz[2]. Der linke Rand der Reproduktion ist derjenige, welcher der Austrittsöffnung am nächsten lag. Von der Mitte dieser Öffnung aus zählen wir im folgenden stets die Winkel, der linke Rand ent-

---

[1] Gelegentlich war es nötig, dem Stäbchen durch einen dünnen Kollodiumüberzug einen festeren Halt zu geben.

[2] Das LiF bildete ein feines Pulver von dem Aussehen gebrannter Magnesia.

spricht also 9°. Man sieht wie auf dem Film kreisförmige, scharfe Interferenzstreifen auftreten, welche sich gegen 90° hin gerade strecken, um nach Überschreiten von 90° und bei Annäherung an 180° allmählich wieder in kreisförmige Linien mit wachsender Krümmung überzugehen. Die Aufnahme zeigt also, wie vom LiF-Stäbchen als Mittelpunkt aus besonders große Intensitäten auf einzelnen Kegeln in den Raum hinausgestrahlt werden. Der Durchschnitt eines dieser Kegel mit dem zylindrisch gebogenen Film entspricht einer solchen photographierten Linie. Der Winkelbereich, in dem von einem Raumelement des Stäbchens merkliche Strahlung ausgeht, muß jeweilig sehr gering sein. Tatsächlich ist nämlich die Breite der photographierten Linien bei genügend harter Strahlung identisch mit der Dicke des zerstreuenden Stäbchens und kann also für spektroskopische Zwecke wesentlich verringert werden.

Bei diesem Versuch stammte die Primärstrahlung von einer Kupferantikathode.

In Fig. 2 ist das Photogramm reproduziert, welches mit demselben LiF-Stäbchen jetzt aber bei Bestrahlung mit Röntgenlicht von einer Platinantikathode erhalten wurde.

Die Fig. 3 zeigt eine Reproduktion der Zerstreuung von 'amorphen Silizium' mit Kupferstrahlung erhalten.

Ganz ähnliche Erscheinungen lieferten auch die anderen oben genannten Substanzen.

Bekanntlich hat zuerst Friedrich[1] bei der Durchstrahlung von Wachs und Paraffin Ringe um den Durchstoßungspunkt der Primärstrahlen mit der Platte photographiert. Es blieb damals unklar, wie dieselben zustande kamen. Es fehlen zwar bei Friedrich entscheidende Experimente, auch ist die Konstitution der benutzten Substanzen unbekannt. Trotzdem glauben wir nach der Schärfe der Ringe in der Reproduktion urteilend, daß auch dort regellos gelagerte Kriställchen für die Erscheinung verantwortlich zu machen sind.

§ 2. *Die Theorie*

Die Laue-Braggsche Theorie der Kristallinterferenzen führt bekanntlich zu den folgenden beiden Hauptsätzen:

I. Fällt ein Röntgenstrahl auf einen Kristall auf, dann können 'reflektierte' Strahlen entstehen, die man so konstruieren kann, als ob sie durch gewöhnliche optische Reflexion an den Netzebenen desselben hervorgebracht wären.

---

[1] *Phys. Zs.* **14** (1913) 317 [this Vol. preceding paper].

II. Die in dieser Weise konstruierten Strahlen sind indessen nur dann wirklich vorhanden, wenn die an zwei zugehörigen aufeinander folgenden Netzebenen reflektierten Strahlen einen Gangunterschied aufweisen, der ein ganzzahliges Vielfaches der auffallenden Wellenlänge ist.

Die Formeln, welche diese Sätze ausdrücken, schreiben wir hier nur für das reguläre System hin[1]; sie lauten dann folgendermaßen: Nennt man den Winkel des monochromatisch vorausgesetzten einfallenden Strahles mit der gerade betrachteten Netzebene φ, die (ganzzahligen) Indizes dieser Ebene h_1, h_2, h_3, die Wellenlänge λ und die Seitenlänge des elementaren Kubus des kubischen Raumgitters a, dann sind zunächst nur solche Winkel möglich, welche die Bedingung

$$\sin \varphi = \frac{\lambda}{2a} \sqrt{(h_1{}^2 + h_2{}^2 + h_3{}^2)} \tag{1}$$

erfüllen. Dabei betrachten wir die Indizes 3, 2, 1 z.B. als verschieden von 6, 4, 2 oder 9, 6, 3, damit in Formel (1) zugleich die Interferenzen höherer Ordnung mit enthalten sind.

Läßt man nun auf ein regelloses Gemisch von kleinen Kristallen monochromatische Röntgenstrahlung auffallen und greift eine bestimmte Netzebene h_1, h_2, h_3 heraus, dann wird nach (1) dieselbe nur dann reflektieren können, wenn sie so orientiert ist, daß der Winkel zwischen dieser Ebene und dem einfallenden Strahl den aus (1) folgenden Wert φ hat. Nun kommt es aber natürlich auf die absolute Orientierung der Kriställchen im Raum nicht an; die an h_1, h_2, h_3, reflektierten Strahlen erfüllen deshalb einen Kegel, dessen halber Öffnungswinkel ϑ den Wert 2φ hat, da die Achse des Kegels von dem einfallenden Strahl und jede Erzeugende von einem reflektierten Strahl gebildet wird. (...) [Intensity discussion not reproduced, its essentials being given in Vol. I pp. 196–197. Ed.]

Ordnet man nun die Indizestriplets nach steigender Quadratsumme, dann entspricht jeder folgenden dieser Summen nach (1) ein Kegel mit größerer Öffnung. Unsere Behauptung geht dahin, daß die von uns photographierten Linien die Schnittlinien solcher Kegel mit unsrem kreisförmig gebogenen Film darstellen.

Ist dem so, dann muß es möglich sein, nach Ausmessung der den Einzellinien entsprechenden Kegelöffnungen 2ϑ, festzustellen, daß die aufeinander folgenden Werte von $\sin^2 (\vartheta/2)$ in Übereinstimmung mit (1) sich wie ganze Zahlen verhalten. Unsere Aufnahmen erfüllen, wie im nächsten Paragraphen gezeigt wird, tatsächlich diese Forderung.

---

[1] Der allgemeine Fall läßt sich ebenfalls ohne Mühe erledigen. Es ist dazu nur eine geeignete Kombination derjenigen Formeln nötig, welche M. v. Laue in der *Enz. d. math. Wiss.* Kap. 24, 467 ff. zusammengestellt hat.

Über dieses hinausgehend zeigt sich aber, daß eventuell gewisse Kombinationen von ganzen Zahlen in den nach steigender Quadratsumme geordneten Indizestriplets als Linien auf dem Photogramm vollständig fehlen. Das ist jeweilig nur auf Grund einer speziellen Bauart des Strukturfaktors zu verstehen, der dann für solche Richtungen verschwinden muß. So ist man imstande, aus den fehlenden Linien auf den Bau von S und damit auf die gegenseitige Lage der als strahlende Zentren angenommenen Atome zu schließen. Letzteres ist im Wesen identisch mit der Braggschen Schlußweise. Der skizzierte Weg führte z.B. zu der Feststellung (vgl. den nächsten § 3), daß LiF ein Gitter bildet wie NaCl, KCl usw., dessen Eckpunkte abwechselnd mit Li-Atomen und F-Atomen besetzt sind, während das sogenannte amorphe Silizium eine Gitterstruktur aufweist, welche mit der von Diamant identisch ist. Diese letzte Feststellung ist besonders deshalb interessant, weil Si im periodischen System unmittelbar unter C steht.

§ 3. *Diskussion dreier Beispiele*

a) Fig. 1 zeigt eine Reproduktion des bei Bestrahlung eines aus äußerst fein gepulvertem LiF gepreßten Stäbchens erhaltenen Films in natürlicher Größe. Die Primärstrahlung bestand wesentlich nur aus der K-Serie der Cu-Antikathode. Auf dem Original waren im ganzen 16 Linien zu erkennen. Die Ausmessung der Photographie ergab für den jeweiligen halben Öffnungswinkel ϑ des zugehörigen Kegels die in Spalte 2 der Tabelle I in Grad angegebenen Werte. Die erste Spalte enthält die Bezeichnungen s.s. = sehr schwach, s. = schwach, m. = mittel, st. = stark, welche eine ungefähre Schätzung der Linienschwärzung bedeuten sollen. Die Hauptintensität der Cu-Strahlung wird bekanntlich nach Moseley geführt von der α-Linie (Wellenlänge $1{,}549 . 10^{-8}$ cm), die zu derselben Serie gehörige zweite Linie an Intensität ist die β-Linie mit einer Wellenlänge von $1{,}402 . 10^{-8}$ cm. Der erste Schritt zur Entwirrung der in Tabelle I gegebenen Werte von $\sin(\vartheta/2)$ bestand nun darin, daß nach Verhältnissen von der Größe $1{,}402/1{,}549$ gesucht wurde, denn wenn α- und β-Linie an der gleichen Netzebene reflektiert vorkommen, müssen sich nach (1) die zugehörigen Verhältnisse von $\sin(\vartheta/2)$ wie die Wellenlängen verhalten.

Das führte unter Berücksichtigung der angegebenen Intensitätsverhältnisse zu einer Ausscheidung von 6 Linien. Die übrigen in der Tabelle I großgedruckten Zahlen wurden nun als zur α-Linie allein gehörig betrachtet. Jetzt galt es, für diese die Indizes der reflektierenden Ebenen zu finden. Es zeigte sich bald, daß nur ein Gitter abwechselnd mit Li- und F-Atomen

DEBIJE AND SCHERRER

Tabelle I. (LiF, Kupferstrahlung).

Schwärzung	ϑ in Grad	$\sin \dfrac{\vartheta}{2}$	h_1, h_2, h_3	$\dfrac{\sin \vartheta/2}{\sqrt{(h_1{}^2, h_2{}^2, h_3{}^2)}}$	Zahl der Ebenen	Intensitä
s.s.	30,0	0,259	1,1,1.	0,150	8	—
s.	3,38	0,290	1,1,1.	0,168	8	—
st.	37,8	0,323	1,1,1.	0,187	8	3,85
st.	44,2	0,377	2,0,0.	0,189	6	10,2
s.	56,2	0,472	2,2,0.	0,167	12	—
st.	63,8	0,528	2,2,0.	0,187	12	10,2
s.s.	67,4	0,554	3,1,1.	0,167	24	—
s.s.	71,4	0,583	2,2,2.	0,168	8	—
m.	76,6	0,620	3,1,1.	0,187	24	3,15
m.	80,8	0,647	2,2,2.	0,187	8	4,51
m.	97,8	0,753	4,0,0.	0,188	6	1,86
s.	111,0	0,824	3,3,1.	0,189	24	1,82
st.	116,0	0,848	4,2,0.	0,190	24	8,10
st.	137,6	0,932	4,2,2.	0,190	24	6,75
s.s.	153,2	0,973	4,4,0.	0,172	12	—
st.	166,6	0,993	$\left\{\begin{matrix}3,3,3.\\5,1,1.\end{matrix}\right\}$	0,191	$\left\{\begin{matrix}8\\24\end{matrix}\right\}$	1,71

in konstanten Abständen besetzt paßte. Die Li-Atome sowie die F-Atome eines solchen Gitters gehen nämlich je aus der Parallelverschiebung eines Würfels mit besetzten Mitten der Seitenflächen hervor. Von den Li-Atomen sind also als Grundstock im Elementarwürfel 4 Exemplare anzunehmen, für welche das Schema gilt

$$p_1 = 0, \; q_1 = 0, \; r_1 = 0,$$

$$p_2 = 0, \; q_2 = \tfrac{1}{2}, \; r_2 = \tfrac{1}{2},$$

$$p_3 = \tfrac{1}{2}, \; q_3 = 0, \; r_3 = \tfrac{1}{2},$$

$$p_4 = \tfrac{1}{2}, \; q_4 = \tfrac{1}{2}, \; r_4 = 0.$$

Entsprechend gilt für die F-Atome

$$p_1 = \tfrac{1}{2}, \; q_1 = \tfrac{1}{2}, \; r_1 = \tfrac{1}{2},$$

$$p_2 = \tfrac{1}{2}, \; q_2 = 1, \; r_2 = 1,$$

$$p_3 = 1, \; q_3 = \tfrac{1}{2}, \; r_3 = 1,$$

$$p_4 = 1, \; q_4 = 1, \; r_4 = \tfrac{1}{2}.$$

Bildet man nun mit Hilfe dieser Angaben den Strukturfaktor S, dann findet man

$$S = (A_{\text{Li}} + e^{i\pi(h_1 + h_2 + h_3)} A_{\text{F}})$$
$$\{1 + e^{i\pi(h_2 + h_3)} + e^{i\pi(h_3 + h_1)} + e^{i\pi(h_1 + h_2)}\} \tag{3}$$

wenn A_{Li} die vom Li-Atom und A_{F} die vom F-Atom zerstreute Amplitude bedeutet.

Aus der angegebenen Form von S folgt:

α) Netzebenen mit gemischten Indizes reflektieren nicht,

β) Netzebenen mit ungeraden Indizes reflektieren eine Intensität, welche proportional $16 (A_{\text{Li}} - A_{\text{F}})^2$ ist,

γ) Netzebenen mit geraden Indizes reflektieren eine Intensität, welche proportional $16 (A_{\text{Li}} + A_{\text{F}})^2$ ist.

Nun ist A_{Li} sicher wesentlich kleiner als A_{F}, da die Atomgewichte von Li bzw- F: 7 bzw. 19 betragen. Die unter β) und γ) genannten Ebenen werden also beide merklich reflektieren.

Ordnet man nun die Indizestriplets nach steigenden Quadratsummen und läßt mit Rücksicht auf α) die gemischten Indizes fort, dann bekommt man die in Tabelle I in der vierten Spalte groß gedruckten Zusammenstellungen. Ist das Modell richtig, dann muß jede beobachtete Linie denselben Wert von

$$\frac{\sin (\vartheta/2)}{\sqrt{(h_1{}^2 + h_2{}^2 + h_3{}^2)}}$$

liefern. Daß dem tatsächlich so ist, zeigen die großgedruckten Zahlen der fünften Spalte. Der kleine Gang der Zahlen von 0,187 bis 0,191, der übrigens nur etwa 2 Proz. ausmacht, ist außerdem leicht erklärlich durch den Umstand, daß das Stäbchen in der Kamera nicht genau im Mittelpunkt gestanden hat.

Eine Kontrolle liefert die Ausführung derselben Rechnung an den in kleinem Druck angegebenen auf die β-Linie bezüglichen Zahlen der Tabelle. Auch sie liefern eine gute Konstanz des oben genannten Verhältnisses und bestätigen damit das Modell von neuem. Die zu alleroberst in der Tabelle I aufgeführte, sehr schwache Linie bildet die einzige Ausnahme. Sie ist nicht mit Sicherheit reell. Wir führen sie trotzdem mit auf, weil die Tabelle dem tatsächlichen Gang der Beobachtungen entsprechen soll, bei welcher zuerst der Film ausgemessen wurde, ohne eine Tabelle der erwarteten Gesetzmäßigkeit zur Hand zu haben, während nachher die Gesetzmäßigkeiten an Hand der Zahlen festgestellt wurden, ohne die Aufnahme weiter zu berücksichtigen.

Tabelle I wird vervollständigt durch eine 6. Spalte, in der für jede Linie die Zahl der mitwirkenden Netzebenen angegeben ist. In der 7. Spalte

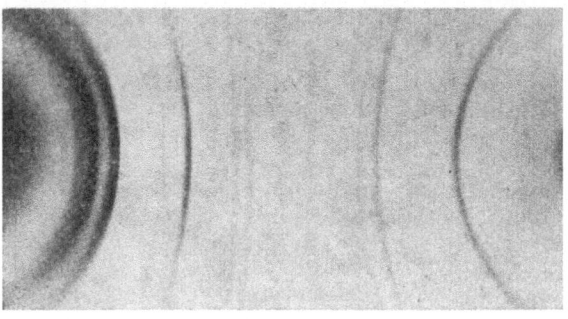

Fig. 1.

Fig. 2.

Fig. 3.

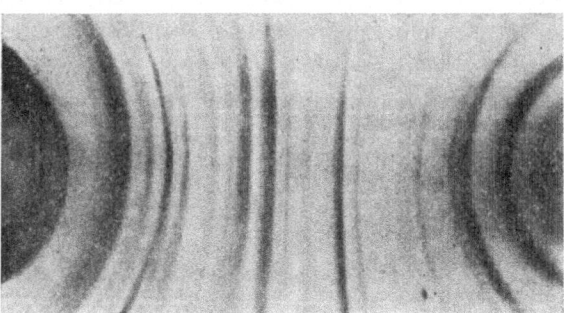

Fig. 4.

stehen die mit Hilfe dieser Zahl und mit Rücksicht auf das im vorigen Paragraphen hervorgehobene Resultat über die Abhängigkeit der Intensität von $h_1{}^2 + h_2{}^2 + h_3{}^2$ für dieselbe ausgerechneten Werte. Da es sich nur um Werte handelte, welche den rohen Intensitätsschätzungen der ersten Spalte gegenüber zu stellen sind, wurde der Einfluß der Wärmebewegung außer acht gelassen. A_{Li} wurde gleich 7; A_F gleich 19 gesetzt. Die Zahlen widerspiegeln nur im Groben den Gang der Angaben aus Spalte 1.

Aus dem auf die α-Linie bezüglichen Mittelwert der Spalte 5 folgt unter der Annahme $\lambda_\alpha = 1{,}549 . 10^{-8}$ cm aus (1) für die Seitenlänge des Elementarkubus des LiF:

$$a = 4{,}11 . 10^{-8} \text{ cm}.$$

Aus den β-Linien der Aufnahme folgt mit $\lambda_\beta = 1{,}402 . 10^{-8}$ cm

$$a = 4{,}17 . 10^{-8} \text{ cm}.$$

Beiden Größen stimmen innerhalb der zu erwartenden Beobachtungsfehler miteinander überein.

b) Die zweite Figur der Tafel zeigt eine Aufnahme der Zerstreuung an demselben LiF-Stäbchen, nunmehr aber mit Platinstrahlung bestrahlt. Man sieht, wie jetzt die Linien näher zusammengerückt und in größerer Zahl als bei der vorigen Aufnahme vorhanden sind.

c) Fig. 3 zeigt eine Aufnahme mit 'amorphem' Silizium als strahlende Substanz. Schon rein äußerlich betrachtet zeigt sie den Zusammenhang der photographierten Interferenzen mit der Kristallstruktur durch das Auftreten kleiner stärker geschwärzter Pünktchen, welche auf den Linien zerstreut sind. Diese rühren offenbar von etwas größeren Kriställchen her, die in ihrer zufälligen Lagerung gerade richtig orientiert waren, um die Cu-Strahlung reflektieren zu können. So liefert diese Aufnahme schon qualitativ betrachtet eine Bestätigung unserer Ansicht.

d) Fig. 4 zeigt eine Aufnahme von Graphit, (mit Cu-Strahlung), aus welcher hervorgeht, daß derselbe trigonal kristallisiert. Mit 12 Atomen im rhomboëdrischen Elementarbereich, dessen Seitenlänge sich zu $4{,}69 . 10^{-8}$ cm ergibt.

Göttingen, Phys. Inst. 28. *Mai* 1916
(Eingegangen 31. Mai 1916)

128. Pure iron was then investigated in the form of very fine powder, obtained by reduction of the oxide with hydrogen. A narrow beam of rays from a tungsten target passed through the powder and formed on the photographic plate a kind of generalized Laue photograph, in which every possible plane in the crystal structure had an equal opportunity of reflecting, and reflected all wave-lengths present. What was actually observed was the position of the K lines, which with the tube running at 110,000 volts, stood out very clearly on the continuous background. The reflection of these lines in different planes appeared on the plate as concentric, nearly circular, lines, whose distance from the center should be inversely proportional, approximately, to the spacing of the planes. The distance of these lines from the center can be measured and compared with the values calculated for the assumed crystal structure. If the assumed structure is correct, every calculated line must be present, and no more, and the intensity must fall off in the manner predicted.

Table II.

Indices of Plane	Distance of Line from Center		Intensity of Line			
	Obs.	Calc.	Obs.[1]	Calc.		
				A.	B.	C.
110	.703	.71	1.00	1.00	1.00	1.00
100[2]	1.00	1.00	.46	1.00	.60	.54
211	1.23	1.22	.54	1.00	.60	.46
110(2)	1.42	1.41	.24	1.00	.60	.39
310	1.59	1.58	.18	1.00	.40	.13
111(2)	1.74	1.73	.16	1.00	.60	.29
321	1.89	1.87	.22	1.00	.60	.24
100(2)[4]	–	2.00	–	1.00	.20	.0003
⎰411 ⎱110(3)	2.15	2.12	.12	2.00	1.40	.41
210(2)	2.26	2.24	.03	1.00	.60	.12
332	2.38	2.35	.02	1.00	.60	.15
211(2)	2.50	2.45	.02	1.00	.60	.11
⎰431 ⎱510	2.58	2.55	.10	2.00	1.40	.27
521	2.74	2.74	.02	1.00	.60	.04

The observed and calculated values are given in Table II. In the calculation it was first assumed that the atoms were arranged on a centered cubic lattice, and that the scattering electrons in each atom were concen-

---

[1] The intensity of the first seven lines was measured with a photometer. The rest were estimated.

trated at its center. The spacings calculated on this assumption are given in column 3, and the intensities in column 5, under *A*.* They agree remarkably well with the observed values, but fail to account for three facts, viz.: (1) That the intensity of the lines falls off continuously with increasing distance from the center. (2) That the first order 100 reflection is much too weak for its position. (3) That the second order 100 reflection is entirely lacking.

It is very difficult to conceive of any arrangement of point atoms which will satisfy these conditions and still give all the observed lines. We are forced, I think, to look for the explanation in the internal structure of the atoms. If it is assumed that eight of the 26 electrons in each atom are arranged along the cube diagonals at a distance from the center equal to one-fourth the distance to the nearest atom, calculation gives the values of intensity shown in column 6 under *B*. It is seen that condition (2) above is satisfied, and (3) is nearly accounted for, but not (1). If all the electrons are displaced from the center of the atom along the cube diagonals in 4 groups of 2, 8, 8, 8, at distances 1/32, 1/16, 1/8 and 1/4, respectively, of the distance to the nearest atom, all the observed facts are accounted for within the limit of experimental error (*C*, column 7).

This is obviously only a rough approximation to the correct position of the electrons in the atom, and more accurate data will enable us to place them more exactly. The excellent agreement with experiment, however, indicates that it is a step in the right direction, and gives promise that we shall be able, by this method, to determine the positions of all the electrons in the atom.

It may be noted that the above arrangement of electrons and atoms gives to iron the correct valence, corresponding to its position in the periodic table, and suggests a very interesting mechanism for ferromagnetism.

129. Vor einiger Zeit[1] habe ich eine Methode angegeben, die Meßgenauigkeit bei Debye-Scherrer-Aufnahmen auf sehr einfache Weise zu erhöhen. Der dabei angewandte Kunstgriff besteht darin, daß man den Film in der Kamera derart einspannt, daß nicht, wie meistens, die Linien niedrigerer Ordnungen in der Mitte des Films auftreten, sondern vielmehr die der höheren Ordnungen. Aus dem Abstand zwischen zwei zusammengehörigen Linien kann man dann mit großer Genauigkeit den Atom-

---

* [Note that in this first paper by Hull on the powder diffraction all intensity factors are missing, except an atomic scattering factor; even the plane number factor! Ed.]
[1] *Physica* **6** (1926) 64; *ZS. f. phys. Chem.* (Cohen-Festband) (1928) 100.

abstand berechnen, da ein Meßfehler in dem Linienabstand wegen der großen Dispersion in der Nähe von $\theta/2 = 90°$ den Atomabstand nur wenig beeinflußt.

Selbstverständlich ist diese Methode nur dann geeignet, wenn die Gitterkonstante des Materials und die Wellenlänge des verwendeten Röntgenlichts so zusammenpassen, daß eine Reflexion unter einem Winkel auftritt, der nur wenig von 90° abweicht.

Im allgemeinen wird dies aber nicht der Fall sein; man kann dann aber sehr leicht eine Korrektion anbringen.

Die Fehler, die in den berechneten Atomabständen auftreten, stammen aus zwei Quellen. Erstens ist der effektive Kameradurchmesser meistens nicht sehr genau auszumessen, da man z.B. die Filmdicke nicht gut bestimmen kann, und der Film meistens auch nicht glatt an der Kamerawand anliegt. Sodann ist es im allgemeinen sehr schwierig, das Präparat genau in den Mittelpunkt der Kamera einzustellen. Der Kameraradius wird also mit einem Fehler Δr behaftet sein.

Die Ausmessung der Linienabstände wird durch den Umstand erschwert, daß der Film während der Entwicklung seine Länge ändert, und diese Änderung wird für verschiedene Filme verschieden sein. Wohl dürfen wir annehmen, daß diese Änderung der Filmlänge proportional ist. Der Linienabstand l wird also mit einem systematischen Fehler $\Delta l = Cl$ behaftet sein. Sei d die Gitterkonstante, dann gilt in unserem Fall

$$d = \frac{\lambda}{2\cos(\theta/2)} \sum h^2, \qquad \frac{\theta}{2} = \frac{l}{4r}$$

$$\Delta d = \frac{\lambda}{2\cos^2(\theta/2)} \sqrt{\sum h^2} \sin\frac{\theta}{2} \cdot \frac{l}{4r}\left(\frac{\Delta l}{l} - \frac{\Delta r}{r}\right) =$$

$$= d.\operatorname{tg}\frac{\theta}{2} \cdot \frac{l}{4r}\left(\frac{\Delta l}{l} - \frac{\Delta r}{r}\right).$$

Da aber bei einer Aufnahme d, r, $\Delta l/l$ und Δr alle den gleichen Wert für verschiedene Reflexionslinien haben, so ist $(\Delta l/l - \Delta r/r) =$ konstant und die d-Werte, die man aus verschiedenen Reflexionslinien berechnet, werden folglich mit einem Fehler behaftet sein, der $l\operatorname{tg}(l/4r)$ proportional ist. Für kleine l-Werte wird der Fehler dem Quadrat des Linienabstandes proportional. Korrigiert man also die verschiedenen d-Werte mit Hilfe der Formel

$$\Delta d = Cl \operatorname{tg}(l/4r),$$

so ist das Resultat von allen systematischen Fehlern befreit, ausgenommen denjenigen, die noch in den Wellenlängen der verwendeten Strahlung stecken.

Den Kameradurchmesser braucht man nicht einmal genau auszumessen, und selbst mit einer sehr einfachen Kamera, bei der das Präparat eventuell noch etwas exzentrisch eingespannt sein kann, findet man noch sehr genaue Werte für d, da in Δr auch die Exzentrizitätsfehler enthalten sind.

Auch für die Bohlinkamera, in der von W. F. de Jong[1] angegebenen Ausführung, kann man dieselbe Korrektion anbringen.

Als Beispiel wählen wir zuerst eine Aufnahme mit einer gewöhnlichen Kamera mit dem Radius $r = 4{,}2$ cm. Diese Kamera hat verschiedene Fehler; erstens ist der Präparathalter nicht genau in der Kameramitte justiert, zweitens wird der Film mit Federn gegen die Innenwand angedrückt, so daß man niemals sicher ist, ob der Film tatsächlich überall glatt anliegt.

Spannt man jetzt den Film so ein, daß die austretende Strahlung durch die Filmmitte geht, dann ist die Übereinstimmung der d-Werte, berechnet aus den verschiedenen Reflexionen, nicht sehr genau.

Nehmen wir z.B. eine Wolframaufnahme ($Cu K_{\alpha_1 \alpha_2}$-Strahlung). Wir finden die d-Werte unter d_I angegeben.

Indizes	d_I	d_{II}	Korr.	d_{III}
(110)	3,11	3,062		
(200)	3,14	3,126		
(220)	3,17	3,131		
(310)	3,17	3,138	24	3,162
(222)	3,17	3,144	16	3,160
(321)	3,17	3,149	9	3,158
(400)	3,17	3,157	2,5	3,159^s
			Mittel	3,160 ± 0,002

Welcher Wert der wahrscheinlichste ist, wissen wir vorläufig nicht: denn die ersten sind durch die Abmessungen des Präparates und die eventuelle Exzentrizität des Präparates gefälscht, die letzten Werte durch Fehler die dadurch entstehen, daß der Film seine Länge in nicht genau bekannter Weise geändert hat.

Legen wir den Film so ein, daß die eintretende Strahlung durch die Filmmitte geht, dann wissen wir jedenfalls, daß die d-Werte, aus höheren Ordnungen berechnet, die besten sind, da sowohl die Fehler von den Präparatabmessungen als auch diejenigen, die von der Filmänderung herrühren, desto kleiner sind, je mehr sich der Reflexionswinkel 90° nähert (d_{II}).

[1] *Physica* **7** (1927) 23.

Auch jetzt sind noch große Abweichungen vorhanden: Wir wissen jetzt aber, daß der *d*-Wert, wie er aus der Reflexion (400) folgt, der beste ist. An den vier letzten Werten z.B. können wir jetzt noch die Korrektion anbringen. Wir finden dann die unter d_{III} angegebenen Werte.

Mit sehr einfachen Hilfsmitteln finden wir also den Atomabstand mit einer Genauigkeit, die man sonst nur z.B. durch Eichung mit einer Normalsubstanz erreichen kann.

Eindhoven, Natuurkundig Laboratorium der
N.V. Philips Gloeilampenfabrieken

ROTATION DIAGRAM

130. L'enregistrement photographique continu fournit simultanément plusieurs spectres dus à des reflexions sur des plans réticulaires différents qui sont entraînés par la rotation du cristal. Il est facile de déterminer les caractéristiques de ces plans ainsi que de mesurer la dispersion qui leur est propre et d'avoir ainsi d'un seul coup les valeurs relatives de leur densité en centres de diffraction.

Il est facile de voir qu'en faisant tourner le barillet sur lequel est placé le cristal de telle façon qu'à l'origine le rayon soit parallèle à la face réfléchissante, d'abord de zéro à α, puis de 180° − α à 180°, on produit sur une plaque sensible normale au faisceau incident deux spectres symétriques par rapport à l'axe de ce faisceau. Cette disposition est précieuse pour déterminer l'angle d'incidence correspondant à une raie parce qu'elle réduit l'opération à faire sur la plaque à la mesure de la distance entre deux raies fines symétriques, ce qui donne le double de l'écart de chacune, au lieu de se servir comme zéro de la trace toujours diffuse du faisceau incident.

131. Es ist bereits mehrfach darüber berichtet worden, daß die natürlich gewachsenen Cellulosefasern sowie Seidenfäden eigenartige Röntgendiagramme liefern, die einen gewissen Einblick in das Wesen der Faserstruktur gewähren[1]. Namentlich findet man, wenn Durchleuchtung im monochromatischen Lichte vertikal zur Richtung des ausgespannten Faserbündels erfolgt, daß statt der zu erwartenden Debye-Scherrer-Kreise ein Diagramm erscheint, das aus Punkten bzw. Streifen besteht, die in eigen-

[1] Herzog u. Jancke, *Ber. d. D. chem. Ges.* **53** (1920) 2162. Herzog, Jancke, Polanyi, *Ztschr. f. Phys.* **3** (1920) 343. P. Scherrer in Zsigmondy, *Kolloidchemie*, 3. Aufl. (1920) 408. Herzog. u. Jancke, *Festschrift der Kaiser-Wilhelm-Gesellschaft*, S. 118.

artig symmetrischer Anordnung den Durchstoßpunkt des Röntgenstrahles umgeben. Die Fig. 2 zeigt eine Platte, die man auf diese Weise mit Ramiefasern erhält.

Zur Erläuterung der Methode, die zur Analyse solcher Faserdiagramme dient, soll zuerst mit ihrer Hilfe das Röntgendiagramm für einen bekannten Fall konstruiert werden, nämlich für die Anordnung von Debye und Scherrer, bei der ein feines Kristallmehl von monochromatischem Röntgenlichte getroffen wird. In dem Häufchen Kristallmehl, das als Beugungszentrum dient, heben wir irgendeine Netzebenenart hervor (z.B. die Netzebenen 132) und sprechen fortab von der Gesamtheit dieser Netzebenen als von den Netzebenen N.

Die abgebeugten Strahlen verhalten sich nun nach Bragg bekanntlich so, als würden sie von den Netzebenen reflektiert—wobei nur die unter einem bestimmten Gleitwinkel (γ) auftreffenden Strahlen reflektiert und alle anderen durchgelassen werden. Dabei ist

$$\gamma = \arcsin (\lambda/2D) \tag{1}$$

(λ = Wellenlänge des Röntgenlichtes, D = Gitterkonstante der reflektierenden Netzebene).

Bei Konstruktion des Röntgendiagrammes können also alle einer bestimmten Ebene parallel gelegenen Netzebenen N durch eine einzige gleichgerichtete Netzebene N ersetzt werden, wobei die Lage dieser (je eine Parallelschar repräsentierenden) Netzebenen innerhalb des Beugungszentrums beliebig angenommen werden kann. Die Rolle der Netzebenen N kann somit durch eine im Beugungszentrum konstruierte Halbkugelfläche übernommen werden—deren Tangentialebenen ja alle möglichen Richtungen haben—, vorausgesetzt, daß die Halbkugelfläche nur unter jenem Gleitwinkel reflektiert, der den Netzebenen N nach (1) zukommt.

Die reflektierende Halbkugel, für die wir diese Bedingung einführen wollen, sei die dem Röntgenstrahle zugewendete Hälfte der in Fig. 1 als 'Netzebenenkugel' bezeichneten Kugel.

Man sieht leicht, daß die Gesamtheit der Punkte auf dieser Halbkugel, denen der Gleitwinkel γ zukommt, einen Kreis bilden, dessen Ebene vertikal zur Strahlrichtung steht und dessen Abstand vom zugehörigen Pole gleich 90°—γ ist. Die Punkte auf der Netzebenenkugel, die diesen Kreis bilden, ersetzen also vollkommen die Netzebenen N. Wir nennen ihn daher den 'Reflexionskreis' dieser Netzebenen (Kreis R in Fig. 1). Ein Blick auf die Fig. 1 zeigt, daß die am Reflexionskreise reflektierten Teile des Strahles eine Kegelfläche (vom Öffnungswinkel 2γ) bilden, also die bekannten Debye-Scherrer-Diagramme liefern. Insbesondere kommt man für eine

133

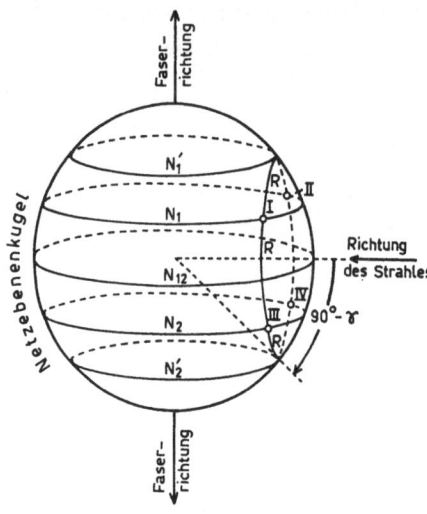

Fig. 1.

senkrecht zur Strahlrichtung stehende Platte zu kreisförmigen Interferenz-
linien.

Wie können nun im Sinne dieser Konstruktion Punktdiagramme entstehen, wie sie die Fasern liefern? Welcher Art ist das geometrische Gebilde, das an Stelle der Netzebenenkugel zu treten hat, wenn die Netzebenen N statt Kreisen Punkte erzeugen sollten?[1]

Erstens muß es nach dem bisher Gesagten möglich sein, dieses Gebilde
auf die Fläche der Netzebenenkugel aufzutragen, denn diese umfaßt ja
alle möglichen Netzebenenlagen, zweitens muß gefordert werden, daß das
gesuchte Gebilde Teile des Reflexionskreises umfaßt—denn nur Punkte,
die auf diesem Kreise liegen, können reflektieren. Hieraus findet man das
fragliche Gebilde, indem man noch bedenkt:

1. Daß es den Reflexionskreis nur in einzelnen Punkten schneiden (also
nicht etwa auf ganzen Stücken überlagern) darf, da im Diagramm bloß
Punkte auftreten sollen,

2. daß diese Schnittpunkte bei Drehung der fraglichen Gebilde um eine
zum Strahl vertikale Richtung (die Faserrichtung) unverrückt bleiben müssen—da das Diagramm sich bei Drehung der Faser um seine Achse nicht
ändert.

Zufolge 1 müssen die gesuchten Gebilde Linien (und nicht etwa Flächen)
sein, zufolge 2 müssen diese Linien Kreise sein, deren Ebenen quer zur

---

[1] Der Eingangs erwähnte Umstand, daß neben Punkten auch Streifen auftreten, ist
für das Prinzip der Erklärung belanglos und kann daher bei der folgenden Ableitung
unberücksichtigt bleiben.

Faserrichtung liegen. An Stelle der Netzebenenkugel treten also Netzebenen*kreise*.

Es ist leicht zu zeigen, daß die Darstellung der Lagenmannigfaltigkeit der
Netzebenen in der Faser durch solche Kreise die charakteristischen Symmetrieverhältnisse der Punktdiagramme voraussehen läßt:

Ist N_1 in Fig. 1 ein Netzebenenkreis, so kann man auch den gleichgroßen
Parallelkreis N_2 als solchen betrachten, da dieser dieselbe Lagenmannigfaltigkeit darstellt. Auf diesen zwei Kreisen sind insgesamt vier reflektierende Punkte vorhanden: die vier Punkte, in denen sie den Reflexionskreis schneiden.

Das entstehende Punktdiagramm ist also eine Abbildung dieser vier
Schnittpunkte (*I, II, III, IV* in Fig. 1) und wird demnach zwei Symmetrie-
achsen haben, eine parallel zur Faser und eine quer zu dieser. Man findet
dies bestätigt an der Ramiefaseraufnahme in Fig. 2: Jeder Interferenzpunkt
tritt in vier symmetrischen Lagen auf,—nur jene Punkte, die gerade auf
eine Symmetrieachse fallen, treten (diesem Umstande entsprechend) bloß
zweimal auf[1].

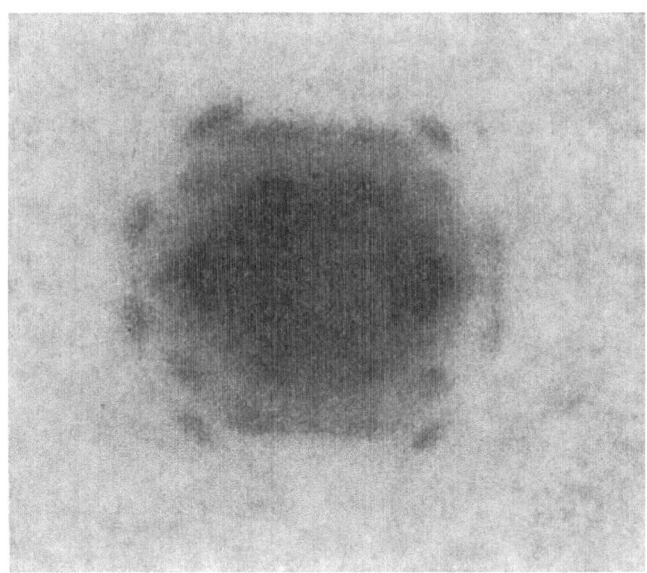

Fig. 2.

---

[1] Eine dem Wesen nach ähnliche Ableitung der Punktdiagramme findet sich bei Herzog,
Jancke, Polanyi *l.c.*

Für unsere weiteren Überlegungen müssen wir uns über die Bedeutung dieser Doppelpunkte im Sinne unseres Konstruktionsschemas klar werden. Dabei zeigt sich, daß diese Bedeutung eine sehr verschiedene ist, je nachdem die Doppelpunkte entlang der Faserachse liegen oder auf der Querachse auftreten. Laut unserem Konstruktionsschema bedeutet der erste Fall, daß die Netzebenenkreise, statt den Reflexionskreis zu schneiden, diesen nur berühren (siehe N'_1 und N'_2 in Fig. 1). Letztere—übrigens ziemlich seltene—Erscheinung ist physikalisch bedeutungslos, was man schon daraus sieht, daß sie gegebenenfalls dadurch zum Verschwinden gebracht werden kann, daß man eine größere Wellenlänge einstrahlen läßt, wodurch der Polabstand des Reflexionskreises vergrößert wird, so daß dieser die Kreise N'_1 und N'_2, die er zuvor bloß berührt hatte, nun durchschneidet.

Dagegen ist es physikalisch bedeutungsvoll, wenn Doppelpunkte auf der Querachse auftreten. Solche entstehen, wenn die Netzebenenkreise Größtkreise werden, wodurch beide notwendig zusammenfallen, und zwar in einem Kreise, der die Lage von $N_{1,2}$ in Fig. 1 hat. Die Netzebenen, deren Lagenmannigfaltigkeit durch einen Größtkreis dargestellt ist, liegen alle parallel zur Faserachse, gehören also einer Zone an, die in der Faserrichtung liegt. Ein besonders häufiges, intensives Auftreten von Doppelpunkten auf der Querachse des Diagrammes (wie etwa am Ramiediagramm in Fig. 2 sichtbar) bedeutet also, daß eine besonders flächenreiche Zonenachse der Kristallite in der Faserrichtung liegt.

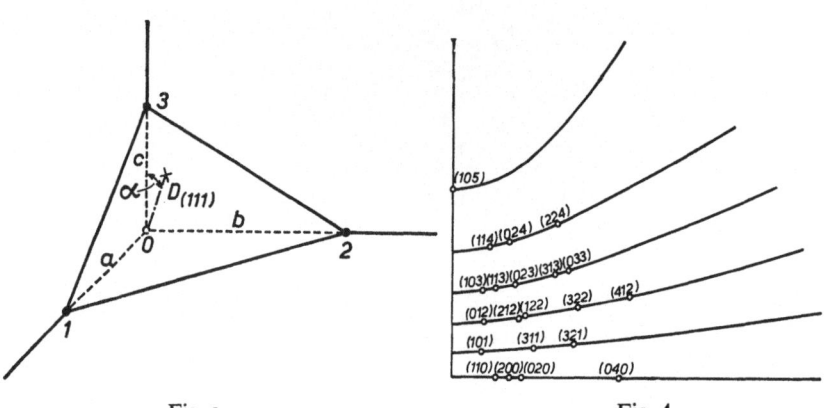

Fig. 3.   Fig. 4.

Nun kommen als solche Zonenachsen natürlich vor allem die Haupt-achsen in Betracht, und man wird daher untersuchen, ob es nicht eine Hauptachse der Kristallite ist, die in der Faserrichtung steht. Das Ergebnis ist positiv: sowohl Flachs wie Seidenfasern erweisen sich in diesem Sinne als nach einer Hauptachse geordnet.

Um dies festzustellen, war es nötig, die charakteristischen Merkmale eines Röntgendiagrammes, wie es von einer nach einer Hauptachse ge-ordneten Faser erzeugt werden muß, abzuleiten, was nach folgendem Prin-zip geschah.

In einem beliebigen (etwa rhombischem) Raumgitter mit den Grund-perioden a, b, c sei eine Netzebene durch ihre drei kristallographischen Indizes festgelegt, z.B. sei es Einfachheit halber die Netzebene (111).

Sind in Fig. 3 die Punkte 0, 1, 2, 3 Gitterpunkte und die Entfernungen a, b, c die drei Grundperioden, so ist die Ebene, die durch die Punkte 1, 2 und 3 geht, eine (111)-Ebene. Die Entfernung des Punktes 0 von dieser Ebene ist also der Netzebenenabstand $D_{(111)}$ der Netzebene (111). Denkt man sich nun z.B., die Faser sei nach der c-Achse geordnet, so sieht man sofort, daß der Winkel α, den die von 0 aus auf die Ebene (111) gefällte Normale mit c einschließt, gleich dem Winkelabstand des Netzebenenkreises (111) von der Faserachse wird.

Ferner ist aus Fig. 3:

$$D_{(111)} = c \cos \alpha, \tag{2}$$

woraus mit Hilfe von (1) der Glanzwinkel $\gamma_{(111)}$ der Ebene (111) sofort berechenbar ist. Damit ist nun auch der Polabstand des zur Ebene (111) gehörigen Reflexionskreises (der einfach $90° - \gamma_{(111)}$ beträgt), bekannt, und damit auch dieser Kreis festgelegt.

Die Auffindung der zur Netzebene (111) gehörigen Interferenzpunkte ist nun eine rein rechnerische Aufgabe. Man hat nur die Lage der Punkte, in denen die Netzebenenkreise den Reflexionskreis schneiden, auszurechnen und die Strahlen, die an diesen Punkten reflektiert werden, bis zur Platte zu verfolgen.

Eine rechnerische Durchführung dieses Gedankenganges unter Verall-gemeinerung desselben auf beliebige kristallographische Indizestripel hat zum Ergebnis geführt, daß sämtliche Interferenzpunkte einer nach einer Hauptachse geordneten Faser auf einer Schar von Hyperbeln liegen müs-sen, wie sie in Fig. 4 dargestellt ist. Die Abbildung zeigt bloß einen Qua-dranten dieser Hyperbeln. Die auf den Kurven liegenden Kreise deuten die Lage der Interferenzpunkte der Flachsfaser an, die Zahlen sind die wahrscheinlichsten kristallographischen Indizes der reflektierenden Netz-ebenen.

Man sieht, daß alle Punkte auf den einzelnen Hyperbeln an der (in der Faserrichtung gelegenen) *c*-Achse gleiche Indizes haben und daß diese von Kurve zu Kurve je um Eins wachsen.

Es liegt auf der Hand, daß dieser Umstand die Berechnung jener Elementarperiode, die parallel zur Faserrichtung liegt, sehr erleichtert. Sobald man die Hyperbeln durch das Diagramm durchgezogen hat, kann man die Grundperioden *c* aus denselben geradezu ablesen.

Hat man eine der drei Grundperioden auf diese Weise festgelegt, so ist hierdurch die Berechnung der beiden anderen wesentlich erleichtert, da das Problem damit viel von seiner Unbestimmtheit verliert.

Für die röntgenographische Strukturbestimmung eines Stoffes, ist es also vorteilhaft, wenn seine Kristallite nach einer Hauptachse geordnet sind, besonders ist dies bei organischen Körpern der Fall, die recht unscharfe Bilder geben und wo außerdem noch die Interferenzen erster Ordnung häufig fehlen, so daß die Anwendung der Runge-Johnsen-Töplitzschen Rechenmethode zu falschen Resultaten führt.

Unter Verwendung dieses Umstandes ist eine Verbesserung der Debye-Scherrerschen Methode entstanden, die darin besteht, daß man die zu untersuchende Substanz bei sehr hohem Druck bis zum Eintreten des 'Fließens' preßt, wodurch die Masse eine nach einer Hauptachse geordnete Struktur erhält[1]. Man erhält so Streifendiagramme, deren Auswertung nach den oben angedeuteten Prinzipien viel sicherer ausführbar ist als jene der gewöhnlichen Debye-Scherrer-Diagramme.

132. Die Untersuchungen M. Polanyis[2] gehen aus von der Auffindung neuartiger monochromatischer Röntgendiagramme durch Herzog, Jancke[3] und Scherrer[4]. Diese Diagramme werden durch eine Faserstruktur der verwendeten Substanzen erzeugt, die auf einer Gleichrichtung von Einzelkriställchen parallel zu einer oder mehreren ausgezeichneten kristallographischen Achsenrichtungen besteht, wobei im übrigen die Kristallite beliebig gegeneinander verdreht sind. Die geometrische Auswertung der Diagramme in den Arbeiten von Polanyi zeigt, daß die Struktur des Einzelkristalles in der Anordnung der Beugungsflecke zum Ausdruck gelangt, so daß in weiterer Verfolgung der charakteristischen Merkmale umgekehrt

---

[1] Becker, Herzog, Jancke, Polyani, *Ztschr. f. Phys.* nächste Nummer.
[2] *ZS. f. Phys.* **7** (1921) 149; *Die Naturwiss.* **9** (1921) 288, 337.
[3] K. Becker, R. O. Herzog, W. Jancke und M. Polanyi, *ZS. f. Phys.* **5** (1921) 61.
[4] P. Scherrer, *Ber. d. D. Chem. Ges.* **53** (1921) 2162; *ZS. f. Phys.* **3** (1920) 343.

Schlüsse auf die Kristallstruktur gezogen werden können, wie es z.B. auf Grund der Diagramme von Cellulose versucht worden ist[1].

Der Verfasser hält es daher für nicht unangebracht, auf eine Methode hinzuweisen, die seit 1919 im Leipziger mineralogischen Institut angewendet wird, und deren Ergebnisse er einer kristallographischen Deutung unterzogen hat[2]. Das Verfahren beruht auf einer Drehung des Kristalles oder Kristallsplitters während der Bestrahlung mit monochromatischem Röntgenlicht. Die Netzebenen des Kristallgitters erzeugen dabei in Reflexionsstellung auf der photographischen Platte ein Diagramm, das mit einem charakteristischen Faserdiagramm eine unverkennbare Ähnlichkeit besitzt (abgesehen von der Form der Interferenzflecke).

In der Tat, es ist für den Röntgeneffekt gleichgültig, ob die Netzebenen sich wie bei den Stoffen mit Faserstruktur durch die Parallelstellung der Kristallite von vornherein in der eigentümlichen Anordnung befinden, oder ob sie durch eine Drehung des Kristalles um die gleiche bevorzugte Richtung nacheinander in die passende Lage versetzt werden. Infolgedessen sind auch die seinerzeit von mir abgeleiteten geometrischen Beziehungen die gleichen, wie sie von M. Polanyi für eine kristallinische Substanz mit einfach idealer Faserstruktur angegeben worden sind. Der Fall mehrfach idealer Faserstruktur kann angesehen werden als eine Überlagerung von Diagrammen mit jeweils verschiedener kristallographischer Orientierung der Drehachse.

Gerade die bei der Debye-Scherrerschen Methode recht schwierige und bei organischen Substanzen mit niedriger Symmetrie fast unmögliche Bestimmung der Indizes der reflektierenden Gitterebenen ist mit Hilfe der von mir Zonenkurven genannten Kurven mit relativer Leichtigkeit ausführbar. Schon der erste Anblick eines Drehspektrogrammes lehrt, daß sich die einzelnen Spektren in nahezu geradlinigen horizontalen und vertikalen (oder auch schräg dazu liegenden) Kurven nach Art eines optischen Kreuzgitterspektrums anordnen. Die nähere Untersuchung zeigt, daß die horizontale Kurvenschar, wie auch von Polanyi gefunden wurde, aus Hyperbeln, die andere die Hyperbeln nahezu rechtwinklig (bzw. schräg) schneidende Schar aus Kurven vierter Ordnung gebildet wird. Es lassen sich leicht entsprechende Indizesschemata aufstellen, und es zeigt sich, daß jede Raumgruppe bei der Drehung um bestimmte Zonen des Kristalles eine charakteristische Verteilung der Spektren ergibt, die mit Hilfe passender Tabellen Schlüsse auf den Feinbau gestattet.

[1] R. O. Herzog, W. Jancke und M. Polanyi, *ZS. f. Phys.* **3** (1920) 343.
[2] Vgl. F. Rinne, Röntgenographische Feinbaustudien, *Abh. d. math.-phys. Klasse der sächs. Akademie der Wiss.* **38** (1921) Nr. 3. Eine ausführliche Mitteilung über die Drehmethode wird demnächst erscheinen.

Verbindet man die Spektren der verschiedenen Ordnungen von ein und derselben Netzebene jeweils durch einen kontinuierlichen Linienzug, so entstehen lemniskatenähnliche Kurven vierter Ordnung, die in der Nähe des Primärfleckes fast geradlinig verlaufen[1]. Bei Einstrahlung eines kontinuierlichen Spektrums während der Drehung des Kristalles lassen sich die genannten Kurven physikalisch realisieren: Die Netzebenen liefern Abbilder des kontinuierlichen Spektrums in Form von radialen Streifen, deren zentrale Begrenzung durch die untere Grenzwellenlänge gegeben ist. Derartige, von Polanyi als 'Radialdiagramme' bezeichnete Spektrogramme treten auch beim gewöhnlichen Laueverfahren infolge Verbiegungen, Krümmungen mechanischer Art oder infolge chemischer Umsetzungen bei Kristallen auf.

In Ansehung obiger Beziehungen dürften die von mir abgeleiteten und in einer demnächst erscheinenden Arbeit näher ausgeführten Erörterungen dazu dienen, Gebilde mit Faserstruktur leichter zu entziffern.

Das Hauptverwendungsgebiet liegt nach meiner Meinung in der Bestimmung der Feinstruktur von Stoffen, die in gut ausgebildeter Kristallform erhältlich sind. Sie versagt auch nicht bei der Untersuchung von Kristallen mit niedriger Symmetrie, z.B. des rhombischen, monoklinen und triklinen Systems. Bei solchen bringt die Entzifferung der Debye-Scherrerschen Diagramme einmal wegen der Überlagerung der Spektren, der geringen Intensität, sodann infolge der Kompliziertheit der zu bestimmenden quadratischen Form größte Schwierigkeiten mit sich. Auch die Braggsche Methode führt nur durch Verwendung einer großen Zahl von verhältnismäßig ausgedehnten Kristallplatten zum Ziel, während die Methode der vollständigen Spektraldiagramme unter Benutzung der von H. Seemann angedeuteten und von mir ausgearbeiteten allgemeinen geometrischen Grundlagen der kristallographischen Indizesbestimmung auch bei Verwendung fast mikroskopisch kleiner Kristalle oder Kristallsplitter mit wenigen Diagrammen unter Drehung um verschiedene Zonen Ergebnisse liefert.

Leipzig, Mineralogisches Institut der Universität, Januar 1922

133. Die oben genannte röntgenographische Methode hat in letzter Zeit eine gewisse Verbreitung erlangt und wird auch in der Literatur mehrfach erörtert. Da die historischen Darstellungen dabei nicht unwesentlich auseinandergehen, wollen wir hier den Standpunkt angeben, zu dem uns die Durchsicht des gesamten verfügbaren Materials geführt hat.[1]

[1] In letzter Zeit hat K. Becker (*ZS. f. Phys.* **17** (1923) 352) Andeutungen darüber gemacht, daß die in den Arbeiten von M. Polanyi und K. Weissenberg gegebene historische

Die ersten photographischen Aufnahmen des vollständigen Beugungs-
bildes, welches bei Einstrahlung eines Linienspektrums während der kontinuierlichen Drehung des Kristalles entsteht, wurden gelegentlich rönt-
genspektroskopischer Versuche von de Broglie[1] erhalten; E. Wagner[2] hat
das Auftreten schräger Spektren als störenden Begleitumstand aufgefaßt
und seine Beseitigung mit experimentellen Mitteln versucht. Die Ausnut-
zung zur Kristallstrukturbestimmung geschah erstmalig durch H. Seemann[3],
der durch Einstrahlung eines monochromatischen, stark konvergenten
Primärstrahlbündels auf den feststehenden Kristall 'vollständige Spektraldiagramme' bei Steinsalz, Rohrzucker und Kaliumplatincyanür erhielt und
die Möglichkeit erörterte, die auftretenden 'Nebenspektren' zur Kristall-
analyse zu verwerten. Über eine praktische Verwendung seiner Methode
ist uns nichts bekannt. Unabhängig hiervon gab E. Schiebold[4] im gleichen
Jahr ein Verfahren zur Kristallstrukturbestimmung unter Verwendung der
Nebenspektren an, das eine Kombination von Lauediagramm und Dreh-
spektrogramm darstellt.

Im Jahre 1919 und 1920 ist diese Methode im Leipziger Institut weiter
entwickelt und in einer Reihe von Dissertationen angewendet worden.
Herr Schiebold hat über diese im Frühjahr 1921 am Mineralogentag vor-
getragen[5].

Zu gleicher Zeit erschien die erste Arbeit von M. Polanyi[6]. Letzterer
zeigte, daß das Beugungsbild natürlich gewachsener Fasern prinzipiell mit
einem Drehkristalldiagramm übereinstimmt und gibt (unter Anwendung
auf die Zellulose) die einfachste Form der Schichtlinienbeziehung an, welche
nachfolgend mit K. Weissenberg[7] in folgender, allgemeiner Form aufgestellt und als Strukturbestimmungsmethode ausgearbeitet worden ist.

Die Interferenzstrahlen eines rotierenden Kristalls liegen auf Kegel-
mänteln, die mit der Drehachse koaxial sind und deren Öffnungswinkel

---

Darstellung, namentlich in bezug auf die Arbeiten von E. Schiebold, falsch sei. Dem-
gegenüber stellen wir gemeinsam fest, daß die von Polanyi und Weissenberg seinerzeit
(*ZS. f. Phys.* **10** (1922) 44) gegebene Schilderung der Publikationslage (die übrigens im
Einvernehmen mit Herrn Schiebold erfolgte) vollkommen korrekt ist.

[1] M. de Broglie, *C. R.* **157** (1913) 924 u. 1413 [this Vol. paper **130**]; **158** (1913) 177, Fig. 1.

[2] E. Wagner, *Ann. d. Phys.* **46** (1915) 868.

[3] H. Seemann, *Phys. ZS.* **20** (1919) 55 u. 169.

[4] E. Schiebold in F. Rinne, *Einführung in die kristallograph. Formenlehre usw.*, 3. Aufl.,
S. 198–200. Leipzig 1919.

[5] Eine erstmalige Publikation der Methode erfolgte erst im Jahre 1922 in vorläufiger
Form, *ZS. f. Phys.* **9** (1922) 2, 180 und eingehender bei Rinne, *Kristallograph. Formenlehre*, 4. u. 5. Aufl., S. 237–242, 1922 u. *ZS. f. Kristallograph.* **57** (1923) 579.

[6] *Die Naturw.* **9** (1921) 337 [this Vol. paper **131**]; *ZS. f. Phys.* **7** (1921) 149.

[7] *ZS. f. Phys.* **9** (1922) 123.

$\mu_1, \mu_2, \ldots \mu_n$ mit der Identitätsperiode a in Richtung der Drehachse nach der Gleichung

$$n\lambda/a = \cos \mu_n + \cos \beta \qquad (2)$$

(n Ordnungszahl des Kegels, β Winkel* zwischen Strahl und Drehachse, λ Wellenlänge, a Identitätsabstand in Richtung der Drehachse) zusammenhängen. Diese Formel gestattet ohne Kenntnis der Indizierung der einzelnen Diagrammpunkte, lediglich durch Vermessung der Kegelöffnungswinkel die Identitätsperiode in Richtung der Drehachse zu bestimmen. Die Messung der Identitätsperioden in Richtung der Kristallachsen gibt die Kantenlänge des Elementarkörpers.

Die Drehkristallmethode ist also zuerst von E. Schiebold in den erwähnten Dissertationen zu Strukturbestimmungen verwendet worden. Dabei hat er bereits die Konstanz des Schichtlinienindex benutzt, die unabhängig davon von Polanyi mitgeteilt wurde. Von letzterem stammt auch die Errechnung von Identitätsabständen aus dem Schichtlinienabstand, was im allgemeinen von ihm gemeinsam mit Weissenberg ausgearbeitet wurde.

134.   *Introductory*

The applications of X-ray diffraction to problems in pure and applied science have reached such large dimensions that it becomes profitable at this stage to treat of instruments made for this specific purpose rather than to leave to each investigator the burden of creating special apparatus for his own needs. There is no reason why universal instruments should not be used, for all but very special problems, to give results as reliable and easily obtainable as those, say, of the spectroscope. The present papers are an attempt to describe first of all the general problem of the requirements and capabilities of apparatus for the different methods of X-ray crystallography, and secondly one particular universal instrument in detail with an account of its adjustment and operation.

X-ray Crystallography

The problems of X-ray crystallography divide themselves naturally into two parts:

(*a*) the determination of the inner atomic structure of the crystals of a pure substance;

---

* [See footnote p. 158. Ed.]

(b) the determination of the mutual arrangement of small crystals in aggregates and mixtures.

To these may be added what is strictly not crystallography at all:

(c) the determination of the wave-lengths of X-rays from different sources.

Of these (a) is mainly important in theoretical and (b) in applied investigations, covering as it does the whole range of materials, metals, alloys, textiles, building materials and ceramics. (c) again is more of theoretical importance on account of the picture it gives of the atomic interior, although it will have a growing importance, second to if not greater than the spectroscope, in delicate qualitative analysis. These papers will concern themselves mainly with the analysis of single crystals, although the other aspects will be touched on in so far as the apparatus to be described has been designed to include them.

Experimental Data of X-ray Crystallography

The experimental data necessary for a complete crystal analysis may be summarised as follows:

(1) Measurement of spacings of planes of known indices leading to the determination of the size and shape of the unit cell.

(2) Measurement of crystal density from which with (1) the number of molecules per cell can be found.

(3) Determination of the symmetry class by the methods of ordinary crystallography.

(4) Determination of the indices of absent reflections leading with (3) to the determination of the space group; and this combined with (2) giving the molecular symmetry.

(5) Measurement of the intensities of reflections from planes of known indices, and of the correcting factors necessary to obtain the structure factors, from which, together with (4), the complete structure may be derived.

Experimental Methods

So far we have considered the necessary data for crystal analysis without thinking of how they may be obtained experimentally. The broad divisions of experimental methods depend on the way the emergent beams of X-rays are observed and measured. Of the three methods of detecting X-rays, fluorescent screen, photographic plate and ionisation chamber, the first, owing to the low power of X-ray tubes and the inefficiency of screens is not, for the present, a practicable method, although it has possibilities. Of the other two, the photographic method has the advantage of being able to register a great number of reflections at once in a way that permits of accurate measurements of distances, whereas the ionisation method is greatly superior in its ability to measure reflection intensities directly. The methods of X-ray crystallography may also be considered from another standpoint, that of the position of the crystal itself. Referring to the expression of Bragg's Law

$$\sin \theta = \lambda/2d$$

we see that while d is a constant for any particular crystal plane, either λ or θ may be made to vary.

The Laue Method

In the first case the crystal is fixed, but the beam of X-rays is a white one, i.e. contains all values of λ within a certain range. Under these circumstances each crystal plane reflects only that portion of the beam that has a wavelength satisfying Bragg's Law for the particular value of its glancing value θ. This is the original method of Laue. The method of registering the emergent beam is here nearly always photographic. It can be seen that as the values of λ are not known, d cannot be found and the shape of the pattern produced depends only on the symmetry of the crystal and its position relative to the incident beam. The densities of the spots however do give some measure of the intensities of the reflections, but are subject to three large corrections: for the unequal distribution of intensity in the incident radiation and for the different absorption of different wave-lengths both in the silver bromide film and in the crystal itself, in addition to the ordinary corrections for extinction, etc. This, combined with the uncertainty resulting from the superposition of different orders of reflection makes its value for intensity measurements mainly qualitative. On the other hand the Laue photograph is often essential for determining the symmetry, and it is unsurpassed in the number of different reflections registered in one exposure.

Rotation Methods

If instead of fixing the crystal and varying λ we use a source of monochromatic X-rays and turn the crystal round some axis, usually perpendicular to the beam, then a certain number of planes will come in turn into the reflecting positions for the particular wavelength of the incident rays. The reflected rays may be received on a photographic plate or film as in the so-called Rotation Method or they may be followed by a movable ionization chamber which is the Bragg Ionization Spectrometer method. In both methods, as λ is known and θ measured, d can be found. The Rotation Method can deal with more reflections and can give more accurate spacing measurements than the Ionization Spectrometer. Also, for reasons which will be explained, it is far more reliable than the present forms of Ionization Spectrometer in the determination of cell size and space group. On the other hand, though superior to the Laue method, it falls far behind the Ionization Spectrometer for measurements of intensity, requiring, as all photographic methods must, some form of Photometer. For intensity measurements the Ionization Spectrometer is the standard to which all other methods are referred. It has however the disadvantage of requiring fairly large crystals which limits its use to substances which crystallise well

Powder Methods

Instead of varying θ by turning the crystal in the beam of X-rays we may use a crystal powder which is in effect a very large number of minute crystals orientated in all directions at random. Only those crystals reflect which offer planes satisfying Bragg's Law and the reflected beams spread out in cones whose angle is 2θ and depends only on d. The reflected beams may be received on a plate or in a movable ionization chamber. This is the powder method of Debye and Scherrer and of Hull. It has the great advantage of being applicable to almost all solid substances and especially to those that cannot be obtained even in minute ($> .01$ cm in dimensions) single crystals. Moreover it is free from the troublesome secondary extinction corrections that apply in all other methods. On the other hand, there is nothing but the spacing to indicate what are the indices of any plane, and, except where the symmetry is high or where the cell size has been found by other methods, it is generally impossible to assign indices and the photograph is meaningless. Even when indices can be assigned the reflections after the first three or four overlap except in the simplest crystals, thus making intensity measurements useless.

Comparison of Methods

The range of utility and the accuracy of the various methods of X-ray crystal analysis can be conveniently summarised as in the table on p. 147.

145

Combination of Methods

It will be seen that no single method gives satisfactory results in all classes of measurement, so that while it is possible and sometimes inevitable that only one method should be used, the full value of an analysis cannot be realized without employing all methods to amplify and check each others results. It may be said roughly that the Laue method and the rotation method are necessary for the preliminary work of determining the cell and the space group while the painstaking and accurate intensity measurements of the Ionization Spectrometer are necessary to establish the details of the structure; the Powder method being used as an auxiliary capacity or in those cases, which are much rarer than is supposed, where single crystals cannot be procured. Luckily the use of five different methods of analysis does not require as many forms of apparatus. All the photographic and all the ionization methods can be combined into two instruments: the X-ray Photogoniometer and the Ionization Spectrometer. The purpose of these papers is to give the principles of construction of the former instrument and an account of its use for Rotation Photographs in particular. The technique of Laue Photography is dealt with in Wyckoff's 'The Structure of Crystals', Ewald's 'Krystalle und Röntgenstrahle' with a convenient summary of papers by Schiebold, Z. Physik. **28** (1924) 355, while the classical 'X-rays and Crystal Structure' gives the best account of the Ionization Spectrometer. For technical applications and descriptions of apparatus Mark's ‚Die Verwendung der Röntgenstrahle in Chemie und Technik' is invaluable.

The X-ray Photogoniometer

The Photogoniometer consists essentially of three parts:

(1) A system of apertures for limiting the breadth and angular divergence of the incident beam of X-rays.

(2) A goniometer head for holding the crystal in the beam of X-rays, for adjusting it in different angular positions and for allowing it to be turned uniformly about an axis perpendicular to the beam through a complete rotation or only through limited angles. This involves some form of steady running electric or clockwork motor.

(3) A photographic plate placed perpendicular to the incident beam behind the crystal which registers all the reflected beams within a fairly small solid angle. When a larger solid angle is required, a film is employed surrounding the crystal supported in a cylindrical camera whose axis is the axis of rotation of the crystal.

	Laue Photograph	Rotation Photograph	Ionization Spectrometer	Powder Photograph	Ionization Spectrometer (powder)
Applicability to crystals of different sizes	Medium 0.1–.01 cm	Medium 0.1–.01 cm	Large > 0.1 cm with developed or ground faces	Small < .01 cm	Small < .01 cm
Applicability to crystals of different symmetry	All	All	All, but troublesome with mono- and triclinic crystals	Only cubic tetragonal, trigonal and hexagonal crystals	Only cubic tetragonal, trigonal and hexagonal crystals
Number of reflections observed	Very many, usually hundreds	Many, up to two hundred	Any number, but troublesome and slow, usually about forty	Usually not more than forty, very crowded	Very few
Determination of indices of reflecting planes	Simple and certain	Troublesome, but certain with care	Certain, but troublesome, especially for general planes	Almost impossible, except for planes of low indices	Almost impossible, except for planes of low indices
Measurement of spacing	Impossible	Highest obtainable accuracy	Accurate	Fairly accurate but for overlapping	Fairly accurate but for overlapping
Measurement of cell size	Only indirectly from other methods	Simple and certain	Liable to errors in complex crystals	Fairly good, where indices can be found	Fairly good
Determination of symmetry class	Best X-ray method	Poor indirect method	Possible from intensity measurements	Impossible	Impossible
Determination of space-group	Only indirectly, liable to error	Best method	Certain only if enough planes be observed	Possible if enough indices can be determined	Possible
Accuracy of intensity measurements	Very poor, too many corrections necessary	Better than Laue method	Standard method, but extinction corrections necessary	Best photographic method	Most accurate, less corrections, but limited application

147

Besides these essential features others may be added for special purposes. The most important are a flat crystal plate holder, taking the place of the goniometer head, which converts the instrument into an X-ray spectrometer; and a collimator and telescope to assist in setting the crystal and which converts the instrument into an optical goniometer. It is sometimes useful but by no means necessary to be able to have the incident beam not perpendicular to the axis or to have the plate not perpendicular to the incident beam.

Theory of the Rotation Method

When such an instrument is used for the rotation method a beam of X-rays from a tube with a copper (iron or molybdenum, etc.) anticathode to produce a monochromatic radiation (really K_{α_1}, K_{α_2} and K_β) falls on the crystal which is kept rotating continuously. Each time a plane comes into the reflecting position (this happens in general four times in a revolution) it makes a trace on a certain spot of the photographic plate and in the course of a sufficient number of revolutions leaves a visible spot on the plate. In general each plane gives rise to four spots which, in the usual case of the beam being perpendicular to the axis, are symmetrically distributed about lines on the plate parallel to and perpendicular to the axis. The photographic density of any spot is simply proportional to the total time of exposure and by lengthening this every reflection however faint will be registered. There is however a limit below which a spot will not appear against a background fogged by general scattering and white radiation but this difficulty can be overcome by the oscillation method (see p. 154). In the pattern of spots that makes up a rotation photograph each spot corresponds to a reflection from a plane of the crystal, and it is the first task of analysis to determine the indices of the planes corresponding to every spot (usually referred to as the indices of the spot). The determination of the unit cell is found to be one step in this process.

The Reciprocal Lattice

The interpretation of rotation photographs is enormously simplified by the use of the mathematical device of the Reciprocal Lattice, first introduced by Ewald[1].

[1] *Z. f. Kryst.* **56** (1921) 129 [Vol. I p. 115; the reciprocal lattice vectors by Bernal are λ times those in the paper cited, after Ewald's first version of the reciprocal lattice, *Phys. Zs.* **14** (1913) 465. Ed.]

The advantage of using the reciprocal lattice for rotation photographs is that the reflection from any plane corresponds both to a spot on the photographic plate and to a point of the reciprocal lattice and that the distribution of spots on the plate is closely related to the distribution of points in the reciprocal lattice.

Rotation Diagram Coordinates

For rotation photographs the most convenient coordinate system to which to refer the points of the reciprocal lattice are the cylindrical coordinates ξ, ω, ζ; ξ being the distance of any point from the axis, ζ its height above the equatorial plane through the origin, and ω its azimuthal angle with respect to some arbitrary fixed line. Of these only ξ and ζ can be found directly from the photograph as there is nothing to show exactly at what angle any plane reflects. We have of course $\rho^2 = \xi^2 + \zeta^2$. In the crystal lattice ξ corresponds to the reciprocal of the distance of the trace of the corresponding plane in the equatorial plane from the origin, and ζ to the reciprocal of the intercept of the corresponding plane on the axis (see Fig. 1).

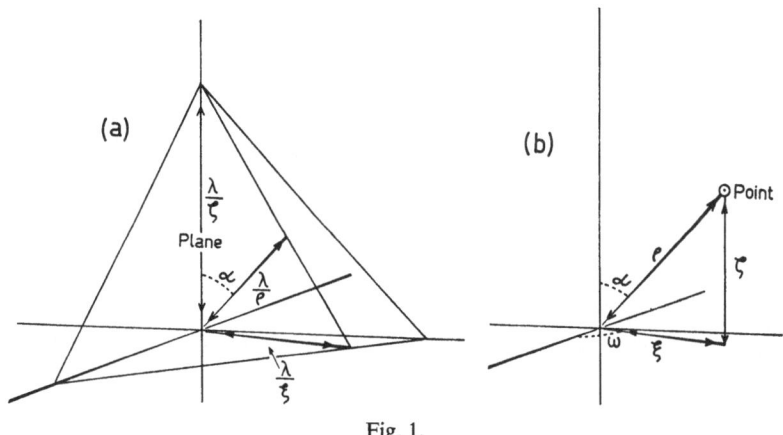

Fig. 1.

A diagram on which all the ξ and ζ values of the reciprocal lattice are plotted is called a rotation diagram. It may be considered to be produced by rotating the reciprocal lattice about the axis and marking the traces of each point on any meridian plane.

149

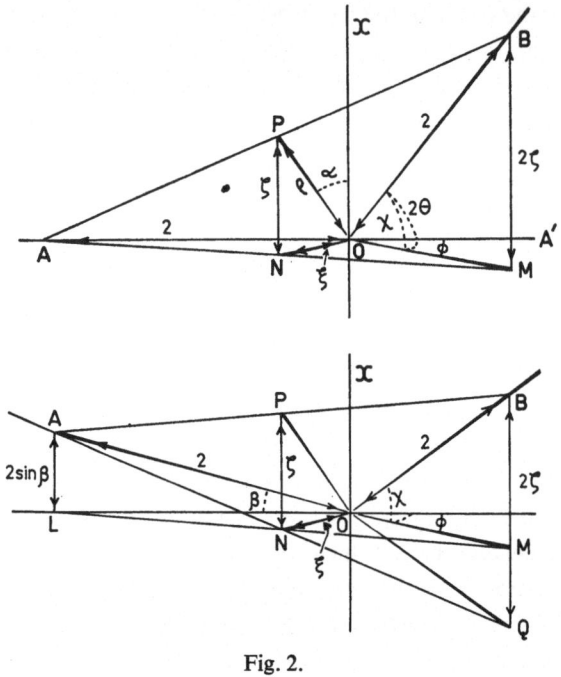

Fig. 2.

Transformations

The coordinates of points on the rotation diagram are related to the angular coordinates χ and ϕ of the reflected beams by the equations

$$\xi = \{1 - 2 \cos \chi \cos \phi + \cos^2 \chi\}^{\frac{1}{2}},$$

$$\zeta = \sin \chi,$$

where χ is the angle measured from the equatorial plane and ϕ that measured from the principal plane, i.e. the meridian plane passing through the incident beam (see Fig. 2). The above equations only hold when the incident beam is perpendicular to the axis. In the more general case, where it is inclined at an angle $\pi/2 - \beta$, we have

$$\xi = \{\cos^2 \beta - 2 \cos \beta \cos \chi \cos \phi + \cos^2 \chi\}^{\frac{1}{2}},$$

$$\zeta = \sin \beta + \sin \chi.$$

In order to be able to interpret the rotation photographs, ξ and ζ must be given in terms of quantities directly measurable. The most suitable are the Cartesian and polar coordinates of the spots on the plate or on the unrolled cylindrical film. For the former we have, for a plate distance D from the crystal,

$$\xi = \left\{2 - \frac{2}{\{1 + r^2/D^2\}^{\frac{1}{2}}} - \frac{y^2/D^2}{1 + r^2/D^2}\right\}^{\frac{1}{2}},$$

$$\zeta = \frac{y/D}{\{1 + r^2/D^2\}^{\frac{1}{2}}}.$$

For a cylindrical film of radius R

$$\xi = \left\{1 - \frac{2 \cos (x/R)}{\{1 + y^2/R^2\}^{\frac{1}{2}}} + \frac{1}{1 + y^2/R^2}\right\}^{\frac{1}{2}},$$

$$\zeta = \frac{y/R}{\{1 + y^2/R^2\}^{\frac{1}{2}}}.$$

Interpretive Charts

From these equations or their equivalents ξ and ζ may be calculated and tables for this purpose are given in the author's paper[1], but the work is rather laborious because of the number of spots involved, and by the use of two charts prepared by the author the necessity of calculation is done away with, except where great accuracy is required. The photograph is simply projected on to the chart by means of an enlarging lantern until the image corresponds to a plate distance of 10 cms for the plane chart or for a camera radius of 5 cms for the cylindrical chart, and ξ and ζ can be read off directly in either case. This method achieves a great saving of time and trouble without serious loss of accuracy. From the ξ and ζ values thus obtained a rotation diagram is constructed and compared with the theoretical rotation diagram derived from the reciprocal lattice. In this way the indices of the reflecting planes can be immediately determined. The detail of the method is explained in the author's paper, but the general principles are as follows.

Measurement of Cell Size

If a crystal is rotated about any zone axis, i.e. along a line joining any two points of the lattice, this is equivalent to rotating the reciprocal lattice about an axis perpendicular to a corresponding plane of points (see Fig. 3).

[1] *Proc. Roy. Soc.* A **113** (1926) 117.

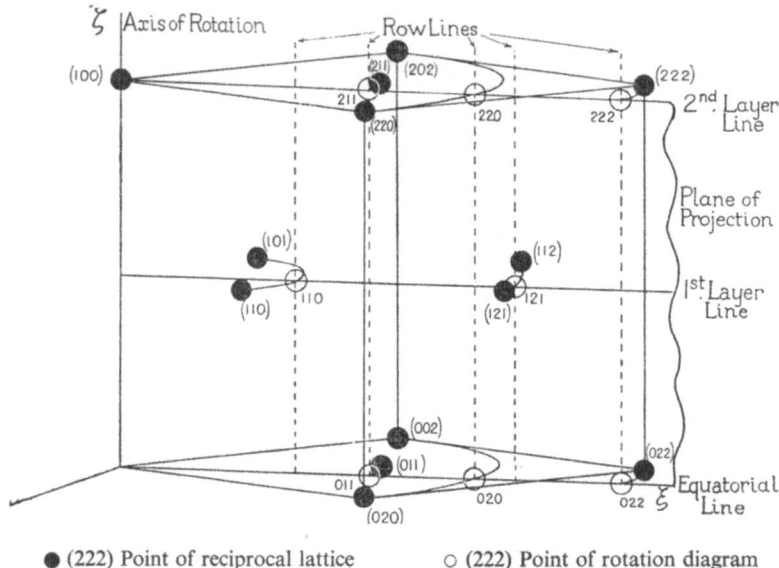

● (222) Point of reciprocal lattice   ○ (222) Point of rotation diagram

Fig. 3. Formation of Rotation Diagram from a face-centred cubic reciprocal lattice, corresponding to a body-centred crystal.

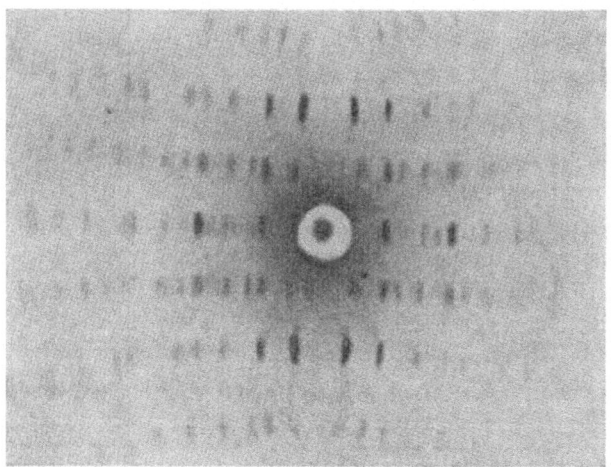

Fig. 4. Typical zone axis rotation photograph. The original was taken by rotating a monoclinic needle-shaped crystal about the *b* axis at 4 cms. from the plane photographic plate in a pencil of X-rays from a copper anticathode.

It is easy to see that the ζ of the reflected beams will have values which will all be multiples of the perpendicular distance between planes parallel to the equatorial plane. Each set of reflections corresponding to a layer of points in the reciprocal net will lie on a line on the plate. These lines are called 'layer lines' (see Fig. 4). On the plane plate they are hyperbolas; on the cylindrical film, straight lines.

The constant ζ differences between adjacent layer lines give immediately the length of the primitive translation, i.e. the distance τ between lattice points in the direction of the axis by the simple relation

$$\tau = \lambda/\zeta.$$

If we take the three principal axes a, b, c of the crystal in turn as axes of rotation, their length and consequently the exact size of the unit cell may be found from this relation. Some difficulties exist in monoclinic and triclinic crystals owing to the arbitrary way in which crystallographers are obliged to assign axes. But these can easily be overcome. From these measurements the size and shape of the cell of the reciprocal lattice can be found even more directly, for in orthogonal crystals a^*, b^*, c^* are of course given by the three ζ intervals. The ξ and ζ values for the plane (hkl) of any orthogonal crystal rotated about its axis are

$$\xi = \{\kappa^2 b^{*2} + l^2 c^{*2}\}^{\frac{1}{2}},$$

$$\zeta = ha^*.$$

Here the reflections of all planes which have the same first index h, lie on the same layer line, while the reflections of planes with the same kl indices lie on the lines of equal ξ the socalled 'row lines'. These values may be calculated and compared plane by plane with the observed values of ξ and ζ or graphical methods may be used throughout and in either way the indices of the reflecting planes may be determined. In monoclinic and triclinic crystals the procedure is more complicated as no row lines appear on photographs taken about crystallographic axes. And here the method of taking photographs not about axes but about normals to crystal faces can be used to advantage. In this case row lines appear and not layer lines.

Resolution of Spots

If the spots on the photograph could be reduced to geometrical points the methods outlined above would be sufficient to determine the indices of all spots on a rotation photograph but unfortunately the spots are of finite size and often overlap except in the more symmetrical classes, i.e. cubic, tetragonal, trigonal and hexagonal. To escape the uncertainty introduced in this manner, which is analogous in two dimensions to the overlapping of Debye rings in powder photographs, it is possible by various methods, such as reducing spot size, use of more accurate adjustment, methods of mis-setting crystals to spread spots, etc., to reduce the overlapping to a minimum. Nevertheless with complex organic crystals with large cells all these methods fail because the spots are so numerous and close together, and it is necessary to make use of the angular position of the crystal on reflection in order to identify the spots. This is equivalent to introducing the third coordinate ω of a reciprocal lattice.

The Oscillation Method

If instead of rotating the crystal through a complete revolution the rotation is limited to a uniform back and forth movement through 5 or 10 degrees, only a limited number of planes will reflect. The indices of planes which reflect for any particular range of oscillation can be calculated, or much more easily found by a very simple geometrical construction described in the author's paper[1]. In such a limited range the chance of overlapping almost disappears and by going over the crystal section by section in this manner every spot can be identified. The usefulness of the oscillation method is much wider than its ability to resolve confused spot patterns. By its use the relative length of exposure of each photograph is much shortened so that it brings up spots which are too faint or hidden in the fogged background to appear in ordinary rotation photographs, and it is also extremely useful for setting crystals without developed faces.

SUMMARY

The operations necessary for the determination of cell size and indices by the method of rotation photograph may be summarised as follows:

(1) Three photographs are taken with complete rotation about each of the three principal axes of the crystal in turn.

(2) The ζ values of the layer lines in each photograph are measured and axial lengths a, b, c of the crystal and a^*, b^*, c^* of the reciprocal lattice are determined.

[1] *Proc. Roy. Soc.* A **113** (1926) 150.

(3) The ζ and ξ values of each spot on the photograph are calculated or found directly from a chart. They are compared with the theoretical values calculated for planes of known indices of the reciprocal lattice. Alternatively, the comparison may be made graphically between the observed rotation diagram and that derived from the reciprocal lattice. In this way the indices of each spot are determined, but if all the spots are not separate it is necessary to proceed to (4).

(4) A series of photographs are taken, each one being a rotation or oscillation about a principal axis of the crystal through a limited angle of 15, 10 or 5 degrees according to the complexity of the crystal. The indices and ξ, ζ coordinates of the planes which should reflect in each of these oscillations are found graphically or by calculation and compared with those obtained from the photograph. In this way the indices both of the planes appearing on the photographs and of those which do not appear are found.

Davy Faraday Laboratory

135.   WEISSENBERG DIAGRAM

Durch eine mit der Kristalldrehung zwangsläufig gekoppelte Bewegung der Aufnahmevorrichtung (Drehung oder Verschiebung eines Filmzylinders oder Platte) kann die Kristalldrehung zwischen zwei Reflexionen experimentell bestimmt und daraus eindeutig die Winkel aller reflektierenden Ebenen zueinander und ihre Identitätsabstände berechnet werden, wodurch auch bei Fehlen kristallographischer Indizien eindeutige Gitterbestimmung und Orientierung von Kristallsplittern gewährleistet ist.

Die Aufgabe, alle Winkel zu messen, welche je zwei beliebige Netzebenen in einem Kristall miteinander bilden, läßt sich durch die folgende Modifikation des Drehkristallverfahrens mit Hilfe des Röntgengoniometers eindeutig und streng lösen.

Wir gehen dabei so vor, daß wir jede Netzebene für sich relativ zu einem festen Koordinatensystem vermessen und daraus die Winkel zwischen je zwei Netzebenen berechnen.

Fig. 1 zeigt ein schematisches Bild des Röntgengoniometers.

Der Kristall ist in der Figur durch eine Lagenkugel ersetzt; diese ist mit einer Drehachse $F\bar{F}$ starr verbunden, welche unter dem Winkel β zum einfallenden Strahl steht und die Lagenkugel in ihren Polen $F\bar{F}$ durchstößt. Der Schlitz einer Blende B umgibt als Parallelkreis in der Polhöhe μ die Lagenkugel; es werden also von allen abgebeugten Strahlen nur diejenigen

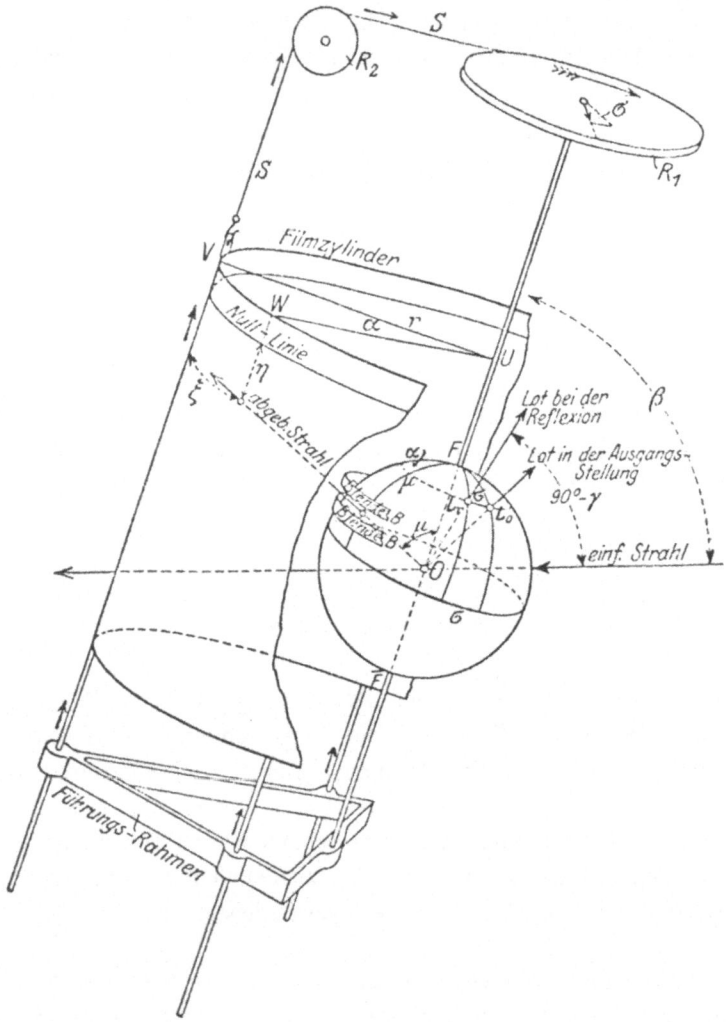

Fig. 1.

durch den Schlitz treten können, deren Winkel mit der Drehachse gleich μ ist. (In der Figur ist der Übersichtlichkeit halber nicht die ganze Blende, sondern nur ein Stück derselben eingezeichnet.)

Die Aufnahmevorrichtung[1] besteht aus einem zur Drehachse koaxialen Filmzylinder, welcher bei Drehung der Kristallachse zwangsläufig axial verschoben wird; durch die Führungen im Rahmen ist die Richtung der axialen Verschiebung, durch die beiden Rollen R_1R_2 und die darüber laufende Schnur S ist die zwangsläufige Koppelung angedeutet.

In der Ausgangsstellung (Nullstellung) des Kristalls durchstößt das Lot OL_0 einer Netzebenenparallelschar die Lagenkugel im Punkt L_0; diese Lage der Netzebenenparallelschar mit dem Identitätsabstand D wird gemäß dem Braggschen Reflexionsgesetz:

$$n\lambda = 2D \sin \gamma_{D\lambda} \qquad (1)$$

den einfallenden monochromatischen Röntgenstrahl von der Wellenlänge λ im allgemeinen nicht reflektieren können. Wird nun der Kristall gedreht, so bewegt sich L_0 auf dem in Fig. 1 angedeuteten Parallelkreis, und der Winkel zwischen einfallendem Strahl und Lot ändert sich; sobald dieser Winkel den nach Bragg gegebenen Wert $90 - \gamma_{D\lambda}$ erreicht, tritt Reflexion an der Netzebenenparallelschar ein; der Durchstoßpunkt des Lotes durch die Lagenkugel sei bei der Reflexion mit L_r, der Winkel, um den der Kristall gedreht werden mußte, um aus der Ausgangslage das Lot in reflexionsfähige Lage zu bringen, sei mit σ bezeichnet. Durch die zwangsläufige Koppelung der Kristalldrehung σ mit einer axialen Filmverschiebung η kann σ leicht aus dem Diagramm bestimmt werden. Aus den Koordinaten $(\xi\ \eta)$ des zu einer Netzebene gehörigen Interferenzpunktes lassen sich in Verbindung mit dem durch die Schlitzblendenstellung gegebenen Winkel μ und dem bekannten β (Winkel zwischen einfallendem Strahl und Drehachse) die Lage der Netzebenenparallelschar im Kristall und ihr Identitätsabstand bestimmen (...).

Anwendungen des Röntgengoniometers

Das Röntgengoniometer ist in erster Linie als Hilfsmittel zur Strukturbestimmung gedacht und bildet insofern einen Abschluß der Drehkristallmethode, als es gestattet von jeder überhaupt reflektierenden Netzebenenparallelschar so viele Parameter zu messen, als zu ihrer eindeutigen Bestimmung erforderlich sind. Ein Debye-Scherrerdiagramm liefert nämlich für jede Netzebene *einen* experimentellen Parameter (Radius des Debye-Scherrerkreises), das Schichtliniendiagramm durch Vermessung der zwei Koordinaten des Interferenzpunktes auf dem Film *zwei* Parameter (μ, α) und das Röntgengoniometer *drei* Parameter (μ, α, σ); eine Netzebenenparallelschar hat aber im Bravaisgitter nur drei Freiheitsgrade (nämlich den Identitätsabstand und zwei Freiheitsgrade, welche durch die Richtung der Normalen gegeben sind), somit ist sie durch drei unabhängige Parameter eindeutig bestimmt; prinzipiell ist somit diese Methode abgeschlossen.

157

Einige spezielle Aufgaben, welche sich mit Hilfe des angegebenen Röntgengoniometers streng lösen lassen, seien nachfolgend kurz skizziert.

I. Gitterbestimmung und Orientierung eines Kristallsplitters von dem keinerlei kristallographische Daten gegeben sind. (...)

II. Analyse von Schichtliniendiagrammen. Wesentlich einfacher gestaltet sich das Vorgehen, wenn eine rationale Richtung im Kristall bekannt[1] ist und als Drehachse eingestellt werden kann; das gewöhnliche Drehdiagramm um diese Richtung ergibt ja ohne Indizierung der Netzebenen, gemäß der Schichtlinienbeziehung

$$n\lambda/J = \cos \mu_n + \cos \beta, * \qquad (19)$$

sowohl die Identitätsperiode J in Richtung der Drehachse als auch die für die Einstellung der Blende des Röntgengoniometers gesuchten Winkel μ.

Die vollständige Goniometrierung des Kristalls kann nun mit Hilfe des Röntgengoniometers streng durchgeführt werden, indem die Schlitzblende der Reihe nach auf die nach Gleichung (19) bestimmten μ_n-Werte eingestellt wird, werden bei jeder Aufnahme alle Schichtlinien bis auf eine weggeblendet und diese eine allein analysiert. Jede solche Aufnaheme liefert die goniometrische Vermessung und somit die Indizierung aller Netzebenen der eingestellten Schichtlinie (n) (also aller Ebenen, deren Index bezüglich der Drehachse n ist). Bei vielen organischen Kristallen ist diese Goniometrierung zur Strukturbestimmung notwendig, da die Grundperioden so groß sind, daß eine eindeutige Indizierung der Netzebenen auf den Schichtlinien des gewöhnlichen Drehdiagramms nur bezüglich der Drehachse möglich ist; die beiden anderen Indizes bleiben unsicher, da zu ihrer Bestimmung nur eine experimentelle Größe (der Netzebenenabstand D) und das Rationalitätsgesetz der Indizes (quadratische Form) benutzt werden kann.

Kaiser Wilhelm-Institut f. Faserstoffchemie, Berlin-Dahlem

136. Es wird eine Ausführungsform des von K. Weissenberg vorgeschlagenen Röntgengoniometers angegeben und die Anwendung an einem Beispiel erläutert.

K. Weissenberg hat in Bd. 23, S. 229 dieser Zeitschrift** eine Modifikation des Drehkristallverfahrens angegeben, die es ermöglicht, die bei diesem

[1] M. Polanyi, *Die Naturwiss.* **9** (1921) 337 [this Vol. paper **131**]; *Z. Physik* **7** (1921) 149, und M. Polanyi u. K. Weissenberg, *Z. Physik* **9** (1922) 44.

* [This equation differs from the usual form $n\lambda/J = \cos \mu_n - \cos \beta$ by the fact that β 'Winkel zwischen einfallendem Strahl und Drehachse' actually denotes the supplement of this angle—see Fig. 1—Ed.]

** [This Vol. preceding paper.]

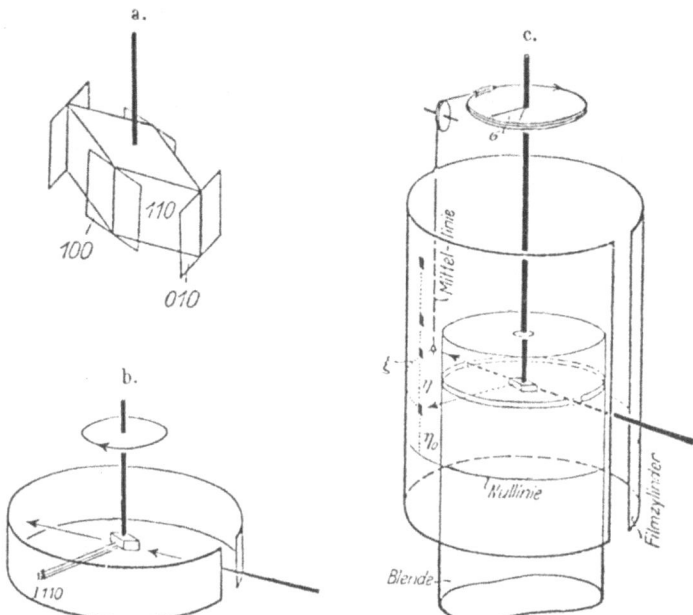

Fig. 1.

Verfahren auftretenden Schichtlinien gesondert bezüglich des Zustandekommens der einzelnen Interferenzflecke zu analysieren. Dies wird erreicht durch eine Registrierung der Aufeinanderfolge der Interferenzen vermittelst einer 'mit der Kristalldrehung zwangsläufig gekoppelten Bewegung der Aufnahmevorrichtung'.

Das Prinzip und die Vorteile dieses Verfahrens mögen am c-Achsendiagramm eines rhombischen Kristalls (Fig. 1a) erläutert werden.

Beim gewöhnlichen Drehkristallverfahren (Fig. 1b) wird ein und derselbe Interferenzpunkt I der Netzebene (110) während einer Umdrehung viermal nacheinander belichtet. Bewegt man den Filmzylinder während der Kristalldrehung in der in der Fig. 1c gezeichneten Weise, so kommen diese vier Interferenzen auf dem Filmzylinder nebeneinander zur Abbildung, und zwar liegen sie auf einer Geraden, die im Abstand ξ der Mittellinie des Films parallel läuft. Ihr gegenseitiger Abstand η liefert den Winkel σ, um den sich der Kristall zwischen zwei reflexionsfähigen Lagen der Netzebene (110) gedreht hat. Der Abstand der ersten Interferenz von der Nulllinie η_0 liefert den Winkelabstand der Ebene von der Ausgangsstellung des Kristalls.

Zur konstruktiven Ausführung habe ich die in Fig. 1c angedeutete Anordnung gewählt. Fig. 2 zeigt das fertige Instrument. Der horizontal gelagerte zylindrische Filmträger F wird vermittelst des Wagens W auf dem

159

Fig. 2.

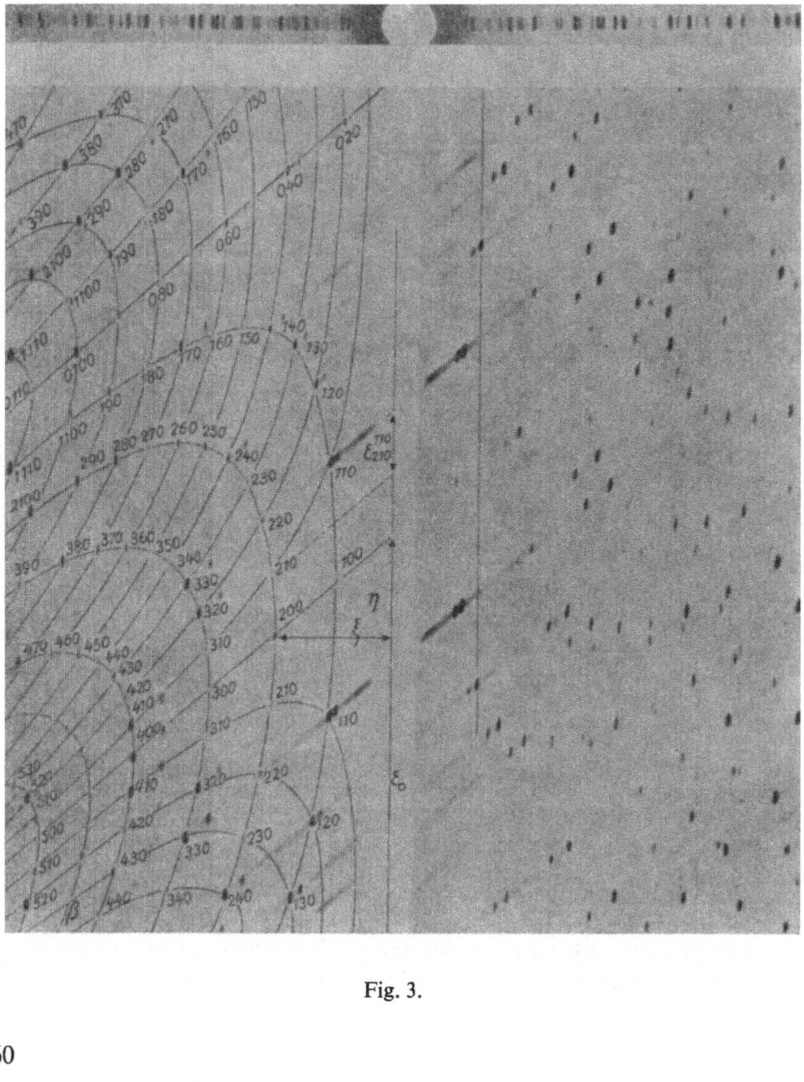

Fig. 3.

Schienenpaar *S* hin und her bewegt. Die Kopplung wurde so ausgeführt, daß einer Bewegung um 180 mm eine gleichzeitige Drehung des Kristallträgers *G* um 180°, also 1° Kristalldrehung = 1 mm Filmbewegung entspricht. (...)

Fig. 3 zeigt oben den Äquator einer gewöhnlichen Drehaufnahme des rhombisch kristallisierenden Diaspor (*c*-Achse). Darunter ist dessen Auflösung vermittelst des Röntgengoniometers abgebildet. Man erkennt leicht die Gesetzmäßigkeiten, die ein solches Diagramm beherrschen.

Die Interferenzflecke ein und derselben Art liegen auf vertikalen Geraden übereinander, die Interferenzen höherer Ordnung ein und derselben Netzebene auf den schrägen Geraden, deren Neigungswinkel eine Konstante des Apparates ist. Man kann sofort das Intensitätsverhältnis der verschiedenen Ordnungen ablesen und sehen, ob die betreffende Röntgenperiode normal ist oder nicht. So sind beispielsweise bei (100) die geraden Ordnungen stark, bei (110) umgekehrt die ungeraden Ordnungen, während bei (120) ein annähernd gleichmäßiger Abfall der Intensitäten statthat. Es ist also sozusagen eine ganze Serie von Braggschen Aufnahmen nebeneinander zur Abbildung gekommen.

Die Interferenzen ordnen sich zu zweierlei Kurvenscharen $hn0$ und $nk0$, und zwar bleibt für jede Schar auf den einzelnen Kurven der eine der beiden Indizes h, k konstant, während n die Reihe der ganzen Zahlen durchläuft. Für die Fälle $h00$ und $0k0$ entarten die Kurven zu Geraden, Asymptoten der zugehörigen Kurvenscharen. Die β-Linien bilden ähnliche Kurven.

Kaiser Wilhelm-Institut für physikalische Chemie
und Elektrochemie in Berlin-Dahlem

137. Bei dem Weissenbergschen Röntgengoniometer[1], das auf dem Drehkristallverfahren beruht, wird durch Anwendung einer 'Schichtlinienblende' und einer mit der Drehung des Kristalls gekoppelten axialen Verschiebung des Filmzylinders eine eindeutige Indizierung jedes auftretenden Interferenzflecks ermöglicht. Man kann mit Hilfe der von Weissenberg angegebenen Formeln aus den Daten der Filmvermessung für jeden Interferenzfleck Lage und Netzebenenabstand der interferierenden Fläche bestimmen. Diese rechnerische Auswertung ist besonders für Aufnahmen von Kristallen niederer Symmetrie verhältnismäßig zeitraubend. Nachstehend wird ein graphisches Verfahren mitgeteilt, das sich des reziproken Gitters bedient. Alle für den Interferenzvorgang charakteristischen Größen können im rezi-

---

[1] K. Weissenberg, *Z. Phys.* **23** (1924) 229.

proken Gitter einer einfachen geometrischen Konstruktion entnommen werden, die besonders deshalb für die Auswertung von Röntgenaufnahmen jeglicher Art geeignet erscheint, weil die Erweiterung der Konstruktion zur gleichzeitigen Darstellung *aller* Interferenzen einer Aufnahme verhältnismäßig einfach gelingt. So ist eine graphische Auswertung von Drehkristallaufnahmen vor kurzem von J. D. Bernal[1] aus der gleichen Konstruktion abgeleitet worden. In der zitierten Arbeit wird eine Begründung dieser Konstruktion ausführlich angegeben. Bernal erörtert daselbst auch die Eigenschaften des reziproken Gitters, das er durch Einführung einer Konstanten k etwas allgemeiner definiert als es P. P. Ewald[2] getan hat. Durch zweckmäßige Wahl der Konstante k kann dann die beabsichtigte Konstruktion besonders einfach gemacht werden. Hier sollen reziprokes Gitter und Hilfskonstruktion so kurz dargestellt werden, wie es zum Verständnis der nachherigen Anwendung auf Röntgengoniometeraufnahmen ausreicht. Um den Vergleich mit Bernal zu erleichtern, verwende ich die von ihm gewählte Buchstabenbezeichnung.

Das reziproke Gitter wird definiert durch

$$\rho_{(hkl)} \cdot d_{(hkl)} = k^2, \tag{1}$$

wobei $\rho_{(hkl)}$ der Abstand des Punktes (hkl) im reziproken Gitter vom Nullpunkt,

$d_{(hkl)}$ der Netzebenenabstand der Ebene (hkl) im wirklichen Gitter und

k eine Konstante ist.

Dabei steht der Fahrstrahl vom Nullpunkt zum Punkt (hkl) im reziproken Gitter senkrecht auf der Netzebene (hkl) im wirklichen Gitter. Durch eine vollständige Angabe des Translationsgitters ist das reziproke Gitter gegeben und umgekehrt; jeder Netzebene des reziproken Gitters entspricht eine darauf senkrechte Translationsrichtung im wirklichen Gitter.

Die Interferenzkonstruktion erhält man gemäß folgender Vorschrift:

Man konstruiere das reziproke Gitter in richtiger Orientierung zum Translationsgitter. Um zu ermitteln, ob eine gegebene Richtung des Primärstrahls zum Gitter reflexionsfähig ist, lege man einen Strahl in dieser Richtung durch den Ausgangspunkt des reziproken Gitters. Man konstruiere eine Kugel, die den Ausgangspunkt des Gitters berührt, und deren Mittelpunkt auf dem gegebenen Strahl mit dem Abstand k^2/λ vom Ausgangspunkt liegt. (λ = Wellenlänge des Röntgenstrahles.) Eine Netzebene des Translationsgitters reflektiert nun den Primärstrahl dann und nur

[1] J. D. Bernal, *Pr. Roy. Soc.* **113** (1926) 117.
[2] P. P. Ewald, *Z. Krist.* **56** (1921) 129.

dann, wenn ihr zugehöriger Punkt im reziproken Gitter auf der Oberfläche dieser Kugel liegt. (Die Kugel wird weiterhin Reflexionskugel genannt.) Die Richtung des abgebeugten Strahls ist gegeben durch die Richtung vom Mittelpunkt der Reflexionskugel zu dem zugehörigen Punkt des reziproken Gitters.

Die Ableitung dieser Vorschrift soll hier nicht gegeben werden; sie geht vom Braggschen Reflexionsgesetz aus. Für die Anwendung macht man zweckmäßig $k^2 = \lambda$. Dadurch wird der Radius der Reflexionskugel gleich 1. Die Punkte des reziproken Gitters beziehen wir, ebenso wie Bernal, auf ein Zylinderkoordinatensystem mit den Koordinaten ω, ξ und ζ (Fig. 1); Zylinderachse ist dabei die Drehachse. Die Koordinaten des abgebeugten Strahls bezeichen wir, wie bei Drehkristallaufnahmen üblich, mit μ und α, den Winkel, den der Primärstrahl mit der Drehachse bildet, mit $(90° - \beta)$.

Da nun eine Röntgengoniometeraufnahme eine Drehkristallaufnahme ist, bei der alle Interferenzstrahlen, die nicht eine durch die Stellung der Schichtlinienblende vorgeschriebene Koordinate μ besitzen, abgeblendet werden, durchstechen alle zur Aufnahme gelangenden Interferenzstrahlen die Reflexionskugel auf einen Breitenkreis mit dem Radius $(90° - \mu)$, der senkrecht zur Drehachse steht. Es können daher nur diejenigen Punkte des reziproken Gitters zur Aufnahme gelangende Interferenzstrahlen erzeugen, die in der Ebene des Breitenkreises liegen. Um möglichst viele Punkte zur Abbildung zu bringen, ist es daher zweckmäßig, dicht bezetzte Netzebenen des reziproken Gitters senkrecht zur Drehachse zu stellen und die Schichtlinienblende auf eine der Netzebenen einzustellen. Dies bedeutet, daß man eine möglichst kurze Translationsperiode als Drehachse wählt und die Schichtlinienblende auf eine Schichtlinie einstellt. Von dieser Schichtlinie liefert nun das Röntgengoniometer dadurch eine Analyse, daß mit der Drehung des Kristalls eine Verschiebung des Filmzylinders längs seiner Achse gekoppelt ist, und infolgedessen die Lage des Interferenz-

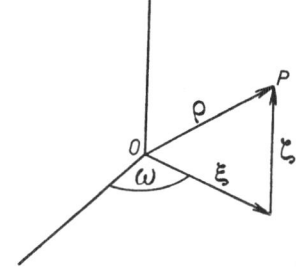

Fig. 1. Koordinaten des reziproken Gitters.

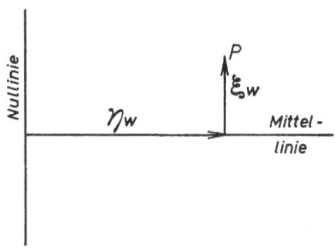

Fig. 2. Schema für die Vermessung der Röntgengoniometeraufnahmen.

163

flecks auf dem Film neben dem Braggschen Glanzwinkel noch die Stellung des Kristalls im Augenblick der Reflexion angibt. Mißt man nämlich, wie in Fig. 2, die den aufgerollten Film zeigt, die Koordinaten ξ_w und η_w des Interferenzflecks P, so liefern diese nach Weissenberg

$$\alpha = \xi_w \frac{360}{2\pi r_F} \tag{2}$$

$$\sigma = \eta_w \frac{360}{\eta_{360}}, \tag{3}$$

dabei ist

α die geographische Längenkoordinate des Interferenzstrahles,

σ der Winkel, um den der Kristall gedreht werden mußte, um von der Ausgangsstellung in die Reflexionsstellung zu gelangen,

r_F der Filmradius,

η_{360} die Verschiebung des Filmzylinders für eine Drehung des Kristalls um 360°.

Zur Auswertung der Aufnahme soll aus den Koordinaten des Interferenzflecks auf dem Film die entsprechende Netzebene des reziproken Gitters konstruiert werden. Dies geschieht mit Hilfe der eingangs gebrachten Kugelkonstruktion, von der nur der in Betracht kommende Schnitt senkrecht zur Drehachse zu zeichnen ist (Fig. 3). Die Figur enthält zunächst

D den Durchstoßpunkt der Drehachse,

S_k den Schnittkreis der Reflexionskugel mit dem Radius r und

M' dessen Mittelpunkt.

Dabei ist, wie aus Fig. 4 hervorgeht,

$$r = \cos \mu, \quad \overline{M'D} = \cos \beta.$$

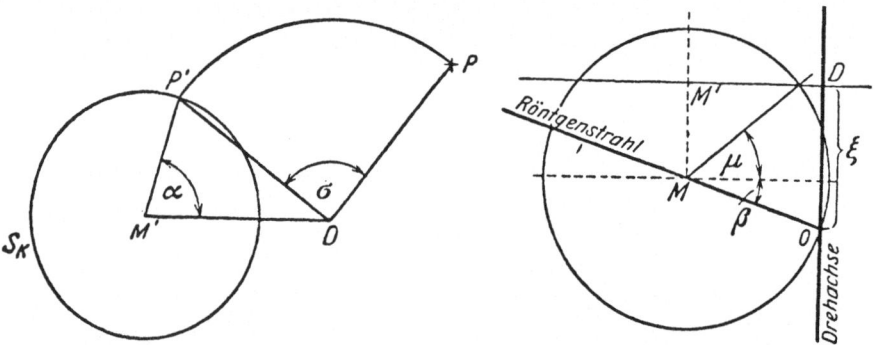

Fig. 3. Konstruktion der Punkte der reziproken Netzebene.

Fig. 4. Querschnitt längs der Drehachse und durch den Mittelpunkt der Reflexionskugel. [In the Fig. ξ should read ζ. Ed.]

Es bedeutet ferner P die Lage eines reziproken Gitterpunktes bei der Ausgangsstellung des Gitters und P' seine Lage im Augenblick der Reflexion. Daher ist $\sphericalangle\,P'DP$ gleich dem Winkel σ; und $\sphericalangle\,DM'P'$ gleich dem Winkel α. Die Vorschrift zur Konstruktion der Netzebene lautet danach:

Man trägt zunächst den Winkel α ab und findet P'; dann dreht man $\overline{DP'}$ um den Winkel σ und findet so P. Auf diese Weise sind die Ausgangslagen für alle zur Interferenz gekommenen Punkte aufzusuchen.

Da die Punkte auf Scharen paralleler Geraden liegen müssen, haben wir in der dann vorliegenden Netzebene für jede weitere Auswertung eine geeignete Grundlage. Es ist leicht, bei bekanntem Translationsgitter die Punkte zu indizieren, bei unbekanntem Translationsgitter ist es leicht, für die Zahlen, die die Aufnahme liefert, Mittelwerte zu erhalten.

Die Fig. 4 ergibt auch den Abstand der betrachteten Netzebene vom Nullpunkt, der gleichzeitig die ζ-Koordinate aller in dieser Ebene liegenden Gitterpunkte ist, zu

$$\zeta = \sin\beta + \sin\mu. \tag{4}$$

Für die Aufnahmen von Äquatorschichtlinien und, wie wir später sehen werden, bei geeigneter Einstellung des Apparates, auch für die Aufnahmen höherer Schichtlinien läßt sich das obige Verfahren noch etwas zweckmäßiger gestalten. Äquatorschichtlinienaufnahmen, die man dadurch erhält, daß man bei senkrechten Eintritt des Primärstrahls die Schichtlinienblende auf $\mu = 0°$ einstellt, enthalten viele Interferenzen, die in mehreren Ordnungen vorkommen, im speziellen ist zu jedem Punkt (hkl) der Gegenpunkt $(\bar{h}\bar{k}\bar{l})$ vorhanden. Die verschiedenen Ordnungen einer Interferenz —damit auch Punkt (hkl) und Punkt $(\bar{h}\bar{k}\bar{l})$—liegen dabei auf schrägen Geraden, die mit der Mittellinie des Films den Winkel v bilden. (In Fig. 5 ist Punkt P' Gegenpunkt zu P.) Dabei ist, wie unten begründet werden wird

$$\mathrm{tg}\,v = \frac{4\pi r_F}{\eta_{360}}. \tag{5}$$

Es ist zweckmäßig, solche Aufnahmen nach den Koordinaten ε und φ zu vermessen, wie in Fig. 5 angedeutet ist. Die Formel (2) ist dann zu ersetzen durch die folgende

$$\alpha = 2\varphi\,\frac{360\cos v}{\eta_{360}}. \tag{6}$$

Der Vorteil der Vermessung der Strecke 2φ besteht darin, daß sie als Abstand Punkt—Gegenpunkt leicht und genau meßbar ist. Von ε dagegen soll jetzt bewiesen werden, daß es direkt proportional der Zylinderkoordi-

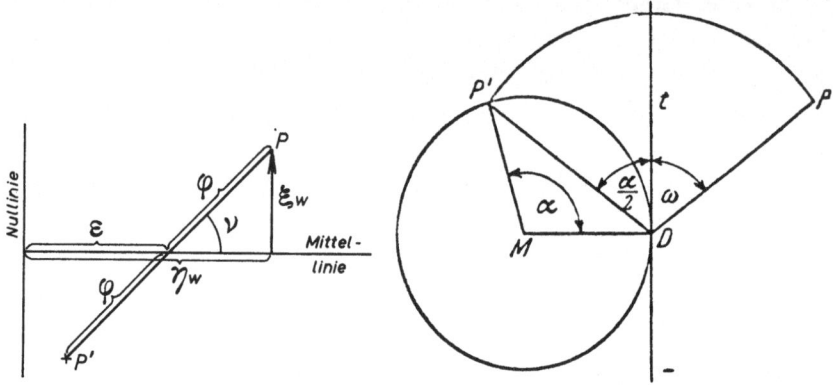

Fig. 5. Schema für die Vermessung der Röntgengoniometeraufnahmen bei Einstellung $\beta = \mu$.

Fig. 6. Konstruktion der Punkte der reziproken Netzebene bei Einstellung $\beta = \mu$.

nate ω (Fig. 1) des reziproken Gitterpunktes ist. Wenn wir Gleichung (5) als Definition des schrägen Koordinatenkreuzes annehmen, so ist nach Fig. 5

$$\varepsilon = \eta_w - \xi_w \frac{\eta_{360}}{4\pi r_F},$$

daraus erhält man

$$\varepsilon \frac{360}{\eta_{360}} = \eta_w \frac{360}{\eta_{360}} - \xi_w \frac{360}{4\pi r_F},$$

mit Hilfe von Formel (2) und (3) ergibt sich

$$\varepsilon \frac{360}{\eta_{360}} = \sigma - \frac{\alpha}{2}. \tag{7}$$

Wir betrachten jetzt die Konstruktion im reziproken Gitter, dargestellt in Fig. 6. Die Äquatorschichtlinie entspricht der durch den Nullpunkt gehenden Ebene des reziproken Gitters. Diese Ebene schneidet aus der Kugel einen Großkreis aus, der den Nullpunkt (in der Fig. mit D bezeichnet, da er auch Durchstoßpunkt der Drehachse ist) berührt. Als Hilfslinie wird hier die Tangente t gezogen. Diese teilt den Winkel σ in die beiden Winkel $\overline{P'D} : t$ und $t : \overline{DP}$. Wie leicht einzusehen, ist $\overline{P'D} : t$ gleich $\alpha/2$; ferner ist $t : \overline{DP}$ die Zylinderkoordinate ω des Punktes P (t als Ausgangsrichtung angenommen). Als Gleichung formuliert hat man

$$\omega = \sigma - \frac{\alpha}{2},$$

berücksichtigt man Gleichung (7), so erhält man

$$\omega = \varepsilon \, \frac{360}{\eta_{360}}.$$
(8)

Man hat dann eine zweite Möglichkeit, das Netz zu konstruieren. Man ermittelt $\overline{DP}'$ aus α, wie oben, dreht dann $\overline{DP}'$ zunächst in die Richtung t und dann weiter um den Winkel ω. Es ist dabei von Vorteil, daß man den Winkel ω für alle Punkte von derselben Ausgangsrichtung abzutragen hat. Es empfiehlt sich nun, um den ersten Teil der Konstruktion ganz zu ersparen, $\overline{DP}$ auf andere Weise zu ermitteln. Es ist, wie aus Fig. 6 folgt, und weil bei Äquatorschichtlinienaufnahmen der Kreisradius gleich 1 ist:

$$\overline{DP} = 2 \sin \frac{\alpha}{2}.$$
(9)

Gleichung (9) mit Gleichung (6) vereinigt, ergibt $\overline{DP}$ als Funktion von 2φ, die man sich am besten als Kurve darstellt.

Die endgültige Art der Auswertung von Äquatorschichtlinienaufnahmen ist dann folgende: Die Filmvermessung ergibt 2φ und ε. $\overline{DP}$ erhält man aus 2φ mit Hilfe der Kurve. ω erhält man aus ε mit Hilfe eines bekannten Proportionalitätsfaktors. $\overline{DP}$ und ω aber sind die Polarkoordinaten der reziproken Gitterpunkte.

Aus obigen Ergebnissen ist auch abzuleiten, warum höhere Ordnungen, sowie Punkt (hkl) und Gegenpunkt $(\bar{h}\bar{k}\bar{l})$ auf den durch Gleichung (5) bestimmten Geraden liegen. Da nämlich diesen Punkten jeweils derselbe Wert ω zukommt (bei Punkt und Gegenpunkt mit entgegengesetzt gleichem ζ), und da ε nur von ω abhängt, so haben die Interferenzflecken auch die Koordinate ε gemeinsam; ihr geometrischer Ort ist eine schräge Gerade wie in Fig. 5.

Geht man zu höheren Schichtlinien über, so wird diese Anordnung im allgemeinen verschwinden (s. die Reproduktionen von Aufnahmen bei A. Gerstäcker, H. Möller, A. Reis: Über den Kristalbau des Pentaerythrit-Tetraazetates und -Tetranitrates[1]). Als Ursache kommt sowohl in Frage, daß Punkte mit gleichem ω in dieser Netzebene des reziproken Gitters nicht mehr gesetzmäßig vorkommen, oder daß ε nicht mehr von ω allein abhängig ist. Während die Entscheidung über die erste Bedingung durch die Symmetrie der als Drehachse benutzten Kristallkante unabänderlich gegeben ist, läßt sich die zweite Bedingung durch geeignete Einstellung des Apparates immer erfüllen. Die Einstellung auf höhere Schicht-

---

[1] *Z. Krist.* **66** (1928) 355.

linien kann sowohl geschehen an der Primärstrahlblende (ausgedrückt durch β) als auch an der Schichtlinienblende (ausgedrückt durch μ). Um eine durch den Abstand ζ bestimmte Netzebene aufzunehmen, ist Gleichung (4) maßgebend, die noch unendlich viele Einstellungen zuläßt. Um aber ε proportional ω und nur von diesem abhängig zu machen, fügen wir die Bedingung hinzu, daß $\beta = \mu$ sein soll, d.h. daß der Primärstrahl und die Interferenzstrahlen der betreffenden Schichtlinie die gleiche Neigung zur Achse des Filmzylinders erhalten sollen. Gleichung (4) ergibt für diesen Fall

$$\sin \beta = \sin \mu = \frac{\zeta}{2}. \qquad (10)$$

Es ist damit erreicht, daß, wie an Fig. 4 zu erkennen ist, $r = \overline{MD}$ wird und so der Durchstoßpunkt der Drehachse auf den Schnittkreis fällt. Infolgedessen haben wir die in Fig. 6 dargestellten Verhältnisse und die oben abgeleitete Gleichung (8) ist gültig. Durchsticht nun die Drehachse das reziproke Netz in einem Gitterpunkt[1], so liegen die übrigen Punkte zentrosymmetrisch zum Durchstoßpunkt und eine Aufnahme wird unter diesen Umständen die Anordnung einer Äquatorschichtlinienaufnahme zeigen. Auch die Auswertung geht dann wie bei diesen vor sich. Auf dem Film werden 2φ und ε gemessen. ε ergibt ω, wie oben. 2φ ergibt zunächst α, aus diesem erhält man $\overline{DP}$ nach Gleichung (11), die an Stelle von (9) tritt, da der Schnittkreisradius jetzt $\cos \mu$ beträgt.

$$DP = 2 \sin \frac{\alpha}{2} . \cos \mu. \qquad (11)$$

Umgibt die reziproke Gitterebene den Durchstoßpunkt nicht zentro-symmetrisch, so bedient man sich, trotzdem keine Zuordnung Punkt—Gegenpunkt vorhanden ist, mit Vorteil der schrägen Koordinaten zur Vermessung von ε. Statt φ mißt man bequemer ξ_w, das ja φ proportional ist. Die Anwendung dieser Größen hat wie bei den symmetrisch angeord-neten Aufnahmen zu geschehen.

Zusammenfassung

Die Anwendung des reziproken Gitters eignet sich in vielen Fällen zur Auswertung von Röntgenaufnahmen, insbesondere von Schichtlinienauf-nahmen, da alle Interferenzflecken einer Schichtlinie solchen Punkten des

---

[1] Diese Bedingung ist identisch mit der Forderung, daß im Translationsgitter die Dreh-richtung [uvw] senkrecht auf der Netzebene (uvw) stehe.

reziproken Gitters entsprechen, die in einer Netzebene des reziproken Gitters liegen. Wird eine Schichtlinie mit dem Weissenbergschen Röntgengoniometer aufgenommen, so genügen die Vermessungsdaten des Films —ohne Kenntnis sonstiger Daten des Kristalls—zur zeichnerischen Darstellung des Punktnetzes im reziproken Gitter, die sich wiederum zur weiteren Auswertung der Filmdaten sehr gut eignet. Oben wurde ein allgemeines, auf graphischer Methode beruhendes Verfahren zur Ermittlung dieses Punktnetzes aus den Koordinaten der Interferenzflecken auf dem Film angegeben. Weiter wurde gezeigt, daß Filmvermessung und weitere Auswertung am vorteilhaftesten an solchen Röntgengoniometeraufnahmen geschehen, bei denen Primärstrahl und Interferenzstrahlen die gleiche Neigung zur Achse des Filmzylinders einnehmen. Für diesen Fall wird eine vereinfachte Ausführung des graphischen Verfahrens beschrieben.

Institut für Physikalische Chemie an der Technischen Hochschule, Karlsruhe

B. LONG CHAIN COMPOUNDS

138. The X-ray Investigation of Fatty Acids, by A. Müller (1923)

■

139. An X-ray Investigation of Saturated Aliphatic Ketones, by B. Saville and G. Shearer (1925)

■

140a. The Investigation of the Properties of thin Films by Means of X-rays, by W. H. Bragg (1925)

■

140b. Sur l'interprétation des spectres X d'acides gras, par M. de Broglie et J. J. Trillat (1925)

■

141. Rayons X et Composées organiques à longe chaine. Recherches spectrographiques sur leurs structures et leurs orientations, by J. J. Trillat (1926)

■

142. Crystallography of the Aliphatic Dicarboxylic Acids, by W. A. Caspari (1928), *with an insertion from*

142a. X-ray Investigation of Long Chain Compounds (n. Hydrocarbons), by A. Müller (1928)

■

143. Some Examples of Information obtainable from the long Spacings of Fatty Acids, by S. H. Piper (1929)

■

144. The Connection between the Zig-Zag Structure of the Hydrocarbon Chain and the Alternation in the Properties of Odd and Even number-ed Chain Compounds, by A. Müller (1929)

■

138. It is well known that ordinary paraffin wax gives an extremely strong powder-reflection of X-rays. The first experiment which showed this was described by Friedrich (*Physik. Zs.* **14** (1913) 317) [this Vol. paper **126**]. He sent a pencil of X-rays through a layer of paraffin wax and obtained on a photographic plate a very well-marked reflection ring round the incident beam. The theory of the phenomenon was not known at the time when the experiment was made. Ordinary paraffin wax is known to contain molecules which possess long CH_2-chains. De Broglie (*Compt. rend.* **176** (1923) 738) and Friedel pointed out recently that these long chains could be detected by means of X-rays and described an experiment in which they measured a large spacing (43.5 Å.U.) in a film of sodium oleate. A series of systematic measurements have been undertaken and have been described by Piper.[1] He found the interesting fact that the length of the spacings increases with an increasing number of CH_2-groups. This confirms the theory which chemists gave long ago.

The author has made independent investigations along the same line. The substances used were a series of fatty acids belonging to one class. Here again it is found that large spacings exist in these substances, and the size of the spacings increases with the number of CH_2-groups present in the acid.

The following was the method of procedure. A small amount of the substance was melted on a glass plate and spread out so as to form a strip 2–3 mm wide, 1 cm long, and 0.1–0.2 mm thick. Those fatty acids which were used were found to crystallise in flakes which oriented themselves parallel to the glass surface. They gave extremely good reflections. This is clearly shown by the good definition of the lines which are near the centre line (see Plate 1). It is obviously necessary in these circumstances, to oscillate the crystal-holder; the crystal flakes, being oriented parallel to the plane surface of the glass, act as a single crystal. The glass plate, on which the substance was deposited, was fixed on the table of a small X-ray spectrometer. Following the method usually adopted, reflections were obtained on each side of the primary beam. This avoids errors which would be involved by using the image of the primary beam as a zero line. The substance was oscillated through 10° on either side of the direct beam.

The photographs, the reproduction of which is given on Plate 1, show a group of well-defined lines lying symmetrically on each side of the central beam. The distances between corresponding lines are different for the different substances investigated. They decrease as the number of CH_2-groups increases, that is, the spacing increases with the number of CH_2-groups. These sharp lines were all found to be the various orders of reflection by the main set of planes, having a spacing which may be called d_1. Besides the sharp lines, there are two rather broad lines outside. These lines, which show on all the plates except that of capric acid, are in the same position

---

[1] The paper is to appear shortly in the *Proceedings of the Physical Society of London.*

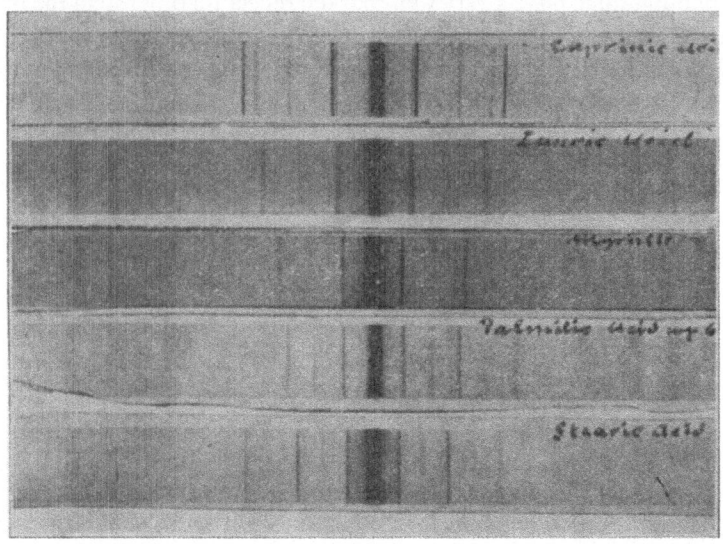

Plate 1.

Table I.

Acid.	M.p.	N.	d_1	d_2	d_3
Capric	—	10	23.2	—	—
Lauric	43·44°	12	27.0	4.11	3.68
Myristic	—	14	32.2	4.12	3.72
Palmitic	62.5	16	34.7	4.08	3.65
Stearic	69–69.5	18	38.7	4.05	3.62
Behenic*	80.5	22	47.8	4.10	3.66

* Not reproduced in Plate I.

for all the substances. The spacings corresponding to these lines may be called d_2 and d_3. The values of the three spacings are given in Table I, where 'M.p.' denotes the melting point, and N the total number of carbon atoms in the chemical formula. The table shows that there is a distinct increase in the spacing d_1 which corresponds to the increase in the number of carbon atoms or CH_2-groups in the substance. The increase of d_1 per CH_2-group in the chemical formula is constant within the limits of error; its value is

$$\delta d_1/\delta N = 2.0 \text{ Å}.$$

The approximate constancy of this increase is shown in Fig. 1.

The existence of the two small spacings, d_2 and d_3, which have the same

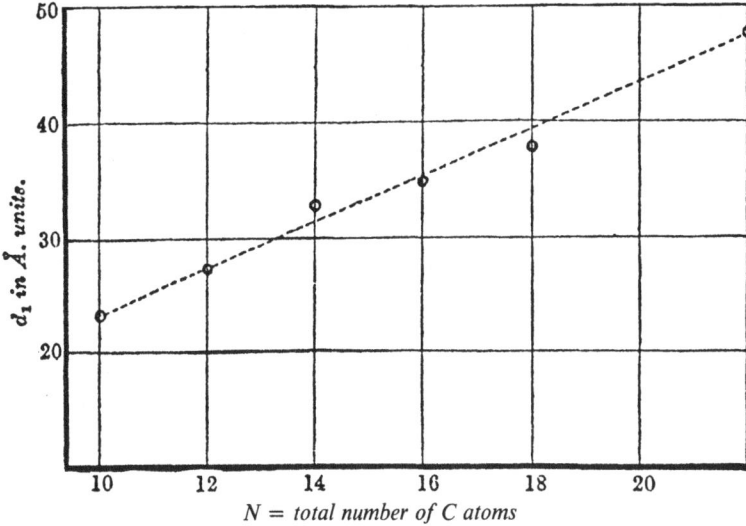

Fig. 1.

value for all the substances investigated, suggests that the unit cell is a long prism which has the same cross-section for all the substances in question. The length of the prism increases proportionally to the number of CH_2-groups in the substance. It is obviously impossible to give the exact structure if the angles between the planes which constitute the cell are not known. If the angles are supposed to be right angles it is possible to calculate the number n of molecules in the unit cell. We have

$$\text{Density} = n \times \frac{\text{Mol. wt.} \times \text{Weight of H-Atom}}{\text{Volume of unit cell}}. \tag{1}$$

For stearic acid, n is found to be very nearly 1.0. But if we calculate the length of the chain which is obtained by piling all the carbons and oxygens one on top of the other in a straight line, we find for that length something like 30 Å.U. The observed spacing is 38.7 Å.U. The difference is far too large to be accounted for by experimental errors. We have therefore to suppose that there is more than one molecule in the unit cell and that the assumed form of the cell is only an approximation. The right length of d_1 could be obtained by putting two or more molecules together in a spiral or zigzag arrangement. How this can be done, and what assumptions are necessary to satisfy equation (1) must be the subject of a more detailed investigation.

It is interesting to note that the first and third orders of the d_1 spacing are very strong on all the photographs, the second and fifth moderately strong, and the fourth in most of the cases very weak.

Adam (*Proc. Roy. Soc.* **101** (1922) 452) has measured the length of these long chains in fatty acids by an entirely different method. He calculated them from the area and the density of unimolecular surface films. His figures are: myristic acid, 21.1; pentadecoic acid, 22.4; stearic acid, 26.2; behenic acid, 31.4 Å.U.

Considering the fact that Adam investigated the substances in a different state, it is obviously impossible that his figures should be in close agreement with the ones given in Table 1. The order of magnitude, however, is the same, and his figures, too, show distinctly the increase of size of the chain with the increase of the number of CH_2-groups in the molecule.

In conclusion, the author wishes to take this opportunity of thanking Professor Sir W. Bragg for his very kind interest in the work. It is well known that the substances used in this investigation are very difficult to prepare in a pure state, and the author gratefully acknowledges the help received from Mr. N. K. Adam, of Sheffield University, from whom most of the specimens were obtained.

139.   *Interpretation of X-ray Results*

In Fig. 1 the results of the investigation of the saturated aliphatic ketones are shown graphically, the spacing being plotted against the total number of carbon atoms in the molecule. It will be observed that the ketones group themselves into two sets represented by the straight lines I and II. In either of these sets the continued addition of CH_2 groups to the molecule results in a uniform rate of growth of the spacing, thus confirming the result,

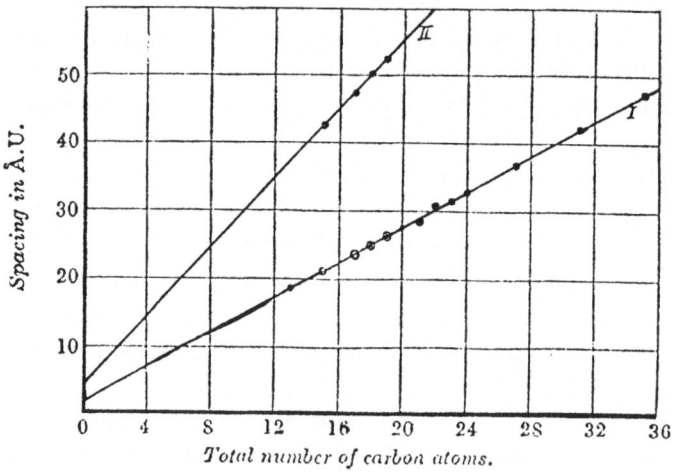

Fig. 1.

previously deduced from the measurements of other series, that the size of this spacing bears a close relation to the length of the molecule. The actual values of the spacings in group I are what we should expect if only a single molecule existed between successive reflecting planes. In group II the spacing suggests a length double that of a single molecule. The condition previously suggested for the existence of a spacing corresponding to the length of a double molecule was that the molecule should possess an active group at its end. The end group of the ketone molecule is CH_3 if the formula be written

$$
\begin{array}{ccccccccc}
\text{H} & \text{H} & \text{H} & & \text{H} & \text{O} & \text{H} & & \text{H} & \text{H} & \text{H} \\
| & | & | & & | & \| & | & & | & | & | \\
\text{H--C--C--C--} & & & \text{--C--C--C--} & & & \text{--C--C--C--H} \\
| & | & | & & | & & | & & | & | & | \\
\text{H} & \text{H} & \text{H} & & \text{H} & & \text{H} & & \text{H} & \text{H} & \text{H}
\end{array}
\qquad (1)
$$

which, as we shall see later, is the most probable arrangement. Such an end group must be regarded as inactive chemically, and therefore, in accordance with our empirical rule, we should expect to find only one molecule between successive planes. Four of the ketones examined give, however, a spacing corresponding to two molecules. These four are not distributed at random, but are all methyl ketones in which there is only one CH_3 group separating the carbonyl group from the end of the molecule. It would appear, therefore, that in these cases the CO group is sufficiently near the end of the molecule for these methyl ketones to be considered as possessing an active end group—$CO.CH_3$. In the ethyl, propyl, etc., ketones, the carbonyl group is too far from the end of the molecule to make its effect felt in the building of the crystal. If this interpretation is correct, it would appear to open up a method for the investigation of the range of action of the oxygen atom.

Further confirmation of this interpretation of the X-ray data as closely representing the lengths either of one or of two molecules comes from Fig. 1. If the spacings of the methyl ketones are halved, the points representing them fall, within the limits of experimental error, on the line representing the other members of the series. They are shown in the figure by circles. Typical photographs of groups I and II are shown in the first two spectra of Fig. 3. The first represents methyl heptadecyl ketone, the second propyl pentadecyl ketone. Both contain 19 carbon atoms, but it is clear that the spacing of the methyl ketone is twice that of the other, as is shown by the fact that the 2nd, 4th, 6th, 8th, etc., orders of the methyl ketone coincide with the 1st, 2nd, 3rd, 4th, etc., orders of propyl pentadecyl ketone. This would suggest that, when the spacing corresponds to the length of two molecules, these are so arranged in the crystal that there is little or no overlapping lengthwise of the two molecules. Further, it will be observed that the intercepts on the vertical axis of the lines I and II are approximately 2 and 4 Å.U. This is just what we should expect from the extra hydrogens on the final carbons; in the case of a single molecule these would contribute about 2 Å.U., in the case of a double molecule 4 Å.U., to the spacing.

It has been pointed out that the position of the carbonyl group does not affect the length of the molecule; it has, however, a very marked influence on the intensity distribution. On certain simple assumptions it is possible to calculate what the intensity distribution should be if the molecule conforms to the proposed structure. The carbon chain may be regarded as a uniform distribution of scattering matter extending over the greater part of the distance between two planes; at the ends where the extra hydrogens occur there will be an almost complete absence of scattering matter, while there will be an excess at that point of the chain where the oxygen atom is situated. Fig. 2 represents diagrammatically such a structure. It is not proposed to enter into a mathematical analysis of the problem, as a full account of the method will be published elsewhere, but the results to be expected may briefly be indicated. If the oxygen occurs at the middle of the chain, as in the di-n-ketones, the odd orders will be strong and the even orders weak;

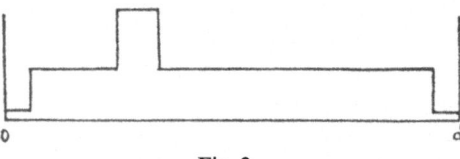

Fig. 2.

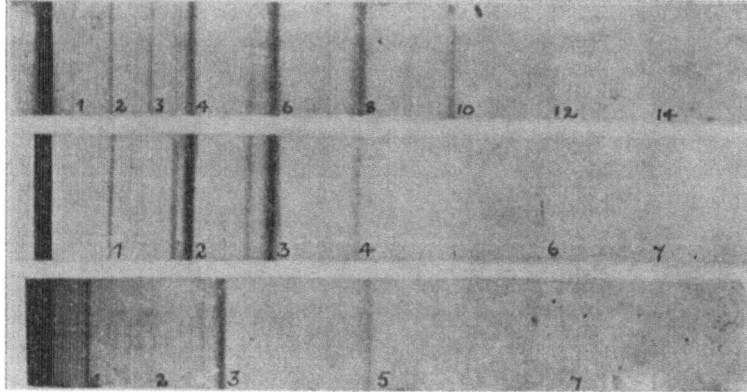

Fig. 3.

if the oxygen is one-third of the way along the molecule, the third, sixth, ninth, etc., orders will tend to disappear; if one-quarter along, the fourth, eighth, etc., orders will be weak, and so on. Half-way between successive minima, positions of maximum intensity will occur. Such predictions are in complete agreement with observation. In Fig. 3, in the third photograph, which represents a di-*n*-ketone, the 1st, 3rd, and 5th orders are well marked, while the even orders are almost non-existent. The second photograph represents propyl pentadecyl ketone, where the oxygen atom is about one-fifth of the way along; the higher orders do not show well in the reproduction, but there is a minimum at the fifth order with a corresponding maximum between the second and third orders. The case of the di-*n*-ketones is of peculiar interest, as the intensity distribution shown by these compounds is found in many other series, e.g., the fatty acids. In all the other cases where such a distribution has so far been observed, the X-ray spacing corresponds not to the length of a single molecule, but to that of two molecules. In a previous paper it was suggested that when two molecules existed lengthwise between two successive planes they were oriented in opposite directions. In the case of these ketones, we have to deal only with a single molecule, but it is clear that a molecule whose formula is $C_nH_{2n+1}.CO.C_nH_{2n+1}$ may be regarded as equivalent from this point of view to two half-molecules, $C_nH_{2n+1}.(\frac{1}{2}CO)$, placed end to end in opposite directions. The fact that we find the same intensity distribution in the di-*n*-ketones and in the fatty acid series appears to be new evidence in favour of the suggestion that the two acid molecules which occur end to end between successive planes have their carboxyl groups oriented in opposite directions. Provided that the oxygen atom is kept at the same fractional distance from the end of the molecule, the intensity distribution is almost independent of the number of carbon atoms in the chain, just as, where the chain is continuous, we find that the intensity distribution remains practically unaltered, although the number of carbons is varied within wide limits.

We see, therefore, that from a study of the X-ray spacing it is possible to determine the number of carbon atoms in the ketones, whilst a consideration of the intensity distribution enables us to locate, within reasonably narrow limits, the position of the carbonyl group.

In conclusion, we have much pleasure in expressing our great indebtedness to Professor Sir W. H. Bragg, F.R.S., for his neverfailing interest in the work and his many helpful suggestions made during its progress.

Davy Faraday Laboratory, Royal Institution

177

140a. The even orders of the fatty acid spectra are very weak compared to the odd orders. Such an effect can be produced in an optical grating by an alternation of white and black lines on a grey ground. An alternation of strong and weak lines gives strength to the even orders: substituting white for one of the blacks is equivalent to changing the sign of its contribution. Gratings can be so made as to illustrate the point. If molecules pointing opposite ways are joined by their carboxyl terminations, then the methyl ends of the molecules are weaker in scattering centres than the general average along the molecule, but the parts where the carboxyl groups join together are above the average in strength.

140b. I. Les spectres X d'acides gras obtenus par la méthode du cristal tournant présentent des particularités qui les différencient complètement des spectres de cristaux ordinaires.

On constate que les ordres impairs sont très intenses par rapport aux ordres pairs, et, de plus, chacune des séries paire et impaire décroît régulièrement en intensité à mesure que l'ordre n de la réflexion augmente. On a là un phénomène qui est exactement l'inverse de celui présenté par le sel gemme dans ses réflexions sur les plans (111). Or, il est absolument impossible d'expliquer une telle répartition des intensités par un raisonnement analogue à celui employé jusqu'ici dans l'analyse cristalline; l'intercalation de plans de diverses densités et le calcul des intensités qui en résulte ne peut jamais donner un tel résultat.

Bragg, dans un article récemment paru, donne une explication assez vague de ce phénomène; nous avons pensé qu'il était intéressant de compléter ces renseignements et de donner une théorie plus générale de ce phénomène.

II. Si le NaCl nous sert de point de comparaison, il s'agit de remplacer les déphasages de $n\pi$ qui existent entre les plans Na et Cl par des déphasages de $(n + 1)\pi$. Un tel système donnera bien l'apparence observée, comme l'on peut s'en assurer aisément.

Pour cela, on considère un milieu matériel contenant des électrons répartis de façon que leur densité en volume soit à peu près constante par rapport à la longueur d'onde employée. On peut supposer ceci dans le cas des acides gras, dont les CH^2 sont, comme on l'a montré, régulièrement espacés entre les plans contenant les extrémités de la chaîne. Un tel milieu homogène ne diffusera que faiblement les rayons X.

Supposons maintenant que ce milieu de densité électronique moyenne Δ soit coupé par une série de plans parallèles équidistants de densité anormalement forte $\Delta + \eta$. Ils agiront comme une série de plans réticulaires et donneront des vibrations réfléchies de forme

$$A_1 \sin(\omega t - \Phi_1) = \Sigma a_1 \sin(\omega t - \varphi_1),$$

en un point A, où la condition de Bragg est réalisée.

Si l'on imagine, au lieu de plans de densité anormalement forte, des plans de densité anormalement faible $\Delta - \eta$, on aura le même phénomène, et, au même point A, on aura des vibrations réflechies

$$A_2 \sin(\omega t - \Phi_2) = \Sigma a_2 \sin(\omega t - \varphi_2).$$

Superposons les deux structures; A sera le siège d'une vibration

$$A_1 \sin(\omega t - \Phi_1) + A_2 \sin(\omega t - \Phi_2)$$

Comme le milieu est devenu homogène et de densité 2Δ, on n'a plus qu'une vibration diffussée négligeable. D'où

$$A_1 \sin(\omega t - \Phi_1) + A_2 \sin(\omega t - \Phi_2) = 0,$$

c'est-à-dire

$$A_1 = A_2 \quad \text{et} \quad \Phi_1 - \Phi_2 = \mp \pi.$$

Donc, dans un milieu homogenè Δ, une série de plans équidistants où la densité est anormalement faible: $\Delta - \eta$, diffusent comme s'ils possédaient une densité anormalement forte: $\Delta + \eta$, mais les phases diffèrent de π.

III. Les acides gras sont constitués par un milieu homogène (chaînes de CH^2) coupé par des plans COOH à grande densité électronique, et des plans à densité faible ou nulle séparant les CH^3 de deux molécules voisines. Ces plans sont répartis alternativement et séparés par la longueur de la chaîne, qui est ainsi la moitié de l'équidistance de Bragg (distance séparant deux plans de même nature). D'après le raisonnement précédent, les plans de densité faible peuvent être considérés comme des plans à forte densité, à condition d'introduire une différence de phase π.

Dans ces conditions, on peut voir facilement que le spectre présente l'apparence inverse de celui de NaCl; tous les ordres pairs sont affaiblis, et tous les ordres impairs renforcés, ce qui correspond à l'expérience.

IV. Cette théorie vérifie donc le fait que deux molécules d'acides gras se soudent bout à bout par leurs CH^3 terminaux. De plus, elle est confirmée par l'examen des spectres des diacides et des carbures saturés; les premiers ne contiennent que des plans anormalement denses (COOH), les seconds que des plans anormalement faibles (compris entre les CH^3 terminaux). Tous deux doivent donc fournir des raies d'ordres successifs à intensités régulièrement décroissantes, ce qui prouve aussi l'expérience.

141.   SOMMAIRE

Les travaux de Müller, Shearer, Piper, Gibbs, etc., ont montré que de nombreux corps organiques à longue chaîne (acides gras, carbures, alcools, cétones, etc.), fondus ou pressés sur du verre, s'orientent de façon à présenter aux rayons X une série de plans réticulaires à grande équidistance. La méthode du cristal tournant permet d'enregistrer photographiquement des spectres qui prouvent que ces équidistances croissent régulièrement avec le nombre des CH^2 de la chaîne et peuvent servir à caractériser ces substances.

J'ai repris ces travaux et les ai poussés dans deux directions différentes:

a) Côté physique: étude des facteurs qui influent sur l'orientation; orientation par le verre, les supports cristallisés ou amorphes, les métaux —étude physique des spectres;

b) Côté chimique: étude de nouvelles séries, étude des savons, étude de mélanges et applications analytiques—étude de réactions chimiques.

Influence de l'épaisseur des couches

Si l'on prend des spectres X avec des couches de plus en plus épaisses, on constate le fait suivant: lorsque la couche est très mince (depuis 1 μ jusqu'à 1/50 mm environ) on n'obtient absolument que les raies correspondant aux grandes distances, c'est-à-dire aux longueurs de chaînes. A mesure que l'épaisseur croît, apparaissent des raies floues, plus éloignées de la tache centrale, et correspondant à de petites distances réticulaires; l'intensité de ces raies augmente tandis que celle des raies d'orientation diminue; ces dernières disparaissent même tout à fait pour une épaisseur de 1 mm environ.

On peut illustrer ceci à l'aide d'une expérience directe: pour cela on prend une lame de verre et on fond une petite quantité d'acide palmitique, qu'on laisse se solidifier sans chercher à l'étendre. On observe qu'il commence par se former une couche très mince et transparente sur laquelle flotte une certaine quantité d'acide fondu. Celui-ci se rétracte progressivement, et finalement se résorbe en une grosse goutelette; le phénomène ressemble à celui signalé par Marcelin pour les couches d'huile et les 'yeux du bouillon'.

La préparation présente en coupe l'aspect ci-dessous (fig. 7): une goutte d'aspect amorphe, fixée sur une couche mince, transparente, biréfringente. Si on prend le spectre, on obtient des raies fines d'ordre 1 et 3, correspondant à l'emplacement des couches minces et montrant que celles-ci sont par-

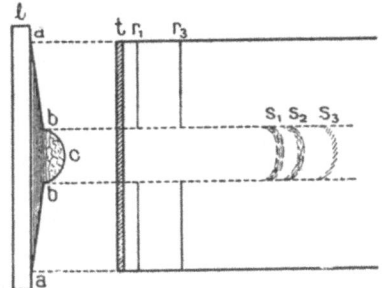

Coupe de la goutte   Spectre correspondant

l—lame de verre.

ab-ab—couche mince bien orientée.

bcb—goutte à structure confuse.

t—tache centrale.

r_1, r_3—raies fines d'ordre 1 et 3 correspondant aux plages *ab-ab*.

s_1, s_2, s_3—anneaux flous correspondant à la région *bcb*.

Fig. 7.

faitement stratifiées. En deçà de ces lignes, et seulement au niveau de la grosse goutte, on observe une série d'anneaux diffus provenant de la diffraction par des microcristaux d'orientation désordonnée; les raies fines ont disparu. Cette expérience montre que les premières assises sont uniquement stratifiées, tandis que les assises supérieures subissent de moins en moins l'action d'orientation due à la lame de verre. Les cristaux ne s'alignent plus, se placent au hasard, d'où apparation des petites distances réticulaires par un processus analogue à celui qui préside à la formation des anneaux par transmission.

On peut donc conclure que l'action d'orientation se localise dans une couche assez mince, pour décroître ensuite progressivement. Il n'est pas possible de fixer une limite précise à la distance jusqu'où s'exerce cette action; on peut tout au plus avoir un ordre de grandeur en étudiant des préparations de plus en plus épaisses, et notant le moment où les raies floues commencent à apparaître, ce qui indique le début d'une structure confuse et non stratifiée. Cette épaisseur est de l'ordre de 1/100 de mm; la stratification se poursuit donc au moins jusqu'à 1/200 mm, ce qui correspond à un empilement d'environ 3.000 molécules ou de 1.500 feuillets.

181

Acides gras saturés

L'étude des acides gras saturés par la méthode spectrographique, commencée par Müller en 1923 [this Vol. paper **138**] a pu être complétée dans ce travail par l'étude d'un certain nombre de termes supérieurs ainsi que de termes inférieurs. Les acides à nombre élevé d'atomes de carbone, ont fourni *les plus grandes distances réticulaires qui aient jamais été mesurées jusqu'à ce jour* (maximum pour l'acide C^{32}: 74.0 A), ce qui présente un intérêt à la fois d'ordre chimique (longueur des molécules) et d'ordre physique (spectrographie des rayons X de grande longueur d'onde).

ACIDES À NOMBRE PAIR ET À NOMBRE IMPAIR D'ATOMES DE CARBONE.—Le premier fait important c'est que la loi de proportionnalité des distances réticulaires au nombre d'atomes de carbone, énoncée par Müller et Shearer, semble ne plus exister, si l'on considère la suite naturelle des acides gras. Mais si on classe ces corps en acides ayant un nombre pair d'atomes de C et en acides ayant un nombre impair d'atomes de carbone, on retrouve alors

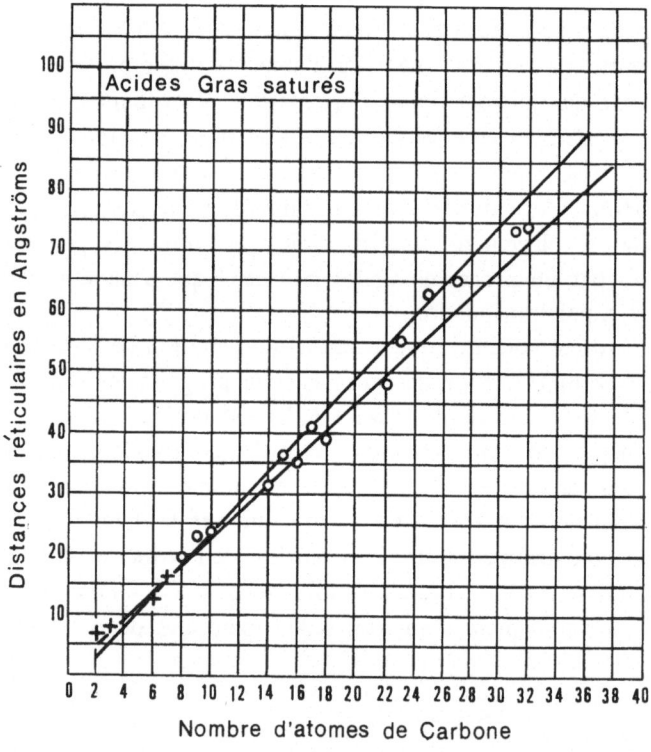

Fig. 12.

que pour chaque série, la distance réticulaire croît régulièrement avec le nombre d'atomes de carbone, *l'accroisement par groupe de deux CH²* *ajoutés étant légèrement différent pour chacune d'elles.*

Si l'on traduit les résultats précédents à l'aide d'une courbe (fig. 12), le phénomène apparaît très nettement: les acides à nombre impair d'atomes de carbone se placent sur une droite située presqu'entièrement au-dessus de celle des acides pairs.

Il est très remarquable que la loi se conserve aussi rigoureusement depuis les premiers termes jusqu'aux termes les plus élevés, malgré les différences de propriétés physiques observées; c'est seulement l'étude complète de la série entière qui a permis de mettre en évidence cette différenciation entre les acides pairs et les acides impairs.

Pour le chimiste organicien, il existe d'ailleurs plus d'une preuve que les acides d'une série homologue manifestent des propriétés différentes, suivant que le nombre de leurs atomes de carbone est pair ou impair: il en est ainsi pour les acides bibasiques de la série oxalique comme pour les acides gras monobasiques.

En effet, les acides pairs sont communs dans la nature, alors que les acides impairs sont en général très rares. De plus on a pu constater la facilité avec laquelle les nombres pairs de la série prennent naissance par oxydation directe de substances appropriées, alors que le rendement des nombres impairs est très faible, excepté dans certaines conditions, soigneusement déterminées, la substance étant oxydée jusqu'au carbone pair immédiatement inférieur. Il paraît donc qu'il y a une plus grande facilité à fixer deux atomes de C qu'un seul à la chaîne d'un acide pair; cela explique la difficulté éprouvée pour préparer certains acides impairs à partir des acides pairs qui l'encadrent (voir acides daturique et margarique).

Enfin—et c'est là un fait frappant pour les acides gras saturés—cette différence suivant la parité se retrouve dans l'alternance des points de fusion. La même régularité se retrouvant dans l'alternance des points de fusion et des distances réticulaires montre combien sont liées profondément ces deux constantes physiques.

On peut donc en résumé énoncer la loi suivante qui complète celle de Müller et Shearer:

Loi: Les distances réticulaires (et donc les longueurs des molécules) des acides gras varient proportionnellement au nombre d'atomes de carbone de leur chaîne, *à condition de grouper ces acides en deux séries, l'une ne contenant que les acides à nombre pair d'atomes de carbone, l'autre que les acides à nombre impair d'atomes de carbone.*

Di-acides pairs et impairs

On constate qu'ici aussi, malgré le petit nombre de mesures, la différence signalée pour les acides gras pairs et impairs, se retrouve également: les diacides à nombre impair d'atomes de carbone ne s'intercalent pas au milieu des diacides à nombre pair (Fig. 14).

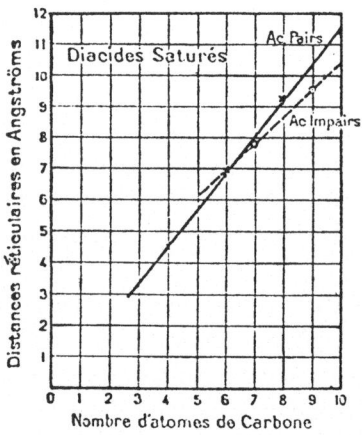

Fig. 14.

La distance réticulaire des premiers est en effet plus voisine de celle de l'acide pair homologue inférieur que de celle de l'acide pair homologue supérieur.

A la différence des acides gras, la droite représentant les acides impairs est plus inclinée que la droite représentant les acides pairs, il faut en conclure, suivant l'hypothèse que nous avons faite, que la molécule des premiers est plus inclinée sur la normale au plan de clivage que la molécule des seconds. En outre il existe une différence fondamentale: la molécule étant symétrique et se terminant des deux côtés par des COOH, il n'y a plus possibilité d'accolement de molécules se tournant le dos; celles-ci se rangeront simplement les unes à la suite des autres. Dans ces conditions, les plans qui nous intéressent seront uniquement chargés en COOH, et la distance *d* mesurée sera celle qui correspond à deux COOH consécutifs, c'est-à-dire à la longueur *d'une seule molécule* au lieu de deux.

Étude et analyse des mélanges

J'ai pu appliquer la méthode spectrographique à l'étude de certains mélanges (acides gras et graisses) et j'ai montré que moyennant certaines précautions il était possible d'obtenir les spectres de chacun des constituants. Dans le cas des triglycérides notamment, des mélanges de trois de ces corps ont pu être aisément décelés, il y a donc là une très intéressante application analytique dans une partie de la Chimie organique où ce genre de recherches est particulièrement délicat et aléatoire.

Application aux phénomènes de lubrification[1]

Les phénomènes d'orientation moléculaire et les applications spectrographiques qui en résultent peuvent fournir des indications nouvelles et assez inattendues sur la lubrification.

Bragg avait en effet émis l'hypothèse que l'origine du graissage devait être recherchée dans la structure feuilletée que certaines substances peuvent acquérir par simple pression; il était intéressant de vérifier cette hypothèse en s'adressant à des graisses et à des mélanges divers employés pratiquement comme agents de lubrification.

Les substances étudiées (suifs, graisses blanches, graisses 'consistantes', graisses à robinets, etc.) présentent un aspect amorphe et possèdent une consistance pâteuse. Une petite quantité de ces produits est placée sur une lame de verre ou de métal, puis recouverte d'une autre lame de verre ou de métal à qui l'on imprime ensuite un mouvement de va-et-vient, en même temps qu'on presse les deux pièces l'une sur l'autre. Dans ces conditions, on constate que la résistance observée au début quand la graisse commence à s'étaler, diminue progressivement à mesure que la couche s'amincit et que la graisse s'étale. On sépare alors par glissement les deux lames qui se trouvent enduites d'une mince couche graisseuse; c'est cette préparation que l'on place sur le barillet du cristal tournant et que l'on radiographie (anticathode fer; 40.000 volts, 8 à 12 milliampères).

Dans la plupart des produits essayés, on constate ainsi l'apparition de spectres d'orientation dont certains sont très intenses. La présence de ces spectres prouve d'une façon certaine que la couche graisseuse s'est stratifiée par suite du frottement des pièces l'une contre l'autre. Si l'on introduit de

---

[1] J.-J. Trillat, *C.R.*, (29 mars 1925) p. 843. Recherches sur les phénomès de lubrification au moyen de la spectrographie par les rayons X.

la graisse dans un moyeu de roue par exemple, il se produit d'abord par la rotation de la roue un laminage qui a pour effet de faire passer la substance à cet état stratifié, par suite de l'orientation des molécules; c'est à ce stade que correspond la résistance qu'offrent au début de leur mouvement des pièces fraichement graissées. Après quoi, la stratification s'opérant, les feuillets formés glissent les uns sur les autres à la manière de cartes à jouer.

Structure des couches de graisse

Les stratifications observées pour les graisses de graissage se produisent sur une épaisseur notable, car pour donner lieu à un phénomène de diffraction visible, il faut au moins 4 à 500 empilements ou feuillets. De plus, il est probable que les principaux constituants du produit donnent leur stratification propre, comme le montrent certains cas où l'on observe deux spectres.

Il paraît certain que l'orientation moléculaire, qui est à la base du phénoméne, s'amorce à la surface du verre ou du métal, absolument comme dans le cas des acides gras fondus sur des métaux; or j'ai montré antérieurement que cette orientation était favorisée par la présence de groupes actifs dans la molécule (acides gras saturés ou non, glycérides)[1]: on peut peut-être expliquer ainsi la raison d'être des mélanges de matières grasses d'origine animale ou d'acides gras[2] aux huiles minérales, les premiers de ces corps étant susceptibles de s'attacher au métal, en s'orientant, et par suite en présentant à l'extérieur une surface garnie de CH^3 sur laquelle les molécules d'hydrocarbures glissent très aisément.

Enfin il y a lieu de signaler que les phénoménes de stratification présentés par les graisses pressées offrent un intérêt analytique: il est possible d'avoir en effet des renseignements sur la composition de la substance, la présence de graissés d'origine animale ou minérale, par le calcul des distances réticulaires et l'examen de la répartition des intensités dans les spectres; on peut également, en utilisant successivement des lames de verre ou de métal, déceler les constituants acides qui se manifestent par l'apparition d'un spectre nouveau sur lame métallique, et étudier enfin l'altération des surfaces graissées par ce procédé de radiométallographie extrèmement sensible.

Travail fait au Laboratoire privé de M. Maurice de Broglie

[1] J.-J. Trillat, *Thèse de Doctorat*, Paris, mars 1926.
[2] P. Woog, *Comptes rendus* **181** (1925) 772.

142. The straight-chain saturated aliphatic dicarboxylic acids, besides being all solid at the ordinary temperature, lend themselves better to crystallographic study than the corresponding monocarboxylic acids, in that their crystals are harder and less liable to distortion. From C_4 onwards they all crystallise in the monoclinic-prismatic class. The length of the unit cells of these crystals, along c, depends on the number of carbon atoms in the molecule, as established by Shearer and Müller for other long chain compounds.

For obtaining crystals of adequate quality, ethyl acetate was found in most cases to be the best solvent. With all the acids, cooling of hot aqueous solutions leads to crystals which are well shaped except for extreme thinness upon (001), whereby they are rendered useless.

The crystals were subjected to X-ray analysis by means of rotation photographs of single crystals about their three axes, upon glass quarter-plates. When required for closer investigation of particular reflexions, oscillation photographs were also taken. The β-angles and cell-dimensions (in Å.U.) of the several crystals are summarised in Table I.

Table I.

Acid.	a	b	c	β	Mols. in cell.
C_6, Adipic	10.27	5.16	10.02	137° 5′	2
C_7, Pimelic	9.93	4.82	22.12	130° 40′	4
C_8, Suberic	10.12	5.06	12.58	135° 0′	2
C_9, Azelaic	9.72	4.83	27.14	129° 30′	4
C_{10}, Sebacic	10.05	4.96	15.02	133° 50′	2
C_{13}, Brassylic	9.63	4.82	37.95	128° 20′	4
C_{18}, Hexadecanedicarboxylic	9.76	4.92	25.10	131° 10′	2

As regards the c dimensions of the several unit cells, it is seen that the crystals fall into two groups, according to the odd or even number of carbon atoms in the molecule. The c axes of the even-carbon series are nearly proportional to the number of carbon atoms, whilst in the odd-carbon series they are proportional to twice that number. In the latter series the heights of the cells are doubled and must be taken up by two molecules end to end; correspondingly the cells contain the substance of four instead of two molecules.

This difference in crystal structure between even-carbon and odd-carbon acids cannot be unconnected with the well-known alternation in their melting points. Ultimately the melting point must depend on the ease with which molecules are torn out of the crystal lattice, and it is plausible

187

to suppose that the double cells of the odd carbon acids (having relatively lower m.p.'s) are less stable in this respect than the more compact cells of the even-carbon acids. A full explanation, however, must be deferred until not only the exact structures of the crystals but also the degree of molecular association in the molten substances are better known.

An alternation in most of the data in Table I is apparent as the homologous series is ascended, and is especially striking with the β-angles, which are always considerably smaller for odd-carbon than for even-carbon acids.

Orders of (001) attained at most only moderate intensity and were represented by both odd and even orders, beginning with (001), in the even-carbon acids, but by even orders only, beginning with (002), in the odd-carbon acids. By a simple application of the structure factor of intensity, this disappearance of the odd orders indicates a division of the unit cell across c into halves of nearly equivalent reflecting power, and is what would be expected from the structure (Fig. 2) [p. 190] of odd-carbon acids. A minor subdivision of the c heights is caused by successive layers of the carbon atoms themselves in the plane of (001). The effect of this, again by operation of the structure factor, is well brought out in the c axis photographs, where remote hyperbolas (layer-lines) show a certain intensity after an interval of weak or unrepresented hyperbolas. Thus, in adipic acid the 3rd and 4th hyperbolas are relatively strong, in suberic the 4th and 5th, in sebacic the 6th, and in the C_{18} acid the 8th and 9th. In pimelic acid the 8th hyperbola emerges strongly, in azelaic the 10th, and in brassylic the 14th.

142a. A systematic discussion of these facts can best be done with the aid of the geometrical structure factor. We shall first discuss the properties of the simplest possible model of a long chain crystal.

This hypothetical crystal is assumed to have only one molecule to the unit cell (axes a, b, c) the molecule itself consisting of 'n' identical atoms or scattering centres of scattering power 1. These atoms are placed at equal distances along one axis of the crystal, say, the 'c' axis. If 's' is the distance between two successive atoms, and c_0 the co-ordinate of the first atom —this last quantity being expressed as a fraction of the 'c' axis—then the co-ordinates of the pth atom are:—

$$x_p = 0; \; y_p = 0; \; z_p = c_0 . c + (p - 1).s.$$

The geometrical structure factor for this single chain is then:—

$$S_n = \left| \sum_{p=1}^{p=n} \exp \left[2\pi i (x_p h/a + y_p k/b + z_p/c) \right] \right|$$

$$= \left| \sum_{p=1}^{p=n} \exp\left[2\pi i\{c_0 + (p-1).s/c\}.l\right] \right| =$$

$$= \left| \exp\left[\pi i\{2c_0 + (n-1).s/c\}.l\right] . \frac{\sin \pi n \, s/c.l}{\sin \pi \, s/c.l} \right|$$

$$= \left| \frac{\sin \pi n.s/c.l}{\sin \pi \, s/c.l} \right| . \tag{1}$$

s/c and 'n' being constant for a given arrangement of the crystal, this S_n is a function of the index 'l'. This function shows sharp maxima provided 'n' is large, as we assume it to be. These maximum values of S_n appear when $s/c.l$ is an integer or in the neighbourhood of an integer; 'l' can only vary in steps of a unit. We get therefore a very marked increase of the numerical value of our function at constant intervals of 'l',

It is interesting to note that for the two reflections on either side of a maximum a small variation of s/c produces a comparatively large change in the ratio of the intensities calculated for these reflections.

In other words, a relatively rough estimate of the intensity ratio is quite sufficient to give an accurate numerical value for our fundamental constant s/c.

142. On scrutinising all the planes represented by X-ray reflexions, it was found that in no case is there a general halving in the (hkl) series; hence the space-lattice is the simple lattice Γ_m. On the other hand, such planes as (100), (101), (300), etc., did not appear, but (200), (205), (402), etc., were frequent; that is, ($h0l$) is halved when h is odd. Lastly, since (020) appears but never (010), we may conclude that the space-group is C_{2h}^5.

If the molecule is asymmetrical, four differently oriented molecules must constitute a unit cell. Such is the case with odd-carbon acids: their molecules therefore must be without symmetry. In the case of even-carbon acids there are only two in the cell, and the molecule must have a centre of symmetry of its own which it contributes to the symmetry of the cell. From the structural formulae, it is easily seen how the one series of acids can have central symmetry and the other none, thus:

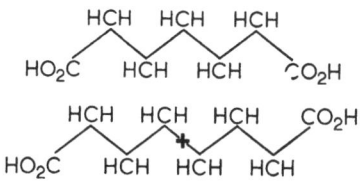

Any straight line drawn through the centre of symmetry at × cuts equivalent atoms or groupings at opposite boundaries of the molecule when the carbon-atom number is even, but not when it is odd. Similar considerations apply, for example, to hydrocarbons; and, indeed, it is found by Müller (*loc. cit.*) that octadecane, $C_{18}H_{38}$, has two molecules to the cell, but $C_{29}H_{69}$ has four. All monocarboxylic acid molecules may be expected to be asymmetrical and to form four-molecule cells.

The lengths of the c axes in Table I would not permit a row of carbon atoms of diameter 1.54 Å.U. to lie with their centres in a straight line; these atoms must therefore lie in zigzag formation along the chain. The unit cells of adipic and pimelic acids, on this hypothesis, are shown diagrammatically in Fig. 2.

The molecules are shown as flat zigzag chains of carbon atoms, the angle of zigzag being not far from the tetrahedral angle. There is as yet no evidence as to the arrangement of the methylenic hydrogen atoms or of the groups of hydrogen and oxygen atoms which cluster at each extremity, and they are not shown in the figure; in the case of even-carbon acids, however, we know that all these atoms must be disposed in harmony with the centro-symmetry of the molecule.

Those molecules which lie half-way across the cell are drawn in dotted lines. Molecules B, in the case of even-carbon acids, are so disposed as to lie symmetrically both across a glide-plane and about a screw-axis

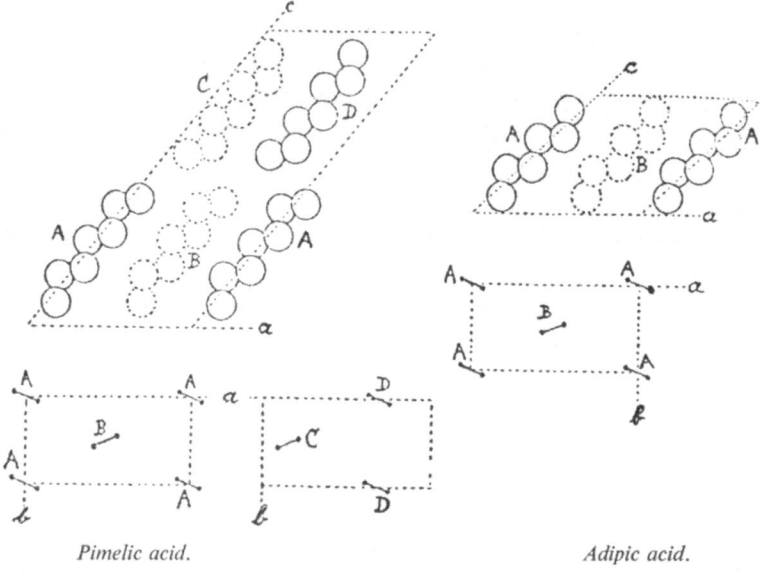

Pimelic acid.   Adipic acid.

Fig. 2.

towards molecules A. In the odd-carbon cell, molecules B are reflexions of A, molecules C are derived by rotation from A and molecules D are reflexions of C. The upper half of the cell is provisionally shown slightly displaced against the lower half in the a direction, on the analogy of the structure of $C_{29}H_{60}$ as worked out by Müller (*Proc. Roy. Soc.* A **120** (1928) 437), but space-group considerations alone afford no information as to this. The planes of the zigzag chains will not lie exactly parallel to (010) or (100), because the central molecule B would then be almost identical with A, in which case we should expect halvings of the (*hkl*) planes in the X-ray spectra. In Fig. 2 the projections of the molecules upon the base are shown inclined at an undetermined angle, which angle, or its symmetrical counterpart, must be the same for all molecules.

The author desires to express his indebtedness to Prof. J. F. Thorpe, F.R.S., and to Mr. N. K. Adam for courteous gifts of specimens, and to Sir William Bragg, F.R.S., for the interest he has taken in this work.

Davy Faraday Laboratory, The Royal Institution

143. Müller showed[1] that the hydrocarbons have two modifications which depend on temperature. It has been shown in this laboratory[2] that even-numbered fatty acids are still more complex, having at least three forms which depend, not only on temperature, but to a certain extent on previous treatment, and de Boer[3] has published similar results for odd acids. Thus in the figure, which is an extension of that given by de Boer, the dots show the spacings of fatty acids prepared by Dr. Malkin. It will be seen that the even acids fall on three lines A, B, C and the odd on four A', B', C', D'. Inspection of this graph shows that a spacing of 40 Å units may be ascribed to an acid of 15, 17, or 18 carbon atoms. Therefore in order to identify the acid it is necessary to determine which particular form it has assumed. This uncertainty can be avoided by photographing the potassium salt instead of the acid[4], but it will be shown below that it is possible to obtain unequivocal results from the acid graph.

Identification of a pure acid is best made by taking a photograph at a few degrees below its melting-point, since it will then definitely show a C spacing if even, and C' if odd. A preliminary photograph of the crystalline

[1] Müller and Saville, *J. Chem. Soc.* **127** (1925) 602.

[2] Piper, Malkin and Austin, *J. Chem. Soc.* (1926) 2310.

[3] de Boer, *Nature* **119** (1927) 634.

[4] Piper, *J. Chem. Soc.*, in the Press.

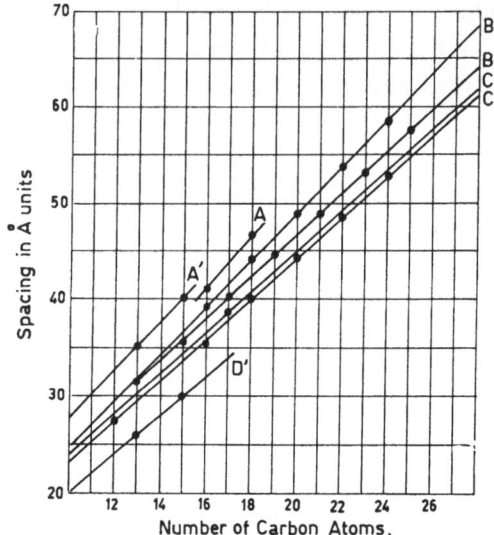

C or C′ spacings occur at high temperatures; A, B, or B′ at low temperatures.

Fig. 1.

powder at room temperature is an additional help as this will usually give a confirmatory spacing of one of the lower temperature modifications. Thus an acid showing a spacing of 40 Å units at room temperature will show near its melting-point a spacing of 34.4 if of 15, 38.8 if of 17, and 40 if of 18 carbon atoms.

144. X-ray investigation[1] of a single crystal of n-$C_{29}H_{60}$ has shown that the CH_2-groups of the chain molecule lie equally spaced on two parallel rows, the lines between successive centres thus forming a zig-zag. Analysis of X-ray photographs of similar substances leaves no doubt that the same chain persists not only through a whole series of hydrocarbons, but is also present in a large number of other carbon chain compounds. The object of this paper is to show how a number of observations on their physical properties are connected with the zig-zag structure of the chain molecule.

The following Fig. 1 represents a section through a crystal. X-rays show that the chains lie side by side, the parallel lines in the drawing being the chain axes. The ends of the molecules, marked by small circles, arrange themselves in equidistant parallel layers. In the following deductions we consider molecules in which the two end groups are chemically identical, such as hydrocarbons, dibasic acids, etc.

[1] *Roy. Soc. Proc.* A **120** (1928) 437.

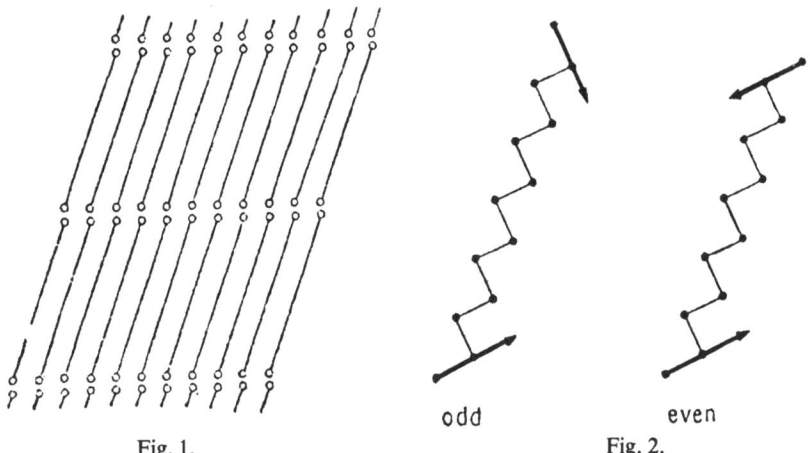

Fig. 1.   odd   even

Fig. 2.

It is impossible to see from this simple model why the odd and even numbered substances should have alternating physical properties. The difficulty, however, disappears when the zig-zag chain structure is introduced into the model.

The following Fig. 2 shows the two types of an odd and an even numbered chain molecule. The CH_2-groups are marked by black circles. The lines connecting the two last groups in each drawing (marked by arrows) are parallel in the even, and not parallel in the odd numbered chain. This is an important point to remember.

How the molecules link together at their ends is the question which next arises. X-ray investigation of the hydrocarbon $C_{29}H_{60}$ has shown that the two nearest molecules join so as to have a centre of symmetry. It is assumed here that this type of linkage exists in all chain compounds both odd and even. This must be correct provided the crystals have chains of such length that there is no appreciable interaction between the end groups of each molecule, and provided moreover the chain axes are at right angles to the end layers. For short chains and inclined axes we no longer expect to find a strictly centrosymmetrical linkage, but at least something approaching it.

It is interesting to see that this centrosymmetrical linkage is to be expected if the end groups are electric—or magnetic dipoles. If two dipoles join as shown in Fig. 3, the arrangement has centrosymmetry.

Fig. 4 shows the arrangement of even and odd numbered molecules according to the principles given above. The horizontal lines indicate the positions of the successive end layers.

The rings at the ends of each chain represent the end groups. The drawing shows that there is an essential difference between the odd and even numbered arrangements. The first pattern repeats itself every second molecule

193

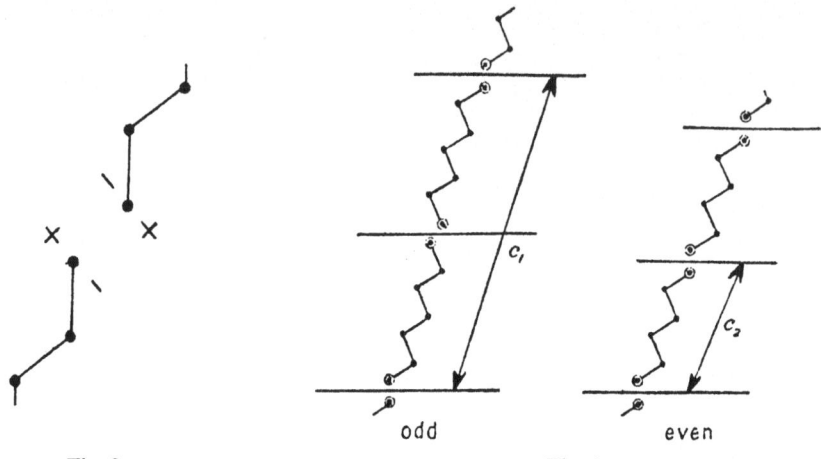

Fig. 3.   Fig. 4.

as we move in the direction 'c'. In the second pattern all the successive molecules are identically situated. The period of repetition is marked by the long arrows c_1 and c_2. In the actual crystal one expects therefore to find two molecules to lie along the 'c' axis in those crystals which are built of odd numbered chains, and only one molecule in crystals containing even numbered molecules.

This conclusion agrees with the results of a recent investigation on the dibasic acid series by Dr. Caspari[1].

Certain features of the crystal habit of these long chain compounds will next be discussed with the aid of the model. The following Fig. 5 shows the arrangement of the molecules in the two types of crystals.

The chain axes are here represented by fine straight lines, the arrows at the ends of the chains are end groups. The diagram shows again the essential difference between the odd and even substances. In the even series all the end layers are identically situated, i.e., they can be brought into coincidence by a parallel shift in the direction B_2B_3. This no longer holds for the odd number series. Here there are two types of layers. The first layer I in the diagram is identical with III; V ... and the second layer II is identical with IV; VI ... ; I and II cannot be made to coincide by a parallel shift. Therefore all the odd members of a series and all the even members of a series are similar amongst themselves. Between the odd and the even members, however, there exists a distinct structural difference which has its origin in the peculiar structure of the chain. This difference must be responsible for the alternations of the properties between the odd and the even numbered substances.

[1] *J. Chem. Soc.* (1928) 3235 [this Vol. paper **142**].

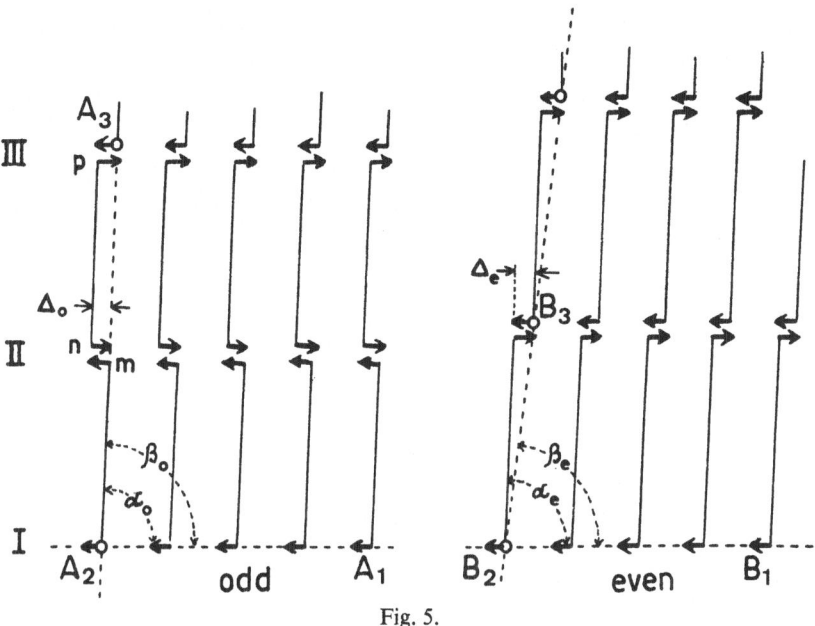

Fig. 5.

An explanation of the behaviour of the β-angles can now be suggested.

In Fig. 5 the angles β_o and β_e are quantities which can actually be observed, and measured either directly with a crystalgoniometer, or indirectly by using X-ray data. Supposing we deal first with substances whose chains are so long that the mutual influence of the end groups at each end of the molecules is negligible, and whose β-angles are nearly $90°$. The angles α_o and α_e between the chain axes and the basal planes A_1A_2, B_1B_2 are then no longer dependent upon the chain length, and according to the diagram we expect to find that the β_o angle in the odd number series to be constant, i.e., independent of the chain length. This does, however, not hold for the even series, and the reason for this can be seen in the diagram. The chain axes are separated by an amount $\varDelta_o$ or $\varDelta_e$ as shown in the drawing. Supposing we start from A_2 and follow the line A_2mnpA_3. A_2 and A_3 are identical points in the lattice, and the angle β_o measures therefore directly the inclination of the chain axes relative to the basal plane, i.e., it is identical with α_o. In the second drawing which refers to the even numbers β_e is no longer identical with α_e and depends both upon the chain length and the quantity $\varDelta_e$. The β-angles must therefore behave differently in the odd and the even series. As the chains become shorter we expect to find a gradual change in both α_o and α_e to take place. Superposed on this will be the effect which has just been discussed. We expect that the β-angles in both the odd and the even series will undergo a gradual change as we

195

ascend the series but they will alternate, i.e., they must lie on two separate smooth curves.

This conclusion again describes exactly the observations in the dicarboxylic acid series (*loc. cit.*).

It has been known for a long time that the melting points of chain compounds alternate as shown in the following Fig. 6. It is now quite easy to give a reason for this.

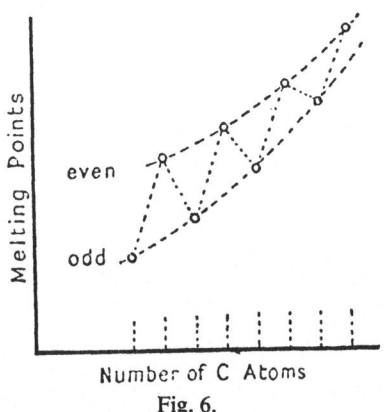

Fig. 6.

If the structures of the odd and the even substances differ from each other as has been shown in this paper, then their corresponding lattice energies will also be different, and their melting points will behave as indicated in the above diagram. Similar things will happen for other physical constants. The molecular volumes and the heat of crystallisation alternate in the fatty acid series. This has been shown by W. E. Garner[1] in his very interesting work on the properties of chain compounds.

A numerical calculation of the lattice energies is at present not possible. Too little is known about the forces which hold these substances together. A more detailed experimental study will perhaps help to see deeper into the nature of the forces. An account of some new observations will appear shortly.

In conclusion, the writer wishes to thank Sir W. Bragg for taking a friendly and encouraging interest in this work.

---

[1] *J. Chem. Soc.* **125** (1923) 881 and **127** (1925) 720.

C. FIBRE STRUCTURES

145. Was sind die Ursachen der eigentümlichen Dehnbarkeit des Kautschuks? I: Über die Änderung der Röntgenspektrums des Kautchuks bei der Dehnung, by J. R. KATZ (1925)

▰

146a. X-Ray Studies of the Structure of Hair, Wool and Related Fibres, I—General, by W. T. ASTBURY and A. STREETS (1932)

▰

146b. X-Ray Studies of the Structure of Hair, Wool and Related Fibres. II—The Molecular Structure and Elastic Properties of Hair Keratin, by W. T. ASTBURY and H. J. WOODS (1934)

▰

145.   1. *Problemstellung*; *Röntgenspektrogramme des ungedehnten Kautschuks*

Die Eigenartigkeit der Dehnbarkeit des Kautschuks harrt noch immer ihrer Erklärung.

Ich habe nun versucht, mit der Röntgenspektrographie diesem alten Problem näher zu kommen.

Leider gibt Kautschuk—wie 1920 zuerst P. Scherrer[1] festgestellt hat—kein Linienröntgenogramm (wie kristalline Substanzen es tun), sondern bloß einen sog. 'amorphen Ring', so wie Flüssigkeiten und glasartige Substanzen ihn aufweisen. Die mittlere Identitätsperiode des amorphen Ringes beträgt ±4,5 Å.-E., ist also auffällig klein, wenn man bedenkt, daß beim Kautschuk sicher ein hochmolekularer Körper vorliegt, während die Identitätsperiode dennoch von der gleichen Größenordnung ist wie bei den einfachst gebauten Flüssigkeiten. Was die Identitätsperiode solcher 'amorphen Ringe' bedeutet, weiß man eigentlich noch gar nicht; einzelne Forscher sehen darin bekanntlich den mittleren Abstand der Schwerpunkte der benachbarten Moleküle oder Atome, andere eine intramolekulare Identitätsperiode. Für beide Auffassungen gibt es Argumente, aber zurzeit kann man noch nicht sicher entscheiden.

Diesem Umstand—daß der Kautschuk bloß einen amorphen Ring gibt, und daß man dessen Bedeutung noch nicht versteht—mag es wohl vor allem zu verdanken sein, daß andere Forscher nicht schon vor mir ver-

[1] R. Zsigmondy, Lehrbuch der Kolloidchemie. 3. Aufl. (Leipzig 1920).

sucht haben, diesem Problem mit röntgenspektrographischen Untersuchungen näher zu kommen.

2. *Röntgenspektrogramme des gedehnten Kautschuks*

Fäden aus dem vorhin genannten, ganz amorphen *Kautschuk, drei- bis sechs-
fach gedehnt, zeigen neben dem noch sichtbaren amorphen Ring ein Linien-
spektrogramm,* wie es ein aus vielen Kriställchen bestehender Körper
aufweist, dessen Einzelkristalle alle mit einer Achse der Dehnungsrichtung
parallel liegen. Also: neben dem 'amorphen Ring' ist ein 'Faserdiagramm'
aufgetreten. Die Interferenzen liegen sehr schön auf hyperbelartigen Kurven
(sog. Schichtlinien), wie es bei einem Faserdiagramm (auf einer Platte aufgenommen) sein sollte.

Diese überraschende Tatsache—wofür es soweit mir bekannt keine
einzige Parallele gibt—wurde dann ausführlich weiter verfolgt. Beim nicht
vulkanisierten Hevea-Kautschuk waren die zuerst sichtbaren Streifen des
Faserdiagramms (welche übereinstimmen mit den zweiten Aequatorial-
interferenzen—alles bei reiner K_α-Strahlung des Kupfers) etwas außerhalb
des amorphen Ringes eben sichtbar bei einer Dehnung von etwa 100
Proz.; sie waren dann gleich intensiv als dieser Ring. Je stärker man jetzt
dehnt, um so intensiver werden die neuen Streifen, verglichen mit der
Intensität des amorphen Ringes, bis bei sehr starken Dehnungen (und
üblichen Belichtungen) dieser Ring noch eben schön sichtbar ist, die Kristallinterferenzen aber sehr stark sind. [Subjoined illustrations taken from

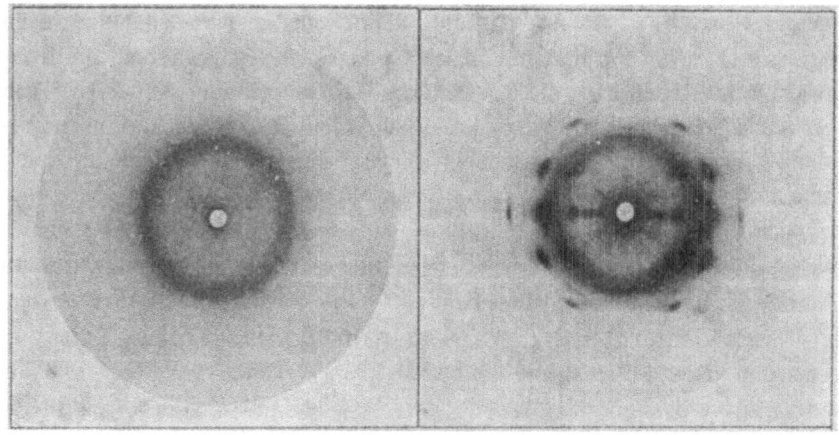

X-ray diagrams of india-rubber, unstretched and stretched, resp.

W. T. Astbury, The Structure of Fibres, *Ann. Rep.* **28** (1931) p. 322. Ed.]
Bei Aufnahmen auf einer Platte sind, auch bei langen Belichtungen, nirgends
Verschmierungen des Faserdiagramms zu Debye-Scherrer-Kreisen sichtbar.
Es liegen also die Kristallite, welche das Liniendiagramm sichtbar machen,
vom Anfang der Dehnung an (100 Proz. Dehnung) einander weitgehend
parallel. Die Lage der einzelnen Interferenzen des 'Faserdiagramms' und der
Diameter des 'amorphen Ringes' scheinen sich—soweit die Genauigkeit
meiner jetzigen Messungen reicht—bei der Dehnung nicht oder nicht viel
zu ändern; doch sind Präzisionsmessungen darüber noch im Gange.

In einer zylindrischen Kamera untersucht—gedehnter Faden in der Achse
des Zylinders—zeigt es sich, daß die Interferenzen nicht über die ganzen
360° verteilt sind wie man das erwarten sollte, sondern schon bei einem
kleinen Ablenkungswinkel ganz aufhören (15–40°).

Warum verschwinden hier alle peripheren Interferenzen? Liegt das an
einer teilweisen Unordnung des Gitters? Das scheint die wahrscheinlichste
Annahme. Und wird der um so kleiner, je stärker die Dehnung? Man
denkt dabei unwillkürlich an den sog. Debye-Faktor der Wärmebewegung
der Moleküle im Kristallgitter.

3. *Die 'kristallinischen' Teilchen des gedehnten Kautschuks*

Aus dem Faserdiagramm läßt sich leicht die Identitätsperiode des Gitters
der Kristalle ablesen in der Richtung der Achse[1], also hier in der Dehnungsrichtung. Sie beträgt ± 8 Å.-E. Die Berechnung der Identitätsperioden in
den beiden anderen Richtungen ist viel heikler, wenn keine größeren, gut
ausgebildeten Kristalle vorliegen. Da aber die innersten Aequatorialinter-
ferenzen einem Gitterebenenabstand von etwa $6\frac{1}{2}$ Å.-E. entsprechen,
scheint auch bei dieser hochmolekularen Substanz wiederum ein kleines
Gitter vorzuliegen, und nicht ein großes, wie man erwarten sollte.

*Was kann es bedeuten, daß wir eine so kleine Elementarzelle finden bei
einer Substanz, die so sicher hochmolekular ist wie Kautschuk?* Bekanntlich
haben R. O. Herzog[2] und seine Mitarbeiter M. Polanyi[3] und R. Brill[4]
wiederholt die Theorie verteidigt, daß die Elementarzelle nicht kleiner sei
als das Molekül der Substanz, und daß man daher aus demselben einen
oberen Wert ableiten könne für die Größe des Molekulargewichts. Die

[1] Siehe M. Polanyi und K. Weißenberg, *Z. Physik* **10** (1922) 44.
[2] Zum letzten Mal noch in seinem Referat auf der Faserstoffsitzung der Naturforscherversammlung in Innsbruck.
[3] M. Polanyi, *Naturwiss.* **9** (1921) 288.
[4] R. Brill, Diss. (Berlin 1923).

Anwendung dieses Prinzips führte zu dem paradoxalen Ergebnis, daß eben diejenigen Substanzen, die man bis jetzt allgemein als die typischen hochmolekularen angesehen hatte, wie Polysaccharide (Zellulose), Eiweißkörper (Seidenfibroin), es nicht sein sollten! Beide Substanzen sollten höchstens ein Molekulargewicht von 600 haben! Diese Auffassung stieß auf lebhaften Widerstand bei den präparativen Chemikern, die—von ihrem Standpunkt mit vollem Recht—auf ein viel höheres Molekulargewicht schlossen. Gegen diese These—daß die röntgenspektrographische Elementarzelle einen oberen Wert ergeben sollten für die Größe des Molekulargewichts—wurde dann von J. R. Katz Protest eingelegt in der Diskussion zu dem Vortrag von R. O. Herzog auf der Innsbrucker Naturforscherversammlung.

Er führte an, daß eben die quellbaren, hochmolekularen Körper einen eigentümlichen Molekularbau zu besitzen scheinen; ihre Moleküle könnten aus Grundkörpern von niedrigem Molekulargewicht aufgebaut sein, die durch Nebenvalenzen oder anderswie zusammenhängen[1]. Und *die röntgenspektrographische Elementarzelle messe*—wenn diese Elementarbausteine regelmäßig in drei Dimensionen geordnet liegen—*bei solchen Körpern möglicherweise nur den Elementarbaustein des Moleküls—den Grundkörper—nicht aber das ganze, viel größere Molekül.* Jedenfalls läßt sich feststellen, daß die Grundbausteine des Kautschuks—mögen es eigentliche Moleküle oder Grundbausteine desselben sein—sich unter dem Einfluß des Zuges ordnen, und zwar so, daß viele kleine Körperchen vorliegen, in denen die Elementarbausteine nach drei Richtungen regelmäßig geordnet liegen, und die alle einander und der Dehnungsrichtung parallel liegen. Besonders ist zu betonen, daß die Elementarzelle sich in drei Richtungen regelmäßig wiederholt. Entstehen die 'Kristalle' tatsächlich erst durch die Dehnung? Dann würde sich daraus eine neue Auffassung der Dehnung des Kautschuks ergeben, welche die Kautschukforschung in ganz neue Bahnen lenken würde. Sie gibt z.B. eine ungezwungene Erklärung vom Joule-Effekt des Kautschuks; d.h. von der Tatsache, daß Kautschuk bei der Dehnung sich erwärmt (statt sich abzukühlen wie andere Substanzen). Denn jede amorphe Substanz, die kristallisiert, entwickelt bekanntlich Wärme. Eben diesen Joule-Effekt konnte man bis jetzt so schwierig erklären.

Diese Untersuchungen wurden ausgeführt im Staatsseruminstitut zu Kopenhagen (Dir. Dr. Th. Madsen). Prof. N. Bohr und Dr. H. A. Kramers danke ich für ihr dauerndes Interesse bei der Ausführung dieser Versuche.

[1] Wie schon vorher H. Pringsheim für Polysaccharide, E. Abderhalden für Eiweißkörper angenommen hatte.

146a. *Introduction*

It is now some ten years since it was first realised that, in common with natural and artificial cellulose fibres, animal fibres with a protein basis are in many cases sufficiently crystalline to yield a pronounced interference figure when examined with monochromatic X-rays. Such 'X-ray fibre diagrams' were reported in 1921 by Herzog and Jancke[1] for muscle, nerve, sinew, and hair, and in 1924 similar photographs from human hair were obtained by one of the present writers[2].

The photographs show that without doubt a considerable proportion of the structure of hair[3] is crystalline or pseudo-crystalline, and that this constituent is common to all the fibres examined, in the sense that substantially the same X-ray photograph is always obtained, from the finest Merino wool to such large-scale structures as quills. In fact, the interference figures given by human hair and the tip-end of a procupine quill, respectively, are practically indistinguishable. Presumably, we are dealing with a photograph of crystalline keratin, or of one of the forms of keratin, if indeed there is more than one fundamental keratin. It is immediately noteworthy that the dimensions of the photograph are not such as would be expected from a substance of very high molecular weight, of the order usually associated with proteins; rather are we reminded of the case of cellulose, in which it seems clear that a comparatively simple unit may be repeated, 'via' primary valency bonds, an indefinite number of times.

The second important result of the present investigation has been to show that the X-ray photograph of unstretched hair is quite different from that of stretched hair. On stretching the hair, the α-photograph, as we shall call it, fades away and is gradually replaced by the β-photograph, the interferences of which first become prominent at about 30 per cent extension. The α-photograph is finally lost sight of at about 60 per cent extension, at which point the β-photograph is almost as well defined as it is possible to obtain it by this method, since very soon after, in the neighbourhood of 70 per cent extension, the fibres break[4].

---

[1] 'Festschrift der Kaiser Wilhelm-Gesellschaft,' 1921.
[2] W. T. A. For a series of lectures on 'The Imperfect Crystallisation of Common Things,' delivered by Sir William Bragg at the Royal Institution.
[3] We shall use the word 'hair' in its most general sense, and when necessary state the kind of hair being referred to.
[4] J. B. Speakman, *J. Text. Inst.* **17** (1926) 457.

The Low Tension Photograph

It will be shown how the action of steam on the β-phase brings about a change which prevents its return to the α-phase and gives rise to the pheno-menon known in the textile trade as 'permanent set'.

We wish to suggest here that *the basis of the unstretched fibrous keratins is a series of hexagonal ring systems linked along the fibre axis by 'bridge atoms'*, in a manner analogous to what is generally accepted for the cellulose and related structures. In point of fact, there is a considerable body of evidence in favour of such a hypothesis.

In this connection it may be mentioned that the ordinary peptide chain

may be considered to be built up of sections each of length equal to about $3\frac{1}{2}$ Å.U., which is a little larger than the spacing, 3.32 Å.U., found in β-hair. In view of the fact that a spacing does not necessarily equal a chief intra-molecular distance, it is possible that the spacing, 3.32 Å.U., may be evidence of the existence of peptide chain in β-hair.

The Contents of the Unit Cell. From the evidence as yet available, we cannot fix the dimensions of the unit cell of either α- or β-hair with any degree of finality. We may only suggest those of a cell which represents to a first degree of approximation the repeating characteristics of the structure without actually being the strict unit of pattern. Such a cell, in the low-tension form of hair, appears to be defined by a length of 5.15 Å.U. along the fibre axis, and two lengths of 27 Å.U. and 9.8 Å.U. at right angles to the fibre axis and to each other. There is not the least doubt that the true minimum unit of pattern is greater than these magnitudes imply; still, the smaller unit is repeated approximately within the larger and must be the expression of some chemical grouping which is the structur-al basis of the completed protein molecule. Similarly, in the β-form, the X-ray diffraction figure focuses our attention on a pseudo-unit defined by a translation 6.64 Å.U. along the fibre axis, and translations 9.3 Å.U. and 9.8 Å.U. at right angles to the fibre axis and to each other.

202

If we are justified in attaching any significance to these dimensions, then it ought to be possible to link them up with the reversible transformation which exists between the α- and β-forms.

{Note added:

We have been able to show that by suitable treatment with steam the load/extension curve of hair may be permanently smoothed out, and that elasticity of form may be demonstrated *in cold water* over a range of extensions from about −30 per cent to about +100 per cent. The whole fibre then behaves as an elastic chain whose length may be almost exactly doubled without rupture occurring. Without entering into detailed discussion, we may say that this observation suggests strongly that the full α − β transformation of hair is accompanied by an elongation of the keratin complex of approximately 100 per cent, and that some relation on this basis must exist between the respective dimensions and features of the α and β X-ray photographs. If now we assume that the two strong meridian reflections, 5.15 Å.U. and 3.32 Å.U., that characterise the reversible transformation, are linked by the relation

$$5.15 \rightleftarrows (3 \times 3.32)$$

we have at once a quantitative explanation of the maximum extension and also *the clue to the nature of the keratin chain*. According to this scheme the transformation may be represented as follows:

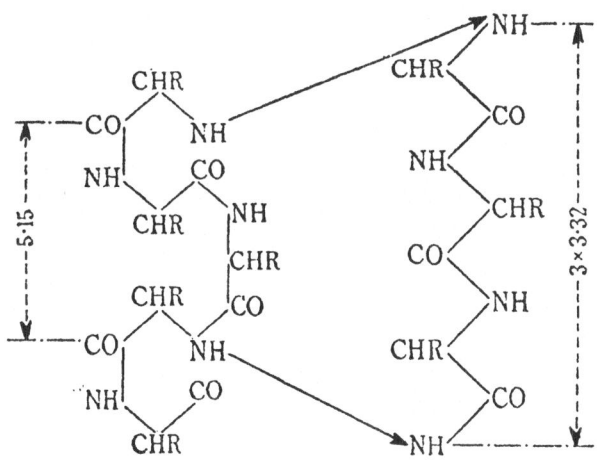

W.T.A

Textile Physics Research Laboratory, The University, Leeds.

146b. In a previous communication[1] an account was given of a preliminary exploration, chiefly by X-ray methods, of the problem of the molecular structure of animal hairs. The present paper is a natural continuation of the record, in which earlier tentative suggestions are either confirmed or rejected, and an attempt is made to lay bare the general structural principles underlying the properties of the protein, *keratin*. It will be unnecessary here to outline once more the historical development of the subject; we shall proceed at once to the main point of this introductory section, which is to give what appears to be the solution of the problem before setting out in detail the experimental facts and arguments leading up to it. Such a procedure is advisable because of the complex nature of the properties under discussion; such a long series of experiments have been involved in their elucidation, that without some sort of preliminary statement of the chief conclusions, the issue is apt to grow confused.

Briefly, the whole argument rests on the discovery[2] that the X-ray 'fibre photograph' which appears to be common to all mammalian hairs, human hair, wool, whalebone, nails, horn, porcupine quills, etc., and which is undoubtedly the diffraction pattern of crystalline, or pseudo-crystalline, keratin, the common fibre substance of all these biological growths, is changed into a quite different fibre photograph when the hair is stretched. The change is a reversible one, recalling that previously discovered by Katz[3] in rubber, because when the hair is returned to its initial unstretched length, the normal keratin photograph reappears. It is clear that the X-ray effects give a diffraction record of a reversible transformation involving not merely an internal slipping of the fibre substance or a rotation of 'micelles' into stricter alignment.

The β-form is thus represented by fully-extended peptide chains in which each amino-acid residue takes up, on the average, a length along the fibre-axis of 3.4 A, while the α-form is represented by a series of pseudo-diketopiperazine rings which follow each other according to a pattern of length 5.1 A. The unfolding of the rings is clearly accompanied by an elongation of 100%, and the suggested pattern offers an explanation of both the characteristic meridian reflection of the α-form (5.1 A) and of the decrease of resistance of the β-form, as compared with the α-form, to the action of reagents such as steam, etc.

Only a part of the elastic properties of hair are to be interpreted by the application of this principle of intramolecular unfolding; many of its

[1] Astbury and Street, *Phil. Trans.* A **230** (1931) 75 [this Vol. preceding paper]; referred to later as I. Cf. also Astbury and Woods, *Nature* **126** (1930) 913.
[2] Astbury, *J. Soc. Chem. Ind.* **49** (1930) 441; *J. Text. Sci.* **4** (1931) 1.
[3] *Chem. Z.* **49** (1925) 353; *Naturwiss.* **13** (1925) 411.

most striking characteristics are to be referred to the nature and distribution of the side-chains denoted above by the general symbol R. Though there does not appear to be any sharp discontinuity in the physical and chemical properties of the keratin complex as a whole, we have to recognize that both the form and the limits of the load/extension curve may be varied over a wide range simply as a result of the changes which take place in the configuration of the side-chains only. As the most convenient example of such side-chain disturbances it will serve for the present to quote the preferential attack of steam which is to be noticed in β-photographs for extensions of 50% and upwards, and which is undoubtedly the cause of the increased capacity for extension. Other changes which are less clear from the X-ray point of view, but which nevertheless are very obvious when examined by more familiar physico-chemical methods, are the freeing of certain side-chain restrictions so as to give rise to the phenomenon of super-contraction, and the 'permanent set' of the β-form on prolonged steaming of the fibre in the stretched state. This latter transformation evidently involves the building-up of new side-chain linkages which fix the β-form in the stretched state and preclude once and for all the possibility of ever regaining the normal α-photograph.

The X-ray photograph of β-keratin, I, is most conveniently referred to an orthogonal cell of dimensions, $a = 9.3$ A., $b = 6.7$–6.8 A., and $c = 9.8$ A., of which b is the most prominent period along the molecular chains, while a and c are 'side-spacings'. With regard to the latter two points emerge, (i) that the equatorial 'spot' nearer the centre which gives the c-spacing is preserved more or less unchanged when the α-photograph is transformed to the β-photograph, and (ii) that the transformation calls into existence on the equator a very strong spot of spacing $a/2$, i.e., 4.65 A. From a study of existing X-ray data on proteins the interpretation of these results seems clear, that, in fact, the spacing 9.8 A. common to both α- and β-photographs arises from the lateral extension of the side-chains (the R-groups of the general formula given above), while the spacing 4.65 A [1] represents the distance of approach of the main-chains on those sides free from side-chains. The controlling factor in this closest approach of neighbouring 'backbones' is most probably attraction between ($=NH$) and ($=CO$) groups, whereby the chains are grouped in pairs, thus:

---

[1] In I attention was drawn to the fact that this spacing is practically equal to the chief spacing in the X-ray photograph of cystine, the most abundant amino-acid in hair; but in the light of subsequent evidence, we wish now to withdraw the suggestion that the two spacings have anything more than a numerical relationship.

Such an arrangement accounts readily for the fact that the *a*-dimension of the simplest orthogonal cell given above is not 4.65 A., but 9.3 A., represented on the equator by an intense second order (200).

The strongest evidence that the equatorial spacing, 9.8 A. (the reflection (001)), must be associated with the lateral extension of the side-chains, comes from an X-ray study of water adsorption and the action of steam. 'Quadrant photographs' of porcupine quill, both α and β, brought first to 0% R.H. by prolonged drying over phosphorous pentoxide and then to 100% R.H., show that though the bulk of the water adsorbed by animal hairs leaves the X-ray photograph unchanged, I, some of it does actually penetrate the crystallites in such a way as to increase the spacing, 9.8 A., by a few per cent. The action of steam, however, is even more striking. An X-ray photograph of human hair stretched in steam to twice its original length, shows a marked 'spreading' of certain spots along the hyperbolae ('smear lines'). The only reflections in the photograph of β-keratin which are unaffected by the action of steam belong all to the zone [001], from which it follows that the spacing disturbance is confined to the zone-axis, of this zone, i.e., to the direction of the spacing, 9.8 A., which we have associated with the lateral extension of the side-chains.

Textile Physics Research Laboratory, The University, Leeds

CHAPTER IX

Defect Structures

Mixed Crystals, Random Structures, Rotating Groups, Alloys

Only slightly more than two decades have passed since the first atomic arrangement determination was made by X-ray crystal structure analysis. The first revolutionary effect of this first determination was to show that the chemists picture of chemical molecules has no significance in crystals of the NaCl type. The analysis of other atomic arrangements in crystals rapidly followed. These earlier analyses were made with a working hypothesis known as the 'structure theory', which all proposed structures had to strictly fulfill. The 'structure theory' required that each chemically different sort of atom had to be assigned to a specific set of equivalent positions which it alone occupied and completely filled[1]. As a consequence of this requirement only a 'selected' but large group of atomic arrangements were until lately determined, which in view of more recent work represented examples of well defined structure types, or to be more exact, of rather ideal solids. The atomic arrangement of many crystals therefore could not be determined, since no structure for them was possible which satisfied the 'structure theory'. To such crystals belonged many well crystallised and important substances such as many alloys, solid solutions having large homogenity ranges, and other intermetallic compounds, as

[1] In his monograph on the 'Geometrische Kristallographie des Diskontinuums', Leipzig 1919, p. 551–52, P. Niggli called attention 16 years ago to the fact, that a considerable departure from the structure theory as here defined may occur.

well as many substances whose very remarkable chemical and physical properties, as for example, mixed crystal formation between completely different chemical compounds, self-diffusion, and ionic conductivity made a knowledge of their structures very desirable.

Within the past few years however, many workers have concentrated their attention onto those compounds for which no structure is possible, which satisfies the 'structure theory', and have gradually succeeded in finding atomic arrangements for them, which not only lead to agreement between calculated and observed intensities, but also offers a ready and plausible explanation of their characteristic physical and chemical properties.

L. W. STROCK
Z. f. Krist. **93** (1936) 285

MIXED CRYSTALS

147a. Die Konstitution der Mischkristalle, by L. VEGARD and H. SHELDERUP (1917)

147b. Die Konstitution der Mischkristalle und die Raumfüllung der Atome, by L. VEGARD (1921)

148. Röntgenographische Untersuchung der Kristallstrukturen von Magnetkies und verwandten Verbindungen, by N. ALSÉN (1925)

149. Die Bau-Zusammenhänge innerhalb der Kristallstrukturen II, by F. LAVES (1930)

150. Vacant Positions in the Iron Lattice of Pyrrhotite, by G. HÄGG (1933)

151. Spinel Structures with and without variate atom equipoints, by T. F. W. BARTH and E. POSNJAK (1932)

152. The Crystal Structure of γ-Fe$_2$O$_3$ and γ-Al$_2$O$_3$, by E. J. VERWEY (1935)

153. Die Kristallstruktur des Hochtemperatur-Jodsilbers α-AgJ, by L. W. STROCK (1934)

154. Solid Solutions whith a Varying Number of Atomes in the Unit Cell, by G. HÄGG (1935)

147a. Mit Hilfe der experimentellen Mittel, die uns die Röntgenstrahlenanalyse zur Verfügung gestellt hat, ist es indessen jetzt möglich die Struktur der Mischkristalle näher zu erforschen.

Bauten sie sich durch eine Aufeinanderlagerung dünner homogener Schichten der beiden Stoffe auf, so müßte eine solche Konstitution durch den Charakter der Röntgenspektra dadurch erkennbar sein, daß jeder Stoff seine eigenen Maxima mit der für ihn charakteristischen Intensitätsverteilung geben müßte. Das Spektrum des Mischkristalls wäre als eine Superposition der Spektra der beiden Komponenten zu betrachten.

Die andere Möglichkeit dagegen, daß die Mischkristalle als einheitliche Kristalle reflektieren, würde bedeuten, daß die Atome der beiden Komponenten in dasselbe Atomgerüst hineingelagert wären.—In diesem Falle wäre es möglich, daß für gewisse Mischungsverhältnisse die Einlagerung von einer solchen Regelmäßigkeit wäre, daß für gewisse Netzebenen ganz neue Reflexionsmaxima auftreten könnten.

Mit der Absicht die Konstitution der Mischkristalle aufzuklären, haben wir im Physikalischen Institut in Christiania eine Reihe von Versuchen angefangen, und wollen im folgenden einen kurzen Bericht über die Bestimmung der Konstitution einiger Systeme geben. Bei diesen Versuchen haben wir die Braggsche Reflektionsmethode in Anwendung gebracht.

Bis jetzt haben wir die folgenden vier Mischkristalle untersucht:

    I. 3 Mol. K Br, 97 Mol. K Cl

    II. 50   ,,   ,, , 50   ,,   ,,

    III. 89   ,,   ,, , 11   ,,   ,,

    IV. 67,5 ,,   ,, , 32,5 ,, NH_4Br.

Eine Zusammenstellung der beobachteten Spektra ist in der folgenden Tabelle gegeben. Zum Vergleich sind auch die berechneten Winkel für die reinen Substanzen angeführt:

Substanz	Kristall-ebene		Intensitäts-verteilung
K Cl	(100)	5°34′	normal
3 K Br, 97 K Cl	(100)	5°34′	66, 27, 9, 2,6
	(100)	5°27′,5	16, 6, 2
50 K Br, 50 K Cl	(110)	7°39′	13,5, 3,5
	(111)	4°42′	2, 22, 5, 0,5
89 K Br, 11 K Cl	(100)	5°18′	24, 7,5 1,8
K Br	(100)	5°18′	normal
67,5 K Br, 32,5 NH_4Br	(100)	5°17′	28, 12, 4, 1,5
NH_4Br	(100)	5°19′,5	normal

Aus den Zahlen der Tabelle sehen wir, daß die *Mischkristalle wie einheitliche Kristalle reflektieren.*

Die Flächen (100) und (110), für welche die reinen Komponenten eine normale Intensitätsverteilung geben, zeigen auch bei den Mischkristallen 'normale' Spektra. Die Anordnung der Atome im Mischkristall muß also eine solche sein, daß die Netzebenen (100) und (110) gleichwertig und äquidistant sind.

Von besonderem Interesse ist ein Vergleich zwischen den Spektren des Mischkristalls II (Zusammensetzung: 50 Mol. K Br, 50 Mol. K Cl) und den Spektren der beiden Komponenten.

Der Verschiedenheit der Molekularvolumina von K Br und K Cl entspricht ein Unterschied der Reflexionswinkel, der im Vergleich mit dem möglichen Fehler der Winkelbestimmung ziemlich groß ist. Wäre der Mischkristall aus mechanisch aufeinandergelagerten Schichten der beiden Komponenten zusammengesetzt, müßte man also in seinem Spektrum jedes der beiden Spektra der Komponenten erkennen können. Für den Mischkristall würde man nicht einfache, sondern doppelte Maxima erhalten.—In den beobachteten Spektren finden wir aber keine doppelten, sondern nur einfache Maxima mit Reflexionswinkeln, die zwischen denjenigen der beiden Komponenten liegen.

Nun wäre es indessen denkbar, daß die beobachteten einfachen Maxima dadurch zustande kämen, daß die Maxima der einzelnen Komponenten nicht die für eine Trennung beider genügende Schärfe hätten.—Daß aber eine solche Deutung der Versuchsergebnisse ausgeschlossen ist, geht aus der Figur hervor. Die Kurven geben die Variation des Ionisationsstroms als

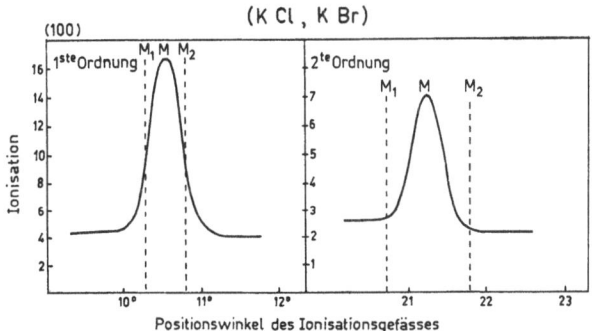

Funktion des Reflexionswinkels in der Nähe der Maxima von erster und zweiter Ordnung des Spektrums der Fläche (100). Die punktierten Linien geben die Lage, welche die Spektra der reinen Substanzen haben würden. M_1 und M_2 sind die Lagen der Maxima von K Br bzw. K Cl.—Wir sehen sofort aus der Figur, daß die Maxima für den Mischkristall einfach sind. Wir können also das Spektrum des Mischkristalls nicht als eine Superposition der Spektra der beiden Komponenten betrachten.

Dies Ergebnis führt uns zur folgenden Vorstellung über die Konstitution der Mischkristalle:

Die Raumgitter der Kristalle der reinen Komponenten bauen sich aus einer bestimmten Anzahl von Elementargittern auf. In unserem Falle haben wir vier Elementargitter von Alkaliatomen und vier Elementargitter von Halogenatomen. In einem Mischkristall haben wir nun dieselbe Anzahl von Gittern jeder Atomgattung; aber die Elementargitter bestehen jetzt nicht mehr aus derselben Atomsorte, sondern einige Atome der einen Komponente sind durch entsprechende Atome der andern ersetzt.—Durch diese Substitution wird sich das Volumen des Elementargitters ändern. Die Molekularvolumina der beiden Komponenten gleichen sich unter der Wirkung der Atomkräfte (Kristallisationskräfte) aus. In unserm Falle muß sich also das Gitter von KCl ausdehnen, dasjenige von KBr dagegen sich zusammenziehen. Das Gleichgewicht bestimmt sich nach den Untersuchungen von Retgers in der Weise, daß sich das Volumen annähernd additiv aus denjenigen der isomorphen Bestandteile zusammensetzt.

Es ist weiter eine Frage, wie sich die eingelagerten Atome anordnen. Es wäre ja denkbar, daß in gewissen Fällen eine solche Regelmäßigkeit der Anordnung entstehen könnte, daß dadurch neue Reflexionsmaxima aufträten.

Wäre dies der Fall, so sollte man erwarten, besonders einfache Verhältnisse in dem Falle zu bekommen, daß im gemischten Gitter eine gleiche Anzahl der beiden Atomsorten vorhanden wäre. Wir haben deshalb den Mischkristall 50 KCl, 50 KBr sorgfältig untersucht, um möglichst neue Maxima, die einer größeren Gitterkonstante entsprächen, zu finden. Solche neuen Maxima haben wir aber vergebens gesucht.

In unserem Falle würde schon die einfachste regelmäßige und kubische Anordnung für gewisse Netzebenen neue Maxima geben.

Diese Anordnung würde darin bestehen, daß in den Punktreihen parallel den Koordinatenachsen (Seiten des Elementarkubus) die Cl- und Br-Atome regelmäßig abwechselten. Wäre die Seitenlänge des Elementargitters einer der reinen Komponenten a, so würde also der Mischkristall durch Elementargitter von annähernd der doppelten Seitenlänge aufgebaut.

Sehen wir von der Änderung des Molekularvolumens ab, so würde die angenommene regelmäßige Substitution von Cl durch Br die Gitterkonstanten aller derjenigen Reflexionsebenen unverändert lassen, für welche die Indizes h, k, l nicht alle unpaarzählig sind, sonst würde durch die Substitution eine Verdoppelung der Gitterkonstanten eintreten.

Von den von uns untersuchten Flächen ist es nur die (111)-Fläche, für welche eine Verdoppelung der Gitterkonstante eintreten sollte. Die dadurch bedingten neuen Maxima haben wir aber nicht beobachten können. Da sie aber der Theorie nach für Mischkristalle von KCl und KBr sehr schwach sein müßten, ist es dennoch nicht ganz ausgeschlossen, daß sie wirklich vorhanden, aber nur zu schwach für die Beobachtung seien. Eine endgültige Entscheidung der Frage, ob für gewisse Flächen der Mischkristalle neue Maxima auftreten können, ist deshalb künftigen Untersuchungen, die wir über andere Mischkristalle vornehmen zu können hoffen, vorbehalten.

Sollte es sich durch diese bestätigen, daß wirklich keine neuen Maxima aufträten, so würde dies darauf hindeuten, daß die Annahme einer regelmäßigen kubischen Substitution aufgegeben werden müßte. Die Substitution müßte nur eine statistisch gleichmäßige sein, derart daß ein jedes ausgewähltes Punktelement (Punktreihe, Netzebene) von Cl-Atomen nach der Substitution durchschnittlich die gleiche Anzahl Cl- und Br-Atome enthielte.

Christiania, Physikalisches Institut

147b. Da die Braggsche Methode gewisse Schwierigkeiten für die Auffindung möglicher neuen Maxima bietet, bin ich bei meinen weiteren Untersuchungen zu der Debye-Scherrerschen Pulvermethode übergegangen.

Eine genaue Berechnung der Gitterkonstante ist durchgeführt und die Ergebnisse sind in Tabelle 2 gegeben.

Tabelle 2.

Kristall	Zahl der Moleküle im Elementarwürfel	a Å.-E.	Δa Å.-E.	V Å.-E.^{3}	ΔV Å.-E.^{3}
K Br	4	6,5968		71,77	
K Br K Cl	4	6,4423	0,1545	66,84	4,93
K Cl	4	6,2903	0,1520	62,23	4,61

Die für das System K Br − K Cl gefundene Seitenlänge (a) des Elementargitters erfüllt mit großer Genauigkeit die Gleichung der Additivität:

$$a_m = \frac{p - 100}{100}\, a_{\text{K Br}} + \frac{p}{100}\, a_{\text{K Cl}},$$

wo p die Anzahl Molekularprozente K Cl im Mischkristall bezeichnet. Welche Allgemeinheit und Strenge diesem Additivitätsgesetz zukommt, müssen weitere Untersuchungen ergeben. Auch die Molekularvolumina folgen annähernd diesem Additivitätsgesetz. Die angegebenen Zahlen aber deuten darauf hin, daß die linearen Dimensionen die Additivität am besten erfüllen.

Deutung des Additivitätsgesetzes

Das gefundene Additivitätsgesetz würde uns unmittelbar die Vorstellung vermitteln, daß jede Atomart einen für das Element charakteristischen Raum verlangt. Den einfachsten Ausdruck für die Raumfüllung eines Atoms bekommt man, wenn man mit Bragg den Atomen Kugelgestalt zuschreibt. Für die Konstitution der Mischkristalle sind dann die folgenden beiden Extremfälle zu betrachten.

1. Die einander ersetzenden Atome haben im Mischkristall denselben Atomdurchmesser wie in der reinen Substanz.

2. Die einander ersetzenden Atome besitzen im Mischkristall denselben Atomdurchmesser. Der Atomdurchmesser muß dann innerhalb gewisser Grenzen änderungsfähig sein.

Im ersten Falle konnten nicht alle Elementargitter dieselbe Größe besitzen, sondern mußten innerhalb gewisser Grenzen schwanken, oder.die für den Mischkristall gefundene Seitenlänge des Elementargitters ist als ein Durchschnittswert zu betrachten. Mischkristalle würden dann eine gewisse 'Mikrozerstörung' besitzen. Eine vorhandene 'Mikrozerstörung' sollte einerseits eine mit wachsendem Ablenkungswinkel zunehmende Unschärfe der Linien verursachen. Andererseits würde die Intensität mit wachsender Ablenkung herabgesetzt werden.

Da die Mischkristalle wenigstens annähernd ebenso scharfe Linien wie die reinen Komponenten geben, und da auch kein merklicher Intensitätsabfall mit wachsendem Ablenkungswinkel zu beobachten ist, können wir schließen, daß die 'Mikrozerstörung' recht gering sein wird, oder die einander ersetzenden Atome müssen sich im Mischkristall auf annähernd denselben Atomdurchmesser einstellen.

Christiania, Physikalisches Institut

148. Der Schwefelüberschuss des Magnetitkieses kann entweder durch eine Substitution von Fe- gegen S-Atome, oder durch die Eindringung von Schwefelatomen in die Zwischenräume des Gitters entstehen. Wenn auch die spez. Gewichtsbestimmungen von Magnetkies und Troilit grosse Fehler aufweisen, scheint es doch offenbar, dass das spez. Gewicht des Troilits grösser als das des Magnetkieses ist. Da nebenbei die c-Achse bei grösserem Schwefelgehalt abnimmt, während die a-Achse sich konstant hält, müssen die S-Atome die Fe-Atome substituieren. Es ist sehr wahrscheinlich, dass die umgekehrte Substitution in eisenreichen Verbindungen vorhanden ist.

Stockholms Högskolas Mineralogiska Institut

149. Es ist schon lange bekannt, daß Magnetkies bei der Analyse einen Schwefelüberschuß ergibt, und man war der Ansicht, der Schwefel befinde sich im FeS als feste Lösung. Von Alsén sind daraufhin mehrere Präparate untersucht worden, wobei sich unter anderem folgendes ergab:

Tabelle 51.

	a	c	c/a
FeS ungefähr dem Verhältnis 1 : 1 entsprechend	3,43	5,79	1,69
FeS + S	3,43	5,68	1,66
FeSe ungefähr dem Verhältnis 1 : 1 entsprechend	3,61	5,87	1,62
FeSe + Se	3,51	5,55	1,58

Aus dem Umstand, daß beim FeS durch Schwefelüberschuß oder beim FeSe durch Selenüberschuß die Elementarzelle, wie aus Tabelle 51 ersichtlich, verkleinert wird, schloß Alsén, es sei unwahrscheinlich, daß die S- bzw. Se-Atome statistisch verteilt zwischen das der festen Verbindung FeS ententsprechende Gitter eingelagert seien, da sonst eine Vergrößerung der Elementarzelle zu erwarten wäre. Er nahm daher an, daß entsprechend dem Schwefelüberschuß die Fe-Atome durch S-Atome ersetzt seien. Gleiches gilt natürlich auch für die Se-Verbindungen.

Aber auch dieser Deutung scheinen gewisse Schwierigkeiten entgegenzustehen, die sich besonders bei den FeSe-Verbindungen bemerkbar machen. Aus Tabelle 51 geht hervor, daß sich bei Ersatz des Fe durch Se die Elementarzelle verkleinert. Nun scheint aber nach der bisherigen Kenntnis von Atomgrößen Se und S wesentlich größer zu sein als Fe.

Man sollte also eigentlich auch bei einer Substitution des Eisens durch Schwefel oder Selen eine Vergrößerung der Elementarzelle erwarten. Im I. Teil wurde bei Besprechung einiger Kristalltypen, welche nicht der Strukturtheorie gehorchen, auf die Möglichkeit von 'unvollständigen Kristallen' hingewiesen. Vielleicht würde sich mit Hilfe einer derartigen Annahme das merkwürdige Verhalten dieser Verbindungen mit S- bzw. Se-Überschuß erklären lassen. Man hätte dann nur anzunehmen, es würden in den Ketten der Fe-Atome einige Atome fehlen. Würden die unbesetzten

Punkte sich rein statistisch über den gesamten Gitterkomplex verteilen, so ließe sich eine derartige Struktur höchstens durch Intensitätsmessungen von einer der Strukturtheorie gehorchenden unterscheiden bzw. durch genaue Messungen des spez. Gewichtes. Eine derartige Annahme würde nun zwanglos die Verkleinerung der Elementarzelle bei S- bzw. Se-Überschuß erklären, und insbesondere eine stärkere Verkleinerung der *c*-Achse gegenüber der *a*-Achse fordern, entsprechend den experimentellen Daten (vgl. Tabelle 51).

Zürich, Mineralogisches Institut

150. Previous investigations on the solubility of sulphur in iron sulphide (FeS) have led to the conclusion that the solid solutions of sulphur in iron sulphide are formed by substituting some of the iron atoms in the original lattice by sulphur atoms. Assuming that the radius of the sulphur atoms is smaller than that of the iron atoms, this hypothesis explains the fact that the lattice dimensions decrease with increasing sulphur content. It seems, however, doubtful if this relation between the radii of iron and sulphur atoms agrees with reality, and the difficulties are still more increased when one has to explain the analogous (only more pronounced) lattice variations in solid solutions of selenium in iron selenide (FeSe) in the same way, on the assumption that the selenium atoms are smaller than the iron atoms.

There exists, however, a second explanation of the observed effect, which requires no special relation between the sizes of the two kinds of atoms,

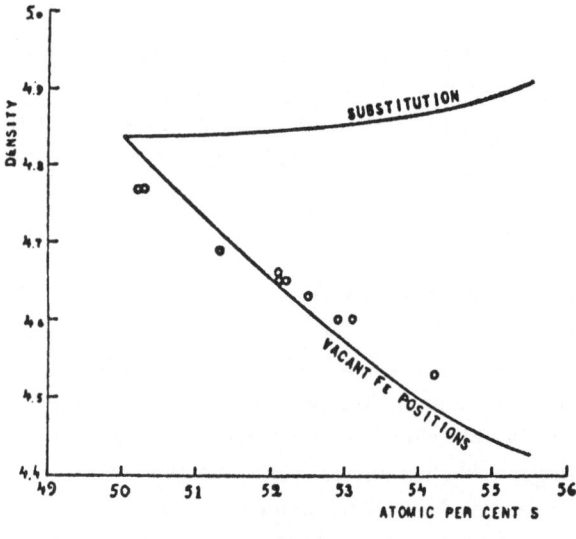

Fig. 1

namely, that the excess of sulphur is caused by an increasing number of vacant positions in the iron lattice. Recent X-ray measurements carried out in this Institute show that the phase of the 'nickel arsenide type', constituting the solid solutions in question, is homogeneous from about 50 to about 55.5 atomic per cent sulphur, and by means of the lattice dimensions observed, the density has been calculated for a solid solution of the substitution type and for the type characterised by vacant iron positions. The curves are shown in Fig. 1 together with previously observed density values, which strongly favour the assumption of such vacant positions.

Theoretically, it should be possible to decide the question also by intensity measurements of the X-ray interferences, but in this case the difference necessitated by the two theories is too small to be detected. A study of the solutions of selenium in iron selenide, where the difference should be more marked, and also where the density measurements probably will give a still more striking difference, has been begun.

No definite facts indicate a regular distribution of the vacant positions, so possibly they might be distributed at random in the iron lattice. Theoretically, the appearance of random vacancies is as natural as the appearance of random substitution, or of a random adding of dissolved atoms into the interstices of a lattice. These vacant positions make a new type of solid solution like the above possible, and the possibility of their appearing should be kept in mind in the determination of any structure where they are not electrostatically impossible.

In this connexion it might be mentioned that the present study has shown that the phase under investigation exhibits a 'superstructure' at the homogeneity limit richest in iron, that is, at the composition FeS. The symmetry is still hexagonal, but the longest basal diagonal of the old unit cell (of the nickel arsenide type) becomes the a-axis in the super cell, while the c-axis of the old cell is doubled. The new super cell contains 12 iron and 12 sulphur atoms. The superstructure disappears rapidly when the sulphur content of the phase increases.

Institute of General and Inorganic Chemistry, University, Stockholm

151. The determination of the crystal structure of spinels by W. H. Bragg[1], and independently by S. Nishikawa[2], was one of the very early accomplishments of crystal analysis. In both cases the work was based on the study

[1] W. H. Bragg, *Phil. Mag.* **30** (1915) 305.
[2] S. Nishikawa, *Proc. Math. Phys. Soc. Tokyo* **8** (1915) 199 [Vol. I paper **19**]

of spinel ($MgO.Al_2O_3$) and magnetite ($FeO.Fe_2O_3$). Since that time a large number of compounds of the general formula XY_2O_4, both minerals and synthetic products, have been examined, and stand on record as having the accepted spinel structure. In most cases, however, no attempt had been made to prove by intensity calculations the correctness of this statement.

In view of some discrepancies of the spinel structure with certain general results of crystal analysis, especially the fact that the law of constant atomic radii seemed to be violated, an examination of this question was undertaken by the present authors, and it was found[1] that in the cases of $MgO.Fe_2O_3$ and $MgO.Ga_2O_3$, which unquestionably are spinels, the intensity of X-ray diffractions could not be reconciled with the accepted arrangement of the spinel structure. The paper just referred to was a preliminary statement of our findings. It was obvious to us that 'it appeared possible to get interatomic distances that lead to more reasonable radii for the cations' (*op. cit.*) by choosing special values for the parameter determining the positions of the 32 oxygen ions, but it was equally obvious that no such adjustment of the parameter could possibly result in a better agreement between observed and calculated intensities of magnesium ferrite and magnesium galliate. For this reason it was not necessary to go through a complete intensity computation, which is very tedious.

In order to account correctly for the observed intensities in the spinels $MgO.Fe_2O_3$ and $MgO.Ga_2O_3$ it was necessary to assume that the 16 equivalent positions ($16c$) of the unit cell of the spinel structure are not occupied by chemically equivalent ions. Instead of having the 'spinel arrangement' with magnesium in $8f$ and iron (or gallium) in $16c$, it was found that 8 iron (or gallium) ions must occupy positions $8f$, and that 8 iron (or gallium) ions + 8 magnesium ions must occupy positions $16c$.

Since then we have continued the study of spinels, and found an additional number of spinels with variate atom equipoints[2]. For the present paper computations showing the effect of the parameter value u on the intensities of the X-ray reflections from the more important planes have been carried out and the calculations based on the atomic F-values[3].

MAGNESIUM FERRITE ($MgO.Fe_2O_3$). The intensities of a number of re-

---

[1] T. F. W. Barth and E. Posnjak *J. Wash. Acad. Sci.* **21** (1931) 255.

[2] In our preliminary paper (*op. cit.*) we had proposed for convenience to call a unit cell in which structurally equivalent positions are occupied by different atoms a cell with variate atom equipoints.

[3] The F-values were taken from the following sources: Ni from R. W. G. Wyckoff, *Physic. Rev.* **35** (1930) 583; Ti from R. W. James and G. W. Brindley, *Phil. Mag.* **12** (1931) 81; Fe, Mg, Al, and O from W. L. Bragg and J. West, *Z. Krist.* **69** (1929) 118. Values for the other elements, Sn, In, Ga, and Zn, were found by interpolation.

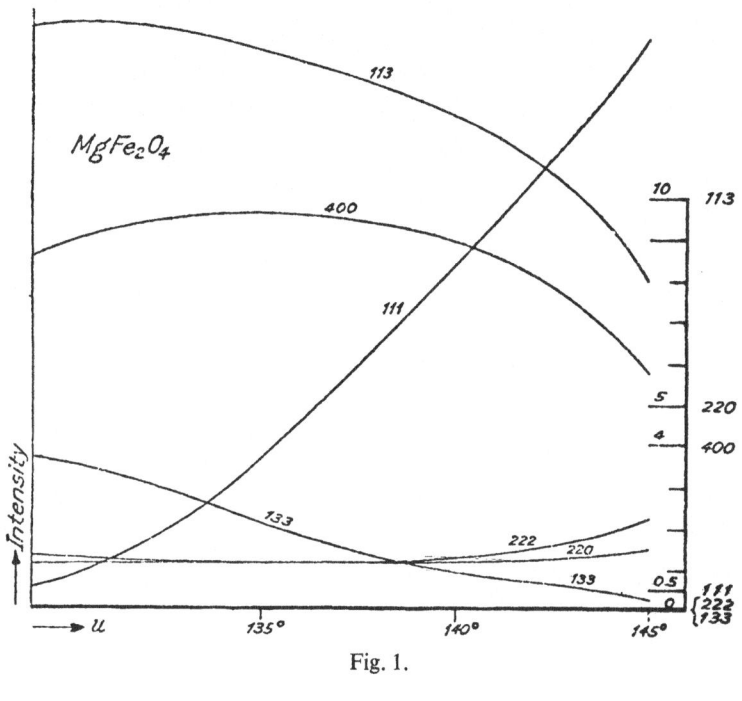

Fig. 1.

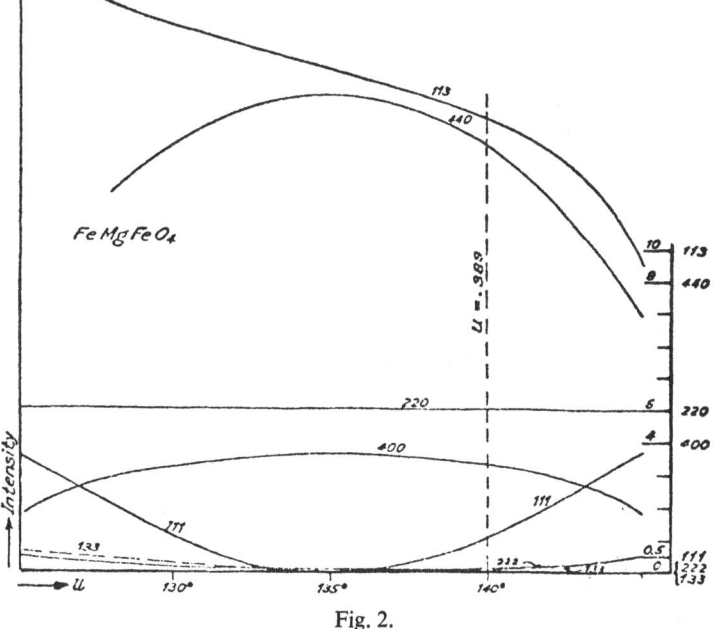

Fig. 2.

Figs. 1 and 2. The effect of the parameter value on the intensities for the normal spinel structure (Fig. 1), and the spinel structure with variate atom equipoints (Fig. 2) of magnesium ferrite.

219

flections from faces with relatively large spacings (which are most sensitive to changes in the parameter value) have been calculated for various values of u, both for arrangement $MgFe_2O_4$, and for arrangement $FeMgFeO_4$. These values were plotted separately for each arrangement against the corresponding u values. The curves thus obtained for each reflection are given in Figs. 1 and 2. On the right-hand side of each diagram is shown on a relative scale the actual observed intensity of these reflections. An examination of Fig. 1 will show conclusively that the normal arrangement $MgFe_2O_4$ cannot be the correct one, as it is obviously impossible to obtain an agreement between observed and calculated intensities regardless of any value of the parameter u. On the other hand, Fig. 2 shows clearly that the observed intensities agree remarkably well with the computations based on the arrangement with variate atom equipoints, and that the best parameter value is somewhat greater than 135°; the value that should be assigned to the parameter is $u = 140° \pm 2°$ (= 0.390 ± 0.006 a). These computations therefore prove conclusively the existence of variate atom equipoints in the magnesium ferrite spinel.

Conclusion

In our preliminary paper it was pointed out that in mixed crystals the existence of variate atom equipoints was probably generally accepted, and we expressed the opinion that there seemed to be no valid reason why the same phenomenon should not be expected to take place in ordinary compounds. Since then we have been able to demonstrate that the cubic modification of lithium ferrite ($Li_2O \cdot Fe_2O_3$) presents another example of an ordinary compound with the variate atom equipoint arrangement. The structure was found to be a face centered one with atomic positions 4c and 4d (that is, a 'sodium chloride' arrangement) in which $2 Li^+ + 2 Fe^{+3}$ occupy together the equipoints 4c. The evidence was very convincing in view of the fact that this is the simplest possible structure with variate atom equipoints. Thus the new structure of spinel in no way presents any longer an exceptional case.

The data given in this paper show, however, that two structural spinel arrangements with positions 8f, 16c and 32b are actually in existence. In one —XY_2O_4—different cations occupy positions 8f and 16c, which was always thought to be the case, and in the other—$YXYO_4$—position 16c is occupied jointly by an equal number of the different cations. The first arrangement was shown to exist in the aluminates: $NiAl_2O_4$, $CoAl_2O_4$, $FeAl_2O_4$, $MnAl_2O_4$, and $ZnAl_2O_4$; the evidence from magnesium aluminate (spinel) does not permit distinction, but by analogy one may be inclined to think that it would have the same arrangement as the other aluminates. All other compounds which were investigated—$FeMgFeO_4$, $GaMgGaO_4$, $InMgInO_4$, $MgTiMgO_4$, $FeTiFeO_4$, $ZnSnZnO_4$—were proved to have the arrangement with variate atom equipoints. Whether, and in what way, electronic configuration governs the differences in the spinel arrangements will be the task of future investigators to work out.

Geophysical Laboratory, Carnegie Institution of Washington

152. Cubic, instable, γ-modifications of Fe_2O_3 and Al_2O_3 were obtained and studied by several authors. The investigation by X-rays gave for these oxides spinellike diagrams.

The structure of γ-Fe_2O_3 has been discussed by Welo and Baudisch[1] and more carefully by Thewlis[2]. The powder diagrams of Fe_3O_4 and γ-Fe_2O_3 are practically identical. Fe_3O_4 has spinel structure. The unit cell contains 32 O-atoms and (8 + 16) Fe-atoms. It is believed therefore that the structure of γ-Fe_2O_3 is given by a spinel unit cell with four additional O-atoms: $Fe_{24}O_{36}$. The positions assumed by Thewlis for these 4 atoms are considered to be more probable than those of Baudisch and Welo with regard to ionic radii. For γ-Al_2O_3 a similar structure has been proposed.

However, Burgers[3] determined the length of the unit cell of γ-Al_2O_3, $a = 7.90$ Å, and pointed out that a structure with 12 molecules Al_2O_3 in the cell ($Al_{24}O_{36}$) would lead to a calculated density 4.1 which is not in accordance with the experimental data: Biltz determined carefully the density of γ-Al_2O_3 and gave 3.42; the actual value may be somewhat higher (the oxide has a very loose structure and may easily contain traces of air) but is probably much smaller than 4.1.

Finally the structure with 4 additional O-atoms is contradictory to all crystallographical work on structures of a related type. In α-Al_2O_3, α-Fe_2O_3, $BeAl_2O_4$ etc. the lattice is built up by a hexagonal close packing of the relatively large O^{2-}ions, containing the cations in the interstices between the anions[4]. According to the principle that the packing of the oxygen-ions determines the structure of many oxides, double oxides, etc., W. L. Bragg and others were able to find the complicated structure of several silicates. In the spinel structure of $MgAl_2O_4$, Fe_3O_4 etc. we deal

[1] Welo, L. A., and Baudisch, O., *Phil. Mag.* **50** (1925) 399.

[2] Thewlis, J., *Phil. Mag.* (7) **12** (1931) 1089.

[3] Burgers, W. G., Claassen, A., and Zernike, J., *Z. Physik* **74** (1932) 593.

[4] Bragg, W. L., and Brown, G. B., *Z. Kristallogr.* **63** (1926) 122.

with a cubic close packing of the oxygen-ions. It seems therefore very improbable that the closely related γ-oxides should contain four additional O^{2-} ions with positions between the close packed O^{2-} ions. In Thewlis's structure of γ-Fe_2O_3 the smallest oxygen-oxygen distance works out to be 2.14 Å, an impossibly low value.

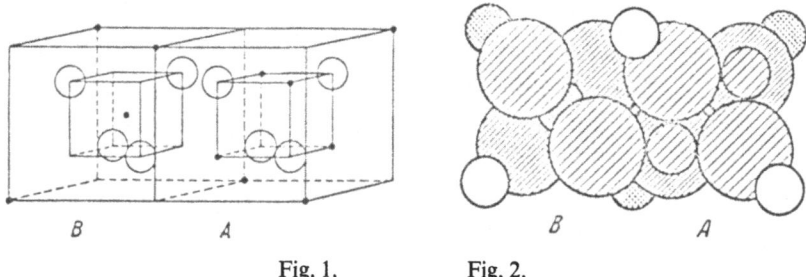

Fig. 1.   Fig. 2.

In the cubic close packing we have for each oxygen-ion one octahedron hole (between six O^{2-} ions) and two tetrahedron holes (between four O^{2-} ions). We can describe the cubic cell of the spinel structure best by subdividing it into 8 small cubes of side-length $\frac{1}{2}a$; the large cell is an alternating packing of two different types of the latter. In Fig. 1 we gave $\frac{1}{4}$ of the unit cell, in Fig. 2 a corresponding packing drawing. In four of the small cubes (type A) we deal with 4 O^{2-} ions and 4 cations with a coordination number 6, say $Al_4^6O_4$ (the high figure denoting the coordination number); these ions have approximately the positions of Na^+ and Cl^- in the unit cell of NaCl, i.e. all octahedron holes are filled up by the cations. The four remaining small cubes (B) contain 4 O^{2-} ions (again in the face centred positions) and a cation in two of the tetrahedron holes: $Mg_2^4O_4$; one cation is in the middle of the small cube, one at one of the corners[1]. The small cubes are arranged regularly in such a way that equal cubes have no faces in common. According to Barth and Posnjak[2] we have to change this picture slightly for a number of spinels. In order to account correctly for the observed intensities in the spinel $MgGa_2O_4$, for instance, they assume that half of the Ga ions occupy the positions $8f$ (i.e., following Wyckoff's notation, the 4×2 tetrahedron holes in cubes B) and $8\,Mg^{2+} + 8\,Ga^{3+}$ are distributed over the 16 equivalent positions $16c$ (the 4×4 octahedron holes in cubes A).

[1] In the figures there are four of the latter, but each of them are contained in common by four B-cubes. These ions can also be considered as belonging to both A- and B-cubes, but since they have tetrahedronhole-positions we will describe them as ions belonging to the A-cubes.

[2] Barth, T. F. W., and Posnjak, E., Z. Kristallogr. **82** (1932) 325 [this Vol. preceding paper].

If we assume for γ-Fe$_2$O$_3$ and γ-Al$_2$O$_3$ the same O^{2-} arrangement as in Fe$_3$O$_4$ or MgAl$_2$O$_4$, it is necessary to take a number of cations out of the cell. This leads to $21\frac{1}{3}$ Fe(Al) + 32 O in the unit cell of spinel. A possible structure might be obtained by the assumption of a still larger cell Fe$_{72}$O$_{108}$, or by a lower symmetry. However, the X-ray diagrams do not give any indication for these assumptions. We will therefore assume that the vacant positions corresponding to the $2\frac{2}{3}$ cations, failing in the spinel arrangement of the γ-oxides, are distributed in a way statistically over the whole lattice; thus a cubic symmetry may be maintained, and the $21\frac{1}{3}$ Fe(Al) ions in the unit cell have to be considered as an average number.

A more detailed description of this structure seems difficult. The difference between the diagrams of Fe$_3$O$_4$ and γ-Fe$_2$O$_3$ is too small for definite conclusions (as has been proved indirectly by the work of Thewlis). But a difference to be observed was that the reflexion (111), in the Fe$_3$O$_4$ diagrams distinctly present, failed for γ-Fe$_2$O$_3$, This cannot be explained by a change of the parameter (after Claassen[1] $u = 0.379$), as other reflexions that are still more sensitive to changes of u, are not perceptibly altered in their intensity. The positions of the oxygen atoms in γ-Fe$_2$O$_3$ are thus practically the same as in Fe$_3$O$_4$, u being $= 0.379$ or almost $\frac{3}{8}$ (corresponding to an ideal close packed oxygen lattice). The structure factor of the reflexion (111) is 8 Fe$_A$ $-$ 5.66 Fe$_B$ (for Fe$_3$O$_4$, with $u = \frac{3}{8}$). Hence its absence indicates that in γ-Fe$_2$O$_3$ preferably the octahedron holes of the cubes A will be vacant on account of the removal of $2\frac{2}{3}$ Fe atoms per unit cell.

A similar comparison can be made for MgAl$_2$O$_4$ and γ-Al$_2$O$_3$ (Mg and Al have almost the same scattering power). We estimated the intensities of a diagram of ordinary spinel and compared them with the intensities of γ-Al$_2$O$_3$. Apparently both diagrams are again practically the same; a small difference is perhaps, that γ-Al$_2$O$_3$ gives a weak (222) reflexion (structure factor 32 O $-$ 16 Al$_A$) but no (111) line, MgAl$_2$O$_4$ only a weak (111) reflexion. The latter would indicate again that preferably octahedron-Al-positions are vacant in γ-Al$_2$O$_3$.

Summary

The cubic spinel structure of γ-Al$_2$O$_3$ and γ-Fe$_2$O$_3$ is described as an averaged structure with regard to the cations. The oxygen arrangement

---

[1] Claassen, A., *Proc. physic. Soc.* **38** (1926) 482.

is the same as in normal spinels. The unit cell contains 32 oxygen ions. $21\frac{1}{3}$ cations and $2\frac{2}{3}$ vacant positions are distributed in a way statistically over the 24 (8 + 16) cation positions of the spinels.

Natuurkundig Laboratorium der N.V. Philips' Gloeilampenfabrieken
Eindhoven

153.   *Kubische Struktur mit Lücken*

Das es nicht gelingt, eine der strengen Strukturtheorie genügende Struktur für α-AgJ zu finden, muss man die Möglichkeit erwägen, dass eine Punktlage nicht vollständig besetzt ist, d.h. dass die Struktur Lücken hat.

Zunächst sei die Möglichkeit betrachtet, dass das Jod die 2 zählige Punktlage $(000; \frac{1}{2}\frac{1}{2}\frac{1}{2})$ besetzt (also eine kubisch raumzentrierte An-ionenpackung), und das Silber eine n zählige Punktlage teilweise ausfüllt. Eine raumzentrierte Punktlage muss gewählt werden, weil, wie oben erwähnt wurde, die Auslöschung einem solchen Gitter entspricht. Für Silber wird man versuchsweise eine raumzentrierte Punktlage derart wählen, dass möglichst wenig Lücken entstehen, z.B. (6e):

$$(000 + \tfrac{1}{2}\tfrac{1}{2}\tfrac{1}{2})(0\tfrac{1}{2}0, \circlearrowright)$$

Eine Intensitätsuntersuchung dieser Struktur zeigt, wie aus Spalte 5 von Tabelle 8 zu ersehen ist, dass eine solche Anordnung für α-AgJ nicht vorliegen kann. Die Intensitätsberechnung wurde nach folgender Formel ausgeführt:

$$J \sim F^2 \cdot \frac{1 + \cos^2 2\theta}{\sin^2 \theta . \cos \theta} . S;$$

darin bedeuten:

$$F^2 = (A^2 + B^2); \quad A = (F_J . \cos \Phi + F_{Ag} . \cos \Phi)$$

$$B = (F_J . \sin \Phi + F_{Ag} . \sin \Phi).$$

$\Phi = 2\pi(hx + ky + lz);$ S Flächenhäufigkeitszahl.

Die Werte der Streufaktoren F_{Ag} und F_J sind der bekannten Arbeit von James und Brindley[1] entnommen.

Auch die nächste Möglichkeit, dass das Silber die Punktlage 12h

$$(000 + \tfrac{1}{2}\tfrac{1}{2}\tfrac{1}{2})(\tfrac{1}{2}0\tfrac{1}{4}, \circlearrowright; \tfrac{1}{2}0\tfrac{3}{4}, \circlearrowright)$$

[1] R. W. James and G. W. Brindley, *Z. Krist.* **78** (1931) 470.

zum Teil besetzt, ergibt keine Übereinstimmung zwischen beobachteten und berechneten Intensitäten (vgl. Spalte 4, Tabelle 8). Jedoch erkennt man, das die Veränderung der Intensitäten gegenüber den für ein reines Jodgitter berechneten (Spalte 3) in der richtigen Richtung erfolgt.

In besonderem Masse gilt dies für das Intensitätsverhältnis von (110), (200) und (211), während die Umkehrung des Verhältnisses (220): (310) bestehen bleibt. Diese wird jedoch in der Anordnung 6e korrigiert, sogar überkorrigiert. Man wird also versuchen, die Anordnungen 6e und 12h zu kombinieren. In der Tat erhält man so überraschend gute Übereinstimmung (Spalte 6). Nur bei grossen Indices berechnen sich der nicht beobachtete Reflex (400) und der sehr schwache Reflex (420) grösser als der schwach beobachtete Reflex (411 + 330). Man wird also versuchen, eine neue Netzebene in $\frac{1}{8}$ des Gitterabstandes einzuschalten. Dies kann dadurch geschehen, dass man auch noch die Punktlage 12n ($u = \frac{1}{8}$) und 12m ($u = \frac{5}{8}$) [= 24j ($u = \frac{1}{8}$): (000 + $\frac{1}{2}\frac{1}{2}\frac{1}{2}$) ($u\bar{u}0$, $\circlearrowright$; $\bar{u}u0$, $\circlearrowright$; $uu0$, $\circlearrowright$; $\bar{u}\bar{u}0$, $\circlearrowright$), Ed.] teilweise besetzt.

Tabelle 8. Vergleich beobachteter und berehcneter Intensitäten für α-AgJ.

1	2	3	4	5	6	7	8
	Ber.		Ber. Intensitäten: Jod in 000; $\frac{1}{2}\frac{1}{2}\frac{1}{2}$				Beob.
hkl	θ	Jod	Silber gleichmässig verteilt in				Intensi-täten
	in Grad	allein					(visuell)
			12h	6e	6e+12h	6e+12h+24j	
110	12°30′	40.1	20.0	20.0	20	20.7	10
200	17 50	8.2	13.7	28.8	17.5	15.5	5$^+$
211	22 01	17.6	29.3	8.8	21.1	22.2	10
220	25 39	5.5	2.7	11.2	6.4	5.7	2$^+$
310	28 57	7.3	3.7	5.3	3.7	4.6	1–2
222	32 01	1.8	0.03	3.7	1.1	1.4	0
321	34 57	8.5	14.1	6.3	10.3	7.8	3
400	37 45	0.8	2.8	1.7	2.8	1.2	0
330	40 30	4.2	2.1	3.0	2.1	1.9 } 5.8	1$^+$
411						3.9	
420	43 12	2.5	4.2	5.2	5.5	2.7	0–1$^-$
332	45 53	2.4	3.9	1.7	2.8	2.9	0–1$^-$

Die völlige Übereinstimmung der für diese Anordnung berechneten (Spalte 7, Tabelle 8) und der beobachteten Intensitäten (Spalte 8) zeigt, dass diese Struktur für α-AgJ möglich ist; sie aber auch die einzig mögliche ist. Denn es gibt keine weiteren Plätze für Ag$^+$ in der Zelle als die

42 in Rechnung gestellten Lücken. Andererseits ist es auch aus Intensitätsbetrachtungen unmöglich, die Plätze von J^- und Ag^+ zu vertauschen, weil (200) sich dann stets stärker berechnet als (110).

Eigenschaften der Struktur

Die Struktur lässt sich beschreiben als raumzentriertes J^--Gitter, in dessen Lücken die beiden Silberionen wahllos angeordnet sind. Das Silber befindet sich ähnlich wie eine Flüssigkeit im Jodgitter.

Diese Struktur gestattet eine zwanglose Deutung der markanten Eigenschaften des α-AgJ (elektrolytische Leitung und Diffusion des Ag^+).

Zum Schluss möchte ich Herrn Prof. Dr. Dr. E. h. V. M. Goldschmidt meinen herzlichsten Dank für seine wertvollen Ratschläge und sein ständiges Interesse aussprechen. Auch Herrn Privatdozenten Dr. F. Laves schulde ich für viele fördernde Diskussionen grossen Dank.

Göttingen, Mineralogisch-petrographisches Institut der Universität

154. In recent years many cases have been observed, where the number of atoms in the unit cell of a solid solution varies with the composition. These cases have been treated from different viewpoints and, therefore, a comprehensive discussion may perhaps be of some use.

The most frequent type of solid solution is doubtless caused by the substitution of atoms of one kind for atoms of another kind. This was also the type of solution first studied and interpreted by means of X-ray methods. Solid solutions were previously known, in which, even without the knowledge later attained with the aid of X-rays, a substitution of atoms as the cause of the variation in the composition could scarcely be accepted. Zambonini[1] showed, for example, that many salts of Ca, Sr, Ba and Pb were able to form solid solutions with the corresponding salts of rare earths, although their chemical formulae were different. Among others he obtained solid solutions in all proportions between $PbMoO_4$ and $La_2(MoO_4)_3$ and also regarded yttrofluorite as a solid solution of YF_3 in CaF_2.

The X-ray studies then revealed many cases where atoms of one kind were added to the interstices of a given lattice. This was shown by Westgren and Phragmén to be the case with the solution of C in γ-Fe.[2] Later, many

[1] Zambonini, F., *Riv. Min. e Crist. Italiana* **45** (1915) 1.
[2] Westgren, A. and Phragmén, G., *Z. physik. Chem.* **102** (1922) 1.

analogous solid solutions have been found, mainly in the systems between transition elements and H, C and N.[1] These phases, which were characterized as 'Einlagerungsstrukturen', seemed to appear only if the dissolved atoms were small enough compared to the metal atoms.

The interstitial phases could also be termed solid solutions, formed by addition of one kind of atoms to a given lattice. Solid solutions were also soon found, which could most conveniently be considered as formed by a subtraction of atoms from a given lattice. The solutions of S in FeS,[2] Se in FeSe[3] and O in FeO[4] are thus caused by the subtraction of varying amounts of Fe atoms from the lattices of the composition FeX. In this way lattice positions, originally occupied by Fe, will be vacant.

It is a matter of convenience if a solid solution belonging to one of the two types last mentioned, is termed a solution formed by addition or by subtraction. The term chosen depends only on which of the two limits of the homogeneity range is considered to represent the solvent. In a great many cases, only one of the limits of the homogeneity range is characterized by a welldefined lattice, and it is quite natural that such a limit is chosen as characteristic of the solid solution. Thus, for example, austenite is most conveniently regarded as a solid solution of the addition type, where C is added to the lattice of γ-Fe. Theoretically, it could also be regarded as a solution obtained by subtracting C from the homogeneous phase richest in carbon, but as neither the lattice nor the composition are well defined at this point, this would be somewhat arbitrary.

Notwithstanding the fact that the terms addition or subtraction solution can thus be valuable in describing certain types, it must be kept in mind that, theoretically, they cannot be distinguished. In this paper both are treated as one type of solid solution, which is probably best characterized as a solid solution where the number of atoms per unit cell varies[5]. The author thinks the term 'interstitial solution' (Zwischenraumslösung) suitable for this type of solid solutions as a whole. In certain cases an interstitial solution can then for convenience be specified as an addition or a subtraction solution.

[1] Hägg, G., *Z. physik. Chem.* (*B*) **12** (1931) 33.

[2] Hägg, G. and Sucksdorff, I., *Z. physik. Chem.* (B) **22** (1933) 444.

[3] Hägg, G. and Kindström, A. L., *Z. physik. Chem.* (B) **22** (1933) 453.

[4] Jette, E. R. and Foote, F., *J. Chem. Physics* **1** (1933) 29.

[5] A variation of the number of atoms per unit cell can also be effected in solid solutions, formed by the substitution of molecules. This possibility, which will be most frequently met in solid solutions of organic substances, will certainly not give rise to confusion.

Structural and Geometrical Conditions for the Formation
of Interstitial Solutions

A necessary condition for the formation of an interstitial solution seems to be a certain stability of the lattice in the interstices of which the dissolved atoms are situated. In other words, the structure must contain a rather rigid lattice skeleton, whose stability is to a certain degree unaffected by a change in number of the interstitial atoms. In many cases this implies that the interstitial atoms must be small compared with the atoms forming the rigid lattice, but in cases where the latter shows large interstices, the dissolved atoms can have comparatively large dimensions.

Rigid lattices which are able to take up other atoms in their interstices are formed by the transition elements. The stability of these lattices is probably connected with the fact that they, in most cases, correspond to the simple lattices, characterizing the pure elements. In these lattices the interstices are small and as a consequence only the smallest atoms known (H, B, C, N) can be dissolved interstitially to any large degree.

Other rigid lattices, playing a part in the formation of interstitial solutions, are found in many oxides, sulphides, selenides and halides. Here the relatively large dimensions of the anions facilitate the construction of a rigid lattice, which seems not to be influenced by a variation in the number of ions, situated in its interstices. These dissolved ions are in most cases cations, certainly owing to the fact that such ions generally are comparatively small. Examples of interstitially dissolved anions are, however, also known but in such cases the interstices of the rigid lattice are especially large.

From the works of W. L. Bragg and others it is well known that, in the silicates, oxygen in combination with silicon (aluminium) forms lattices of a high degree of stability. These skeletons are often only slightly influenced by changes taking place in their interstices, and although these changes generally import substitutions, variations of the number of atoms are also known.

It is to be expected that especially close-packed lattices ought to be very stable and only affected to a very small degree by changes taking place in their interstices. A study of the lattices in which interstitial solutions are known to be possible shows, in accordance with this assumption, that the rigid lattices are close-packed in a majority of the cases.

Conditions of Electrostatical Neutrality

The number of neutral atoms per unit cell can vary without disturbing the neutrality of the lattice but a variation of the number of ions must be compensated by other variations of the ionic charges. In the following the different ways in which electrostatical neutrality is maintained have served as a basis for a classification of the different types of interstitial solutions.

Interstitial Solutions of Neutral Atoms or Groups

The most important examples of this type of interstitial solutions seem to occur in the already-mentioned systems of transition elements with H, B, C and N. They are most conveniently regarded as addition solutions of small neutral atoms in relatively stable metal lattices. These phases have been treated in detail earlier and, therefore, a reference to this treatment[1] seems to be sufficient.

The zeolites, where water molecules etc. are distributed in the holes of a firm framework and can be removed and replaced without destruction of the crystal, form other examples of this type of interstitial solutions.

Interstitial Solutions accompanied by Substitution

A typical example of this type of solution was found by Ketelaar[2] in the system $AgI-HgI_2$. Between 50° and 158° there exists an intermediate phase, termed $\alpha-Ag_2HgI_4$, which shows a large homogeneity range. At the composition Ag_2HgI_4 it can be considered as a structure of the zincblende type where $2\,Ag + Hg$ are statistically distributed over 3 of the 4 positions, belonging to one of the face-centred lattices. This structure can be derived from the zincblende structure of AgI if Ag^{+1} ions are replaced by Hg^{+2} ions and the excess in positive charge is balanced by subtraction of Ag^{+1} ions from the cation lattice. The formation of the interstitial solution in this case is evidently connected with the stability of the lattice, formed by the large close-packed I^{-1} ions, and the fact that the cations are very small compared to the I^{-1} ions. Further, the ions Ag^{+1} and Hg^{+2} possess about the same dimensions, which facilitates their distribution in structurally equivalent interstices in the I^{-1} lattice.

[1] Hägg, G., *Z. physik. Chem.* (B) **12** (1931) 33.
[2] Ketelaar, J. A. A., *Z. Kristallogr.* **87** (1934) 436.

An analogous type of interstitial solution is represented by the Mg-Al-spinels, containing Al_2O_3 in excess of the formula $MgO.Al_2O_3$[1]. Mg^{+2} ions in the original spinel lattice are replaced by Al^{+3} ions and an increase in positive charge is prevented by a reduction of the total number of cations. In this way the homogeneity range of the spinel phase is increased and γ-Al_2O_3 can be considered as its (unstable) limit. The unit cell of $MgO.Al_2O_3$ contains $8\ Mg^{+2} + 16\ Al^{+3} + 32\ O^{-2}$ and the unit cell of γ-Al_2O_3 in mean $21\frac{1}{3}\ Al^{+3} + 32\ O^{-2}$. One ninth of the original cation positions are vacant. It was earlier assumed that the extension of the homogeneity range of the spinel phase was due to a replacement of Mg^{+2} ions by Al^{+3} ions and a simultaneous addition of O^{-2} ions into the interstices of the spinel lattice.

Bruni and Ferrari[2] have recently discussed the solid solutions of $MgCl_2$ in LiCl. The Cl^{-1} ions in the lattice of $MgCl_2$[3] form a cubic close-packing, quite as in the structure of LiCl. If every second cation layer at right angles to one trigonal axis in LiCl is left out, the result will be a lattice of the $MgCl_2$ type. As a consequence, the solutions of $MgCl_2$ in LiCl are to be regarded as solutions with vacant positions in the cation lattice. Bruni and Ferrari connect this with the fact that the structure is determined mainly by the anions, as the cations only occupy a small fraction of the total volume.

The structures of $LiMnPO_4$ (lithiophilite) and Li_3PO_4 which, according to Zambonini and Laves[4], both belong to the olivine type, also show the dominating influence of the anion lattice. The cations Mg^{+2}, Fe^{+2}, Mn^{+2} and Li^{+1} in olivine and the mentioned phosphates, are situated in the largest (octahedral) interstices in the deformed hexagonal close-packing of O^{-2} ions. The number of these interstices, occupied by such cations, seems not to be very important for the stability of the lattice. Interstitial solid solutions accompanied by substitution are certainly possible in these cases.

As has been mentioned above, solid solutions of salts (mainly molybdates and tungstates) of Ca, Sr, Ba and Pb with the corresponding salts of rare earths were reported by Zambonini. Later, they have been examined by means of X-rays by Zambonini and Levi[5]. These studies do not throw any light upon the nature of the solid solutions in question but it can be

[1] Hägg, G., Nature 135 (1935) 874; Z. physik. Chem. (B) 29 (1935) 88.

[2] Bruni, G. and Ferrari, A., Z. Kristallogr. 89 (1934) 499.

[3] Bruni, G. and Ferrari, A., Z. physik. Chem. (A) 130 (1927) 488; Pauling, L., Proc. Nat. Acad. Washington 15 (1929) 709.

[4] Zambonini, F. and Laves, F., Z. Kristallogr. 83 (1932) 26.

[5] Zambonini, F. and Levi, R. G., Rend. Accad. Lincei 2 (1925) 149, 225, 303, 377, 462.

said almost with certainty that they must be of the interstitial type, where vacant positions appear in the cation lattice as the concentration of the trivalent cations rises.

The solutions of YF_3 in CaF_2 are perhaps of another type. The available X-ray and density data are not sufficient for a decision of the type of solution but the analogy with the pair PbF_2 and BiF_3, studied by Hassel and Nilssen[1] seems to indicate that the solution of YF_3 in CaF_2 is accompanied by an addition of F^{-1} ions to the lattice of CaF_2. In any case, the interstices in the CaF_2 lattice are large enough to make such an addition possible.

It is well known that, in the silicates, silicon often is partially replaced by aluminium. The resulting decrease in positive charge is in most cases compensated by a replacement of other cations by cations with a higher charge, but an addition of new cations to formerly unoccupied interstices of the lattice is also known. As an example, Warren[2] has shown that the structure of hornblende can be derived from the structure of tremolite in the way that the silicon is partially replaced by aluminium and alkali ions are added to the lattice. Similar examples have been observed by Barth and Posnjak[3]. These authors have found that the structures of $NaAlSiO_4$ (α-carnegieite) and Na_2CaSiO_4 both are related to the structure of β-cristobalite. In $NaAlSiO_4$ half of the silicon in β-cristobalite is replaced by aluminium and $\frac{1}{4}Na^{+1}$ per oxygen atom added to the lattice. In Na_2CaSiO_4 the still larger decrease in positive charge, resulting from the replacement of silicon by calcium, is compensated by the addition of $\frac{1}{2}Na^{+1}$ per oxygen atom. It is possible to obtain preparations with intermediary compositions which can be regarded as interstitial solid solutions.

Interstitial Solutions accompanied by Changes in Valency

Several cases are known where a change in valency of one kind of atom causes a variation of the number of atoms per unit cell.

The solutions of the respective metalloids in FeS, FeSe and FeO have already been mentioned. The excess of metalloid is caused by a partial rise in charge of the Fe^{+2} ions and in order to maintain the neutrality of the lattice some of the original Fe^{+2} positions become vacant.

In a recent investigation[4], the author has studied the structure of the

---

[1] Hassel, O. and Nilssen, S., *Z. anorg. allg. Chem.* **181** (1929) 172.
[2] Warren, B. E., *Z. Kristallogr.* **72** (1929) 493.
[3] Barth, T. F. W. and Posnjak, E., *Z. Kristallogr.* **81** (1931) 376.
[4] Hägg, G., Nature **135** (1935) 874; *Z. physik. Chem.* (B) **29** (1935) 95.

magnetic ferric oxide, $\gamma\text{-}Fe_2O_3$, which is formed by the oxidation of Fe_3O_4 at low temperatures. The oxygen lattice is not changed during the oxidation and the increase in valency of the Fe^{+2} ions is accompanied by the appearance of vacant positions in the Fe lattice. Thus, $\gamma\text{-}Fe_2O_3$ is related to Fe_3O_4 in the same way as $\gamma\text{-}Al_2O_3$ to $MgO.Al_2O_3$. The only difference is that the increase in positive charge per cation, which makes the vacant positions occur is in the former case caused by a rise in charge of atoms of the same sort and in the latter case by substitution.

Institute of General and Inorganic Chemistry of the University,
Metallographic Institute, Stockholm

B. LAYER STRUCTURES WITH RANDOM STACKING

155. Les rayons X ne donnent pas toujours la véritable maille des cristaux, by CH. MAUGUIN (1928)

▬

156. Die 'Wechselstruktur' von $CdBr_2$, by J. M. BIJVOET and W. NIEU-WENKAMP (1933)

▬

157a. Interferenzerscheinungen an zweidimensionalen Kristallen, by F. LAVES and W. NIEUWENKAMP (1935)

▬

157b. Zweidimensionale Cristobalitkristallen, by W. NIEUWENKAMP (1935)

▬

155. M. G. Friedel, dans une Note récente[1], insiste à nouveau sur le fait que l'analyse des édifices cristallins au moyen des rayons X ne fournit pas toujours la maille cristalline véritable. Je pense avoir trouvé dans la biotite (mica noir) et dans les chlorites des exemples propres à illustrer cette thèse.

1. En général, on engendre le feuillet élémentaire des micas en répétant un motif en O^{12} par les translations d'un réseau de losanges (OA, OB, Fig. 1) dont les dimensions sont à peu près indépendantes de la composition chimique du minéral[2]. Dans le cas de la biotite, on trouve une maille (*oa*, *ob*) trois fois plus petite, avec un motif en O^4. Cette exception est unique dans tout le groupe des micas, et par cela même assez suspecte.

2. Chez les chlorites[3] formées de feuillets très analogues à ceux des micas, on retrouve les translations génératrices (*oa*, *ob*) de la biotite. Mais on est conduit ainsi à adopter un motif cristallin (en O^6) qui comporte un nombre d'atomes d'hydrogène $H^{8/3}$ fractionnaire. On pourrait être tenté d'expliquer la valeur fractionnaire de l'exposant de H en lui attribuant une signification statistique (moyenne entre mailles de composition variable), mais alors on devrait s'attendre à le voir varier d'une chlorite à l'autre, ce qui n'est pas. Force est donc d'admettre que la maille véritable est au moins triple de la maille donnée par l'expérience: motif en O^{18} avec 8 atomes H. L'hypothèse la plus plausible, c'est que les feuillets des micas (y compris la

---

[1] G. Friedel, *Comptes rendus* **186** (1928) 1788.
[2] Ch. Mauguin, *Comptes rendus* **186** (1928) 879.
[3] Ch. Mauguin, *Comptes rendus* **186** (1928) 1852.

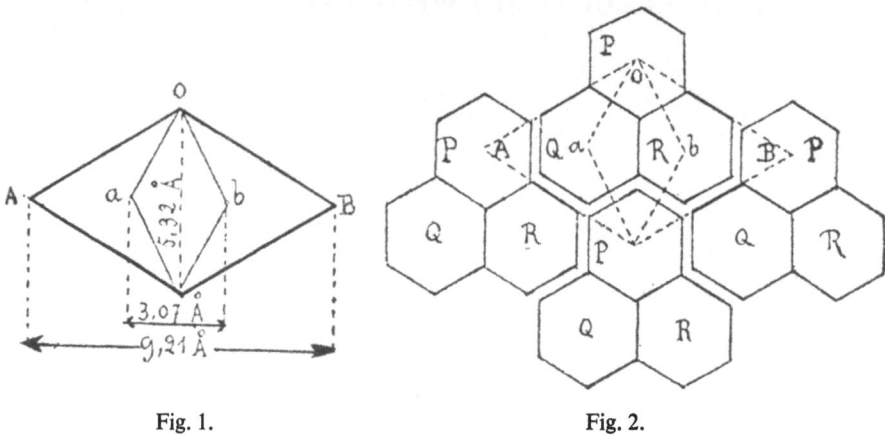

Fig. 1.   Fig. 2.

biotite) et ceux des chlorites admettent les mêmes translations génératrices OA, OB, mais que pour une raison à trouver les rayons X fournissent dans certains cas une valuer trois fois trop petite par la diagonale AB de la maille.

3. Cette manière de voir paraît confirmée par les observations suivantes. La valeur du paramètre AB se déduit des distances qui séparent les rangées horizontales de taches de diffraction sur le cliché (chambre cylindrique) obtenu par rotation du cristal autour de l'axe binaire L_2. Ces rangées sont trois fois plus écartées dans les chlorites (et la biotite) que dans les micas. Mais un examen attentif montre que les rangées de taches qui disparaissent quand on passe des micas aux chlorites (ou à la biotite) sont remplacées par des traînées continues, non résolubles en taches, extrêmement faibles, qui m'avaient d'abord échappées mais que je retrouve sur la plupart des clichés. Il suffirait qu'on consente à considérer ces traînées comme des rangées de taches dégénérées pour retrouver partout le paramètre AB qui paraît bien être le paramètre véritable de tous ces cristaux.

4. Voici un schéma qui permet de se représenter au moins approximativement la façon dont la maille véritable peut être masquée aux rayons X. Supposons le feuillet élémentaire des micas (ou des chlorites) formé de cellules hexagonales (côté $a = \frac{1}{3}\,5{,}28$ Å) où les atomes d'oxygène sont placés exactement de la même manière, mais où les atomes électropositifs prennent trois arrangements distincts que nous désignerons par P, Q, R (Fig. 2). Si le groupe PQR se répète indéfiniment avec la même orientation, les rayons X fourniront OA, OB comme translations génératrices d'un feuillet. Mais admettons que, par suite de macles répétées soit dans un même feuillet, soit au passage d'un feuillet à l'autre, le groupe PQR prenne d'une façon désordonnée trois orientations qui diffèrent de $2\pi/3$

ou $4\pi/3$, chacune avec la même fréquence statistique; tout se passera comme si *oa*, *ob* étaient les translations vraies du système, et l'on trouvera pour le motif la composition moyenne de P, Q, R. Les traînées qui remplacent les rangées de taches sur les clichés semblent indiquer que l'égalité statistique entre les trois orientations n'est pas parfaite.

156. Das Pulverdiagramm von $CdBr_2$ aus wäßriger Lösung führt zu einer Elementarzelle mit $\frac{1}{3}$ Molekül $CdBr_2$. Hieraus ergibt sich eine Struktur, worin sich Schichtenfolgen wie im $CdCl_2$-Typ, und wie im CdJ_2-Typ unregelmäßig abwechseln.

Die Struktur von sublimiertem $CdBr_2$ wurde bestimmt von A. Ferrari und F. Giorgi[1]; sie gehört dem $CdCl_2$-Typ an: rhomboedrische Achsen $r = 6,63$ Å, $\alpha = 34°42'$. Basis Cd (000), Br_I (*uuu*), Br_{II} ($\bar{u}\bar{u}\bar{u}$); $u = \frac{1}{4}$.

Ein Präparat, hergestellt durch Verdunsten einer wässrigen Lösung und Dehydratation im Exsikkator lieferte das Pulverdiagramm der Fig. 1a (cf. Tabelle I). Es zeigt nur einen Teil der Diffraktionslinien der $CdCl_2$-Form (Fig. 1b). Der naheliegende Vergleich mit dem CdJ_2-Typ zeigt ein ähnliches Verhältnis: Unser Diagramm enthält die Linien, welche dem $CdCl_2$- und dem CdJ_2-Typ gemeinschaftlich sind.

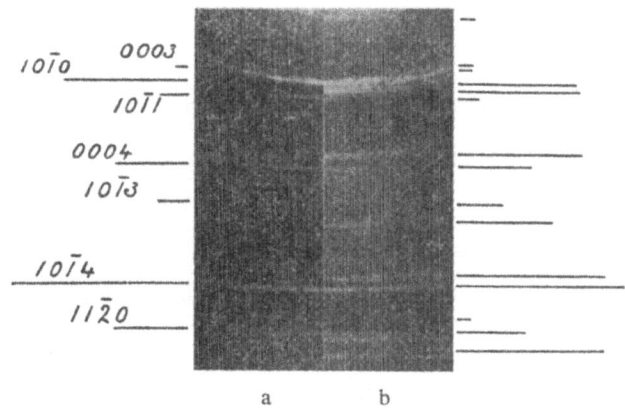

Fig. 1. Röntgenogramme von $CdBr_2$.

a) Aus wäßriger Lösung. b) Sublimiert. Die Diagramme sind auch schematisch gezeichnet. Die Linien in a bilden einen Teil der Linien in b. Die Indizierung der Wechselstruktur ist angegeben.

[1] *Rend. Accad. Lincei* **9** (1929) 1134.

Tabelle I. Pulveraufnahme an $CdBr_2$ aus wäßriger Lösung mit Cu-K$_\alpha$-Strahlen.

Indices	Sin² ϑ ber.	Sin² ϑ exp.	Intensität beob.	Intensität ber.	Inidces	Sin² ϑ ber.	Sin² ϑ exp.	Intensität beob.	Intensität ber.
0001	15	17	St	2	0006	544			$\frac{1}{2}$
0002	60	61	s	$\frac{1}{2}$	$11\bar{2}3$	584			12
0003	136	137	m	2	$20\bar{2}0$	598	601	s	36
$10\bar{1}0$	150	152	m	36	$20\bar{2}1$	618	618	ss	12
$10\bar{1}1$	165	168	s	12	$20\bar{2}2$	658			3
$10\bar{1}2$	210			3	$11\bar{2}4$	690	691	St	72
0004	242	244	St	12	$10\bar{1}6$	693			3
$10\bar{1}3$	286	287	s	12	$20\bar{2}3$	734			12
0005	377	381	s	2	0007	740			2
$10\bar{1}4$	392	394	St	72	$11\bar{2}5$	826	826	ss	12
$11\bar{2}0$	448	451	s	36	$20\bar{2}4$	840	840	St	72
$11\bar{2}1$	464			12	$10\bar{1}7$	889	890	s	12
$11\bar{2}2$	509			3	0008	966	965	m	12
$10\bar{1}5$	527	530	ms	12					

Für die Berechnung der Intensitäten ist $F_{Cd} = 1$ und $F_{Br} = \frac{1}{4}$ angenommen. Die blättrige Entwicklung der Kristalle bedingt das starke Hervortreten der Basis und der benachbarten Flächen.

Strukturbestimmung

Das Pulverdiagramm kann indiziert werden mit

$$\sin^2 \vartheta = 0{,}149_5 \, (h_1{}^2 + h_2{}^2 + h_1 h_2) + 0{,}015_1 \, h_3{}^2$$

für Cu-K$_\alpha$-Strahlen.

In Tabelle I findet man die berechneten und beobachteten $\sin^2 \vartheta$-Werte. Die zugehörige hexagonale Zelle hat die Kanten $a = 2{,}30$ Å, $c = 6{,}23$ Å. Mit der pyknometrisch erhaltenen Dichte 5,0 folgt für die Zahl der Moleküle pro Zelle $0{,}32 \approx \frac{1}{3}$.

Diese zu kleine Zelle läßt sich wie folgt deuten: Die gegenseitige Lage zweier aufeinanderfolgenden hexagonalen Cd-Ebenen ist in unregelmäßiger Abwechslung teils wie in der $CdCl_2$-Struktur—die Translation zwischen ihnen ist eine Rhomboederkante $r(r_1, r_2, r_3)$—, teils wie in der CdJ_2-Struktur—die Translation ist $c = \frac{1}{3} \, (r_1 + r_2 + r_3)$. Gleiches gilt für die Br_I- und für die Br_{II}-Atome: die Translation zwischen zwei aufeinanderfolgenden Br_I- (bzw. Br_{II}-)Ebenen ist r oder c. Wir werden zeigen, daß dann c und $r - c = a$ bei der Diffraktion als Elementartranslationen erscheinen.

Tabelle II. Pulverdiagramme von $CdBr_2$.

			Berechnet für					
a) $CdCl_2$-Typus			b) Wechselstruktur			c) CdJ_2-Typus		
Indices	$\sin^2\vartheta$	νS^2	Indices	$\sin^2\vartheta$	νS^2	Indices	$\sin^2\vartheta$	νS^2
111	15	2	0001	15	2	0001	15	2
						$10\bar{1}0$	50	$\frac{1}{2}$
100	51	3						
110	57	2						
222	60	$\frac{1}{2}$	0002	60	$\frac{1}{2}$	0002	60	$\frac{1}{2}$
						$10\bar{1}1$	65	30
211	77	36						
221	92	6						
						$10\bar{1}2$	110	36
322	132	6						
333	136	2	0003	136	2	0003	136	2
$1\bar{1}0$	150	36	$10\bar{1}0$	150	36	$11\bar{2}0$	150	36
332	157	36						
201	165	12	$10\bar{1}1$	165	12	$11\bar{2}1$	165	12
						$10\bar{1}3$	186	30
						$20\bar{2}0$	200	$\frac{1}{2}$
$11\bar{1}$	201	6						
200	206	2						
321	210	3	$10\bar{1}2$	210	3	$11\bar{2}2$	210	3
						$20\bar{2}1$	215	30
433	217	2						
220	226	36						
311	241	6						
444	242	12	0004	242	12	0004	242	12
443	253	6						
						$20\bar{2}2$	260	36
331	281	6						
432	286	12	$10\bar{1}3$	286	12	$11\bar{2}3$	286	12

Die $\sin^2\vartheta$ sind berechnet für Cu-Kα-Strahlen. Für die νS^2 ist $F_{Cd} = 1$ und $F_{Br} = \frac{3}{4}$ gesetzt worden.

Betrachten wir die Senkrechte durch etwa ein Cd-Atom, mit diesem im Nullpunkt. Von den Punkten c, $2c$, $3c$, ... können wir nur die Möglichkeit einer Besetzung mit Cd-Atome behaupten. Für die Refraktion der Röntgenstrahlen an solch einer unregelmäßig teilweise besetzen Punktreihe, ist die Erfüllung der Refraktionsbedingung der vollständigen Punktreihe hinreichend und notwendig. Hinreichend: wenn die Amplituden aller Punkte sich addieren, gilt dies auch für einen Teil derselben. Notwendig: Wird aus einer größeren Anzahl sich vernichtender Amplituden (wie dies bei Nichterfüllung der Refraktionsbedingung der vollständigen Punktreihe vorliegt) ein Teil abgesondert, ohne bestimmtes Gesetz, so wird dadurch keine merkliche Resultante erhalten.

Für die Br-Atome gilt dieselbe Refraktionsbedingung: Die Punktreihen $u, u + c, u + 2c, \ldots$, bzw. $-u, c - u, 2c - u, \ldots$ sind gleichfalls unregelmäßig teilweise besetzt, und c fungiert wieder als Elementartranslation.

Dieselbe Betrachtung können wir anstellen für Punktreihen nach der Rhomboederkante r. Die Punkte $0, r, 2r, \ldots$ bzw. $\pm u, r \pm u, 2r \pm u \ldots$ sind wieder zum Teil von Cd- bzw. Br-Atomen besetzt. Daher ist r ebenfalls eine Elementartranslation für die Indizierung des Diagramms, und folglich auch $r - c = a$. (Es zeigt somit die Linien, welche sowohl dem $CdCl_2$-Typ wie dem CdJ_2-Typ angehören.)

Die Diffraktion erfolgt also wie an einer hexagonalen Zelle mit den Kanten a und c, und mit

$$\left. \begin{array}{l} \tfrac{1}{3}\ Cd\ in\ 000 \\ \tfrac{1}{3}\ Br_1\ in\ 00u \\ \tfrac{1}{3}\ Br_2\ in\ 00\bar{u} \end{array} \right\} u = 0{,}25$$

Mit dem zugehörigen Strukturfaktor $S = F_{Cd} + 2F_{Br} \cdot \cos uh_3$ sind die Intensitäten (vS^2) der Tabelle I Spalte 5 und der Tabelle IIb berechnet.

Diskussion der Struktur

Unsere Struktur bildet eine Schwankung zwischen dem $CdCl_2$- und dem CdJ_2-Typ. Sie kann gleichfalls als eine wiederholte Zwillingsbildung eines dieser Typen nach der c-Achse gedeutet werden. Es muß dahingestellt bleiben, ob es sich um einen Wiederholungszwilling eines Kristalls des $CdCl_2$-Typ handelt, worin dann vereinzelte Schichtenpaare sich wie im CdJ_2 aneinander legen, oder ob gerade die Mehrzahl der Schichten diese letzte Anordnung aufweist. Die ideale Wechselstruktur wäre ohne Bevorzugung des einen oder des anderen Typus zu denken. Das Röntgenogramm kann jedoch zwischen diesen drei Fällen nicht entscheiden[1].

Längeres Zerreiben im Mörser führt das $CdBr_2$ der gewöhnlichen Form über in die neue Modifikation, wie das geänderte Röntgenogramm zeigt. Die mechanische Beanspruchung hat also eine Verschiebung der Schichten

---

[1] Überflüssig ist wohl die Bemerkung, daß ein Gemenge von—unverzwillingten—Kriställchen der beiden primären Typen ein anderes Diagramm gäbe. Es würde außer den gemeinschaftlichen Linien doch wenigstens auch die stärksten Linien der einzelnen Typen zeigen.

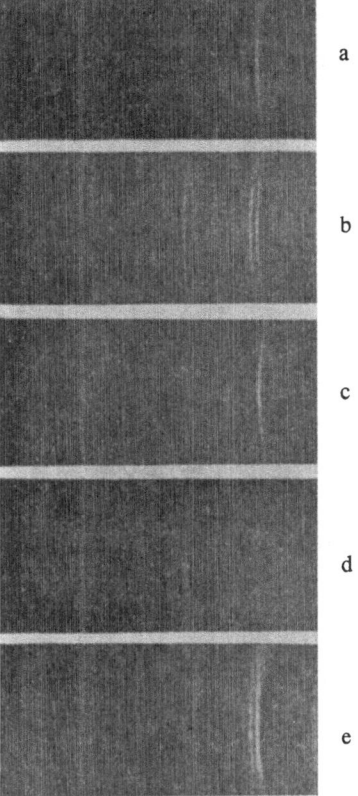

Fig. 2. Röntgenogramme von CdBr₂.

(a) sublimiert und stark zerrieben, (b) sublimiert, (c) aus wäßriger Lösung, (d) id. nach etwa einem Monat, (e) id. nach Erhitzen bis 400°.

zur Folge. Erhitzung auf einige hundert Grad führt einen beschleunigten Übergang von der Wechselstruktur zum CdCl₂-Typ herbei, welcher bei gewöhnlicher Temperatur nur sehr langsam vor sich geht (Fig. 2).

Amsterdam, Laboratorium für Kristallographie der Universität

157a. Einige Aufnahmen, die wir im Göttinger Institut erhielten, und über welche wir in späteren Arbeiten berichten wollen, zeigten, daß zweidimensionale Kristalle öfter in der Natur anzutreffen sind, als es bisher den Anschein hatte. Es schien uns daher von Interesse, die Streueffekte, in welchen die zweidimensionalen Kristalle sich äußern, zusammenzustellen. Es wird sich ergeben, daß nur für Einkristallaufnahmen deutliche charakteristische Merkmale zu erwarten sind, bei Pulveraufnahmen sind die Effekte nicht leicht zu erkennen und schwierig zu deuten. Der letztere Fall wurde bereits von M. v. Laue[1] behandelt. Wir wollen uns daher im folgenden nur mit Einkristalldiagrammen beschäftigen.

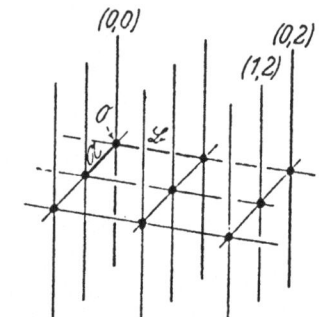

Fig. 1. Atomebene.   Fig. 2. Reziprokes Gitter zu Fig. 1.

Am einfachsten lassen sich die Verhältnisse unter Heranziehung des reziproken Gitters überblicken. Mit M. v. Laue ordnen wir dem Kreuzgitter $\mathfrak{a}$, $\mathfrak{b}$ die reziproken Translationen $\mathfrak{A}$, $\mathfrak{B}$ (in Analogie zu der Ewaldschen[2] Konstruktion für den dreidimensionalen Fall) zu. In den so erhaltenen reziproken Netzpunkten errichten wir eine Schar von Loten, die wir in dieser Notiz Stäbe nennen wollen, dem Vorgehen F. Kirchners[3] folgend. Siehe Fig. 1 und 2. Wir verstehen unter einem Stab (h, k) das Lot, welches im Punkt $[h, k]$ des reziproken Netzes errichtet wurde. Die Beziehungen zwischen Kreuzgitter und reziprokem Gitter werden wie üblich dargestellt durch die vektoriellen Gleichungen:

$$(\mathfrak{a}, \mathfrak{A}) = (\mathfrak{b}, \mathfrak{B}) = 1; \quad (\mathfrak{a}, \mathfrak{B}) = (\mathfrak{b}, \mathfrak{A}) = 0.$$

Wir erhalten die möglichen Reflexionen mittels der von Ewald eingeführten Ausbreitungskugel. Sie hat den Radius $1/\lambda$. Wie im Fall des dreidimensionalen Kristalles legen wir ihre Oberfläche durch den Nullpunkt 0 (siehe Fig. 2). Jeder weitere Punkt P des reziproken Gitters, welcher auf die Kugelfläche zu liegen kommt (hier also jeder Schnittpunkt Stab-Kugel), stellt eine Reflexion dar, deren Richtung gegeben ist durch den Strahl:

[1] Laue, M. v., Kreuzgitterspektren, Z. Kristallogr. **82** (1932) 127.

[2] Ewald, P. P., Z. Kristallogr. **56** (1921) 129 sowie Hdb. d. Phys. **23/2** (1933), Berlin, S. 260 [Vol. I. papers **26** und **27**]. Vgl. auch Bernal, J. D., Proc. Roy. Soc. London (A) **113** (1926) 117.

[3] Kirchner, F., Zerstreuung von schnellen Kathodenstrahlen usw., Ann. Physik **13** (1932) 53.

Kugelmittelpunkt → P. Die Richtung Mittelpunkt → Nullpunkt ist die Richtung des einfallenden Strahls. Da im allgemeinen immer Schnittpunkte von Stäben mit der Kugelfläche vorhanden sein werden, werden für jede Wellenlänge und jede Richtung des Primärstrahles Reflexionen auftreten. (Nur wenn die Wellenlänge sehr groß, größer als 2a und 2b ist, tritt bei streifender Inzidenz keine Reflexion ein.) Mit dieser Vorstellung lassen sich nun die Effekte, welche bei verschiedenen Versuchsanordnungen zu erwarten sind, anschaulich ableiten.

A. Monochromatische Strahlung

1. STEHENDER KRISTALL. Wegen der monochromatischen Strahlung hat die Ausbreitungskugel einen festen Radius, und außerdem, weil der Kristall nicht bewegt wird, eine feste Lage im reziproken Gitter. Um die Schnittpunkte der Stäbe mit der Kugel leicht überblicken zu können, fassen wir jeweils in einer Ebene liegende Stäbe zusammen. Wir wollen dabei zwei Hauptfälle unterscheiden: a) Die Ebenen enthalten den Stab $(0, 0)$, s. Fig. 2; β) die Ebenen enthalten den Stab $(0, 0)$ nicht. Wir behandeln zunächst den Fall a. Eine solche Ebene $E_{(h,k)}$ mit den Stäben ... $(\bar{h}, \bar{k})$, $(0, 0)$, (h, k), $(2h, 2k)$, ... schneidet die Kugel in einem (i. allg. kleinen) Kreise, der durch den Nullpunkt geht. Daher liegen die Richtungen der entsprechenden Reflexionen auf einem Kegelmantel. Im Röntgenogramm werden sie also auf einem Kegelschnitt angeordnet sein wie die Reflexionen einer Zone im gewöhnlichen Lauediagramm. (In unserem Falle ist diese Zone die Atomreihe $[k, h]$ in der Netzebene, Fig. 1.) Wir kommen jetzt zu dem Fall β. Eine Ebene, z.B. mit den Stäben ... $(\bar{1}, k)$, $(0, k)$, $(1, k)$, $(2, k)$ ..., schneidet die Kugel ebenfalls in einem Kreise und deswegen liegen auch hier die Reflexionsrichtungen auf einem Kegelmantel. Der Kreis geht aber nicht durch den Nullpunkt. Im Röntgenogramm werden die Reflexe also auf einem Kegelschnitt angeordnet sein, welcher aber nicht durch den Primärfleck geht. Die Kegelschnitte sind daher nicht den gewöhnlichen Lauekreisen zu vergleichen, sondern sie haben den Charakter von Schichtlinien. Bei der speziellen Versuchsanordnung mit dem Primärstrahl senkrecht zu der Atomebene, also parallel den Stäben des reziproken Gitters, erhält man auf einer senkrecht zum Primärstrahl aufgestellten Platte das bekannte 'Kreuzgitterspektrum', d.h. Anordnung der Reflexe auf radialen Linien (α) und auf Scharen paralleler Linien (β). Es mag nicht uninteressant sein, zu bemerken, daß man bei jeder beliebigen Einstrahlrichtung ein 'Kreuzgitterspektrum' erhalten kann, sofern man nur die Platte parallel der Atomebene aufstellt. (Man sieht dies sofort ein,

wenn man bedenkt, daß die Ausbreitungskugel, wie auch immer der Primärstrahl einfallen mag, von den Stäben des reziproken Gitters immer in einer ähnlichen Weise geschnitten wird.)

2. DREHAUFNAHMEN. Dreht man den zweidimensionalen Kristall um eine beliebige Richtung, dreht sich also das reziproke Gitter um eine beliebige Gerade durch den Nullpunkt, so überstreichen die Stäbe Flächen (i. allg. Hyperboloide, welche zu Ebenen, Kegeln oder Zylindern entarten können). Diese Flächen schneiden die Ausbreitungskugel in Kurven: das Drehdiagramm zeigt daher kontinuierliche Kurven. Wir nehmen im folgenden immer an, daß der Primärstrahl (wie üblich) auf der Drehachse senkrecht steht, und wollen einige Fälle näher diskutieren. a) Ist die Drehachse parallel den Stäben, dann beschreibt jeder Stab (h, k) einen Zylinder. Dessen Schnittkurve mit der Kugel gibt die Reflexionsrichtungen an, welche im dreidimensionalen Fall sämtliche Reflexe $(h, k, 0)$, $(h, k, 1)$, $(h, k, 2)$ ... enthalten würden. Ein zylindrischer Film um die Drehachse zeigt dann Kurven der in Fig. 3 gegebenen Gestalt. Es sind die Schieboldschen[1]

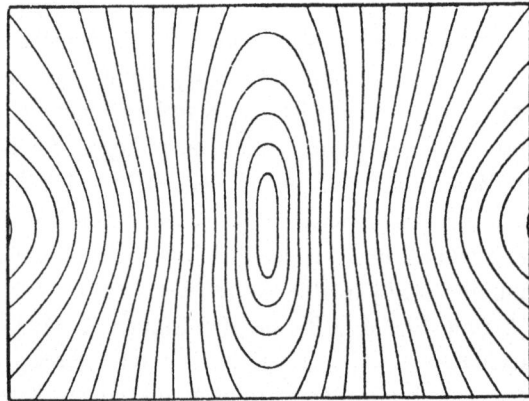

Fig. 3. Schichtlinien 2. Art nach Schiebold. Die Figur wurde mit gewisser Abänderung der oben zitierten Arbeit von J. D. Bernal entnommen.

Schichtlinien 2. Art; sie müssen in unserem Falle als kontinuierliche Kurven auftreten. b) Ist die Drehachse senkrecht zu den Stäben, dreht man also das Atomnetz um eine in ihm liegende Richtung, dann beschreiben die Stäbe Ebenen, welche die Ausbreitungskugel in parallelen Kreisen schneiden. Das Drehdiagramm zeigt daher kontinuierliche Schichtlinien, die horizontal verlaufen. Die horizontale Linie, die einem bestimmten Stabe entspricht, zeigt im allgemeinen einen charakteristischen Intensitätsverlauf.

[1] Schiebold, E., Die Drehkristallmethode, *Forschr. d. Mineralogie usw.* **11** (1927) 167.

Die Intensität steigt gegen die Mitte des Filmes hin (nach kleineren Streuwinkeln) im allgemeinen stark an, um mit einer scharfen Kante plötzlich abzubrechen. (Schneidet nämlich ein Stab die Drehachse nicht, so überstreicht er nicht eine ganze Ebene, sondern die überstrichene Fläche wird nach innen zu von einem umhüllenden Kreise begrenzt. Der Schnittpunkt dieses umhüllenden Kreises mit der Ausbreitungskugel gibt die Richtung der im vorigen Satz erwähnten Kante an.) Der Intensitätsverlauf kann in seinen Einzelheiten mittels Weißenberg-Aufnahmen gut verfolgt werden.

Göttingen, Mineralogisches Institut

157b. Ramdohr[1] unterscheidet drei verschiedene Trachttypen des Cristobalits:

1. Reguläre Oktaeder in gewöhnlicher Ausbildung;
2. sechsseitige Tafeln mit rhomboedrischer Begrenzung;
3. solche, die geometrisch durchaus dem Tridymit gleichen.

Die Kristalle regulärer Tracht geben Drehdiagramme, welche tetragonal zu indizieren sind, mit $a = 7,0_3$; $c = 6,9_0$ Å. Da Reflexe hkl mit $h + k$ ungerade fehlen, ist diese Zelle basiszentriert. Ihre Kanten sind parallel den vierzähligen Achsen des Oktaeders. Die Kristalle sind in der Weise verzwillingt, daß die drei möglichen Orientierungen (die tetragonale c-Achse parallel [100], [010], [001] des Oktaeders) in etwa gleicher Verbreitung vorhanden sind[2]. Das Drehdiagramm zeigt nur scharfe Reflexe, welche alle dem dreidimensionalen Cristobalitgitter entstammen.

Anders verhalten sich die Cristobalite von tridymitähnlicher Tracht. In deren Röntgenogramm finden sich Interferenzerscheinungen wie sie an 2-dimensionalen Gittern zu erwarten sind[3]; durch geeignete Wahl der Versuchsanordnung ist es möglich, jeden in unserer theoretischen Arbeit diskutierten Fall zu verwirklichen (Fig. 1–7). Es läßt sich daher nicht umgehen, in diesen Cristobaliten zweidimensionale Kristallschichten parallel (111) anzunehmen. Die Entstehung dieser Gebilde kann man sich folgendermaßen erklären:

[1] Ramdohr, P., Über die Blaue Kuppe bei Eschwege und benachbarte Basaltvorkommen. Diss. Göttingen (1919). *Jahrb. preuß. geol. Landesanst.* **1** (2) (1919) 284.

[2] Die innige Verzwillingung verhindert zwar die gesonderte Beobachtung einiger Reflexe (hkl und hlk bei k und $l \leqq 2$), beeinträchtigt weiter aber die Brauchbarkeit des Drehverfahrens nicht.

[3] Laves, F., und W. Nieuwenkamp, Interferenzerscheinungen an zweidimensionalen Kristallen, *Z. Kristallogr.* **90** (1935) 273 [this Vol. preceding paper].

Fig. 1–7. Interferenzerscheinungen an zweidimensionalen Cristobalitkristallen[1]. [Fig. 6 and 7 not reproduced.]

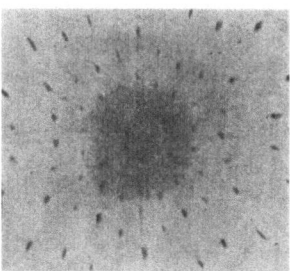

Fig. 1. 'Kreuzgitterspektrum' an Cristobalit. Primärstrahl senkrecht zu (111). Monochromatische Strahlung Cu-K$_\alpha$.

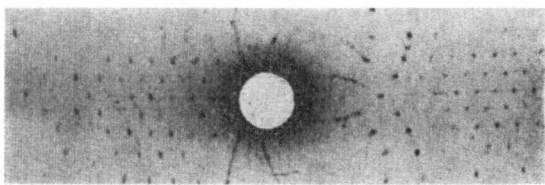

Fig. 2. Primärstrahl unter 45° geneigt zu (111). Zylindrischer Film zur Erhaltung de beiden Kreuzgitterspektren. Für dessen Erklärung siehe Laves und Nieuwenkamp *Z. Kristallogr.* **90** (1935) 273 unter Anordnung A 1. (Die Kontinuität der 'Laue-Kreise' rührt von der Cu-K-Strahlung beigemischter weißer Strahlung her.)

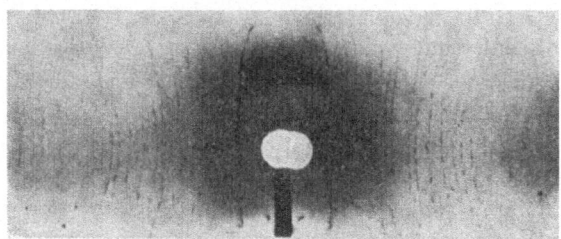

Fig. 3. Drehaufnahme um [111], zeigt kontinuierliche Schichtlinien zweiter Art. (Laves und Nieuwenkamp A 2a).

---

[1] Für weitere Beispiele von Interferenzerscheinungen an zweidimensionalen Atomanordnungen siehe F. Laves, Zweidimensionale Überstrukturen. *Z. Kristallogr.* **90** (1935) 279.

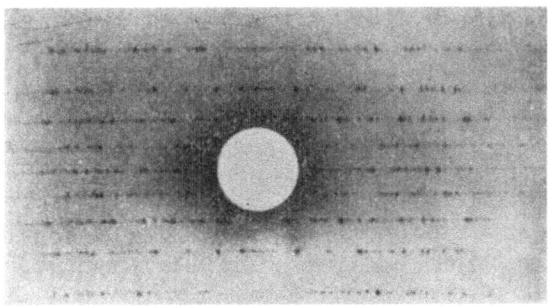

Fig. 4. Drehaufnahme um [110] mit kontinuierlichen Schichtlinien erster Art. (Laves und Nieuwenkamp A 2b).

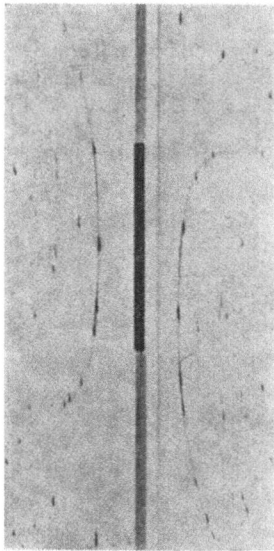

Fig. 5. Weißenberg-Aufnahme des Äquators der vorstehenden Aufnahme.

Die Cristobalite von der dritten Tracht werden von Ramdohr gedeutet als Pseudomorphosen nach Tridymit[1]. Die Strukturen von Cristobalit und Tridymit verhalten sich grundsätzlich wie jene von Zinkblende und Wurzit, oder wie eine kubische und eine hexagonale dichteste Kugelpackung; es

---

[1] Die Richtigkeit dieser Auffassung stellen die Drehdiagramme außer Zweifel; nebst den kontinuierlichen Linien enthalten sie scharfe Reflexe des dreidimensionalen Cristobalits und solche des Tridymits.

245

gibt den beiden Strukturen gemeinsame Schichten parallel (111) in Cristobalit und parallel (0001) in Tridymit. Diese Schichten haben die Dicke eines
Elementarabstands in Tridymit, $\frac{2}{3}$ des Parameters von [111] in Cristobalit.
Die Verknüpfung der SiO_4-Tetraeder ist in beiden Schichten völlig dieselbe,
die Tetraeder haben aber gegenüber der Idealposition des C_9- bzw. C_{10}-
Typus Drehungen erfahren, welche für Tridymit und Cristobalit verschieden sind, entsprechend ihrer verschiedenen Dichte und Lichtbrechung.

Der Übergang Tridymit → Cristobalit innerhalb einer Schicht besteht
daher lediglich in einer kleinen Drehung der SiO_4-Tetraeder, und wird
sich ebenso glatt vollziehen können wie die Übergänge hoch ⇄ tief-Modi-
fikation von Tridymit, Cristobalit oder Quarz.

Die Verknüpfung der Schichten untereinander ist aber in Tridymit eine
andere als im Cristobalit; um aus Tridymit einen dreidimensionalen Cristo-
balitkristall zu erhalten, müssen Bindungen Si-O zerbrochen werden. Hier
steht also dem Übergang eine hohe Energieschwelle im Wege; es ist eben
diese Hemmung, welche die Übergänge Tridymit ⇄ Cristobalit ⇄ Quarz so
stark verzögert, daß sie künstlich noch kaum veranlaßt werden konnten.

Es ist daher anzunehmen, daß in den Pseudomorphosen von Cristobalit
nach Tridymit der Übergang in einzelnen Schichten parallel (111)/(0001)
vollständig erfolgt ist, in den Bereichen zwischen zwei Schichten aber teilweise oder gänzlich gehemmt wurde. Der Kristall setzt sich dann aus einer
Anzahl zweidimensionaler Cristobalitschichten zusammen, die wegen der
unregelmäßigen Zwischenräume bei der Röntgenstreuung keine gesetzmäßige Phasenverknüpfung besitzen: es werden einfach die Intensitäten
der 'Kreuzgitterspektren' addiert.

Herrn Professor V. M. Goldschmidt bin ich für die gastfreundliche
Aufnahme im Institut, sowie für sein stetes förderndes Interesse zu großem
Dank verpflichtet. Die Anregung zu der Untersuchung verdanke ich
Herrn Dr. F. Laves; er stand mir auch während derselben in freundlichster
Weise mit Rat und Tat zur Seite.

C. ROTATING GROUPS

158. Molecular Rotation in the Solid State, by S. B. HENDRICKS (1930)

▬▬

159. An X-ray Investigation of Normal Paraffins near their Melting Points, by A. MÜLLER (1932)

▬▬

160. Gradual Transition in Sodium Nitrate II. The Structure at Various Temperatures and its Bearing on Molecular Rotation, by F. C. KRACEK, E. POSNJAK and S. B. HENDRICKS (1931)

▬▬

161. Molecular Rotation in Solid Sodium Nitrate, by J. M. BIJVOET and J. A. A. KETELAAR (1932)

▬▬

158. *Molecular Rotation in the Solid State*

The determined crystal structures of a number of primary alkyl ammonium halides indicate that in such compounds the carbon atoms are arranged collinearly[1] in a particular group. Thus in the case of primary amyl ammonium chloride[2] the X-ray diffraction data from powders and single crystals can be completely explained by a tetragonal unit of structure containing $2NH_3C_5H_{11}Cl$ with $a = b = 5.01$ A, $c = 16.69$ A. The space group is D_4^2, V_d^3, S_4^1, C_{4v}^1, C_4^1, D_{4h}^7, and the Cl, N and C atoms are at $0\frac{1}{2}u$, $\frac{1}{2}0\bar{u}$, with $u_{Cl} = 0.095$. The absence of reflections in odd orders from planes ($hk0$) with ($h + k$) odd and the intensities of reflections from other planes such as (200) require the carbon atoms of the C_5H_{11} groups to scatter X-radiation as if they are arranged collinearly in each group.

Prof. Linus Pauling, of the California Institute of Technology, has recently suggested to me that the indicated collinear arrangement of carbon atoms might be in error. If the carbon atoms of an alkyl group really have a 'zig-zag' arrangement and the group is rotating[3] about its chain axis independent in phase of other rotating groups, then the result

---

[1] S. B. Hendricks, *Z. f. Krist.* **67** (1928) 106, 475; **68** (1928) 189.

[2] S. B. Hendricks, *Z. f. Krist.* (In press).

[3] See Linus Pauling, *Phys. Rev.* (July 1930).

given above would be obtained. If the temperature should be sufficiently lowered, this complete rotation would be replaced by slight oscillation about some equilibrium positions.

Observations were made on primary amyl ammonium chloride at approximately liquid air temperatures. The density, determined by suspension in mixtures of liquid nitrogen and oxygen, is c. 1.0, probably a little greater than the value 0.953 at 25°C. Diffraction lines on powder photographs (CuK radiation) at liquid air temperatures are similar in spacings to, but markedly different in intensities from, those at room temperatures; some few additional lines requiring a larger unit of structure are also present. It is difficult to determine accurately the structural characteristics of such a complex compound from powder photographs alone. The photographs at liquid air temperatures, however, indicate, from their similarity to photographs at 25°, that the crystals have approximately orthogonal axes and that the atomic arrangement in the unit of structure containing $4NH_3C_5H_{11}Cl$, with the dimensions $a = b = 7.0$ Å, $c = $ circa 16.6 Å, is closely similar to that in the unit of structure previously described. The presence of planes such as (210) and (300), referred to the axes of the larger unit of structure, could be explained by alteration in the structure but probably is best accounted for by absence of the suspected molecular rotation that leads to the fortuitous determination of the unit of structure and atomic arrangement at room temperatures.

The best test for a possible collinear arrangement of the carbon atoms of a C_5H_{11} group is the absence of reflections in odd orders from planes $(hk0)$ with $(h + k)$ odd. (Indices referred to a unit of structure having $a = b = 5.0$ Å, $c = 16.6$ Å). The large axial ratio makes it difficult to distinguish between reflections from $(hk0)$ and (hkl) with l unity on powder photographs. Reflections from (200) and (201), however, are very weak at liquid air temperatures, while reflections from (200) are strong at room temperature. This change in the intensity of (200) could be explained by a great departure of the chain axes of the C_5H_{11} groups from parallelism with the tetragonal axis of the crystal at liquid air temperatures, or, as is most probable, by a departure of the carbon atoms from a collinear arrangement.

The most immediate conclusion is that the carbon atoms in a C_5H_{11} group are arranged in a 'zig-zag' manner and that the characteristics of the X-ray diffraction photographs made at room temperatures from crystals of primary amyl ammonium chloride arise partially from rotation of the C_5H_{11} groups about their chain axes. The configuration of a hydrocarbon chain as deduced from the crystal structure of the primary alkyl ammonium halides is thus probably the same as that first found by Müller and Shearer[1] for some long chain aliphatic compounds. The carbon to carbon separation along the chain axis is *c.* 1.20–1.30 A, and the carbon-carbon distance might well be *c.* 1.54 A.

It thus seems, as Pauling has indicated, that in a crystal containing molecules or molecular groups with small moments of inertia about some axes, these molecules may undergo rotational motion about these axes.

Fertilizer and Fixed Nitrogen Investigations,
Bureau of Chemistry and Soils, Washington, D.C.

159. The present paper is a continuation of previous work[2] on the thermal expansion of normal paraffins.

Experimental

The recording of the lattice dimensions was obtained with the aid of X-ray photographs. It was essential to keep the temperature of the substance constant to a fraction of a degree during the exposure to the X-rays, and it was also desirable occasionally to vary the temperature in small steps of a few tenths of a degree. The small thermostat in which the substance had to be kept during the X-ray exposure was heated to the required temperature by a constant flow of preheated water. The thermostat itself was made from a solid copper rod. It had hollow spaces through which the water circulated and slots through which the incident and the reflected X-ray beams entered and emerged. The flat surface on which the substance was spread went through the axis of rotation of the spectrometer. The spectrometer table carrying the thermostat was oscillated through about 5° during the exposure.

The reflected rays after emerging from the thermostat had to pass through a long narrow slot in a shield behind which was a film drum. This drum was turned through a small angle after each exposure, and in this way a whole series of photographs were recorded on the same film. It was easy to measure the shift of any of the recorded lines relative to a reference line, and this was all that was required for the measurement of the expansion.

[1] *Jour. Chem. Soc.* **123** (1923) 3156; G. Shearer, *Proc. Roy. Soc.* A **108** (1925) 655.
[2] *Proc. Roy. Soc.* A **127** (1930) 417.

A considerable time saving was obtained by the use of an X-ray generator with a rotating anode. The generator had a copper anode and ran as a rule at about 30 KV. and 120–150 milliamps. The average time of exposure was only a few minutes. The temperature adjustment took not more than 5 minutes. The X-rays were, as a rule, filtered by nickel foil and the wavelength used was 1.539 A.

Results

Previous work on the expansion of normal paraffins has shown that the length of the chain axis depends much less upon the temperature than the length of the two other axes. These observations have been confirmed in

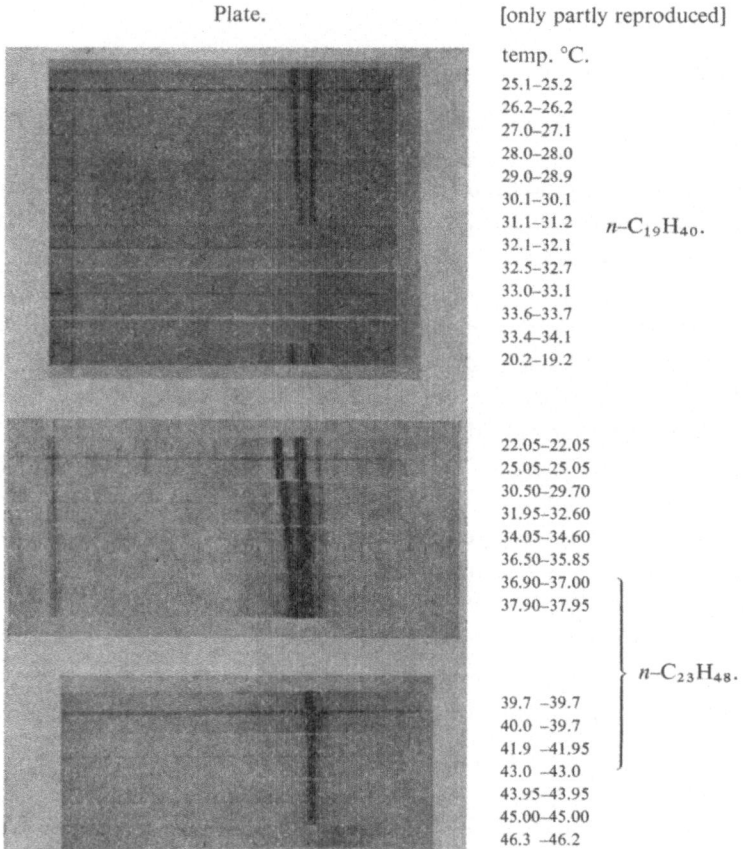

Plate.

[only partly reproduced]

temp. °C.

25.1–25.2	
26.2–26.2	
27.0–27.1	
28.0–28.0	
29.0–28.9	
30.1–30.1	
31.1–31.2	$n\text{--}C_{19}H_{40}.$
32.1–32.1	
32.5–32.7	
33.0–33.1	
33.6–33.7	
33.4–34.1	
20.2–19.2	

22.05–22.05	
25.05–25.05	
30.50–29.70	
31.95–32.60	
34.05–34.60	
36.50–35.85	
36.90–37.00	
37.90–37.95	

$n\text{--}C_{23}H_{48}.$

39.7 –39.7
40.0 –39.7
41.9 –41.95
43.0 –43.0
43.95–43.95
45.00–45.00
46.3 –46.2

the present work. The expansion of the c axis (chain axis) is too small to be measured with the apparatus used in the present work. The investigation is essentially confined to the expansion of the two axes which are in a plane perpendicular to the chain axes, and the discussion of the numerical data will mainly deal with those obtained from the normal form.

The actual photographs of some specimen are reproduced on the Plate. The 110 and the 200 reflection draw closer and closer together and become indistinguishable from each other as the temperature approaches the melting point, showing that the structure changes into hexagonal close packing.

The transition from the less symmetrical form into the hexagonal close packing is a continuous function of the temperature.

$C_{21}H_{44}$ behaves in the same way as $C_{23}H_{48}$. $C_{19}H_{40}$ differs from the two in so far that the substance melts before the symmetrical state is reached, Plate.

Photographs of the substances in the molten state show a diffuse band instead of the well-defined lines characteristic of the solid.

Discussion of Results

The data of the last chapter show that the normal paraffins tend to become hexagonal at the melting points. In the range of C_{21} to C_{29} this state of high symmetry is actually reached. Substances outside this range melt before becoming hexagonal.

X-ray investigation has shown that the CH_2 groups in the paraffin molecule are arranged in a zig-zag chain, and that the chains have only two planes of symmetry intersecting in the chain axis. This is found from measurements made at room temperature. The present work shows that the molecules behave as if they were more symmetrical at the melting point.

The inert character of the paraffin molecule, its heat resisting properties and the fact that the chain length does not depend appreciably upon the temperature, suggests a considerable rigidity of the molecular structure. It seems therefore unlikely that the temperature should produce a radical change of the configuration of the carbon skeleton of the molecule.

The carbon chain is surrounded by hydrogen atoms. This hydrogen shell is more likely to be affected by temperature. It is almost certain that the temperature motion of the hydrogens tends to make the molecule more symmetrical at higher temperatures, but it is impossible to tell at the present stage whether the temperature agitation of the hydrogen molecules accounts

251

for the observed increase of symmetry, or whether the zig-zag structure of the carbon chain is still predominant.

The molecule as a whole must perform oscillations under the influence of the temperature. These oscillations also tend to make the crystal more symmetrical. The moment of inertia of the chain molecule is smallest relative to the chain axis. The amplitudes of the oscillations round this axis may become very large at higher temperatures and the molecules may even perform complete rotations. They would then, on the average, have the symmetry of a circular cylinder and the hexagonal close packing would follow quite naturally.

The writer wishes to express his appreciation to Sir William Bragg, O.M., F.R.S., and the Managers of the Royal Institution for their kind interest in the work.

160. Results of measurements which establish a gradual transition in crystalline sodium nitrate were presented in a previous paper[1]. It was shown that the transition is completed at 275°, but that the changes in heat capacity and volume are distributed over an extended temperature interval.

The hypothesis of molecular rotation within the lattice offers a plausible explanation of the anomalous behavior of several other substances. The consequences of this hypothesis are of great importance in the theory of the solid state; it is accordingly obvious that each new case must be tested for possible contradictions. In sodium nitrate the physical conditions are favorable for a detailed examination in view of the accessibility of the temperature region over which the gradual transition is extended. The structure, in addition to the other criteria of the transition as previously described, can be studied over the whole range of temperatures. The ambiguity encountered in the ammonium salts, due to the low scattering power of the hydrogen atoms, does not enter in this case to vitiate the results by uncertainties in the experimental data.

The Structure of Sodium Nitrate below 185°

Sodium nitrate crystallizes in the ditrigonal scalenohedral division of the hexagonal system. At 25° the unit of structure containing $2NaNO_3$

---

[1] F. C. Kracek, *J. Am. Chem. Soc.* **53** (1931) 2609.

has $a = 6.32$ Å and $\alpha = 47°14'$,[1] corresponding to the crystallographically observed angle of $102°42.5'$. The atomic arrangement according to Wyckoff is

Na atoms at $\frac{1}{4}\frac{1}{4}\frac{1}{4}$; $\frac{3}{4}\frac{3}{4}\frac{3}{4}$

N atoms at 000; $\frac{1}{2}\frac{1}{2}\frac{1}{2}$

O atoms at $u\,\bar{u}\,0$; $\frac{1}{2}-u, \frac{1}{2}+u, \frac{1}{2}$

$\bar{u}\,0\,u$; $\frac{1}{2}+u, \frac{1}{2}, \frac{1}{2}-u$

$0\,u\,\bar{u}$; $\frac{1}{2}, \frac{1}{2}-u, \frac{1}{2}+u$

with $u = \frac{1}{4}$ within close limits.

There are no significant changes in the intensities of the reflections in the region 25 to 185°.

The data obtained from powder photographs taken at temperatures below 185° show satisfactory agreement between observed and calculated intensities of reflections. (*see* Table I).

Table I. Powder diffraction data on sodium nitrate below 185°. MoKα radiation. [Table only partly reproduced.]

	Intensity	
hkl	Obs.	Calcd.
110	2	1.2
211	> 10	27
222	4	4.2
1Ī0	3	3.1
210[a]	5.7	10.5
200	4	3
220	1	1.4
323	7	9.3
321	3	5.6
2Ī0[a]	3	4.4

[a] Planes in which O atoms alone contribute to reflections.

Intensities were calculated on the basis of the equation

$$\rho = c.J.\frac{1 + \cos^2 2\theta}{\sin 2\theta}.F\bar{F}$$

employing the F values published by Bragg and West[2] for sodium and oxygen and assumed F values for nitrogen equal to three-fourths the corresponding values for oxygen.

[1] W. L. Bragg, *Proc. Roy. Soc.* (*London*) A **89** (1913) 468; R. W. G. Wyckoff, *Phys Rev.* **16** (1920) 149.

[2] W. L. Bragg and J. West, *Z. Krist.* **69** (1929) 118.

The Structure of Sodium Nitrate in the Transition Region

Extraordinary changes take place at temperatures above 215° in the intensities of reflections from some planes as compared with 25°. Oxygen atoms alone contribute to odd order reflections from planes with one odd and two even indices, whose sum is not divisible by four (oee; $(e_1 + e_2)/n \neq 4$). At 250° only (120) persists, and at 280°, see Table IV, only uncertain traces of any reflections due to oxygen atoms alone are observable on the films. This indicates that the oxygen atoms at temperatures above 275°, the temperature at which the gradual transition is completed, scatter x-radiation as if they were on the trigonal axes. This observation is in harmony with the postulate that the nitrate ions are rotating about the trigonal axes.

If it is assumed that the nitrate groups, at temperatures above 275°, scatter x-radiation as pseudo-atoms on the trigonal axes, the F values reduce to the following two cases

$$(a) \quad (h + k + l)/4 = n, \quad F = 2NO_3 + 2Na$$
$$(b) \quad (h + k + l)/4 \neq n, \quad F = 2NO_3 - 2Na$$

For large values of $\sin^2 \theta$ only reflections from planes of type (a) would be expected to persist. This is true for planes having spacings smaller than that of (422).

The calculation of intensities of reflections for planes having spacings greater than that of (422) presents a most interesting difficulty. A rotating NO_3 group would be expected to have a time average electronic distribution more closely resembling a torus or a hollow sphere than a sphere. In the case of a torus its scattering power would certainly not be a function of θ alone but would depend markedly upon the orientation of the NO_3 group relative to the reflecting plane. For planes having approximately the same values of θ, the maximum value of F would be expected for that plane most closely perpendicular to the trigonal axes; the minimum for a plane parallel to the trigonal axes. That such an effect plays a prominent role can be ascertained from attempts to calculate intensities of reflections for (110) (222), ($1\bar{1}0$) and (200) on the basis of any assumed monotonic type of F curve for the nitrate group.

We have calculated the intensities assuming that $F(NO_3) = F(N) + F(3O) = F(3\frac{3}{4}O)$—Table IV, case B.

As would be expected, there is no agreement between the observed and calculated intensities of reflections. As a matter of fact, the better agreement for planes of this type is obtained by assuming that the oxygen atoms

254

Table IV. Powder diffraction data on sodium nitrate at 280–290°.
MoKα radiation. [Table only partly reproduced.]

| | | Intensities | |
| | | Calcd. | |
hkl	Obs.	A	B
110	3.2	1.2	10
211	>10	27	27
222	2.5	4.2	1.5
110	3.4	3.1	15.3
210[a]	<0.5	10.5	0.0
200	2.4	3	0.5
220		1.4	7.5
332	7.7	9.3	7.5
321	2.2	5.6	2.5
210[a]	<0.5	4.4	0.0

A, oxygen atoms in fixed positions. B, NO$_3$ group at a point on trigonal axes.

[a] Planes in which oxygen alone contributes to reflections.

occupy fixed positions. Such an assumption is more nearly in harmony with the physically allowable electronic distribution for a torus (three points on the rim) rather than the electronic distribution concentrated at points on the trigonal axes.

We may conclude, then, that there is no serious objection to the hypothesis of molecular rotation as an explanation for the gradual transition in sodium nitrate.

It is significant that during the transition interval the axial angle α decreases as the temperature increases.

The rotational degree of freedom of the nitrate group about an axis normal to the plane of the group is apparently the one excited in the region 150 to 280°. The other two rotational degrees of freedom about axes in the plane of the nitrate group are probably degenerate below the melting point. The moment of inertia is greatest in the first case. It is probable that this characteristic is to be accounted for by a considerably smaller value of V_0, which indeed is suggested by the atomic arrangement at room temperature.

Washington, D.C.

161. It was shown by Kracek and his co-workers[1], that the gradual transition in sodium nitrate at 275° is accompanied by an important change

[1] Kracek, Posnjak and Hendricks, *J. Am. Chem. Soc.* **53** (1931) 3339.

of the intensities of the diffraction lines in the powder diagrams given by this substance. They conclude 'that there is no serious objection to the hypothesis of molecular rotation as an explanation of the gradual transition in sodium nitrate', this conclusion being reached by a qualitative intensity discussion, the quantitative calculation of the rotating model presenting 'a most interesting difficulty'.

Now the scattering power of a ring model has been calculated by Coster[1] and by Kolkmeyer[2] with a view to the possibility of electron binding rings in diamond and by one of us[3] in testing electronic models of lithium. For the case of sodium nitrate we have now performed the intensity calculation along these lines and reached a fair interpretation of the observed intensities, which offers a strong affirmation of the model proposed.

The calculation of the scattering power of a ring of electrons (atoms) can be made in the following way. All points in the same lattice reflection plane have the same phase. A point at a distance d from this plane has a phase difference of $4\pi d \sin \theta / \lambda$ where θ represents the glancing angle.

Let the angle between the reflecting lattice plane and the plane containing the orbit of the N rotating electrons (atoms) be α and $d\varphi$ an element of the orbit containing $N(d\varphi/2\pi)$ electrons. The distance of this element to the lattice plane is then $\rho \sin \alpha \sin \varphi$ (ρ = radius of the orbit).

The diffracted amplitude is

$$\frac{N}{2\pi} e^{4\pi i d \sin \theta / \lambda} \, d\varphi.$$

Integrating this over the circle we find the amplitude of the ring diffraction (phase compared with rays scattered by a point in the plane, e.g., the center of the orbit)

$$\frac{N}{2\pi} \int_0^{2\pi} e^{4\pi i \rho \sin \alpha \sin \theta \sin \varphi / \lambda} \, d\varphi = N J_0 \left(4\pi \frac{\rho}{\lambda} \sin \alpha \sin \theta \right)$$

where $J_0(x)$ represents the Bessel function of the 0th order of the argument (x).

This calculation is based on the very probable supposition that there is no strict phase relation between the rotation in neighboring cells.

In the case of the rotating NO_3 group we substitute $N = 3F(O)$ where $F(O)$ is the atomic scattering factor for oxygenium. The phase is compared with that of the center of the orbit, the nitrogen atom.

We have calculated the intensities in the usual way on the basis of the

[1] Coster, *Verslag. Akad. Wetenschappen Amsterdam* **28** (1919) 391.
[2] Kolkmeyer, *ibid.* **28** (1920) 767.
[3] Bijvoet, *Rec. trav. chim.* **42** (1923) 874.

equation $I = P \dfrac{1 + \cos^2 2\theta}{\cos\theta \sin^2\theta} S^2$ with

$$S = \begin{cases} 2F(\text{Na}) + 2F(\text{N}) + 6F(\text{O})I_0(x) & h + k + l = 4p \\ -2F(\text{Na}) + 2F(\text{N}) + 6F(\text{O})I_0(x) & h + k + l = 2p \\ 0 & h + k + l \neq 2p \end{cases}$$

Here $J_0(x)$ represents the scattering power of a ring, x being equal to $4\pi(\rho/\lambda) \sin\theta \sin\alpha$; ρ, radius of the ring, and α the angle between reflecting and orbit plane.

The atomic scattering factors were taken according to James and Brindley[1]. For nitrogen the factor curve for neutral atoms was taken, as N^{+5} was found not to be in accordance with the observed intensities. For oxygenium also the factor for neutral atoms is used, which only slightly differs from that of O^{-2} and only for small diffraction angles.

Now it is the problem to ascertain whether it is possible to fix a value for ρ which gives good agreement between calculated and observed intensities. As the Bessel function has alternating positive and negative values, it is easy to limit this value. From the fact that $32\bar{1}$ is much stronger than the corresponding neighboring reflections, it follows that the value of ρ lies between 1.1 Å and 1.65 Å or between 2.4 Å and 2.9 Å. The latter value is at once excluded by 211, which is a very strong reflection. The intensities calculated with a radius $\rho = 1.15 \pm 0.05$ Å are in very good agreement with the observed intensities (last columns, Table I).

Table I. [only partly reproduced. Ed.]

hkl	sin θ	α	x	$J_0(x)^a$	½ΣF (O)	½ΣF (N)	½ΣF (Na)	½S	½p	Cont. fact.	I calc.	I obs.
110	0.127	62°	1.61	+0.45	19.95	5.4	−8.85	5.55	3	252	2.3	3.2
211	.161	44°	1.62	+0.44	18.0	4.8	+8.7	21.6	3	150	21.0	>10
222	.170	0°	0.00	+1.00	17.55	4.7	−8.6	13.65	1	134	2.5	2.5
1Ī0	.197	90°	2.86	−0.21	15.9	4.2	+8.2	9.1	3	100	2.5	3.4
210	.214	66°	2.84	−0.20	0.0	0.0	0.0	0.0	6	80	0.0	<0.5
200	.234	75°	3.26	−0.33	14.4	3.8	−7.7	8.65	3	67	1.5	2.4
220	.254	62°	3.24	−0.33	13.65	3.55	+7.4	6.4	3	60	0.7⎫	7.7
332	.254	25°	1.56	+0.48	13.65	3.55	+7.4	17.5	3	60	5.5⎭	
321	.260	48°	2.78	−0.18	13.5	3.5	−7.3	6.25	6	55	1.3	2.2
2Ī0	.302	63°	3.88	−0.40	0.0	0.0	0.0	0.0	3	42	0.0	<0.5

^{a} Jahnke, Emde 'Funktionentafeln'.

Geological Institute University of Amsterdam

Lab. for Gen. and Inorganic Chemistry University of Amsterdam

[1] James and Brindley, Z. Krist. **78** (1931) 470.

D. ALLOYS

162. X-ray Analysis of the Cu-Zn, Ag-Zn and Au-Zn Alloys, by A. WESTGREN and G. PHRAGMÉN (1925)

■

163. The Structure of γ-Brass, by A. J. BRADLEY and J. THEWLIS (1926)

■

164. Researches on the Nature, Properties and Conditions of Formation of Intermetallic Compounds (with Special Reference to certain Compounds of Tin), by W. HUME-ROTHERY (1926)

■

165. The Structure of Certain Ternary Alloys, by A. J. BRADLEY and C. H. GREGORY (1927)

■

166. Zur Chemie der Legierungen, by A. WESTGREN (1932)

■

167. The Electron Theory of Metals, by J. D. BERNAL (1935)

■

> The study of the crystalline structure of metals, alloys and intermetallic compounds is at the same time one of the most important and difficult branches of crystallography. The problem is so full of pitfalls that it is useless to approach it with any but the most carefully organised plans, prepared in collaboration by metallurgist, physicist, chemist, and crystallographer.
>
> W. T. ASTBURY and W. H. BRAGG
> *Ann. Rep.* **21** (1924) 221

162. In the following the authors will try to elucidate the question to what extent X-ray analysis may contribute to a deeper insight into the chemical nature of metallic phases.

W. Rosenhain[1] has recently put forth the hypothesis of the atoms being more intimately combined in intermetallic compounds than in solid solutions, which would bring about that the interatomic distances in the first-mentioned

---

[1] *Nature,* **112** (1923) 832.

substances would differ markedly from the normal ones. Confronted with reliable experimental facts, this view has, however, already turned out to be untenable. The distances between the atomic centres in inter-metallic compounds do not seem to differ perceptibly from those in other crystals. Rosenhain's supposition has evidently its origin in the very common idea that the forces holding together atoms in an intermetallic compound are of another kind, or at least stronger than those acting between atoms in a pure metal or in a solid solution. The validity of this view has, however, never been proved. The bonds probably being non-polar in all cases, there seems to be no reason not to believe that the interatomic forces should be essentially the same in a solid solution as in a compound. They are certainly as 'chemical' in the one case as in the other.

According to the authors' opinion, the fundamental difference between solid chemical compounds and solid solutions lies in their structure. *In an ideal chemical compound structurally equivalent atoms are chemically identical. In an ideal solid solution all atoms are structurally equivalent.* In the latter case some of the atoms of the solvent are replaced by atoms of another kind, which take the place of the former, and which are distributed quite irregularly in the parent lattice.[1]

Of these two extreme types of structures the first one seems to occur comparatively seldom in metallic phases, while the second one is met with frequently. A multitude of metallic phases seem, however, to represent an intermediate stage between these two extreme cases, forming what might be characterized as solid solutions in chemical compounds. (...)

2. X-ray analysis of the Cu–Zn, Ag–Zn, and Au–Zn systems has proved that the five different types of structure occurring in the first-mentioned of these systems [see Fig. on next page] are met with also in the other ones.

The β- and γ-phases of the Cu–Zn system have been investigated in the form of single crystals; in other cases the analyses have been performed by means of the powder method only. Arranged according to rising content of zinc, the structures common to the three systems have turned out to be the following: α, face-centred cubic; β, cubic of CsCl-type; γ, cubic with 52 atoms in the elementary cube; ε, close-packed hexagonal with an axial ratio of 1.55–1.60; and η, close-packed hexagonal with an axial ratio of 1.80–1.90.

Metallographic Institute, Stockholm

---

[1] In exceptional cases this substitution may be of a complex nature (comp. *Journ. Inst. Metals* **31** (1924) 204 and *Journ. Iron & Steel Inst.* **109** (1924) 171.

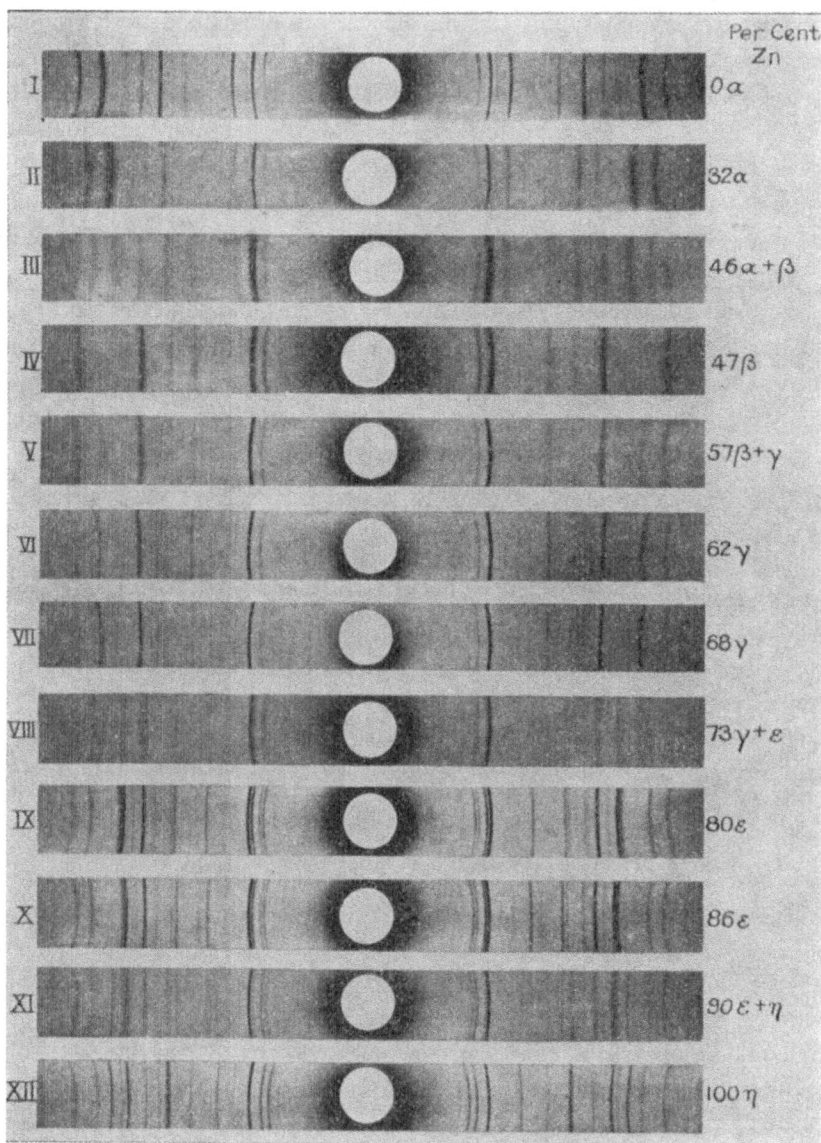

Powder Photograms of Copper-Zinc Alloys.

163.   *Introduction*

Westgren and Phragmén[1] have recently described the results of investigations on the structures of Cu–Zn, Ag–Zn and Au–Zn alloys. As a result of X-ray analysis by the powder method, five different types of structure were found in the case of the Cu–Zn alloys. Structures were successfully assigned to four of these phases, but in the case of the γ-phase a complete elucidation was not attempted. The structure is cubic, and contains 52 atoms to the unit cube.

By way of conjecture, Westgren and Phragmén suggested the formulae Cu_4Zn_9, Ag_4Zn_9, and Au_4Zn_9, as these correspond to compositions coinciding with one of the homogeneous γ-phase ranges. We have found that these formulae are incorrect, the true formulae being Cu_5Zn_8, Ag_5Zn_8, and Au_5Zn_8.

It is at first sight surprising to find such a large number of atoms in the unit, but this is by no means an isolated instance. The γ'-phase of the Cu–Al alloys[2], and even two distinct modifications of the element manganese[3], present the same phenomenon. It is a remarkable fact that α-manganese[4], γ Cu–Zn, γ Ag–Zn, γ Au–Zn, and γ' Cu–Al all have a unit cell of about the same dimensions, and containing about the same number of atoms. Moreover, the intensities of the reflexions from many planes of these structures are found to be extraordinarily alike, whatever substance is examined. In particular, the two strongest lines on a powder photograph invariably occur at the same part of the film, so that for all these substances two interplanar spacings are particularly pronounced. It is significant that these are the spacings of the (110) and (211) planes of a body-centred cubic lattice whose lattice constant is about $a = 2.95$ Å.

There is clearly a fundamental relationship between the structures of the above substances and the simple body-centred cubic structure. The present paper gives the results of an attempt to determine the relationship in the case of the Cu–Zn alloys.

[1] 'X-Ray Analysis of Copper-Zinc, Silver-Zinc and Gold-Zinc Alloys', by Arne Westgren and Gösta Phragmén, *Phil. Mag.* **50** (1925) 311 [this Vol. preceding paper].

[2] Jette, Phragmén and Westgren, *Journ. Inst. Met.* **31** (1924) 193.

[3] Westgren and Phragmén, *Z. f. Physik* **33** (1925) 777; A. J. Bradley, *Phil. Mag.* **50** (1925) 1018.

[4] In order to avoid further confusion the authors have decided to adopt the nomenclature devised by Westgren and Phragmén for the modifications of manganese.

Possible Space-Groups for γ-Brass

The only reflexions found for Cu–Zn and Au–Zn are those from planes (hkl), for which (h + k + l) is even. In the case of the Ag–Zn alloy there are lines which are exceptions to this rule, but these extra lines are all very weak, and are probably either β-lines or are due to impurities. There are no extra lines on the rotation photogram of γ-brass, taken from a single crystal. This may be taken as conclusive evidence that the structure is built up from the space-lattice $Γ'''$. According to Westgren and Phragmén, there are 52 atoms per unit cell, so that the structure contains 26 inter-penetrating body-centred cubic lattices, each of dimensions 8.85 Å.

From a Laue photograph Westgren and Phragmén deduce that the symmetry of γ-brass is either T_d, O, or O_h. The only abnormal spacings are (hkl), where (h + k + l) is odd, which are halved. Three space-groups satisfy these conditions, namely, T_d^3, O^5, and O_h^9. These are therefore the only possible space-groups.

The above are the only conclusions for which there is direct experimental evidence.

Approximate Positions of the Atoms

To test directly every possible arrangement of atoms which would satisfy the above space-group conditions would be almost impossible. We have selected the most probable types of structure, and tested such arrangements of atoms by comparing the intensities calculated for these arrangements with the intensities of the lines as observed by Westgren and Phragmén. In order to determine which arrangements were feasible, use was made of the following criteria:

(1) A clue to the approximate positions of the atoms is obtained by considering the positions and intensities of the strongest lines.

(2) The distance of closest approach of the atoms is not likely to be very different from that in other crystals of a similar type containing the same atoms.

With one exception, the strongest lines on the film of γ-brass are identical in position and intensity with the lines on the β-brass film.

The lattice of β-brass is body-centred cubic, the lattice constant d_{100} is 2.945 Å, the unit cell contains two atoms. The lattice constant of γ-brass (61.7 per cent Zn) d_{100} is 8.85 Å, and each unit cell contains 52 atoms. The side of the unit cube of γ-brass is therefore exactly three times that of β-brass, and its volume is 27 times as great. It follows that if γ-brass had 54 atoms to the unit cell, they might be arranged in exactly the same way as the atoms of β-brass, namely, on a body-centred lattice.

In such a case, γ-brass would only give rise to those lines which appear on the film of β-brass. The existence of extra lines on the film of γ-brass can be accounted for if we suppose two of the 54 atoms to be removed without greatly displacing the remaining 52 atoms.

These considerations show that a possible structure for γ-brass consists of a body-centred cubic arrangement with 1 atom in 27 removed, the remaining atoms being slightly displaced, but necessarily in such a manner that the cubic symmetry is preserved.

Fig. 1 shows a unit cell containing the whole 54 atoms in the correct positions for a body-centred cubic structure. With the exception of one atom at the centre, all the atoms lie on the surfaces of three concentric cubes which are shown in the figure. The diagram illustrates the special case where every atom is situated at exactly the same position as in a body-centred lattice, but since the unit cell contains so many atoms, they need not all be structurally equivalent. Cubic symmetry will be preserved and the space-group requirements fulfilled if the atoms are divided into four or five groups of equivalent positions. The most general case, which corresponds to the space-group T_d^3, is shown in the figure. Structurally equivalent atoms are denoted by the same symbols. The 54 atoms are

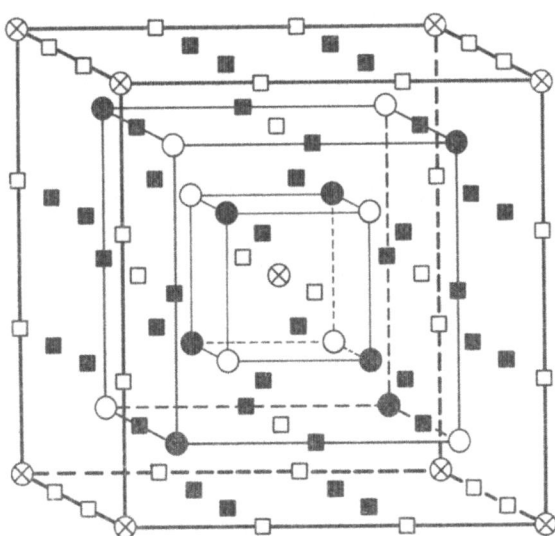

Fig. 1. The Derivation of the Structure of γ-Brass from a Simple Cube-Centred Arrangement of Atoms.

divided between five groups of equivalent positions in the following manner:

Table II.

Type of Atom.	Symbol of Type.	No. of Atoms per Unit Cell.
'X'	⊗	2
'A'	●	8
'B'	○	8
'C'	□	12
'D'	■	24
	Total No.	54

The atoms may be displaced in different ways while still conforming to the space-group $T_d{}^3$, but atoms belonging to the same set must be displaced in the same manner. Fig. 1 will also represent the case of the space-group $O_h{}^9$, the additional condition being imposed that the 'A' and 'B' atoms must be equivalent. Consideration shows that it is not necessary to discuss the case of O^5, since any displacements from the arrangement of Fig. 1 which are consistent with the symmetry elements of O^5 also satisfy the requirements of $O_h{}^9$.

A possible structure of γ-brass must have two atoms per unit cell less than the arrangement of Fig. 1. It is clear that the only two atoms which could be removed, if our general scheme of arrangement is correct, are

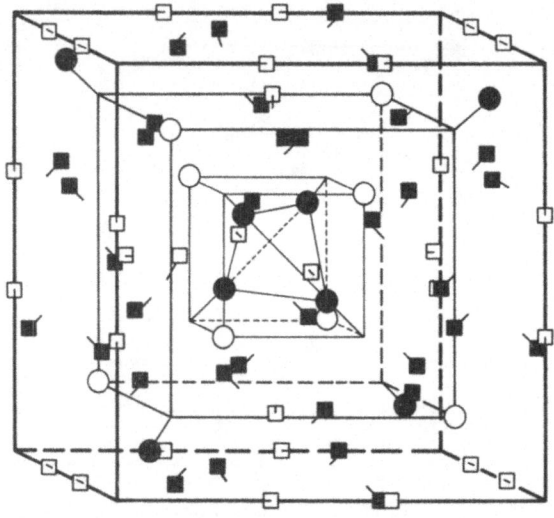

Fig. 2. Structure of γ-Brass.

the 'X' atoms, because there are more than two atoms of each other type in the unit cube.

Fig. 2 represents a unit cell from which the 'X' atoms have been removed. The remaining atoms are slightly displaced from the positions of Fig. 1, the displacement being shown by the short lines. The extents of the displacements actually shown in the figure are those assigned to γ-brass by the present analysis. The displacements have carried some of the D atoms, which were formerly on the boundary faces of the unit cell, entirely into neighbouring cells. These atoms have been inserted in Fig. 2, in order to facilitate comparison with Fig. 1.

Choice of Space-Groups

The arrangement of Fig. 2 would correspond to the space-group O_h^9 only if 'A' and 'B' atoms were made equivalent. As actually drawn it corresponds to the space-group T_d^3. We need not consider space-group O^5, as this requires the atoms to be placed in exactly the same positions as O_h^9, if the structure is to be at all similar to the body-centred cube.

To decide between T_d^3 and O_h^9 requires a determination of the parameters of the atoms, which can be made from considerations of intensities and interatomic distances.

The co-ordinates of the atoms are as follows:

$$(000; \tfrac{1}{2}\tfrac{1}{2}\tfrac{1}{2}) +$$

'A' Atoms: (aaa), $(a\bar{a}\bar{a})$ $\circlearrowright$.

'B' Atoms: $(\bar{b}\bar{b}\bar{b})$, $(\bar{b}bb)$ $\circlearrowright$.

'C' Atoms: $(c00)$ $\circlearrowright$, $(\bar{c}00)$ $\circlearrowright$.

'D' Atoms: (dde) $\circlearrowright$, $(dd\bar{e})$ $\circlearrowright$, $(\bar{d}d\bar{e})$ $\circlearrowright$, $(\bar{d}\bar{d}e)$ $\circlearrowright$.

For space-groups O^5 and O_h^9 $a = b$ and $e = 0$.

At this stage no discrimination will be made between Cu and Zn atoms. The atomic numbers are so nearly equal that Cu and Zn may be considered to scatter X-rays by the same amount. No appreciable error will be introduced into the calculations by this assumption.

In the arrangement shown in Fig. 1, $a = \tfrac{1}{6}$, $b = \tfrac{1}{6}$, $c = \tfrac{1}{3}$, $d = \tfrac{1}{3}$, $e = 0$. If the atoms of γ-brass were situated in these positions, the 'X' atoms being absent, the interatomic distances would be 2.55 Å. This is the same value as the distance of closest approach of the centres of copper atoms in the lattice of pure copper. The zinc atom in metallic zinc has two inter-atomic distances, namely, 2.66 Å and 2.91 Å, whilst in β-brass the inter-atomic distance is again 2.55 Å. One may anticipate that in γ-brass the interatomic distances will again be of the same order of magnitude.

Displacements of the atoms consistent with the symmetry requirements of O_h^9 and T_d^3 may now be tried in order to explain the observed intensities of reflexion. The parameter values were tested by the use of the structure-amplitude formula, comparing the values of S so obtained with the observed intensities. In such a comparison allowance must be made for two other factors which affect the observed intensities. These are the frequency factor N, giving the total number of planes corresponding to the form $\{hkl\}$, and a factor expressing the general falling off in intensity as the angle of reflexion increases. This is due to a number of superimposed effects, and to allow for each separately would bring in many unproved assumptions. The net effect may to some extent be compensated for by the insertion of a factor $\csc^2 \theta$ in the intensity formula. We then obtain the expression $NS^2/\sin^2 \theta$ for comparison with the observed intensities. The use of this purely empirical formula is only justified by its value in affording a rough comparison between reflexions at not too widely differing angles of reflexion. In the present case of a cubic crystal it may be expressed as $NS^2/(h^2 + k^2 + l^2)$.

Table III compares the calculated and observed intensity values for a number of planes selected from the powder photograph data. In the case of the holohedral structure the symmetry requirements, combined with the criterion that the interatomic distance must not depart too widely from 2.55 Å, limit the displacement of the 'A' and 'B' atoms to a small amount, and hardly permit the 'C' and 'D' atoms to be displaced at all. On the other hand, the lower symmetry requirements of T_d^3 allow much larger displacements. In Column III are given values calculated for the holohedral structure with those parameter values which give the best correspondence between the observed intensities and the calculated values. This is so unsatisfactory as to make it clear that considerable readjustments are required in the parameter values. So long as the structure remains holohedral, no appreciable change can be brought about without bringing some pairs of atoms much closer together than 2.55 Å. It may be concluded that no arrangement based upon space-group O_h^9 can satisfy all the requirements, and that it remains to consider arrangements based on T_d^3.

The values shown in Column IV of Table III were obtained by giving the a' and e' parameters the greater displacement which is permitted by T_d^3. The much closer correspondence indicates a closer approximation

Table III.

hkl	Obs. Intensity	NS²/(h² + k² + l²) Holohedral $a' = b' = 52°^1$ $c' = 120°$ $d' = 120°$ $c' = 0$	Hemihedral $a' = 37°$, $b' = 60°$ $c' = 120°$, $e' = 16°$ $d' = 120°$	$d' = 110°$
110	abs.	0.0	0.8	1.4
200	abs.	0.4	0.8	0.1
211	abs.	0.5	1.5	0.6
220	abs.	2.4	0.4	0.9
310	abs.	1.3	0.0	0.7
222	weak	3.2	0.9	2.9
321	weak	3.2	6.1	5.5
400	abs.	1.6	2.0	0.0
330 ⎱ 411 ⎰	strong	⎧ 102.5 ⎨ ⎩ 2.4	66.2 11.4	55.1 36.2
420	abs.	0.5	0.7	0.4
332	mod.	0.5	4.0	9.2
422	weak	0.0	5.5	6.1
521	abs.	0.0	2.8	0.0
440	abs.	8.2	0.9	0.1
433 ⎱ 530 ⎰	abs.	⎧ 1.5 ⎨ ⎩ 3.2	0.6 0.0	1.3 1.8
620	abs.	0.4	0.5	0.7
622	abs.	1.0	1.2	0.9
444	mod.	4.2	2.3	9.2
640	abs.	1.6	1.4	2.5
642	very weak	0.7	1.4	3.3
730	abs.	11.0	1.3	1.6
800	abs.	1.3	0.3	1.2
653	very weak	2.6	0.1	1.4

[1] For convenience in calculation the parameters are expressed here in degrees, so that a' is equal to $360a$.

to the correct structure. These parameter values are actually the maximum displacements allowed by considerations of interatomic distance. The other parameter values were unchanged. A much better agreement was obtained by putting $d' = 110°$, as in the last column of Table III.

Further information with regard to parameter values was obtained by the use of the rotation crystal data. Table IV [not reproduced] shows that the best agreement is obtained by increasing the value of c'.

The best value is about $c' = 129°$. No further change is indicated in the values of the other parameters.

The parameter values, now expressed again as fractions of the side of the unit cell, which give the best agreement between the observed and calculated intensity values, are $a = 0.10_3$, $b = 0.16_7$, $c = 0.35_8$, $d = 0.30_5$, $e = 0.04_5$.

Tables V and VI [not reproduced] give complete lists of observed and calculated intensities from the powder photogram and the rotation photogram.

The agreement between observed and calculated values is sufficiently good to afford strong confirmation of the correctness of the parameter values, and entirely confirms the original supposition that γ-brass has very nearly a body-centred cubic structure.

Table VII.

hkl	$h^2+k^2+l^2$	Observed Intensities			Positions of Atoms which would account for Intensity Changes			
		Cu–Zn	Ag–Zn	Au–Zn	A	B	C	D
211	6	abs.	v.w.	w.	Zn	Cu Ag Au	Cu Ag Au	–
222	12	w.	w.	abs.	Zn	Cu Ag Au	Cu Ag Au	Zn
321	14	w.	abs.	abs.	–	Cu Ag Au	–	Zn
420	20	v.w.	abs.	?	Zn	Cu Ag Au	–	Zn
332	22	m.	w.	w.	Zn	–	–	Zn
422	24	w.	w.	m.	Cu Ag Au	Cu Ag Au	Cu Ag Au	Zn
431 510	26	w.	v.w.	v.w.	Zn	Cu Ag Au	Cu Ag Au	Zn
440	32	abs.	w.	v.w.	Cu Ag Au	Cu Ag Au	Cu Ag Au	Zn
530 433	34	abs.	abs.	v.w.	Zn	–	–	Cu Ag Au
541	42	abs.	v.w.	?	Zn	Cu Ag Au	Cu Ag Au	–
444	48	m.	m.	st.				
550 710 543	50	w.	w.	m.				
642	56	v.w.	abs.	?	Zn	–	–	Zn
644 820	68	w.	w.	st.				
653	70	v.w.	abs.	abs.	Zn	–	–	Zn
822 660	72	st.	st.	m.				

Identification of the Atoms

The scattering powers of copper and zinc are so nearly equal that there was some difficulty in distinguishing the respective copper and zinc atoms, and an indirect method had to be employed for this purpose.

The powder photograms of γ–Cu–Zn, γ–Ag–Zn, and γ–Au–Zn are so similar that there can be no doubt that they have the same type of structure. Table VII gives a list of the lines which have slightly different intensities for the three series of alloys. The differences in intensity are either due to slight differences in parameter values or to the replacement of the copper atoms by the more efficient scatterers, silver and gold. The table shows what distribution of atoms would produce the observed intensity changes. Evidently 'A' and 'D' atoms are zinc and 'B' and 'C' atoms are either copper, silver or gold. There is no indication of any change in the values of the parameters.

Discussion of the Structure

The complete arrangement of atoms in the structure found for γ-brass is shown in Fig. 2. The open squares and circles represent copper atoms, the solid ones represent zinc atoms.

There are 32 zinc atoms and 20 copper atoms in the unit cell. The corresponding formula is Cu_5Zn_8. A confirmation of this formula is given by metallographic data. Several properties of γ-brass alloys exhibit a maximum at compositions corresponding to just over 60 per cent zinc. For this reason metallographers ascribed the formula Cu_2Zn_3 to the alloy. In view of the X-ray data this formula must now be considered untenable. However, Cu_5Zn_8 contains 62.5 per cent Zn, and this accounts equally well for the observed maximum.

Table VIII [not reproduced] gives a list of the interatomic distances found in Cu_5Zn_8. This completely confirms the hypothesis put forward in section 3, that these distances would prove to be very nearly the same as the interatomic distances in Cu, Zn and CuZn.

The copper and zinc atoms are distributed symmetrically throughout the unit so that each atom has the greatest possible number of neighbours of the opposite sort. These neighbours are distributed as follows:

Copper Atoms 'C' atoms have 10 zinc and 3 copper atoms as neighbours.
                'B' atoms have 9 zinc and 3 copper atoms as neighbours.

Zinc Atoms   'A' atoms have 6 zinc and 6 copper atoms as neighbours.
'D' atoms have 5 zinc and 6 copper atoms as neighbours.

The authors desire to express their thanks to Prof. W. L. Bragg, F.R.S., for his kind interest and valuable suggestions during the progress of the work.

The Physical Laboratories, the University of Manchester

164. In the β phases of divalent metals and copper, the β phase is grouped round the composition CuZn, &c. This corresponds to the ratio of 3 valency electrons (1 from the copper atom and 2 from the zinc atom) to 2 atoms.

In the β phases with trivalent aluminium grouped round the composition Cu_3Al, &c., we have 6 valency electrons (1 from each copper atom and 3 from the aluminium atom) to 4 atoms, and the ratio is therefore once again 3 electrons to 2 atoms.

In the β phase of the copper-tin alloys grouped round the composition Cu_5Sn, we have 9 valency electrons (1 from each copper atom and 4 from the tin atom) to 6 atoms, and the ratio is here again 3 electrons to 2 atoms.

It is therefore suggested that the β phases of these alloys of copper, silver, and gold are interpenetrating space lattices of atoms and electrons for which the fundamental ratio is 3 electrons to 2 atoms, with a certain amount of variation on either side, just as the lattice of a pure metal (which as shown later is also a system of interprenetating lattices of electrons and atoms) can still persist in a solid solution where the relative proportion of electrons is no longer the same.

With regard to the structures of these β phases, the X-ray investigations have shown the β phase of Cu–Zn system to possess the body-centred cubic structure. The work on the Al–Cu β phase has not yet been satisfactory, but indirect evidence of the body-centred structure is given.

Note. Since the above theory was put forward a paper has appeared by Westgren and Phragmén[1], in which the β phases of Au–Zn and Ag–Zn alloys are shown to have the body-centred structure in complete agreement with the above theory.

Since the above paper was written, the theory has received substantial confirmation from two papers read at the same meeting of the Institute of Metals. In a paper by Murphy[2] the equilibrium diagram of the system silver-tin was determined, and shown to have a β phase which is grouped

---

[1] *Phil. Mag.* **50** (1925) 311–341 [this Vol. paper **162**].
[2] *J. Inst. of Metals* **35** (1926) 107.

round the composition Ag_5Sn, again giving the ratio of 3 valency electrons to 2 atoms required by the theory detailed above. In a paper by Stockdale[1] some copper-rich, copper-aluminium-tin alloys were examined, and equilibrium diagrams given showing the compositions of four ternary eutectoid points. As shown by the present author in the discussion on Stockdale's paper, the four ternary points determined correspond to a ratio of 2 atoms to 2.95 ± 0.01 electrons. As in the copper-aluminium system the eutectoid point has slightly more copper than that required by Cu_3Al (giving the 3 electrons : 2 atoms), it is clear that the compositions of the ternary eutectoids change so as to keep the ratio of atoms : valency electrons constant, in striking confirmation of the above theory.

Thesis approved for the Degree of Doctor of Philosophy
in the University of London

165. In view of the similarity between the γ-phase of copper-zinc and the γ-phase of copper-aluminium it was decided to test the possibility of forming mixed crystals of the two phases. Alloys corresponding to the formulae Cu_5Zn_8 and Cu_9Al_4 were made up and were then melted together in different proportions. Three such alloys were thus made up, and the structures in each case were identical with the structures of the binary alloys; the alloys were quite homogeneous in character. The similarity between these different structures is independent of the chemical composition but is closely related to the electron distribution. There is in each case the same ratio of valency electrons to atoms, namely, 21 : 13.

Physical Laboratories, University of Manchester

166. Vor einigen Jahren glaubten mein Mitarbeiter Phragmén und ich, daß eine mehr oder weniger vollständige Regelmäßigkeit in der Gruppierung der verschiedenen Atomsorten in bezug aufeinander als Kennzeichen einer sogenannten intermetallischen Verbindung dienen könnte, und wir schlugen deswegen die Definition vor: *In einer idealen intermetallischen Verbindung sind strukturell gleichwertige Atome chemisch identisch*[2]. Es lag nahe, diese Annahme zu machen, weil es sich bei den unmetallischen Verbindungen, die aus Ionen oder Molekülen aufgebaut sind, immer so verhält, daß strukturell gleichwertige Atome identisch sind.

---

[1] *J. Inst. of Metals* **35** (1926) 181.
[1] A. Westgren u. G. Phragmén, *Phil. Mag.* **50** (1925) 311 [this Vol. paper **162**].

In diesen Substanzen kommen offenbar wegen ihrer verschiedenartigen Ladung oder ihrer Teilnahme am Molekülbau den verschiedenen Atomarten im Gitter ganz besondere Funktionen zu. Molekülstruktur ist indessen eine Seltenheit bei den metallischen Substanzen, und wenngleich sie von Ionen, die von frei beweglichen Elektronen zusammengehalten werden, aufgebaut sein dürften, haben ja diese Ionen alle dasselbe Zeichen und können sich deswegen austauschen, ohne daß die Neutralität des Kristalls verlorengeht. Die Verschiedenheit der Atomarten, die eine metallische Substanz aufbauen, bewirkt in manchen Fällen, daß sie sich in gesetzmäßiger Abwechslung von verhältnismäßig niedriger Periodizität in das Gitter einordnen. Wenn diese Regelmäßigkeit vollständig ist, d.h. wenn die Atome auch nicht im geringsten Grade austauschbar sind, muß das Produkt selbstverständlich eine einfache stöchiometrische Zusammensetzung erhalten. Es ist dann der ideale Fall realisiert, den wir oben erörtert haben. Bei festen nichtmetallischen Verbindungen scheinen die einfachen stöchiometrischen Proportionen immer mit einer derartigen Regelmäßigkeit in der Atomgruppierung verknüpft zu sein, und auch eine Anzahl metallische Substanzen sind nach einfachen Atomproportionen zusammengesetzt, weil die verschiedenen Atomarten im Gitter in gesetzmäßiger Weise in bezug aufeinander geordnet sind. Bei den metallischen Stoffen sind aber einfache Atomproportionen nicht *immer* mit einer derartigen Regelmäßigkeit des Kristallbaus verbunden. Sie können unter Umständen ihren Grund in einem ganz anderen Umstand haben.

Einfache stöchiometrische Zusammensetzung von Legierungsphasen als Folge der Anwesenheit von Valenzelektronen in bestimmter Zahl im Kristallgitter

Es gibt ja viele Gründe für die Annahme, daß die metallischen Stoffe als aus positiven Ionen und verhältnismäßig frei beweglichen Elektronen aufgebaut sind. Es hat sich herausgestellt, daß gewisse Gittertypen in den Legierungsreihen dann wiederkehren, wenn die Konzentration der freien Elektronen im Gitter spezielle Werte annimmt. Ich werde im folgenden darauf zurückkommen. Da die im Gitter frei beweglichen Elektronen aus den locker gebundenen Elektronen der zusammengetretenen Atome bestehen, d.h. in den meisten Fällen aus den Valenzelektronen derselben, so ist eine Folge davon, daß Kristalle auch dieser Art nach einfachen Atomproportionen zusammengesetzt sein können. Für das Zustandekommen des Gitters ist für Phasen dieser Art die Konzentration der Valenzelektronen das primär Entscheidende. Die Natur und das Mengenverhältnis der Atomrümpfe hat nur einen sekundären Einfluß auf die Kristallstruktur.

In gewissen Fällen können die verschiedenen Arten der letzteren sich in regelmäßiger Weise in bezug aufeinander im Gitter einordnen. Bisweilen kann diese Ordnung nur partiell sein, so daß gewisse Gruppen strukturell gleichwertiger Punkte chemisch identische Atome enthalten, während die übrigen Gruppen aus zufallsmäßig verteilten Atomarten verschiedener Art bestehen. Endlich kann es auch vorkommen, daß die Atome ganz regellos, d.h. statistisch, auf die Gitterpunkte gestreut sind. Als Beispiele können die Phasen Cu_5Zn_8, Cu_9Al_4, Cu_5Cd_8 und $Cu_{31}Sn_8$ herangezogen werden. In diesen sämtlichen Fällen scheint das Gittergleichgewicht durch das übereinstimmende Verhältnis von 21 Valenzelektronen auf 13 Atome bedingt zu sein. Bradley[1] hat gefunden, daß die beiden ersten dieser Phasen bei Zusammensetzungen nach den angegebenen Formeln sehr nahe dem idealen Grenzfall entsprechen, indem die beiden Atomarten im Gitter in bezug aufeinander vollständig regelmäßig gruppiert zu sein scheinen. Bei Cu_5Cd_8[2] ist die Ordnung nur partiell, und wahrscheinlich gilt das auch für $Cu_{31}Sn_8$. Es ist aber auch möglich, daß man bei dieser letzteren Kristallart mit einer ganz regellosen Verteilung der Kupfer- und Zinnatome auf die Gitterpunkte zu tun hat.

Von diesem letzterwähnten Typus sind die Phasen Ag_3Al, Au_3Al und Cu_5Si. Sie haben alle dieselbe Struktur wie β-Mangan. In Abb. 3 sind

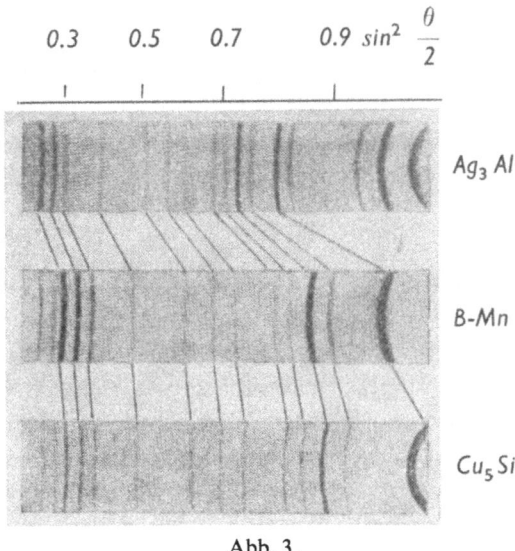

Abb. 3.

[1] A. J. Bradley u. J. Thewlis, *Proc. Roy. Soc., London* A **112** (1926) 678 [this Vol. paper **163**]; A. J. Bradley, *Phil. Mag.* (VII) **6** (1928) 878.
[2] A. J. Bradley u. C. H. Gregory, *Phil. Mag.* (VII) **12** (1931) 143.

die Pulverphotogramme von Ag_3Al und Cu_5Si mit dem des β-Mangans verglichen. Aus der Abbildung ersieht man, daß die relative Intensität der Röntgeninterferenzen dieser drei Stoffe genau dieselbe ist. Silber- und Aluminium- bzw. Kupfer- und Siliciumatome können deswegen nicht in bezug aufeinander geordnet sein, sondern müssen auf die Gitterpunkte statistisch verteilt sein. Ag_3Al hat ein äußerst enges Homogenitätsgebiet; sogar auf röntgenographischem Wege konnte eine Ausdehnung desselben nicht nachgewiesen werden. Diese Phase muß folglich der üblichen Auffassung gemäß zu den chemischen Verbindungen gezählt werden. Orientierende Untersuchungen von Au_3Al deuten darauf hin, daß auch diese Phase einen sehr engen Homogenitätsbereich hat. Auch Au_3Al ist also eine chemische Verbindung. Bei Cu_5Si können die Kupfer- und Siliciumatome sich freilich innerhalb eines Intervalls von ein paar Atomprozenten austauschen[1]; diese Kristallart dürfte trotzdem mit ebenso großem Recht wie FeS als eine chemische Verbindung betrachtet werden können. Das Gemeinsame der drei Phasen ist, wie ihre Formeln angeben, daß in ihnen das Mengenverhältnis zwischen Valenzelektronen und Atomen $3:2$ ist.

Bei den metallischen Stoffen können einfache stöchiometrische Proportionen nicht nur als Folge von regelmäßigen Atomgruppierungen entstehen, sondern sie können auch durch bestimmte Mengenverhältnisse zwischen Valenzelektronen und Atomen bedingt sein. Dies hat selbstverständlich zur Folge, daß die vorher erwähnte, von Phragmén und mir vorgeschlagene Definition einer idealen intermetallischen Verbindung nicht aufrechterhalten werden kann. Die Phase Ag_3Al muß trotz der vollständig regellosen Gruppierung ihrer Silber- und Aluminiumatome nach den üblichen Vorstellungen der Chemiker wenigstens mit ebenso großem Recht wie FeS und FeS_2 als eine chemische Verbindung angesehen werden.

Klassifikation der metallischen Stoffe

Aus dieser ausführlichen und weitläufigen Diskussion dürfte hervorgehen, daß es mit großen Schwierigkeiten verknüpft ist, ja, sogar unmöglich sein dürfte, die metallischen Stoffe unter Anwendung der Begriffe chemische Verbindung und feste Lösung durchgehend zu klassifizieren. Soweit diese Klassifikation bisher durchgeführt werden konnte, hat sie kaum über das hinausgeführt, was wir von den intermetallischen Reaktionsprodukten wissen, daß nämlich 'das Auftreten der Salzvalenzen in den Formeln der Metallverbindungen als zufällig betrachtet werden kann'[2]. Der Begriff

---

[1] S. Arrhenius u. A. Westgren, *Ztschr. physikal. Chem.* (B) **14** (1931) 66.
[2] G. Tammann, Lehrbuch der Metallographie, 3. Aufl., Leipzig (1923) S. 230.

chemische Verbindung ist ja für die Systematik der Chemie von grundlegender Bedeutung gewesen, und es ist leicht verständlich, daß man eifrig versucht hat, auch innerhalb der Legierungskunde Nutzen daraus zu ziehen. Es unterscheiden sich aber doch die metallischen Substanzen mit ihren Koordinationsgittern so radikal von den aus ungleich geladenen Ionen aufgebauten Untersuchungsgegenständen der anorganischen Chemie, sowie von den aus Molekülen zusammengesetzten organischen Verbindungen, daß spezielle Prinzipien zu ihrer Klassifikation herangezogen werden müssen.

Der Zweck einer Systematik muß ja sein, die Entstehungsverhältnisse der Stoffe weitmöglichst aufzuklären. Die Stoffe sollen derart geordnet werden, daß es möglich wird, Auskunft zu erhalten über den Zusammenhang zwischen ihrer Art und den Eigenschaften der sie aufbauenden Komponenten.

Dies scheint nunmehr in der Tat dadurch möglich zu sein, daß wir die metallischen Stoffe ganz einfach nach dem Typ ihrer Kristallstruktur klassifizieren. Wenn auch eine vollständige Systematik der metallischen Stoffe nach diesem Prinzip bisher noch nicht erreicht worden ist, so deutet doch das bisher Erzielte darauf hin, daß es wenigstens möglich ist, auf diese Weise einen Überblick über wichtige Gruppen dieser Substanzen zu gewinnen.

Wiederkehrende Strukturtypen in den Kupfer-, Silber- und Goldlegierungen

Die Legierungen von Kupfer, Silber und Gold dürften die bisher röntgenographisch am vollständigsten untersuchten sein. Es hat sich herausgestellt, daß gewisse Strukturtypen in denselben häufig wiederkehren. Ausführliche Berichte über die beobachteten Analogien sind schon früher veröffentlicht worden[1]. Im folgenden soll nur eine kurze Übersicht darüber gegeben und einige während der letzten Zeit gemachte Beobachtungen mitgeteilt worden.

In den sämtlichen Kombinationen von Kupfer, Silber und Gold mit Zink, Cadmium und Quecksilber, mit Ausnahme des Systems Gold-Quecksilber, das bisher nicht ausführlich untersucht worden ist, sind sogenannte γ-Phasen, die dem γ-Messing analog sind, gefunden worden. Kristallarten dieser Art sind auch in den Systemen Kupfer-Aluminium und Kupfer-Zinn angetroffen worden. Wie schon vorher erwähnt, ist es

[1] A. Westgren u. G. Phragmén, *Ztschr. Metallkunde* **18** (1926) 279; *Metallwirtschaft* **7** (1928) 700; *Trans. Faraday Soc.* **25**, Nr. 98 (1929) 379.

für diese Phasen charakteristisch, daß ihre Homogenitätsgebiete Konzentrationswerte in sich einschließen, bei denen das Mengenverhältnis zwischen Valenzelektronen und Atomen 21 : 13 ist. Es scheint nur γ-Kupfer-Quecksilber eine Ausnahme von dieser Regel zu machen[1]. In vielen Fällen fällt der betreffende Konzentrationswert mit der Grenze des Homogenitätsgebiets zusammen, die der Kupfer-, Silber- bzw. Goldseite des Zustandsdiagramms am nächsten liegt. Die meisten dieser Phasen haben Homogenitätsgebiete, die einige Atomprozente umfassen. $Cu_{31}Sn_8$ hat aber, wie röntgenographisch festgestellt werden konnte, eine ganz konstante, dieser Formel genau entsprechende Zusammensetzung.

Bei mehreren γ-Phasen ist die merkwürdige Erscheinung beobachtet worden, daß die Symmetrie, wenn die Atomsubstitution im Gitter über eine gewisse Stufe hinausgeht, in einer, wie es scheint, kontinuierlichen Weise von kubisch auf eine niedrigere hinabfällt. Die meisten der Interferenzen im Pulverphotogramm werden dabei erst unscharf und teilen sich dann bei noch mehr gesteigerter Substitution in Multipletts auf. Das ist der Fall bei γ-Messing, wenn seine Kupferkonzentration einen gewissen Wert überschreitet. Auch bei mehreren anderen Phasen ist diese Erscheinung beobachtet worden. Im Kupfer-Silicium-System kommt bei höherer Temperatur eine Kristallart dieses entarteten γ-Messingtypus vor; sie erreicht aber nirgends innerhalb ihres Homogenitätsintervalls kubische Symmetrie.

Eine andere Gruppe von Legierungen, die in diesem Zusammenhang von Interesse ist, umfaßt die sogenannten β-Phasen. Ihr Prototyp ist das β-Messing. Für sie alle gemeinsam ist das raumzentriert kubische Gitter, auf dessen Punkte die Atomarten je nach ihrem Mengenverhältnis in verschiedener Weise verteilt sein können (vgl. Abb. 5). Wie W. Hume-Rothery hervorgehoben hat[2], scheint die Gitterstabilität dieser Phasen durch das

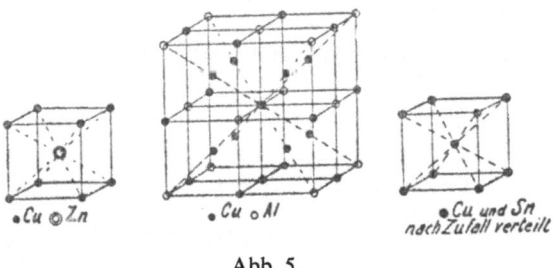

Abb. 5.

[1] N. Katoh, *Ztschr. physikal. Chem.* (B) **6** (1930) 27.
[2] H. Hume-Rothery, *Journ. Inst. Metals* **35** (1926) 313 [this Vol. paper **164**].

übereinstimmende Verhältnis 3 : 2 zwischen Valenzelektronen und Atomen bedingt zu sein. Dieser Strukturtypus ist in etwa zehn verschiedenen Systemen gefunden worden, und in sämtlichen Fällen, außer bei der Silber-Lithium-Phase, die neuerdings von S. Pastorello entdeckt wurde[1], hat sich die Regel von Hume-Rothery bewährt.

In einigen Systemen bei Konzentrationen, wo man nach dieser Regel das Auftreten einer β-Phase erwarten könnte, stößt man aber statt deren auf eine Kristallart von dem viel komplizierteren Strukturtypus des β-Mangans. Solche Phasen sind die oben erörterten Ag_3Al, Au_3Al und Cu_5Si. Das Mengenverhältnis zwischen Valenzelektronen und Atomen ist auch für sie 3 : 2.

Sehr oft findet man in den Kupfer-, Silber- und Goldlegierungen Phasen mit der einfachen Struktur hexagonaler dichtester Kugelpackung. Aus den ersten orientierenden Untersuchungen von Silberliegerungen, die an Ag–Zn, Ag–Al, Ag–Sn und Ag–Sb ausgeführt wurden, schien hervorzugehen, daß auch das Auftreten dieses Strukturtyps mit einem gewissen Verhältnis zwischen Valenzelektronen und Atomen in Zusammenhang stände. Mit steigender Valenz der mit dem Silber kombinierten Komponente schien das Homogenitätsgebiet der Phase hexagonaler dichtester Kugel-Packung in den Zustandsschaubildern dieser Systeme sich in gesetzmäßiger Weise nach der Silberseite hin zu verschieben. Daß die Verhältnisse nicht ganz so einfach sind, hat sich später bei der Untersuchung des Silber-Cadmium-Systems herausgestellt. Hier tritt die Phase hexagonaler dichtester Kugelpackung innerhalb zweier voneinander ganz getrennter Zustandsfelder im Gleichgewichtsdiagramm auf.

Gesetzmäßigkeiten im Aufbau der Legierungen von Übergangselementen

Einige der oben erörterten, in den Kupfer-, Silber- und Goldlegierungen häufig vorkommenden Strukturtypen sind auch in den Kombinationen von Übergangselementen mit Zink, Cadmium und Aluminium wiedergefunden worden.

Vom Typus des β-Messings sind z.B. FeAl, CoAl, NiAl und (Cu, Mn)Al. Phasen, die dem γ-Messing analog sind, kommen nach dem Befund von W. Ekman[2] in Legierungen vor, die aus einem Metall der Eisen-, Ruthenium- oder Osmiumfamilie in Kombination mit Zink oder Cadmium bestehen. Sie können mit den folgenden Formeln bezeichnet werden: Fe_5Zn_{21}, Co_5Zn_{21}, Ni_5Zn_{21}, Rh_5Zn_{21}, Pd_5Zn_{21}, Pt_5Zn_{21} und Ni_5Cd_{21}.

[1] S. Pastorello, *Gazz. chim. Ital.* **60** (1930) 493.
[2] W. Ekman, *Ztschr. physikal. Chem.* (B) **12** (1931) 57.

Im Kobalt-Zink-System hat Ekman auch eine Phase gefunden, die vom β-Mangantypus ist. Ihre Zusammensetzung entspricht etwa $CoZn_3$.

Aus allen diesen Formeln kann der Schluß gezogen werden, daß dieselben Regeln, die für das Zustandekommen der erwähnten drei Strukturtypen in den Kupfer-, Silber- und Goldlegierungen gelten, auch für ihre Entstehung in Legierungen von Übergangsmetallen Gültigkeit besitzen, wenn wir nur die letzteren in diesen Kombinationen als nullwertig betrachten dürfen. Die frei beweglichen Elektronen im Gitter würden demgemäß in diesen Legierungen von Übergangsmetallen mit Zink, Cadmium und Aluminium nur von den letzteren Metallen herrühren*.

Institut für allgemeine und anorganische Chemie der
Universität Stockholm

* [An illustration of the Hume-Rothery rule is given in the subjoined figure taken from A. WESTGREN and W. EKMAN, Structure Analogues of Intermetallic Phases *Arkiv för Kemi, Min. och. Geol.* **10** B No. 11 (1920). Ed.]

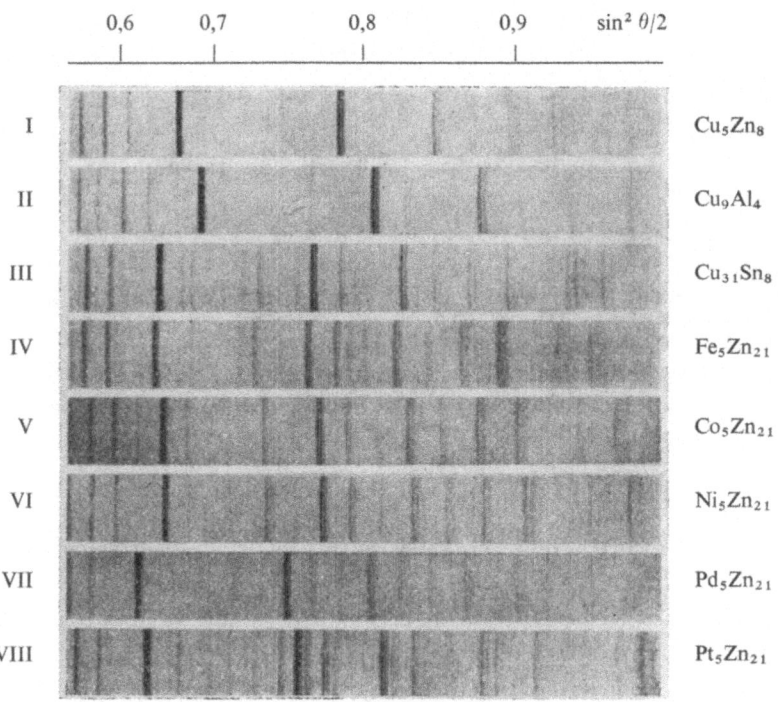

Powder photograms of alloys with γ-brass structure. Fe-*K*-radiation.

Electron:atom ratio 21:13

167.   *The Electron Theory of Metals*

It has long been known that the characteristic structure of certain alloys is determined not so much by its composition as by the ratio of the total number of valency electrons to the total number of atoms in the unit cell. Thus, the characteristic complex structure of γ-brass is shared by the alloys Cu_5Zn_8, Cu_9Al_4, Cu_3Sn_8, Ni_5Zn_{21}, and many others which all have in common an electron: atom ratio of 21 : 13. Similarly, the structure of β-brass corresponds to an electron:atom ratio of 3 : 2, and that of ε-brass to a ratio of 7 : 4. These ratios, first put forward on empirical grounds by Hume-Rothery in 1927, have recently been given a quantum-mechanical explanation by H. Jones[1]. An attempt is here made to give an elementary physical picture which will convey the essential features of the theory.

The state of an electron may be completely defined in terms of the three co-ordinates k_x, k_y, and k_z of its momentum. If we imagine these co-ordinates to be plotted along three rectangular axes, the state of the electron is described by the point (k_x, k_y, k_z) in the momentum space thus defined. If the electron is free, the energy associated with the state is proportional to the square of the momentum, i.e., to the square of the vector joining the point in question to the origin of momentum space. On account of the quantum conditions and the exclusion principle, no two electrons may exist in the same state, so that each electron requires for itself a definite volume of momentum space, and an assemblage of many electrons will occupy a spherical domain whose size is determined by the number of electrons considered. If the electrons are not free, as will be the case in a crystal where we must take into account the effect of the lattice field, it can be shown that the energy of the electron no longer varies continuously with the momentum, but that there are certain planes in momentum space such that when the point describing the state of the electron crosses one of these planes a discontinuous energy change takes place.[2] The electrons in the neighbourhood of a discontinuity are those whose wave-length and direction would lead them to be reflected by the corresponding plane of the crystal. Just below the plane the energy is abnormally depressed; just above it, abnormally raised. Between successive planes the energy varies continuously with momentum and approximates to that of a free electron. It may be shown, further, that the planes in momentum space across which these energy discontinuities take place are parallel to all possible

[1] *Proc. Roy. Soc. London* A **144** (1934) 225.
[2] Brillouin, 'Quantenstatistik', Chap. 8; A. Sommerfeld and H. A. Bethe, 'Handbuch der Physik', XXIV, 2.

crystallographic planes in the crystal, so that in the case of e.g., a cubic substance, the origin is surrounded by a series of concentric zones successively bounded by the faces of the cube, the dodecahedron, the octahedron, and so on. The magnitude of the energy jump across the various zones is not constant, but is proportional to the degree of scattering of electron waves from the corresponding plane in the crystal, and therefore closely related to the intensity of the X-ray reflexion: if all the atoms in the structure have nearly the same atomic number, it is actually proportional to the amplitude of this reflexion.

In the γ-alloys, experiment shows that, of the planes with small indices, only those of the forms {411} and {330} give rise to strong X-ray reflexion. We therefore conclude that the first planes in momentum space outwards from the origin across which the energy shows an appreciable discontinuity are those lying parallel to faces of these two forms. The figure bounded by these two forms is a polyhedron of 36 faces, not very far from spherical in shape.

Let us now imagine that the number of electrons in the structure is gradually increased from zero. At first, the Fermi distribution, i.e., the volume occupied by the states of the electrons in momentum space, will grow very nearly spherically until it comes into proximity with the faces of the zonal polyhedron. When the sphere touches the zone, the distribution will no longer grow in that particular direction, since to do so would correspond to a large energy increment, but rather in other directions in which growth can still take place without the zone being crossed. The distribution will therefore ultimately assume a distorted shape approximating to that of the zone. Jones has calculated that in the case of the γ-alloys, for a unit cell containing 52 atoms, the spherical Fermi distribution inscribed in the zonal polyhedron corresponds to 80 electrons, whereas the whole polyhedron would hold 90. The actual number of electrons in the unit cell is 84, so that in the γ-alloys the surface of the Fermi distribution lies very close to the surface of energy discontinuity. This Jones states to be the distinguishing feature of the γ-structures. The exact filling of the zone corresponds to lowering of the energy of the lattice relative to other lattices with a different zone arrangement, and consequently adds stability to the structure. In terms of the theory of diamagnetic susceptibility given by R. Peierls[1], it accounts for the remarkably large diamagnetic susceptibilities of the γ-alloys, and it also gives an explanation of the large Hall coefficient which these alloys show.

[1] *Z. Physik* **80** (1933) 763.

CHAPTER X

Structure Determinations of Increasing Complexity

A. CONCLUSIONS FROM SYMMETRY ONLY

An X-ray Examination of *i*-Erythnitol, by W. G. BURGERS (1926), *see* paper **110**

▬

CONCLUSIONS FROM CELL-DIMENSIONS ONLY

168. Crystal Structures of Vitamin D and Related Compounds, by J. D. BERNAL (1932)

▬

168. I have had the opportunity of examining by X-rays the crystals of ergosterol and certain of its irradiation products, recently described by a team working at the National Institute for Medical Research. Though the results are only preliminary, they seem of sufficient interest to warrant publication at this stage. Five substances, all of composition $C_{27}H_{41}OH$ or $C_{27}H_{42}OH$, have been examined, with the results shown in the accompanying table.

All have the same b axis of 7.2 A and their a and c axes are simple multiples of 10 A and 20 A respectively. The spacing of the c plane is

Substance	a	b	c	β	$c \sin \beta$ $= d_{001}$	Space Group	No. of Mol. per Cell	No. of Mol. in Asymmetric Unit
Ergosterol	9.75	7.4	39.1	65°	*35.40*	C_2^2–$P2_1$	4	2
α-Dihydroergosterol and ethyl alcohol	30.8	7.4	43.1	53	*34.5*	C_2^3–$C2$	12	3
Calciferol	20.8	7.15	38.5	68	*35.65*	C_2^2–$P2_1$	8	4
Pyrocalciferol calciferol	20.2	7.35	40.0	63	*35.8*	C_2^3–$C2$	8	1
Lumisterol	20.3	7.25	20.4	60	*17.8* $=\frac{1}{2} \times 35.6$	C_2^2–$P2_1$	4	2
Cholesterol	16.4	...	...	...	*33.3*	C_1^1–$P1$	...	...

remarkably constant at 35.5 A. It differs significantly from that of cholesterol, which was examined for comparison. Lumisterol has a halved c spacing, and α-dihydroergosterol, owing to the presence of alcohol of crystallisation, deviates from the others.

From these observations certain conclusions can be drawn:

2. The molecule of ergosterol and its photoderivatives has the approximate dimensions 5 A × 7.2 A × 17–20 A. These form a double layer similar to those of long chain alcohols and acids. Such dimensions are difficult to reconcile with the usually accepted sterol formula

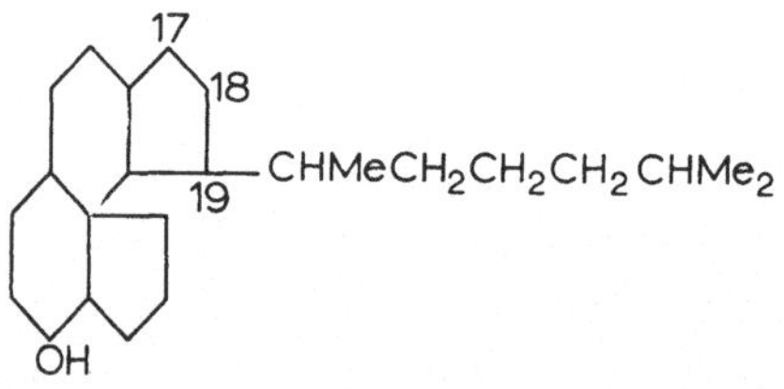

which would lead to a wider and shorter molecule, but agree much better with one where the carbon chain is attached to atom 17 in ring iv,*

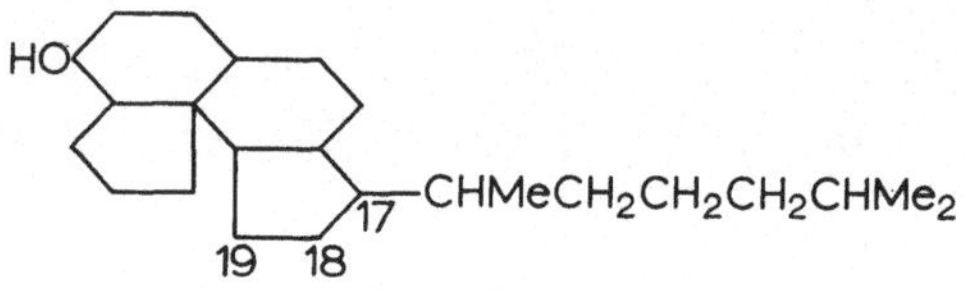

Mineralogical Museums, Cambridge

---

* [This conclusion, confirmed afterwards, is based on the influence of the place of the carbon chain relative to the OH-group on the dimensions of the molecule; the elongated shape of the six-rings in the figures will have been unintentional. Ed.]

B. THE FIRST COMPLETE STRUCTURE
DETERMINATION OF AN AROMATIC COMPOUND

The application of the X-ray methods to the analysis of organic crystals has its own special features both of technique and of interpretation.

The major interest in the attempt to solve an organic structure lies in the determination of the positions of the atoms in the molecule. This is a very difficult problem: but there are clear signs that the difficulties are yielding. The organic molecule usually contains a large number of atoms, and though there is a certain simplicity in the fact that the light elements carbon, oxygen and nitrogen atoms are almost the sole constituents of the structure—the hydrogens are necessarily and quite reasonably neglected at present—yet the number of parameters required to define the relative positons is very great.

If one or two key structures can be completely solved, it is to be expected that the way to the general solution of organic crystals will be opened.

There are two great groups of organic crystals, the aliphatic and the aromatic, based respectively on the chain and the ring of carbon atoms.

The ring has proved much less tractable than the chain. This is partly, perhaps, because its form is variable. It certainly does not possess the hexagonal symmetry which is, naturally enough, given to it in its usual form of representation. Very often it possesses no more than a centre of symmetry: though it has also been found that it can in other cases possess a trigonal axis.

Mrs. Lonsdale *is the first, I believe, to find a crystal—hexamethyl benzene—which it has been possible to solve with satisfactory completeness and certainty. The crystal is triclinic, with but one molecule in the cell. Mrs.* Lonsdale *has been able to place each atom: and to show that the atoms of the molecule all lie in one plane, which plane can be conveniently taken as one of the faces of the unit cell. Thus the atoms lie entirely in a set of parallel planes: the distance apart is 3.67 Å.U. The disposition, therefore, resembles in some respect the graphite arrangement, the chief differences being due to the presence of regularly placed hydrogen atoms. The crystal forms a standard with which others may be compared; its principal plane must give the maximum intensity of reflection to be observed in a hydrocarbon. If a different set of planes in the same crystal or a set in any other similar crystal gives a smaller reflection when the proper allowances have been made it is to be inferred that the atoms in such crystal are not all in one plane.*

Such a procedure is effectively a determination of the F curve of carbon: and it is to be remembered that this curve has already been measured by Debye and by Ponte, using diamond dust. Nevertheless, the difference between the perfect diamond crystal with its prominent co-efficient of extinction, and the imperfect organic crystal with its many peculiarities are so great that the separate determinations of the F curve are most desirable.

The applications of absolute intensity measurements to organic crystals have so far been very few and tentative: it is certain that they must be of great importance in the future, and we may look for a rapid extension of their use now that a beginning has been made. The application to inorganic crystals has been already very successful: it is to be hoped and expected that the study of organic crystals will benefit to a similar degree. In the papers to be presented this afternoon, the relative intensities of reflection by different planes of the same crystal have been measured with good results. The 'absolute' measurements, that is to say measurements in terms of the weight of the crystal and of the intensity of the primary beam will surely be features of the work of the near future.

W. H. BRAGG,
Trans. Far. Soc. **25** (1929) 346

169. X-ray Evidence on the Structure of the Benzene Nucleus, by KATHLEEN LONSDALE (1929)

The Structure of Graphite

In 1924 Hassel and Mark[1] and, independently, Bernal[2] determined the structure of graphite and showed conclusively that, although the unit cell is hexagonal and of the dimensions given by Hull, the carbon atoms are not arranged in puckered hexagons as in diamond but are in flat sheets, as in the Debye-Scherrer structure. They found the carbon atoms to be within 0.1 A of the cleavage plane. Later Ott[3] confirmed their results and

[1] O. Hassel and H. Mark, *Z. Physik* **25** (1924) 317.
[2] J. D. Bernal, *Proc. Roy. Soc.* A **106** (1924) 749.
[3] H. Ott, *Ann. Physik* **85** (1928) 81.

succeeded in showing that the carbon atoms could not, in fact, be more than about 0.03 A from the cleavage plane. The six-carbon rings in graphite are therefore flat and the atoms themselves are only 1.42 ± 0.01 A in diameter, as against 1.54 A for the atoms in diamond. Moreover three of the valency directions of the graphite carbon atom lie in one plane, the fourth exercising a weak attractive influence at right angles or nearly at right angles to this plane. This new result immediately put a new complexion on to the benzene problem. Since, as was mentioned previously, the projections of the graphite plane ring and the diamond puckered ring on to the mean plane of the ring are almost identical, it is not possible to state from the anthracene-naphthalene result that the benzene ring is definitely more like the one than the other. In fact we cannot with certainty say that it is like either; since two types of ring exist, we cannot deny the possibility of the existence of a third having a similar projection.

Statement of the Problem

The most satisfactory determination of the structure of the benzene nucleus is only to be expected from some aromatic compound in which the positions of the atoms themselves can be uniquely found with considerable exactitude, without the help of any assumptions or analogies based on the chemistry of that or other compounds. Such an analysis must try to answer certain very definite questions:

1. Does the organic molecule exist as a separate entity in the crystalline state?

2. Are the carbon atoms in the benzene nucleus arranged in the form of a closed ring?

3. If so, is the ring hexagonal in shape?

4. What are the sizes of the atoms in the ring and the dimensions of the ring itself?

5. Is the ring plane, or are the atoms arranged on more than one plane?

6. What are the positions of the side-chain atoms or groups relative to the nucleus and to one another?

7. What is the arrangement of the valencies in the 'aromatic' carbon atom?

8. How does the carbon atom in aromatic nuclei differ from the 'aliphatic' carbon atom?

9. Is the benzene nucleus always of the same size and shape?

The first question, as to the separate existence of the organic molecule in the crystalline state, has been answered definitely in the affirmative for several aliphatic compounds. Among these are hexamethylenetetramine[1,2], for which a complete structure determination has been given, and certain of the long-chain compounds[3,4]. There is, in fact, very strong presumptive evidence for the separate existence of the molecule in many crystalline organic compounds. The question is necessary, for however likely it may seem that the benzene nucleus should remain as a distinct entity in the crystal, it is better not to have to make that an *a priori* assumption in any structure determination. Nor have suggested theoretical structures always involved such an assumption. Thus in Barlow and Pope's conception of crystalline benzene, a carbon atom selected at random seems to belong to several possible rings. In fact, they themselves remark on this: 'The partitioning of this diagrammatic assemblage into molecular complexes can occur in several different ways'[5]. Huggins[6] has also suggested a theoretical structure for benzene in which molecular boundaries are no longer distinguishable. This he no doubt considered to be reasonable because in diamond, for example, each carbon atom belongs to six distinct rings and in graphite each carbon atom is a member of three rings.

X-ray Evidence from Hexamethylbenzene

$C_6(CH_3)_6$, preliminary results on which were published in *Nature*[7], has provided definite answers to a number of the questions tabulated above. The precise method of structure determination has been described in detail elsewhere[8], but a short analysis may also be useful here.

The unit cell is triclinic and contains one centro-symmetrical molecule of $C_6(CH_3)_6$. Its measured spacings and angles are

$$d_{100} = 7.736 \quad d_{010} = 6.008 \quad d_{001} = 3.694 \text{ Å}$$

$$\sphericalangle \, 010:001 = 129°18' \quad \sphericalangle \, 001:100 = 80°48' \quad \sphericalangle \, 100:010 = 74°0'$$

[1] R. G. Dickinson and A. L. Raymond, *J. Amer. Chem. Soc.* **45** (1923) 22.
[2] H. W. Gonell and H. Mark, *Z. physik. Chem.* **107** (1923) 181.
[3] A. Müller, *Proc. Roy. Soc.* A **114** (1927) 541—gives bibliography of previous work on the subject.
[4] *Ibid.* **120** (1928) 437.
[5] W. Barlow and W. Pope, *Trans. Chem. Soc.* **97** (1910) 2308.
[6] M. L. Huggins, *J. Amer. Chem. Soc.* **45** (1923) 264.
[7] K. Lonsdale, *Nature* **122** (1928) 810.
[8] *Proc. Roy. Soc.* A **123** (1929) 494.

from which can be calculated the lengths of the axes and the axial angles:

$$a = 9.010 \quad b = 8.926 \quad c = 5.344 \text{ Å}$$

$$a:b:c = 1.009:1:0.599$$

$$\alpha = 44°27' \quad \beta = 116°43' \quad \gamma = 119°34'$$

The density calculated from these data is 1.035 grs./c.c.; direct measurement by the suspension method gives 1.042 grs./c.c. at 18.5°C. Measurements of spacing and intensity were made from well over one hundred reflections. The observed intensities had to be corrected for angle of diffraction before they could be compared with one another for the purpose of structure determination. The normal procedure is to use a formula of the type A in order to obtain S, the structure factor of the given plane.

$$I_{\text{obs.}} \propto S^2 . F^2 . \frac{1 + \cos^2 2\theta}{\sin 2\theta} \qquad \text{A}$$

Here θ is the angle of diffraction and F is a factor dependent on θ, which makes allowance for the inner structure of the atoms themselves. It can be deduced from measurements upon simple crystals of known structure (diamond or graphite in the case of carbon)[1] or calculated by some method such as those described by Hartree[2] or Thomas[3]. In formula A the factor F includes the heat factor. The structure factor is far more useful than the observed intensity because it is independent of the spacing of the plane concerned and dependent only on the positions of the atoms relative to that plane. Thus S has a maximum value when all the atoms lie in the given plane and a minimum value when an exact interleaving of the plane occurs.

If $x_1 x_2$, etc., are the perpendicular distances of atoms of scattering power A_1, A_2, etc., from a plane of spacing, d, the structure factor will be

$$S = \Sigma A \exp\left(2\pi i \frac{x}{d}\right) \qquad \text{B}$$

In the case of a centro-symmetrical structure this reduces to

$$S = \Sigma A \cos\left(2\pi \frac{x}{d}\right) \qquad \text{C}$$

Structure factors can be calculated by means of equation C for given configurations of atoms in the unit cell and compared with the structure factors deduced from the observed intensities using equation A.

[1] M. Ponte, *Phil. Mag.* **3** (1927) 195.
[2] D. R. Hartree, *Proc. Camb. Phil. Soc.* **21** (1923) 625.
[3] L. H. Thomas, *Proc. Camb. Phil. Soc.* **23** (1927) 542.

KATHLEEN LONSDALE

Table I.

[Reproduced for a small number of reflections only. Ed.]

hkl	I	S	hkl	I	S	hkl	I	S
100	730	27.5	0$\bar{1}$0	365	24.7	$\bar{1}$10	313	23.6
200	52	14.9	0$\bar{2}$0	21	10.6	$\bar{2}$20	11.2	11.9
300	34	23.6	0$\bar{3}$0	10.5	20.2	$\bar{3}$30	8	19.2
400	30	34.4	0$\bar{4}$0	5.3	24	$\bar{4}$40	4	24
500	0.8	8	0$\bar{5}$0	—	< 17	$\bar{5}$50	—	< 19
600	—	< 12	0$\bar{6}$0	—	< 29	$\bar{6}$60	—	< 35
700	—	< 19	0$\bar{7}$0	—	< 51	$\bar{7}$70	—	< 59
110	11	5.9	1$\bar{2}$0	4.4	6.4	$\bar{2}$10	8	6.4
220	1.5	6.7	2$\bar{4}$0	—	< 9	$\bar{4}$20	—	< 5.8
330	1.2	12	3$\bar{6}$0	—	< 26	$\bar{6}$30	—	< 13
440	0.8	22	[4$\bar{8}$0		< 76]	$\bar{8}$40	—	< 21
340	7	50	4$\bar{7}$0	1.3	58	$\bar{7}$30	6.8	56
001	1000	72.0						
002	95	66.8						
003	19	72.2						
004	3	68.6						
005	<0.8	< 89						
006	<0.8	< 191						
101	120	31.5	01$\bar{1}$	50	28.8	$\bar{1}$11	36	27.4
201	8	11.8	02$\bar{1}$	2.4	11	$\bar{2}$21	1.3	9.0
301	11	20.5	03$\bar{1}$	2	16	$\bar{3}$31	1.7	18
401	8	25	04$\bar{1}$	1.6	24	$\bar{4}$41	—	< 23
501	—	< 11	05$\bar{1}$	—	< 29			

Trial showed that in order to obtain 'observed' factors independent of θ, F values on the behaviour of graphite had to be employed. The measured intensities and the deduced structure factors are given in Table I, the intensities relative to $I_{001} = 1000$ and the structure factors relative to $S_{001} = 72$.

The smallest intensity that could be observed was, in general, about 1/1000 of the (001) intensity. The corresponding structure factor depends, of course, on the angle of diffraction. The smaller the angle, the smaller the value of S giving an observable reflection. For planes from which no reflection could be observed, it is clear that the real structure factor must be less than a maximum value governed by the angle θ. Also, in general, the values of S corresponding to small-spacing (large θ) planes are less accurate than those for large-spacing planes. The accuracy is very roughly indicated in the table.

Several unusual features may be observed by a careful comparison of the structure factors.

(*a*) The structure factors of the set of planes (100) → (110) → (010) are repeated fairly closely throughout the corresponding series $(0\bar{1}0) \rightarrow$ → $(1\bar{2}0) \rightarrow (1\bar{1}0)$ and $(\bar{1}10) \rightarrow (\bar{2}10) \rightarrow (\bar{1}00)$. If one takes any plane (*hk*0) and compares it with $(k\,\overline{h+k}\,0)$ and $(\overline{h+k}\,h\,0)$, the similarity in structure factors (though not as a rule in intensities) is most remarkable. Since *a* is nearly equal to *b*, and the angle between them is nearly $2\pi/3$, it is clear that three such planes as these will intersect the (001) plane at angles of about 60° with one another. The similarities of structure factor between these sets of planes, therefore, show that the [001] zone is hexagonal in structure. The crystal as a whole does not possess hexagonal symmetry, nor does any set of *intensity* measurements nor any kind of Laue photograph show up the symmetry directly, simply because the *c* axis is obliquely inclined to the (001) plane. The lattice (Fig. 3) may be regarded as a series of hexagonal nets displaced somewhat arbitrarily with respect to one another.

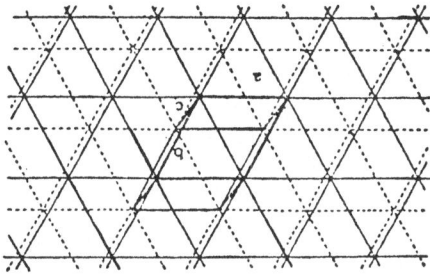

Fig. 3. Lattice projected on to (001) plane.

(*b*) A second unusual feature of the 'observed' values of S is that they are almost independent of the last index '*l*'. This may be seen by taking any plane (*hkl*) and comparing it with the corresponding plane (*hk*0), or better still by comparing sets of planes such as (*hk*0) (*hk*1) $(hk\bar{1})$ (*hk*2) etc. Slight variations from the rule do undoubtedly exist, but they are too small to affect the general argument. Now the only way of arranging atoms so that they occupy the same relative positions with respect to the pairs of planes (*hk*0) (*hk*1) (*hk*2) etc., is by placing them all in the (001) plane, which joins the lines of intersection of all the rest.

291

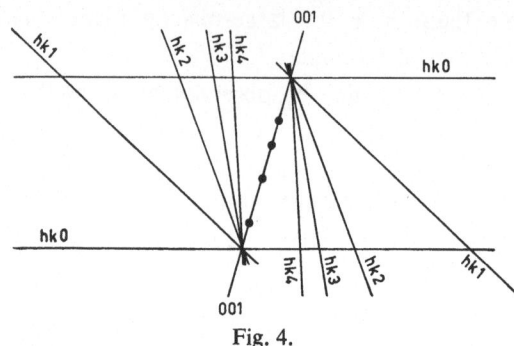

Fig. 4.

The scattering power of the hydrogen atoms is too small (relative to that of the carbon atoms), for these to be included in the argument, but (b) shows that the carbon atoms at least must all lie in or very near the (001) plane, and from (a) we know that they must be arranged approximately hexagonally in that plane.

(c) The closeness of the carbon atoms to the (001) plane at once explains the third remarkable feature of the 'observed' structure factors, which is, that the value of S for (001) is larger than any other in the crystal and is almost independent of the order of reflection (a special case of (b)), so that the observed intensities fall off in much the same way as those from the cleavage plane of graphite.

The structure factors of the planes (340), (4$\bar{7}$0), and ($\bar{7}$30) are also very large, and these are small-spacing planes, so that small movements of the carbon atoms whith respect to them are relatively more important than in the case of a large-spacing plane. The fact that their structure factors are next in magnitude to that of (001) shows that the carbon atoms must also lie very near to these planes. In other words the twelve carbon atoms in the unit cell must be near to the intersections of the planes (340), (4$\bar{7}$0), ($\bar{7}$30), and (001). There are 36 such intersections in the unit cell, but the problem of finding out which of them are occupied by carbon atoms is considerably simplified by the hexagonal structure we know to exist in the (001) plane. If the positions of two out of the twelve carbon atoms can be determined, the remaining ten simply fall in place. It is not difficult to find out which two out of six possible positions must be occupied to give structure factors (calculated from equation C) in passable agreement with those observed (calculated from equation A). Fortunately, only one out of the fifteen possible arrangements does give any such agreement, and the consequent positions of the twelve carbon atoms are shown in Fig. 5. The diameter of the carbon atoms has been taken as the distance between centres of neighbouring atoms. For the sake of completeness the nearest

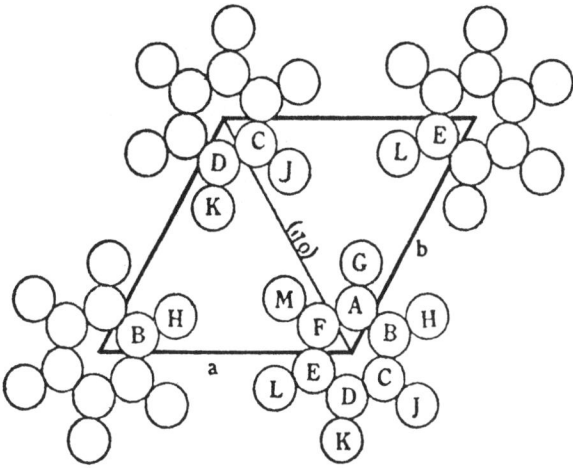

Fig. 5.

atoms in the surrounding unit cells are also shown. Although the above arguments apply simply to the separate atoms, it is clear that *the chemical molecule has emerged as a separate and distinct entity.*

Question (2) and (3) also admit of an immediate answer, as a result of this analysis.

The carbon atoms in the benzene nucleus are arranged in the form of a closed ring, which is not only centro-symmetrical but also possesses pseudo-hexagonal symmetry of the first or second kind. (That is, pseudo-hexagonal or trigonal symmetry together with a centre of symmetry.)

The fourth and fifth questions, as to the size of the atoms and the shape of the ring itself, can only be answered by making small changes in the positions of the atoms and finding how these affect the structure factors, until we obtain a set of factors giving the best agreement with observation. The way in which the size of the carbon atoms and their positions in the (001) plane have been found more exactly has been described elsewhere, so that only the final results are given here. The neglect of the hydrogen atoms prevents the attainment of an exact agreement, but very good agreement follows if the sizes of the atoms are:

Diameter of nuclear carbon atom $= 1.42 \pm 0.03\,\text{Å}$

Diameter of side-chain carbon atom $= 1.54 \pm 0.12\,\text{Å}$

and their co-ordinates, expressed as fractions of the lengths of the crystallographic axes (which are also the axes of co-ordinates):

$$\left.\begin{array}{c}A\\D\end{array}\right\}(\pm0.071\ \pm0.182\ \ 0)\qquad \left.\begin{array}{c}G\\K\end{array}\right\}(\pm0.145\ \pm0.371\ \ 0)$$

$$\left.\begin{array}{c}B\\E\end{array}\right\}(\mp0.109\ \pm0.073\ \ 0)\qquad \left.\begin{array}{c}H\\L\end{array}\right\}(\mp0.222\ \pm0.149\ \ 0)$$

$$\left.\begin{array}{c}C\\F\end{array}\right\}(\mp0.180\ \mp0.109\ \ 0)\qquad \left.\begin{array}{c}J\\M\end{array}\right\}(\mp0.367\ \mp0.222\ \ 0)$$

The comparison is shown in Table II.

Table II.

hk0			Mean S obs.	Mean S calc.	hk0			Mean S obs.	Mean S calc.
100	0$\bar{1}$0	$\bar{1}$10	25.3	26.1	320	2$\bar{5}$0	$\bar{5}$30	15.6	-19.6
200	0$\bar{2}$0	$\bar{2}$20	13.8	-13.3	430			<14	10.6
300	0$\bar{3}$0	$\bar{3}$30	21.0	-22.2	110	1$\bar{2}$0	$\bar{2}$10	6.2	-2.9
400	0$\bar{4}$0	$\bar{4}$40	27.5	-28.5	220	–	–	6.7	11.3
500	–	–	8	5.5	330	–	–	12	10.1
600			<12	-2.5	440		–	22	23.6
700			<19	-21.4	340	4$\bar{7}$0	$\bar{7}$30	55	67.4
710			23	-32.8	230	–	$\bar{5}$20	11	-4.6
610	–	–	18	-22.6	350		$\bar{8}$30	34	40.3
510	–	–	16	21.3	120	2$\bar{3}$0	$\bar{3}$10	21.4	16.5
410			<7	-5.8	240	–	$\bar{6}$20	25	24.3
310	1$\bar{4}$0	$\bar{4}$30	34.6	-30.4	250	5$\bar{7}$0	$\bar{7}$20	39	41.6
620			<20	-16.2			$\bar{4}$10	<5.3	1.1
520			<14	-13.4	140	–	$\bar{5}$10	14	-17.3
210	1$\bar{3}$0	$\bar{3}$20	8.5	-8.8			$\bar{6}$10	<11.5	-1.6
420			<9.2	-3.6			$\bar{7}$10	<18	-9.9
530			<20	-5.4					
						001		72.0	72.0

The agreement is quite spoilt if the diameter of the nuclear carbon atom varies much from 1.42 Å, which is also the diameter of graphitic carbon. It certainly cannot be as big as the carbon atom diameter in diamond, nor as small as 1.30, the value recently suggested by Hendricks[1].

It is important that the effect of 'puckering' the ring upon the calculated structure factors should be recorded. Table III [reproduced for 00*l* and 0*k*0 reflections only] shows a selection of the 'observed' factors and the corresponding calculated factors for the plane ring and for the

[1] S. B. Hendricks, *Z. Krist.* **68** (1928) 189.

Table III.

hkl	Observed S	Calculated S						
		I	II	III	IV	V	VI	VII
001	72.0	72.0	72.0	72.0	72.0	72.0	72.0	72.0
002	66.8	72.0	62.2	62.2	62.2	62.2	51.4	51.4
003	72.2	72.0	47.8	47.8	47.8	47.8	21.1	21.1
004	68.6	72.0	31.7	31.7	31.7	31.7	−12.8	−12.8
005	<89	72.0	16.4	16.4	16.4	16.4	−44.4	−44.4
0$\bar{1}$0	24.7	26.0	26.5	27.1	24.8	29.5	30.2	25.2
0$\bar{2}$0	14.6	−13.6	−18.5	−10.1	−22.9	−3.6	−8.5	−19.7
0$\bar{3}$0	20.2	−23.0	−37.3	−9.0	−15.8	−30.4	−45.7	−0.8
0$\bar{4}$0	24	−28.6	−44.8	−7.8	−16.8	−37.2	−54.7	5.5
0$\bar{5}$0	<17	6.8	0.8	20.2	−2.6	5.5	−0.7	10.8

Column I: plane ring; all atoms in (001).

      II: diamond-like ring; that is, nuclear atoms A, C, E, at perpendicular distance of +0.255 Å, and B, D, F, at −0.255 Å from (001).
Side-chain carbon atoms in (001).

    III: same arrangement inverted in cell.

    IV: nuclear atoms in (001).
Side-chain atoms, G J, L, at +0.255 Å; H, K, M, at −0.255 Å from (001).

     V: same arrangement inverted in cell.

    VI: nuclear atoms A, C, E, at +0.255 Å; B, D, F, at −0.255.
Side-chain atoms G, J, L, at −0.255; H, K, M, at +0.255.

   VII: same arrangement inverted in cell.

diamond-like ring with various amounts of puckering of the side-chain carbon atoms.

Only the plane molecule gives anything like good agreement with the observations. In fact it is found that even a motion of as little as ± 0.1 Å, perpendicular to the (001) plane, of either nuclear or side-chain carbon atoms gives structure factors out of agreement with observation, both in respect of loss of hexagonal arrangement in the [001] zone and in dependency upon the last index 'l'.

The answers to questions (5) and (6) in respect of $C_6(CH_3)_6$ are, therefore, that *the benzene ring is quite flat* and that *the side-chain carbon atoms are attached radially to their respective nuclear atoms and lie in the plane of the ring.*

The ring itself is similar both in structure and in size to the graphite six-carbon ring, and the nuclear (aromatic) carbon atoms, like those in graphite, have three co-planar valencies arranged at 120° to one another.

The distance between centres of carbon atoms in *meta*-positions is 2.46 ± 0.05 Å. Sir William Bragg's brilliant deduction of 2.49 Å as the width of the added ring in the anthracene molecule is, therefore, completely verified. The experimentally determined periodicities, 1.28 and 2.66 Å, along the *c* axis of anthracene, correspond approximately to the distances, 1.23 and 2.46 Å, between the carbon atoms in the benzene nuclei, resolved parallel to the length of the anthracene molecule.

The Fourth Valency Bond in Aromatic Carbon

X-ray structure analysis, however complete, can give very little direct information about the fourth valencies of the nuclear carbon atoms. The only point definitely proved is that they must be arranged so as to give the ring as a whole a centre of symmetry. This quite eliminates the Kekulé static model with its three fixed double bonds, since three points cannot be centro-symmetrical unless, indeed, they are collinear.

Ramanathan[1] has calculated the optical anisotropy of benzene for both a plane and a diamond-like ring, and he finds that the value calculated for a plane ring of graphite size is much larger than the experimental value, even when the hydrogen atoms are ignored, unless the nuclear carbon atoms are assumed to be distinctly anisotropic, with the axis of larger polarisation perpendicular to the plane of the ring. He did not feel justified in making this assumption; but now that the plane benzene ring is established, this hypothesis assumes the form of a consequence.

Apart from this evidence, however, there is no adequate reason to suppose that the fourth valencies in the benzene nucleus are arranged exactly as they are in graphite, that is, normal or nearly normal to the plane of the ring. The way in which the F factors, referred to previously, vary *in* and *normal to* the plane of the benzene ring would no doubt furnish valuable information upon this point, and so would accurate knowledge of the temperature factors. But these are not yet available even for graphite, quite apart from benzene itself.

Question (7), therefore, has only been answered in part.

Relation Between Aromatic and Aliphatic Compounds. That the aromatic carbon atom might resemble that in graphite has frequently been suspected on account of the apparent tervalency of the carbon atom in each case. Debye and Scherrer were the first to put forward the suggestion that graphite is the prototype of aromatic and diamond of aliphatic compounds.

[1] K. R. Ramanathan, *Proc. Roy. Soc.* A **110** (1926) 123.

It is of great importance to know whether the aromatic character of the benzene nucleus is bound up with its graphitic arrangement, or whether that structure can persist even when the aromatic properties are lost. If even the reduced carbon ring is still flat, then the aromatic nature must depend more on the fate of the fourth valency bond than on the configuration of the ring.

In simpler aliphatic compounds the carbon atom is generally tetrahedral or sphenoidal. In the methane molecule, for instance, it is at least tetrahedral[1], in penta-erythrital tetra-acetate the central carbon atom is sphenoidal[2]. Many other examples might be given in which the carbon atom approximates to the diamond type. It is, unfortunately, not possible to determine the positions of the eighteen hydrogen atoms in $C_6(CH_3)_6$, since they are too small in scattering power to be governed by the rules which led to the location of the carbon atoms. If CH_3 groups of neighbouring molecules are to touch one another in the (001) plane, their approximate diameter must be 3 Å. Calculation shows quite clearly, however, that the CH_3 group does not scatter spherically, like a single atom of the same size. The carbon atom belonging to the group is quite definitely at the end nearer to the benzene nucleus, and has a diameter of 1.54 ± 0.12 Å. It may possibly, therefore, be of the same size as the graphite carbon atom, but is much more probably of the diamond size. Even though we cannot tell whether the hydrogen atoms in this particular case are attached at the tetrahedral angle or not, this similarity in size is significant, considered in conjunction with the undoubted tetrahedral arrangement of other carbon atoms in aliphatic compounds. It does, in fact, seem very likely from the X-ray evidence that although carbon atoms in aromatic nuclei are like those in graphite, aliphatic carbon atoms are much more like those in diamond, both in size and in the arrangement of their valencies. The latter, however, can suffer considerable distortion when the added groups are heavy or extensive. More X-ray work is needed before question (8) can be fully answered, and the same is true also of question (9).

Possible Deformation of the Benzene Nucleus. The fact that the nucleus in $C_6(CH_3)_6$ corresponds so closely in structure to the graphite type of ring indicates that very little deformation can have taken place. There seems no reason to suppose that in, at any rate, the simpler fully-substituted derivatives the structure of the nucleus will depart very much from the plane hexagonal form. On the other hand, some symmetry has certainly been lost, for the graphite ring possesses true hexagonal symmetry of the first or second kind, whereas none of the benzene derivatives are more

[1] J. C. McLennan and W. G. Plummer, *Nature* **122** (1928) 571.
[2] I. E. Knaggs, *Nature* **121** (1928) 616.

than centro-symmetrical. Then again, the structural analysis of $C_6(CH_3)_6$ shows that all the side-chain carbon atoms are in the plane of the ring itself, whereas in dichlornaphthalene tetrachloride[1] it seems probable that the two substituted chlorine atoms are not in the plane of the aromatic ring. Whether the valency directions of the nuclear carbon atoms are dependent to a certain extent upon the forces between the attached groups, is a question that must be left open until the structures of more aromatic compounds have been completely analysed.

Acknowledgments. The independent research work which is included in this paper was carried out in the Physics Department of the University of Leeds, with the help of a Scholarship from Bedford College, London. A grant from the Royal Society defrayed the expenses of part of the apparatus employed.

[1] J. M. Robertson, *Proc. Roy. Soc.* **118** (1928) 709.

C. SUCCESSIVE DETERMINATIONS OF THE NAPHTHALENE AND ANTHRACENE STRUCTURES

170. The structure of organic crystals offers a very inviting field of research by the methods of X-ray analysis. To the organic chemist the relative positions of the atoms in the molecule, as also of the molecules in the crystal, are of fundamental importance; and it is with these relations that the X-rays deal in a manner which is new and unique. Moreover, the multiplicity of crystalline forms—and this is true of both organic and inorganic substances—each so precise and invariable, and so obviously related to the atomic and molecular forces, is a sign that if the forces were better understood it would be possible to account for the forms that are known and possibly to build others that are unknown. But in order to acquire such a power we must learn the crystalline structure, so that the physical characteristics of the whole may in the end be referred to the characteristics of the individual atom. Progress has been made with the examination of the structure of some of the simpler inorganic crystals; but organic crystals have been neglected*. Their molecular complexity has been somewhat of a deterrent. Yet, if a way could be found of making determinations of structure, in spite of the complexity, it seems likely that they would quickly be fruitful. The substitutions and additions which are so characteristic of organic chemistry take place in such an immense variety of combinations and grades that the slightest knowledge of the underlying mechanism might lead to useful comparisons and rules.

I have made a careful study of a few crystals, principally naphthalene

* [*see* foot-not p.e 303. Ed.]

and some of the naphthalene derivatives, in order to discover, if possible, some way of handling the complex molecules. The numerical results will be set out later, and may, I think, be taken as sufficiently accurate to make foundations for a theory.

I shall endeavour to show that the results can be explained, so far as can be seen at present, by supposing the benzene ring or naphthalene double ring to have definite form and size, preserved with little or perhaps no alteration from crystal to crystal, and that there are good *a priori* reasons for the supposition. If this principle be accepted the problem is simplified at once. Naphthalene itself is then to be regarded as a structure in which there is but one element, the naphthalene double ring, and no longer as an aggregate of 10 carbon atoms and eight hydrogen atoms of unknown mutual arrangement. A more complex molecule such as either of the naphthols is not to be regarded as an addition of one oxygen atom to these 18, an idea on which nothing can be built, but as a naphthalene double ring of the same size and form as before, except that one particular hydrogen has been replaced by a hydroxyl group. It is then possible to think what changes in the disposition of the molecules might be caused by such a substitution and to compare conceivable solutions with observations on the dimensions of the new crystal. Such a method of procedure is obviously in good agreement with the ideas of organic chemistry.

It is convenient to distinguish the facts regarding crystalline structure which can be obtained by the goniometer and various other means, from the new facts which can be obtained by the use of X-rays. The former are recorded in crystallographic tables such as are given by von Groth in his 'Chemische Krystallographie'. Naphthalene may be taken as an example. In the fifth volume of von Groth's work, on p. 363, a description of naphthalene is given from which the following data are taken:

Monoclinic Prismatic

$a : b : c = 1.3777 : 1 : 1.4364;$

$\beta = 122°49'$

The monoclinic prismatic class has the highest symmetry of which the monoclinic system is capable, having a digonal axis and a plane of symmetry perpendicular to it. The figures give the angular, but not the linear, dimensions of the unit cell of the structure (Fig. 1); the unit cell being the smallest volume, of which, by continual repetition without any change in contents or disposition, the whole crystal can be formed. In any crystal it is possible to choose the unit cell in many ways, but they must all be capable of derivation from one another and must all have the same volume.

300

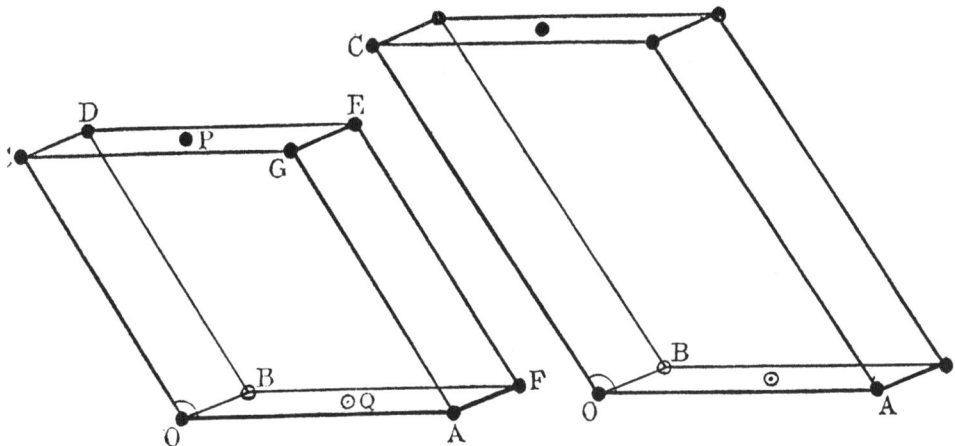

Fig. 1. Unit cells of naphthalene and antrahcene drawn to the same scale.

	$OA = a$	$OB = b$	$OC = c$
Naphthalene	8.34	6.05	8.69
Anthracene	8.7	6.1	11.6

Naphthalene $\alpha = BOC = 90°$, $\beta = COA = 122°49'$, $\gamma = AOB = 90°$

Anthracene $\alpha = BOC = 90°$, $\beta = COA = 124°24'$, $\gamma = AOB = 90°$

The angular dimensions are to be considered as including the angular relations to one another of any pair of planes in the crystal, not merely of the planes bounding the cell.

It is also stated that the (001) face is the cleavage plane; and that, in addition, the faces (110), (20$\bar{1}$), (11$\bar{1}$) are found as bounding planes of the crystal. The angles between various pairs of these faces are also given as observed. The specific gravity is stated to be 1.152. Other information is given by von Groth concerning the optical properties of the crystal; also concerning the methods that have been adopted in growing the crystals from various solutions and the consequent effect on the development of different faces. These facts do not concern us for the moment, but they must be taken into account eventually.

The examination by X-ray analysis gives us the spacings between the planes and, therefore, the linear as well as the angular dimensions of the unit cell. The specific gravity being known, and the actual weight of the molecule, it is possible to find how many molecules are contained in each cell; generally, two or four. In the case of naphthalene, it is found that, assuming the angular dimensions to be correctly given by the crystallographers, the linear dimensions are:

$$a = 8.34, \; b = 6.05, \; c = 8.69.$$

301

These figures are obtained in the following way:

The actual length of the b axis being represented by b, the mass contained in the cell is

$$b^3 \times 1.3777 \times 1.4364 \times \sin 122°49' \times 1.152 \text{ A.U.}$$

(It is convenient in this work to extend the Angstrom system of units so that an A.U. of area is 10^{-16} cm^2, of volume 10^{-24} cm^3, and of mass 10^{-24} gr.)

The mass of the hydrogen atom being 1.662 A.U., the mass of the molecule $C_{10}H_8$ is $128 \times 1.662 = 213$ A.U. Now, from the full results of the X-ray measurements which will be given presently, it is perfectly clear that there are two molecules in each unit cell. Hence, the value b is readily calculated, and the values of a and c also.

Besides these determinations of length, the X-ray method gives also the angle between any pair of planes, whether they form faces or not, provided that a measurable reflection can be found. Also, the relative intensities of the reflection by different faces, as well as the relative intensities of the spectra of different orders given by any one set of planes, yield information as to the distribution of the scattering centres and of the atoms which contain them.

We may now consider what reasons can be put forward for assuming the concrete existence of the benzene and naphthalene rings. If we examine the structure of the diamond we find that the atoms of carbon are tied together so that each is at the centre of gravity of four others. The distance from centre to centre is 1.54 A.U. As I have already pointed out[1] the rigidity of the diamond and the open character of its structure, imply that great force is required to alter the orientation of any coupling with respect to the other three belonging to the same atom. Were it otherwise, all atoms would seek to be surrounded by as many neighbours as possible; the substance would be close packed, and its density would be more than double what it is. The structure of the diamond may also be looked on as consisting of a series of puckered layers parallel to a given tetrahedral plane (see Fig. 6). A sharp blow may cleave the diamond, along one of these layers. If we take a model showing two layers as in graphite, lay hold of the upper layer, and move it to the new position shown in the figure, the structure is now that of diamond. I am following Hull's determination[2]; Debye and Scherrer[3] would flatten out the puckers in the

[1] *Proc. Phys. Soc.* **33** Part 5, August 15 (1921).
[2] *Physical Review* **10** December (1917) 692.
[3] *Phys. Zs.* **18** (1917) 297.

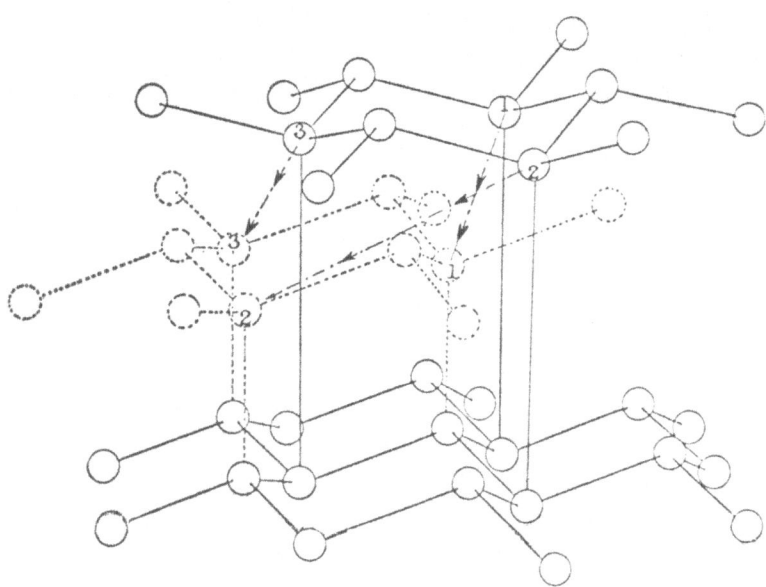

Fig. 6. The fine lines of the diagram show the structure of graphite. By moving the top layer to the position shown by the broken lines the diamond structure is obtained.

planes of graphite*. The point is perhaps not important for our present purpose, but it is necessary, for descriptive purposes at least, to choose one form. According to Hull's measurements, the shortest distance between each pair of atoms lying in the same layer is shortened from 1.54 A.U.—the diamond spacing—to 1.50 A.U. The distance between two successive layers has been increased by 1.35 A.U. A carbon atom in one layer is now at equal distances from its three nearest neighbours in the next layer, the distance being 3.25 A.U.

The bonds between one layer and the next are now greatly weakened; the substance cleaves readily in thin flakes. One layer slides with great ease over the other, though the bonds between the atoms in any layer are at least as strong as before. When all the bonds were of the strong kind, the substance, as diamond, was the hardest thing known. When the one set of bonds has been weakened, the substance, as graphite, is used as a lubricant. Probably its efficiency as such depends both on the weakness of the one set of bonds and the strength of the other. Yet these new bonds are perfectly definite, and the distance between two layers and, therefore, the 3.25 distance between atomic centres is a perfectly constant and de-

* [The reader, when comparing these statements with paper **169** and its introduction, should bear in mind that the present first paper of the anthracene series is of an *earlier* date than that of the paper preceding it in our collection. Ed.]

terminate quantity. It should be of great interest to compare the physical constants of graphite along and perpendicular to the axis, since in the two cases the two kinds of bonds are separately involved. Some of the comparisons would be difficult, but the thermal expansions can probably be compared by X-ray methods.

If the strong bonds between the atoms in a layer remain, and are even drawn a little tighter when the graphite form replaces that of diamond, it seems very reasonable to suppose that the single ring or multiple rings which are so clearly to be distinguished in the network may be separated out as such without loosening the bonds between their component atoms. In fact, these latter bonds might be expected to tighten even a little more. Let us assume that a single ring is a benzene ring, a double ring a naphthalene ring, and so on. Taking the spacings as given by Hull for graphite, the dimensions of the benzene and naphthalene frameworks are as shown in Fig. 7. The figure is constructed to show the arrangement of atoms in the naphthalene crystal, but it will also serve to illustrate the point under discussion. The carbon atoms A to F form a benzene ring, those from A to J the double ring of naphthalene.

The atoms centres A, and G, are 0.71 A.U. above, and the centres D and J the same distance below the plane of the diagram which is supposed to contain all the remaining centres. It should be observed that circles are used to represent the atoms as a convenient method of designation, not as implying that the radius (1.50) may *always* be used in calculating the distance between the centre of any one atom, and the centre of any other atom.

We may now go on to consider individual crystals; and we take naphthalene first. It might have seemed more natural to attack the benzene crystal before the naphthalene; but the latter was chosen because it is a very well-shaped crystal, and is solid at ordinary temperature. Benzene can only be examined under special temperature conditions and then only, with convenience, as a mass of small crystals. The study of benzene and some of its derivatives has been begun, but the greater attention has been given to the naphthalene crystals, and I will describe now the results that have been obtained in their case.

In the next table are set out the results of the linear measurements. They are compared with calculations based on the assignment of *one* molecule to each cell, which is equivalent to supposing each corner of the cell to represent a molecule. The object of making this provisional assumption is to show how the position of the second molecule can be found by comparing the observed with the calculated results.

Table II.

Plane	Calculated spacing	Observed spacing	Nature of reflection
100	7.00	3.46	Strong: indication of spacing 6.92
010	6.05	2.95	Very weak
001 (Cleavage)	7.30	7.30	Very strong: also higher orders
110 (Natural)	4.59	4.55	Strong
11$\bar{1}$ (Natural)	4.70	4.63	Moderate
20$\bar{1}$ (Natural)	4.17	4.12	Strong
021	2.79	2.76	Very weak
10$\bar{1}$	7.51	3.71	Very weak
210	3.04	2.99	Strong
21$\bar{1}$	3.44	3.39	Strong

No reflection obtained from following planes:

011, 012, 10$\bar{2}$, 101, 111, 22$\bar{1}$, 112, 11$\bar{2}$, 221, 21$\bar{2}$, 211, 212.

The table shows that the 100 and 010 spacings are only half what they should be if there were molecules at cell corners only. (N.B. Only one-eighth of a corner molecule is within the cell, and the whole eight count for one whole molecule within the cell.) But the 001 spacing is right. We conclude that there is a molecule at each of the points P and Q (see Fig. 1), each contributing half a molecule to the cell: and that these are in all respects similar to the corner molecules. Molecules placed at P and Q interleave the planes 100, 010, and also 10$\bar{1}$ by other planes of equal density which halve the corresponding spacings. The planes 110, 20$\bar{1}$, 021, are unaffected because they already contain P and Q.

It should be mentioned, however, that the 100 plane seemed to give a small spectrum at half the angle which gave the principal reflection; this would indicate that the second molecule was not quite similar, in orientation or some other particular, to the first, or was not exactly at P and Q. This is also suggested by the fact that 210 and 21$\bar{1}$ give the calculated spacings, whereas half values might be expected.

Although so much information is given in these tables, some of which we have used and some of which we cannot fully use for lack of knowledge, yet it would be hopeless to try to arrange the eighteen atoms of naphthalene on the basis of what has been learnt, without some helpful hypothesis. But we now take the naphthalene double ring as described. Its dimensions are such that it seems quite possible to fit two of them into the cell, if we had some indication as to their orientation thereto.

305

As to this we get a strong hint from a comparison of naphthalene with anthracene ($C_{14}H_{10}$), whose construction shows three rings in a line, as against the two of naphthalene. The crystallographic data of the latter are:

$$a : b : c = 1.4220 : 1 : 1.8781, \quad \beta = 124°24'.$$

$$\text{Specific gravity} = 1.15.$$

The crystals themselves are very small flakes, and it was not possible to find one which could be conveniently treated by the single crystal method. However, by pressing a number of them together against a flat disc, so that all the 001 planes were parallel thereto, however oriented they might be otherwise, it was easy to get a sufficiently accurate determination of the 001 spacing, and therefore the linear dimensions of the unit cell. It appeared that there were two molecules to the cell, as for naphthalene. The dimensions were:

$$a = 8.7, \ b = 6.1, \ c = 11.6.$$

[Renewed determination on minute perfect crystals, W. H. BRAGG, *Proc. Roy. Soc.* **35** (1923) p. 167:

$$a = 8.58, \ b = 6.02, \ c = 11.18, \ \beta = 125°0'$$

specific gravity: from X-rays 1.255, pycnometric 1.250. Ed.]

If these dimensions are compared with those of naphthalene (see Fig. 1), it will be seen that while *a*, *b*, and *β* remain nearly the same, the *c* axis has lengthened considerably, the difference amounting to 2.9 A.U. nearly [(1923)-value: 2.5 A.U. Ed.] Now the extra ring, if of the benzene dimensions, should be responsible for an addition of 2.5 A.U. nearly to the molecule.

It is reasonable to conclude that the molecules in both crystals lie end to end along the *c* axis, and that the structures are similar.

The over-all lengths of the two molecules, without allowance for the hydrogen atoms at their ends, that is to say, in the *β*-positions, are 6.41 and 8.86 respectively. There is, therefore, a vacant space between the ends of two molecules of rather more than 2 A.U., into which two hydrogen atoms have to be fitted. This agrees very well with what might be expected; only it must be remembered that we have no definite indication from studies of crystal structure as to the actual distance between the centres of a carbon and a hydrogen when united by a valency bond, nor between two hydrogen atoms not so united.

We have still to decide in what plane, passing through the *c* axis, the molecule is to be placed, and we have less clear indications with respect to this point than those that have guided us hitherto. On making up a model, however, it is seen that it is much more likely that the plane of the molecule lies nearer to the *ac* plane than the *bc* plane. The molecules lock together much better if that is so. Moreover, if the molecules lie in the *bc* plane, they would be close neighbours in that plane, and at the same time there would be wide gaps between consecutive planes. The plane 100 should, therefore, be prominent, most probably a natural face, perhaps even a cleavage plane: whereas it is neither of these things. But if the molecules lie in the 010 plane the form of the crystal seems much easier to understand, as we shall see later. The β-hydrogens of each molecule lie up against the corresponding hydrogens of the next and the 001 plane passes through them all. It would appear to be the weakest junction in the crystal, and therefore the 001 plane is the cleavage plane.

It must be observed that in the junctions between molecule and molecule there are forces far weaker than the valency forces, which latter unite the atoms of the same molecule. It is the former which bind the molecules into the crystal, nevertheless.

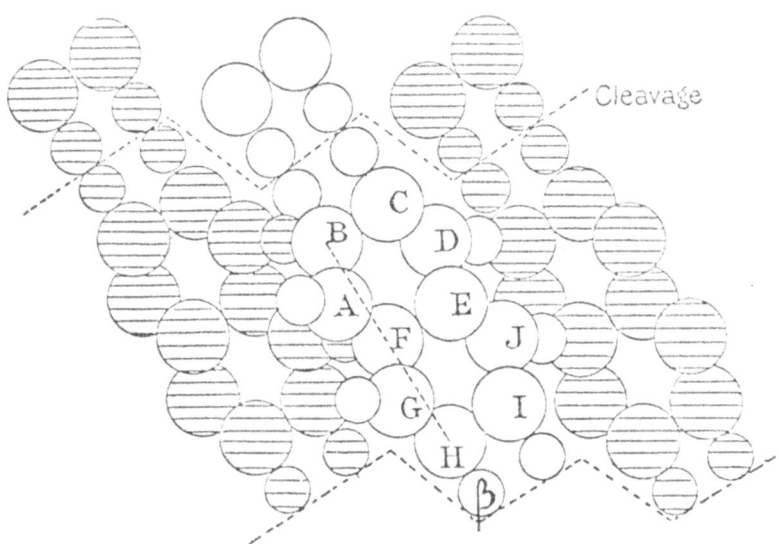

Fig. 7. Showing mutual relations of three naphthalene molecules and parts of others.

The unshaded circles between the two cleavage planes represent a molecule as at *Q* (Fig. 1). The shaded represent molecules *B* and *F* in the same figure. The small circles represent hydrogen atoms, but their size is uncertain.

Diameter of carbon atom = 1.50. *BH* = 4.92. Projection of *AD* on the plane of the diagram = 2.50.

307

When the model is put together in the way now indicated it is found that all the α-hydrogens, those that are attached to the sides of the molecule, lie up against carbon atoms of the next molecule, and that there is an appropriate space waiting for each, of magnitude about 1 A.U., the actual value depending on the orientation of the molecule. The forces exerted at these junctions, though far weaker than valency forces, are stronger than the forces between two β-hydrogens; and therefore if the crystal is ruptured it is the latter which give way first. The forces exerted by the α-hydrogens do so across the 110 and 1̄10 planes, and it is not surprising to find that the latter make natural faces of the crystal and give a strong reflection. The structure now found is a very empty one: it is like lace-work in space. That must be expected, since the specific gravity is so low. The structure is shown in Fig. 7.

The work that I have now described may conveniently be divided into two parts. The first is experimental, and shows, I think, that a large mass of new and valuable information respecting the linear dimensions of organic crystals is to be gained by the methods of X-ray analysis. In the second, I have ventured to suggest and apply a certain principle, viz., that the benzene and naphthalene rings as well as other ring combinations have actual form and dimensions which are nearly, if not quite, the same when they are built into different compounds. Such a principle requires much more illustration before it can be finally established; for the present, however, it shows promise of providing an entry into a very wide field which otherwise seems difficult to break into.

Moreover, the idea fits in extremely well with the work of Langmuir on surface films, and with such results as those described by Adam[1], and with the ideas and experimental results of several other workers. We see clearly that the forces that bind atoms together are of more than one kind. The very strong valency bonds, whether explained as due to electron sharing, or in any other way, are exemplified by all the linkings in diamond and by the linkings in the planes of the graphite flakes. But, besides these are other bonds of much weaker character, such as those extending between an atom in one graphite sheet and its three nearest neighbours in the next. Such bonds as these unite the molecules of the organic compound so as to form the crystal. They are of varying degrees of strength; the cleavage plane shows where they are weakest. They are definitely associated with special points on the molecule as we see from the facts of crystallisation. When a crystal forms in a liquid, or by sublimation, the molecule that attaches itself correctly, and with proper orientation to others already in

[1] *Proc. Roy. Soc.* **99** July (1921).

308

position, is the one that stays there and resists the tendencies of other drifting and thermally-agitated molecules to remove it. It is fixed by the attachment of certain points on its own structure to certain points on the structure of the other molecules. The beautiful exactness of crystal structure is evidence of the precision with which this adjustment is made; and at the same time of the definite molecular form without which precision would be impossible.

171. In a paper published in the *Proceedings of the Royal Society* (A **125** (1929) 542) on the structure of naphthalene and anthracene, J. M. Robertson comes to the conclusion that 'the scattering centres lie nearer the *ac* planes than the *bc* planes, but no simple structure with a plane of symmetry parallel to the *ac* plane is possible', and that the scattering centres lie along a chain structure similar to hydrocarbons. On the other hand, the structure of hexamethylbenzene as determined by K. Lonsdale (*Proc. Roy. Soc.* A **123** (1929) 537) suggests that the benzene rings in aromatic compounds should in all probability be plane structure. This has further support from the plane hexagonal structure of graphite (Ott, *Ann. d. Phys.* **85** (1928) 81). As regards whether the scattering centres are nearer the *ac* plane or the *bc* plane, the optical and magnetic anisotropies which have been measured by S. Bhagavantam (*Proc. Roy. Soc.* A **124** (1929) 545) require that the carbon atoms should lie nearer the *bc* plane than the *ac* plane. The structure proposed by Robertson, however, does not explain the intensities of reflection from many of the crystal planes, which he supposes are due to small glancing angles for those particular reflections. But on evaluating the angle factors for the intensities it is seen that such large discrepancies cannot be explained in that manner.

I made an X-ray investigation into the structure of naphthalene and anthracene, the results of which will be published shortly. It has been found that the best agreement for the intensities of reflections from these crystals is obtained when all the carbon atoms in one molecule are supposed to be in one plane and the planes of the molecules are inclined to the cell faces. The correct positions of the molecules are obtained by first placing them along the *bc* planes, then rotating them through 25° about the *c* axis (the two molecules in the unit cell being rotated in opposite directions), and then rotating them about *b* axes through 12° and 9° for naphthalene and anthracene respectively. The agreement will be best seen by referring to the table appended herewith, where the results for some simple planes are given. Similar agreements were obtained for all the forty planes from

Table I.

Indices	Naphthalene		Anthracene	
	Theoretical Structure Factor	Experimental Structure Factor	Theoretical Structure Factor	Experimental Structure Factor
001	15.3	15.3	13.2	13.2
002	6.0	6.2	8.8	8.4
110	18.2	17.5	27.0	30.3
11$\bar{1}$	5.1	5.9	10.2	8.9
020	6.6	7.0	8.3	7.5
200	15.0	14.8	19.8	18.3
20$\bar{1}$	24.8	23.0	21.0	14.9
20$\bar{2}$	5.2	4.8	9.2	9.9
210	10.0	10.6	14.7	16.2
21$\bar{1}$	9.2	10.0	12.6	14.9

which reflections were observed. It can be easily seen that agreements are much better than those obtained by Robertson.

The intensities of 007, 20$\bar{7}$, 40$\bar{7}$, 60$\bar{7}$ reflections from naphthalene and 00$\bar{9}$, 20$\bar{9}$, 40$\bar{9}$, 60$\bar{9}$ reflections from anthracene, on which Robertson bases his arguments for supposing that the scattering centres lie nearer the *ac* planes, agree qualitatively with experiment as the structure factors for the 40$\bar{7}$ and 40$\bar{9}$ planes respectively come out the highest among the series according to this arrangement of placing the carbon atoms.

Calcutta, Nov. 26, 1929

172. I believe Dr. Banerjee's structure to be essentially correct. It has been clear to me for some time that the last two sections of my paper to which Dr. Banerjee refers must be amended as regards the distribution of the scattering centres in the *a* and *b* directions. During last summer, Sir William Bragg made 'absolute' measurements of the intensities of the reflections from a number of anthracene planes. These measurements were expressed as ratios between the structure factors actually found, and the structure factor to be expected if all the atoms were in the reflecting planes. It was intended that these results and deductions therefrom should be incorporated with my paper, the publication of which was to be delayed for the purpose: unfortunately, owing to my absence from England, there was some confusion during the revision of the proofs and this was not done. Sir William Bragg's figures lead to a structure resembling Dr. Banerjee's so closely that it is interesting to give the following quotation from a letter which he wrote to me. It is in the form of notes upon a table of structure factors:

No. 1: 'A flat molecule, axis along the c axis; plane of molecule making an angle of 25° with the bc plane. This gives good values in the c zone, but not in the b zone; especially the $20\bar{1}$ is far too weak. So next (No. 2) the molecule is tipped over a little more to the upright position (about 6°). This greatly improves the b zone .... In No. 3 a slight buckle is put in, to try to improve the notable 204. The consequences are not very striking. On the whole there is so much agreement that we cannot be very far wrong.'

Plane	S Observed	S Calculated		
		No. 1	No. 2	No. 3
200	0.70	0.68	0.58	0.50
020	0.33	0.32	0.32	0.31
110	0.50	0.47	0.48	0.47
210	0.67	0.58	0.52	0.55
310	0.20	0.23	0.19	0.17
410	0.55	0.67	0.39	0.29
320	0.50	0.55	0.40	0.42
001	0.22	...	...	0.23
002	0.26	...	0.19	0.15
$20\bar{1}$	0.50	0.14	0.43	0.40
204	0.80	0.27	0.50	0.73

Whether the carbon atoms in these molecules lie in one plane as strictly as do the graphite carbon atoms, or those of hexamethylbenzene, can scarcely yet be stated with certainty. But the structure certainly appears to approximate to those types.

Physics Dept., Michigan University, Ann Arbor, U.S.A., Jan. 6, 1930

[*See further* paper **191**. Ed.]

D. SILICATES

The analysis of the silicate structures and the discovery of their natural classification have been the greatest triumph of X-ray crystallography.

J. D. BERNAL and W. A. WOOSTER,
Ann. Rep. **28** (1931) 301

173. The Structure of Silicates, by W. L. BRAGG (1927)

▬

174. The Structure of Diopside, by B. WARREN and W. L. BRAGG (1928)

▬

175. W. L. BRAGG and J. WEST (1928)

▬

176. Zur Frage der Struktur und Konstitution der Feldspate, by F. MACHATSCHKI (1928)

▬

177. The Structure of the Micas and Related Minerals, by L. PAULING (1930)

▬

178. The Structure of Sanidin and other Felspars, by W. H. TAYLOR (1933)

▬

173.[1] At a time when thefundamental conceptions of the structure of matter are being changed so rapidly, when every six months witness the birth of a new analytical method of dealing with the very foundations of our physical science, the study of the crystal patterns of silicates must seem a trivial matter. Yet similar studies have played their part in the extraordinary development of physics in the last decade, and I hope that the refinement of our methods of analysis, such as is represented by the present examples, will in turn prove to be of use.

The results of the particular investigations which I propose to describe are interesting in themselves because the silicates form so large a pro-

[1] Discours delivered at the Royal Institution on Friday, May 20

312

portion of the earth's solid crust, and certain artificial forms are so largely used for technical purposes. I think, however, that it is right to stress another aspect of this study. We are trying to improve the technique of the X-ray examination of solid bodies, to increase the resolving power of our instruments so that we can see finer detail and deal with more complicated structures. It is at present a tedious and difficult task to discover how the atoms are arranged in these bodies; even when some experience in handling them has been obtained, one has to devote much time and concentration to each particular case. Yet every solution makes the next problem easier to attack, and when we look back on the last few years' progress, I think a very real increase in power of analysis is evident as the result of the efforts of many workers in this field.

It is this advancement of technique in which, personally, I am particularly interested. In trying to improve our instrument we examine with it from time to time a new type of solid body—these silicates being an example—and we describe what we see. I am not competent to discuss the discoveries we make; I can only hope that the casting of light at a new angle may be useful to those who have made a life-long study of these particular types of compound. I feel that my main contribution must be a demonstration of what knowledge it is possible to attain by careful X-ray examination.

The silicates present a highly interesting series of problems for X-ray analysis. The numerous crystalline forms have been carefully studied because of their importance to the mineralogist, and they show most interesting relationships and wide variation in composition. It is estimated that oxygen, silicon, aluminium, and iron by themselves compose about 87 per cent of the earth's crust, and if we add four other elements, calcium, sodium, potassium, and magnesium, they amount together to 98 per cent. These are the elements that build up the compounds we are considering, and their relative proportions are a reflection of the fact that most of the earth's crust is composed of compounds of metals with silicon and oxygen.

The silicates occur as solid bodies, the atoms of which are arranged in crystalline patterns. These patterns are often very complex, and it would be difficult to attack them by general methods unless some guiding line could be followed through the intricate maze. Such a guiding line may perhaps be found in the peculiar part which the oxygen atoms play in building up the structure. Not only are the oxygen atoms the most numerous, but also they appear to be the most bulky of the units of which the pattern is woven, so that their predominant size and number make them force the other atoms to conform to certain simple and characteristic arrangements

313

of oxygen atoms which we find occurring again and again as an underlying *motif*, throughout the range of such silicates as have yet been analysed. A few simple examples may help to illustrate this point.

Four spheres, packed together as tightly as possible, assume a tetrahedral arrangement with one sphere standing on three others (Fig. 1*a*). Six spheres when packed together take up a form where three of them lie on top of three others (Fig. 1*b*). Alternatively we may regard this arrangement as four spheres at corners of a square, with one above and one below this square (Fig. 1*c*), the second arrangement being identical with the first regarded from a different view point.

These very simple groupings of oxygen atoms, with an atom of some other element at the centre of the group, occur again and again in the silicates and in many other compounds. In many cases the group not only has a characteristic shape, but also a characteristic size, the distance between the oxygen atoms having a value of about 2.5 Å.U. to 2.7 Å.U. In the list of common elements given above it is only sodium, calcium, and potassium which appear to break up the regularity of this group, and to force the oxygen atoms apart if placed at the group centre. The other elements appear to fit comfortably into the interstices of the oxygen grouping. Since there is a common distance throughout between oxygen atoms, certain atoms of one group can at the same time form part of the next group, and so a continuous structure is built up which may be thought of as a pattern of oxygen atoms with the metal and silicon atoms in its interstices.

This predominance of the oxygen atoms greatly simplifies the analysis of the structures, and makes it easier to visualise the relationships between different types. All these bodies build up a crystalline pattern repeated again and again to space. Now the simplest types of pattern are those such as children are taught to make, when blocks of the same shape and

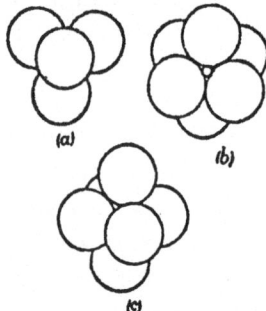

Fig. 1. Groups of four, and of six, oxygen atoms. Such groups, with atoms of silicon or one of the metals at the centre, are repeated indefinitely in the silicate structure.

314

size, but coloured differently, can be stacked together in geometrical designs. The silicates present rather a fascinating analogy to these kindergarten patterns, the oxygen atoms being the blocks and the other atoms the colouring agent. It is as if a complex pattern were embroidered by the other atoms upon a simple underlying fabric of oxygen atoms.

The two types of group, four-fold and six-fold, illustrated in Fig. 1 are composed of oxygen atoms equidistant from their neighbours. We have seen that they may be regarded as spheres compressed into the smallest possible space. Such an arrangement can be continued indefinitely, and if it is done regularly one or other of two characteristic groupings of spheres is the result. These are the well-known forms of cubic and hexagonal close-packing.

This close-packing of the oxygen atoms is a very simple arrangement, and it is interesting to find that a number of silicates are based on it. Its existence in a silicate can be surmised by noting how much volume there is to each oxygen atom in the whole structure. If the oxygen atoms are packed together closely, with 2.7 Å.U. between their centres, it is easy to calculate that each atom occupies a minimum volume of 14 (Å.U.^{3}). (If magnesium and iron are present, they expand this volume slightly by an amount for which allowance can be made.) Further, the refractive index, if ideal close-packing exists, must not be less than 1.7, as the oxygen atoms have a high refractivity.

Using these tests, it appears probable that certain compounds are based on one of the forms of close-packing, and we have examined some of these cases. The close-packing of the oxygen atoms is, however, exceptional, and although in most compounds the fourfold and sixfold groups form the basis of the pattern, this pattern may be of an open type. The various forms of silica, to the investigation of which the Royal Institution has made so large a contribution, are beautiful examples of open patterns built of the four-fold groups, and other examples of these lace-work open

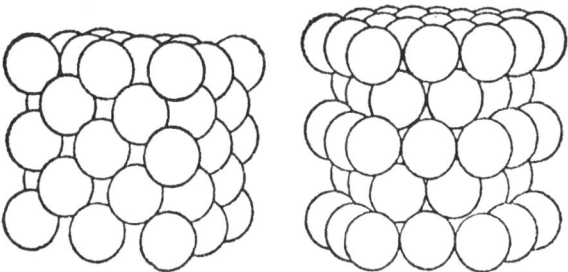

Fig. 2. Spheres in cubic and hexagonal closest-packing.

patterns have been analysed. The closely woven patterns have been chosen for description here because their basis is the more simple.

In order to understand these frameworks of oxygen atoms, with metal atoms packed in the interstices, it is necessary to bear in mind some of the geometrical laws of pattern-making. If a repeated pattern be formed by stacking together blocks of the same size and shape, but different in colour, the unit of the pattern will be composed of a whole number of blocks. It is always possible to outline the pattern with a series of unit cells, each of which just contains one complete example of the pattern and no more. The unit cell may be a large and distorted one if the pattern is complex, but it must always obey one condition. If one corner is placed at the centre of a block, all the other corners will also be at centres. The crystals which are based on one of the close-packed arrangements of oxygen atoms must have a unit cell related to the close-packed arrangement in this way.

Disthene or cyanite, Al_2SiO_5, is an example of such a crystal. Its unit cell could not well be more irregular. Its edges, and the angles between them, are all unequal, as shown in Fig. 3.

Yet its high refractive index (1.72) and small volume per oxygen atom (15.0 Å.U.^{3}) hint that the oxygen atoms are in one of the forms of close-packing. An X-ray examination has proved this to be the case. The packing turns out to be of the cubic type, and the way the disthene cell and the cubically arranged oxygen atoms 'fit together' is shown in Fig. 3. The cell of disthene contains twenty oxygen atoms, and the cell outlined in the

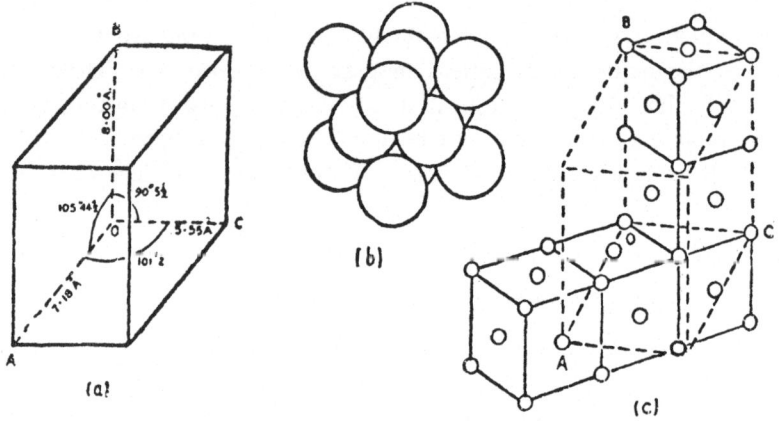

Fig. 3. (a) shows the unit cell of disthene; (b) is a portion of the array of spheres in cubic close-packing; (c) shows to the same scale a more extended portion of this array, the centres of the spheres alone being indicated. The unit cell of disthene is also a unit cell (containing twenty points) of the cubic lattice; this is proved by the way in which OA, OB, OC correspond in length and direction in (a) and (c).

right-hand figure contains exactly twenty close-packed spheres. This very complex pattern of disthene has to be woven into a basis of twenty oxygen atoms, since a multiple of five is demanded by the chemical formula. The irregular disthene cell is the way chosen by Nature of blocking out suitable groups of twenty oxygen atoms from the very simple cubic structure.

When the scattering of X-rays by the crystal is examined, the close-packed arrangement of oxygen atoms shows up strongly. We can consider the effect on the X-rays as composed of that due to the oxygen atoms on one hand, combined with that due to the atoms of metal and silicon on the other hand. The former leads to a simple and intense diffraction pattern, to be expected from a straight-forward cubic face-centred lattice. The aluminium and silicon atoms, which form a complex embroidery on a large scale woven into the oxygens, give a complex diffraction pattern within that due to the oxygen atoms alone. Fig. 4 illustrates the point.

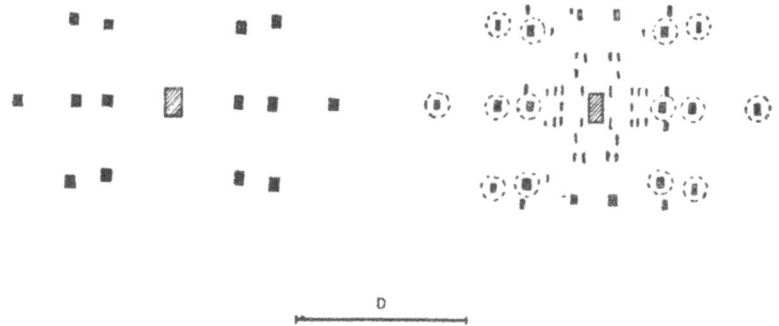

Fig. 4. Rotation photograph around b axis of disthene (right) compared with ideal rotation photograph around cube edge for close-packed atoms (left).

The complex inner pattern contains the information necessary to tell us where the aluminium and silicon atoms are. It is a difficult matter to unravel its story, but it is by no means so formidable as it would have been had we not known that the oxygen atoms are nearly in this simple arrangement, and that the aluminium and silicon atoms are somewhere within the four-fold or six-fold groups of oxygen atoms.

As another example of this pattern-weaving, the series of compounds Mg_2SiO_4, MgO_2H_2 $(Mg_2SiO_4)_2$, MgO_2H_2 $(Mg_2SiO_4)_3$, MgO_2H_2 $(Mg_2SiO_4)_4$ may be examined. The dimensional relationships between these crystals (the chondrodite series) have long aroused interest. If we measure the unit cells of the compounds we find that two axes, outlining one face, remain constant throughout all four crystals. The thickness of the cell measured perpendicularly to this face increases in regular steps in the last three compounds, as if blocks of magnesium silicate were being

317

added on in a regular way. With the aid of X-rays this process can be followed out in detail, and some finer points of it are not without interest. All the compounds prove to be based on hexagonal close-packing, and with the aid of this guiding feature, Mr. West and Mr. Taylor, who have been working on these crystals in my laboratory, have, in my opinion, succeeded in elucidating the approximate positions of all the atoms in these complex patterns. Their results are shown in Fig. 5.

These patterns are very formidable unless one has made a special study of them. However, the figures may perhaps make clear the main points. We have to explain the way in which the unit cell varies in size from compound to compound, and we have also to explain a curious complication. Whereas Mg_2SiO_4 and MgO_2H_2 $(Mg_2SiO_4)_3$ have a rectangular cell (orthorhombic), the other two compounds are built on a slant with the

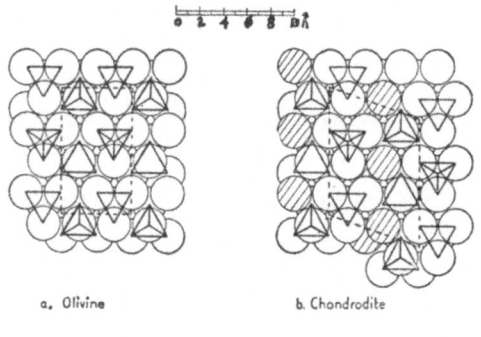

a. Olivine   b. Chondrodite

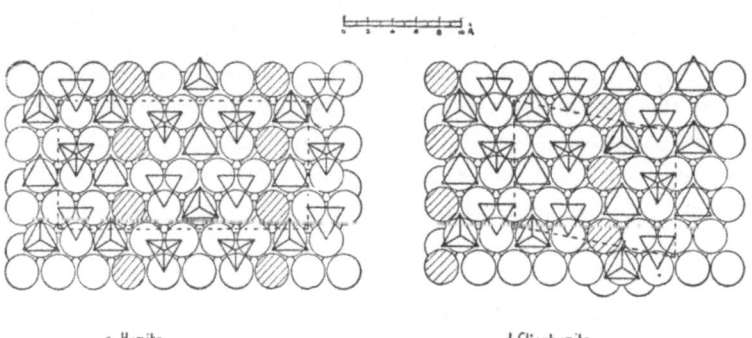

c. Humite   d Clinohumite

Fig. 5. Atomic arrangement of the chondrodite series of minerals (after Taylor and West). Two layers of oxygen atoms in hexagonal close-packing are shown, projected on the plane (100) of each crystal. The a axes of the crystals are perpendicular to the plane of the diagram, and are practically identical for all the crystals. The diagram shows the identity of the vertical b axis in all the crystals, and the relationships in direction and length between the c axes.

318

type of symmetry called monoclinic. The blocks are not clapped directly on top of each other, but are stepped sideways. It is fascinating (to the enthusiast) to see how naturally the results follow from the arrangements shown in the figure.

The circles in Fig. 5 represent the oxygen atoms which form the framework on which our whole structure is based. Two layers of such atoms are shown, representing two sheets of spheres packed closely, one lying on the other. Some of the oxygen atoms are linked together to form a group SiO_4 with a silicon (not shown) at its centre. These groups are shown as tetrahedra in the diagram, and six or three edges of each tetrahedron are drawn, depending on whether its apex is turned towards or away from the observer. The magnesium atoms are left out of the diagram for the sake of simplicity; the shaded pairs of oxygen atoms are to be labelled 'OH', since they must belong to the hydroxyl groups.

Each crystal is a series of alternate strata of magnesium silicate and magnesium hydroxide. That part of it which is magnesium silicate is arranged exactly like the pure magnesium silicate shown in Fig. 5a. The layers of hydroxide cement together the blocks of magnesium silicate. In order to outline the unit cell of any of these patterns, we join up four points in the pattern which are exactly alike. The diagram will show that the measured unit cells, shown by dotted lines, are exactly those cells which satisfy these conditions. Chondrodite, MgO_2H_2 $(Mg_2SiO_4)_2$, must have a slanting cell, the next compound, humite, a long rectangular one, and the last, clinohumite, again a slanting cell. We were surprised in making our X-ray examination to find such a curious difference in the shape of the unit cell between humite on one hand and chondrodite and clinohumite on the other hand. When the pattern was put together, however, it was clear that such a difference followed naturally from the relative numbers of silicate and hydroxide strata.

Though very few examples of silicates have yet been analysed, the temptation to draw conclusions from them is irresistible. The most curious feature about the natural silicates is the immense variety in composition which occurs within one and the same mineral species. Stupendous chemical formulae have to be assigned in order to explain even approximately the relative proportions of the different elements in some well-known types.

It seems to me that the position becomes clearer when we consider the compounds as an embroidery of the metal atoms upon an oxygen framework. We may compare the oxygen framework to the steel girder system of a large ferro-concrete building in course of construction, which is intended to be divided into sets of flats. Before the girder system is filled in, its configuration is very simple indeed. Then certain blocks of it are partitioned

319

off into sets of rooms, each comprising a flat; these flats are the more complex units of pattern in the crystalline structure. The arrangement of the rooms in each flat corresponds to the selection in our oxygen framework of certain spaces in which to put the metal atoms. We can go one step further, and rent our flats to families of slightly different composition, so that a census of the whole building shows a bewildering proportion of types of inhabitants difficult to represent by a definite family formula. This is really the state of affairs with which the chemist is confronted when he attempts to give a formula to many of the silicates.

Some of the very earliest structures which were analysed caused us to revise our ideas of what was meant by the 'molecule' of the chemist. In sodium chloride there appear to be no molecules represented by NaCl. The equality in numbers of sodium and chlorine atoms is arrived at by a chessboard pattern of these atoms; it is a result of geometry and not of a pairing-off of the atoms. This is, of course, not universally true, for this absence of the molecule in solids is in general only found in inorganic compounds. It would appear, however, that the silicates are of this non-molecular type, and that in seeking to assign formulae to them, and to the hypothetical acids of silicon on which they are based, it should be borne in mind that they are really extended patterns. The relative numbers of their constituent atoms are characteristic of the extended pattern, and essentially a result of their solid state, so that it is doubtful whether a grouping of the atoms into molecules has in this case a meaning. It will be very interesting to see what further light the X-ray results can cast on the relationships in this fascinating series of compounds.

174. Diopside is a typical crystal of the large Pyroxene group of minerals. The present investigation has, we believe, determined the positions of all the atoms in the structure, and has brought to light an interesting arrangement of the silicon and oxygen atoms in this metasilicate. The silicon atoms are surrounded by four oxygen atoms as in other silicates, but two oxygen atoms of each tetrahedral group are held in common with neighbouring groups in accord with the three to one ratio of oxygen to silicon atoms. The tetrahedra thus linked together by shared oxygen atoms form endless chains parallel to the c axis of the crystal; they lie side by side and are held together by calcium and magnesium atoms. All three cleavages of the crystal (110), (100), (010), are parallel to these chains, evidence of their relatively great strength. The pyroxene minerals are closely related to another group, the amphiboles, in which many types of fibrous crystals, such as asbestos, occur. It is therefore highly interesting to discover in a typical pyroxene these chains of silicon and oxygen atoms forming a 'grain' in the structure, for this feature may prove to be common to a number of minerals.

The structure is shown in Fig. 8. It is determined by fourteen parameters, and the analysis affords practical examples of methods described in a paper 'A technique for the X-ray analysis of crystals with a large number of parameters' by Mr. J. West and one of the authors (W.L.B.). In order to illustrate these methods, the steps by which the structure was unravelled are described in detail in sections 4, 5, 6 and 7 [not included. Ed.]; the reader who is not interseted in the details of analysis will find a general summary in section 3. The evidence for the correctness of the structure is examined in section 8, and its features are discussed in section 9.

1. *The space-group and dimensions of the unit cell*

The crystals and crystal sections used in our investigations were kindly supplied to us by Professor Hutchinson from the Mineralogical Museum of Cambridge University. The material came from de Kalb, New York, and the composition is reported to be

SiO_2	55,12
Al_2O_3	0,40
FeO	1,12
MgO	18,15
CaO	25,04
Na_2O	0,45
K_2O	0,02
Ignition	0,17
	100,47 total.

The space-group and cell dimensions have been conclusively determined by Wyckoff and Merwin[1], and our observations are in complete agreement with their results. The crystal is monoclinic, the axial angle β being 74°10'. Our values of the axial lengths agree within our order of accuracy with their estimates $a = 9,71$ Å, $b = 8,89$ Å, $c = 5,24$ Å. We find that only such reflexions appear as satisfy the following conditions:

Indices $(0kl)$, k even,
Indices $(h0l)$, h even and l even,
Indices $(hk0)$, $h + k$ even.

[1] *Amer. Jour. Sci.* 9 (1925) 379.

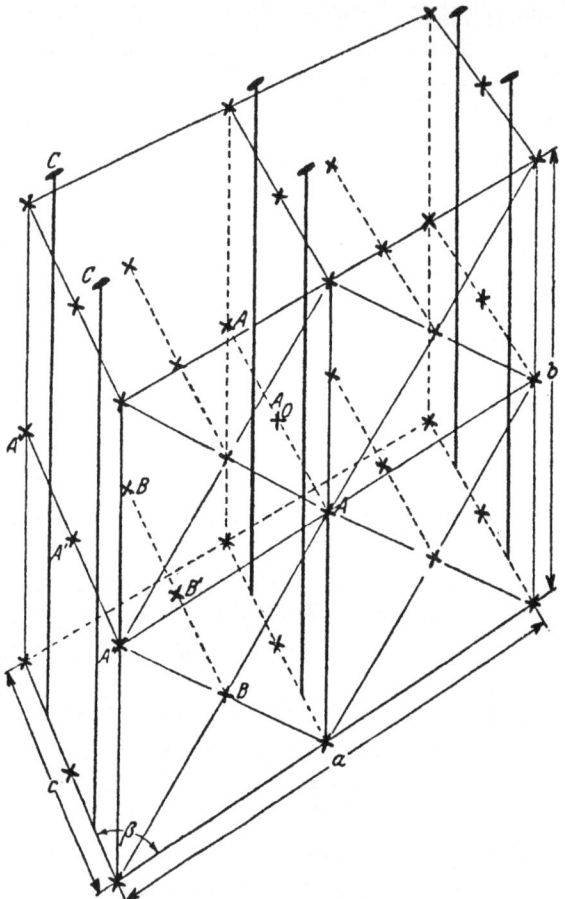

Fig. 1. Space-group 2*Ci*-6.

Assuming holohedral monoclinic symmetry, this is in agreement with their determination of the space-group C_{2h}^6 (2*Ci*-6 in Wyckoff's notation) [C2/c, Ed.]. The unit cell contains four molecules of $CaMg(SiO_3)_2$.

The cell and symmetry elements are shown in Fig. 1. There are four independent sets of symmetry centres, and four of each type in the unit cell. The two-fold axes are shown in the figure. The symmetry plane perpendicular to the *b* axis is a glide plane with translation *c*/2. The (001) face is centred.

An atom in the general position is multiplied by the symmetry operations into eight within the unit cell. Since there are only four calcium atoms and four magnesium atoms within the cell, these must lie at symmetry centres or on twofold axes. No such restriction is necessarily imposed on the positions of the silicon or oxygen atoms.

2. *Experimental measurements*

Our object in the experimental measurements was the determination of a quantity F for a large number of sets of indices (hkl). F measures the contribution of the whole unit cell to the amplitude of a wave reflected from the planes (hkl), expressed in terms of the contribution by a single electron as calculated by the classical formula of electrodynamics. Each value $F(hkl)$ is a non-dimensional ratio, characteristic of the crystal. The values of F provide the data for the determination of the structure.

In order to obtain these values we measured directly or indirectly the value of the integrated reflexion ρ for about a hundred sets of indices (hkl). The direct measurements were made on the a, b and c faces of the crystal. The integrated reflexions of several orders from these faces were compared with the reflexion (400) from a standard crystal of rocksalt, for which the value $\rho = 1{,}09 \times 10^{-6}$ was assumed. Rhodium K_{α} radiation was used, for which $\lambda = 0{,}614$ Å.

The values of ρ for other indices were measured with sections of crystal cut perpendicularly to the a, b and c axes of thickness 0,064 cm, 0,107 cm and 0,210 cm. These sections were kindly prepared for us by Professor Hutchinson. Each was used to measure a series of reflexions around the zone axis perpendicular to it. For instance, the section perpendicular to the b axis was used for reflexions corresponding to indices $(h0l)$. The section was mounted vertically on the crystal table of the spectrometer. It was turned in its own plane until a plane $(h0l)$ normal to the section became vertical and X-radiation traversing the crystal was reflected from this plane. By bringing various planes $(h0l)$ in turn into the vertical position, and setting crystal section and ionization chamber at the correct angles, comparisons between the intensity of reflexion by the different planes could be made very rapidly. The values of ρ cannot be deduced directly from these observations, since certain unknown factors come in. However, these factors vary in a continuous way with glancing angle, and since a number of reflexions round the zone of type $(h00)$ and $(00l)$ had been previously determined by using the (100) and (001) faces of the crystal, these reflexions could be used to calibrate the whole series of observations.

The experimental measurements were carried out for us by Mr. J. West to whom we wish to express our indebtedness. The value of $F(hkl)$ is deduced from the corresponding integrated reflexion ρ by the formula

$$\rho = \left(\frac{Ne^2 F}{mc^2}\right)^2 \cdot \frac{\lambda^3}{2\mu \sin 2\theta} \cdot \frac{1 + \cos^2 2\theta}{2} = \frac{Q}{2\mu}. \tag{1}$$

The number N of unit cells per c.c. is the reciprocal of the volume 435×10^{-24} c.c. of each unit cell. Substituting for other constants

$$\rho = 2{,}42\left|\frac{F^2}{\mu} \cdot \frac{1 + \cos^2 2\theta}{\sin 2\theta} \cdot 10^{-8}\right. \tag{2}$$

The value of the linear absorption coefficient μ was found by direct experiment to be $12.8 \, \text{cm}^{-1}$. This is somewhat lower than the value $14{,}23 \, \text{cm}^{-1}$ which may be calculated for Diopside from the figures in Windgårdh's[1] tables. We have generally found the measured coefficient somewhat smaller than that calculated from the tables by using the additive law for atomic absorption coefficients.

The integrated reflexion ρ in the formula is that which would be observed with an ideally imperfect or mosaic crystal, and the observed values (which will be denoted by ρ') must be corrected for extinction. If we write the formula (1) in the form

$$\rho = \frac{Q}{2\mu}$$

then Darwin has shown that the effect of extinction is to reduce ρ to a smaller value ρ' given by

$$\rho' = \frac{Q}{2(\mu + gQ)} \tag{3}$$

where g is a constant characteristic of the crystal.

We have assumed (3) to be true, and that g is the same for all the specimens of crystal we used. This can only be approximately so, and the formula must in any case cease to hold when gQ is not small compared with μ. The whole question of extinction is very difficult, and it cannot be corrected for satisfactorily in such experiments as these. We must be content with an approximate correction, and regard those cases where our correction for extinction is large as yielding uncertain values of F. Fortunately the correction to the majority of F values is not important.

The value of g was estimated by a preliminary analysis of reflexions around the zone [010] of the crystal. It was found possible to determine the positions of the atoms when projected on the (010) plane although the effect of extinction was uncertain. It was clear however that extinction was considerable, because values of $F(hkl)$ calculated for the structure were smaller than the values deduced from experimental observation, in those cases where ρ' was large. A value of g was found which brought the observed

[1] K. A. Windgårdh, *Z. Phys.* **8** (1922) 363.

values up to the level of calculated values, and this value was used to correct all observations of ρ'. In this way corrected values of $F(hkl)$ were found which made a closer analysis of the projection on (010) possible, and enabled the analysis of other coordinates to be completed.

Formula (3) can be written

$$\rho' = \frac{\rho}{1 + 2g\rho}$$

or

$$\rho = \frac{\rho'}{1 - 2g\rho'}. \tag{4}$$

The value adopted for g was $1{,}06 \times 10^4$. With this value, ρ is twice as great as ρ' when ρ' is 24×10^{-6}. Up to this point the correction is probably not far from the truth, but as one goes to larger values of ρ' it must be regarded as increasingly uncertain.

The values of $F(hkl)$ can now be calculated from the experimental results by the formula

$$\frac{\rho'}{1 - 2g\rho'} = 2{,}42 \cdot \frac{F^2}{12{,}8} \cdot \frac{1 + \cos^2 2\theta}{\sin 2\theta} \cdot 10^{-8}. \tag{5}$$

Values of ρ' and F are given in Tables 1, 2, 3, 4. [Tables 1 and 2 *see* next page; Table 3, values of $F(h0l)$, and Table 4, values of $F(hk0)$, not inserted. Ed.] The values F (calculated) are based on the structure we propose, and are referred to in section 8.

In analysing the structure, we must find positions for the atoms which explain the observed values or $F(hkl)$. In order to compare values calculated for a structure with observed values, F curves for the individual atoms are needed. The atomic F curves used in this analysis are given in Table 5; they were arrived at empirically as the result of work with other silicates, and are referred to in the accompanying paper by Bragg and West.

In the last line we have given the F value for all the atoms in the unit cell when scattering waves with the same phase. It is interesting to compare these figures with observed F's at corresponding values of $\sin \theta$. It will be seen how greatly interference cuts down the resultant amplitude in the majority of cases.

3. *Summary of method of analysis*

When the positions of the atoms depend on a large number of parameters, the main difficulty in analysis is to discover some means of

WARREN AND W. L. BRAGG

Table 1. Values of $F(h00)$, $F(0k0)$, $F(00l)$ derived from direct measurements of ρ'.

$\sin\theta$	Indices	$\rho'\times 10^6$	$\pm F$ (obs.)	F (calc.)	$\sin\theta$	Indices	$\rho'\times 10^6$	$\pm F$ (obs.)	F (calc.)
0,0658	(200)	–	–	– 11	0,3290	(10 0 0)	11,5	56	+56
0,0691	(020)	9,0	19	+ 26	0,3455	(0 10 0)	12,5	65	+66
0,1219	(002)	60?	> 140?[1]	–175	0,3657	(0 0 6)	3,4	29	–48
0,1316	(400)	3,0	15	+ 16	0,3948	(12 0 0)	–	–	– 5
0,1382	(040)	–	–	– 7	0,4146	(0 12 0)	3,0	29	+36
0,1974	(600)	31,0	100	+114	0,4606	(14 0 0)	8,0	58	+56
0,2073	(060)	28,0	94	– 91	0,4876	(0 0 8)	3,7	38	+47
0,2438	(004)	34,0	136	+107	0,5264	(16 0 0)	–	–	+ 5
0,2632	(800)	21,0	76	+ 69	0,5922	(18 0 0)	1,0	22	+21
0,2764	(080)	–	–	– 23					

[1] Value uncertain owing to very high extinction·

Table 2. Values of $F(0kl)$.

$\sin\theta$	Indices	$\rho'\times 10^6$	$\pm F$ (obs.)	F (calc.)	$\sin\theta$	Indices	$\rho'\times 10^6$	$\pm F$ (obs.)	F (calc.)
0,0922	(021)	21,4	44	–42	0,3022	(082)	3,9	28	+30
0,1400	(022)	28,4	75	+59	0,3123	(025)	–	–	+10
0,1513	(041)	29,6	82	–96	0,3200	(064)	–	–	–17
0,1844	(042)	24,0	72	+56	0,3316	(083)	–	–	–10
0,1942	(023)	w.	small	–10	0,3345	(045)	3,1	26	–36
0,2189	(061)	14,0	50	+52	0,3685	(065)	3,2	29	+24
0,2292	(043)	13,1	50	+51	0,3688	(084)	w.	small	–27
0,2406	(062)	25,0	88	+87	0,3714	(026)	–	–	+ 2
0,2534	(024)	w.	small	+11	0,4115	(085)	–	–	+ 4
0,2766	(063)	9,3	44	–35	0,4200	(066)	w.	small	+21
0,2800	(044)	–	–	– 9	0,4483	(047)	2,2	26	+30
0,2832	(081)	3,5	25	+30					

Table 5. Atomic F values, $\lambda = 0,614$ Å.

$\sin\theta$	0,1	0,2	0,3	0,4	0,5	0,6
Calcium	15,1	10,6	7,7	5,9	4,5	3,6
Silicon	10,2	7,6	5,6	4,0	2,8	2,0
Magnesium	9,3	6,6	4,3	2,8	1,9	1,2
Oxygen	6,6	3,3	1,8	0,9	0,5	0,3
4 $CaMgSi_2O_6$	338	209	136	88	60	42

commencing the attack by measuring some of these parameters independent-ly of the others. Once this has been done, and definite information about certain important parameters has been obtained, it is not difficult to find the values of the remainder.

In the present case, it was clear from the outset that an attempt to measure the coordinates parallel to the *a* and *c* axes held out the most promise of successful commencement of the analysis. This will be seen from a consideration of Fig. 1. Since there are only four calcium and four magnesium atoms in the unit cell, each of these atoms lies at one or other of the symmetry centres A, A', B, and B', or on the digonal axes marked C. Let the atomic arrangement be projected upon the plane (010), the base of the cell in Fig. 1. If the base be divided into four quadrants each of side $a/2$, $c/2$, the projected atomic arrangement in each quadrant is the same owing to the nature of the symmetry (cf. Fig. 8). We need therefore only consider one of these quadrants as shown in Fig. 4, and we know that Ca and Mg atoms lie at one or other of the positions A, B, or C in that figure. This feature was of the greatest assistance in the analysis which proceeded in the following stages.

a) A series of reflexions was considered, with indices ($h0l$) where h and l are multiples of 4. Values of F for (400), (404), (004), (800), (804) (40$\bar{4}$), (1200), (80$\bar{4}$), (408), (008), (808), (16 00) had been measured. Successive reflecting planes pass through all the points A, B, C, D as will be clear from Fig. 4, and therefore both Ca and Mg atoms make a full positive contribution to these planes wherever they are situated. Each observed value of $F(h0l)$ (of unknown sign) is to be explained by contributions from calcium, magnesium, silicon and oxygen atoms and we know the value of the first two contributions. Consideration of the values showed that if the silicon atoms were situated in certain regions it would be impossible to explain the observed values wherever the oxygen atoms were placed. For instance, if the resultant $F(h0l)$ is very small and the calcium and magnesium atoms by themselves outweigh any negative contribution the oxygen atoms could make even if they all scattered waves in phase, it is clear that the silicon atoms cannot be making a positive contribution to $F(h0l)$. We can thus draw a series of strips parallel to the planes ($h0l$) which represent forbidden areas for the projected silicon atoms. Such strips for a number of planes are drawn and they exclude the whole area of the cell with the exception of the small regions marked 1, 2, 3, 4 in Fig. 4. *We therefore know that the silicon atoms are in one or other of these regions.* The argument is given in full in section 4.

b) Since the reflexions so far considered have indices ($h0l$) where h and l are multiples of 4, there is no way of distinguishing between positions

327

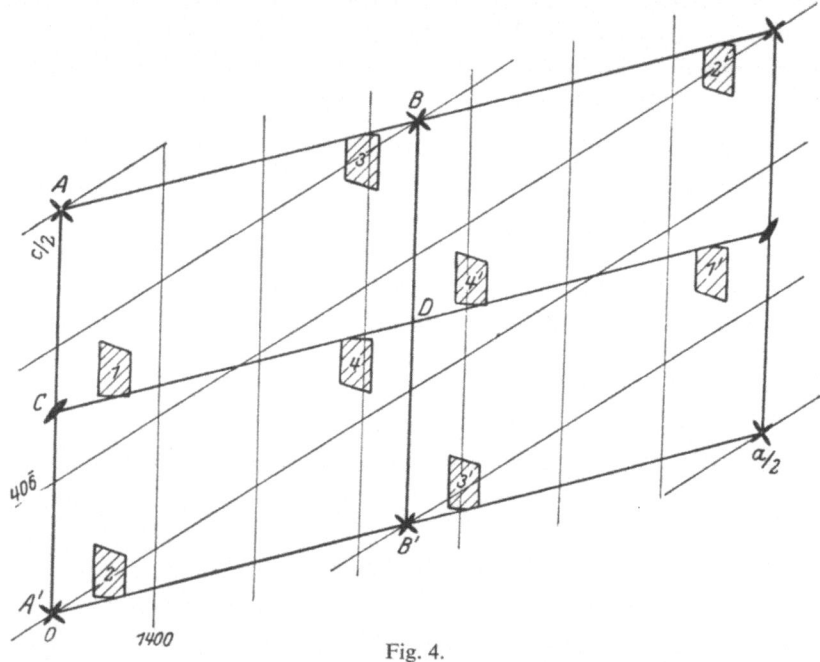

Fig. 4.

A, *B* and *C* for the calcium and magnesium atoms, and positions 1, 2, 3, 4 for the silicon atoms in Fig. 4. By symmetry the silicon atoms must occur in pairs such as those marked 4, 4′, one pair in each parallellogram of side $a/2$, $c/2$.

By now considering the values of F for planes ($h0l$) where h and l are not multiples of 4, we can find the relative positions of the silicon pair, the calcium atom, and the magnesium atom. We may show that (in the projection) *calcium and magnesium atoms occupy the same position, which is arrived at by going a distance* a/4 *from the centre of the silicon pair in a direction parallel to the* a *axis*.

The conclusion is reached by such considerations as the very strong values of $F(hkl)$ for (40$\bar{6}$) and (14 00) which have nearly maximum values. The planes are drawn in Fig. 4. It will be clear for instance, that if the silicons are at 4 and 4′, calcium and magnesium atoms must be at *C*, for any other position would reduce either (40$\bar{6}$) or (14 00). The argument depends of course for its validity upon the actual figures given in full in section 5.

c) So far no consideration of the positions of the oxygen atoms has been necessary, but a useful indication of their position is given by the very small value ± 15 of $F(400)$. Successive planes (400) are represented by AA', BB' in Fig. 4, and calcium, magnesium, and silicon atoms make a strong positive contribution to $F(400)$. This must be almost completely balanced by the oxygen atoms, and consideration of the figures shows that *all oxygen atoms must be approximately half-way between such planes as AA' and BB'* (see final structure in Fig. 8). This involves all oxygen atoms being in the general position, so that they are of three types each with three coordinates.

It is now possible to complete the analysis of the plane (010). The coordinates of the oxygen atoms parallel to the *a* axis are known approximately, and 'trial and error' methods were used to find the three coordinates parallel to the *c* axis. The only arrangement which gave agreement for all planes around the [010] zone was found to have two atoms close together in the projection (O'_1 and O_2 in Fig. 8) and the third in the position O_3 of Fig. 8.

d) We next prove that *the Ca and Mg atoms must lie on rotation axes.* It is shown that when they are placed alternatively at the symmetry centres, it is impossible to find any position for the silicon atoms which can account for certain values of $F(hk0)$. This argument is given in full in section 6 [not inserted, Ed.]. Now that the position of the atoms in relation to the rotation axes in Fig. 4 is known, it is seen that each oxygen atom is at a distance of 1,5 Å from an axis.

e) The structure now begins to assume form. We will associate each oxygen atom with that rotation axis to which it is nearest; it will have a companion on the other side of the axis. The unit cell contains four rotation axes, and each of these will have a calcium atom, a magnesium atom, and three oxygen pairs threaded on it in some such way as that shown in Fig. 7. The arrangement on two of the axes will be reversed in the *b* direction compared with that on the other two.

The group consisting of calcium atom, magnesium atom, and three pairs of oxygen atoms on each digonal axis is repeated at a distance of 8,9 Å, the length of the *b* axis. The very suggestive fact now appeared that, using the normal distances for oxygen to oxygen, calcium to oxygen, and calcium to magnesium, there is just room to stack these atoms within a length of 8,9 Å, as shown diagramatically in Fig. 7. There are only two alternatives. Either the calcium and magnesium atoms occur together, followed by a group of three oxygen pairs, or they occur in the order shown in the figure Mg-2O-2O-Ca-2O-(Mg) which is of course reversed in alternate rows by the symmetry. The former alternative is so improbable that we did not test it.

We used the diagram of Fig. 7 to suggest approximate relative positions of magnesium, calcium, and oxygen pairs along any one digonal axis. The whole structure now resolves into one determined by two parameters. The

329

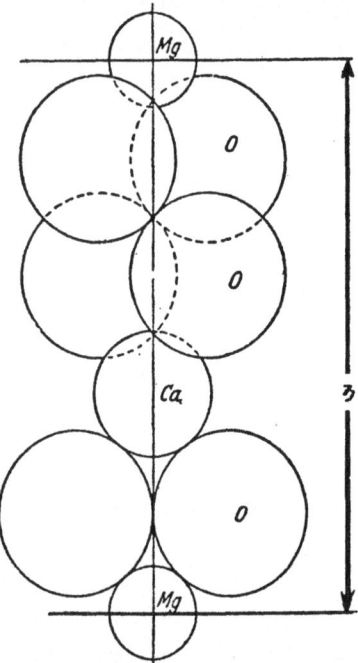

Fig. 7. Type of arrangement round digonal axis.

first fixes the position of some one atom, say calcium, along the twofold axis. The second determines the coordinates parallel to the *b* axis of the silicon atom, its other two coordinates having been already determined. Two coordinates can easily be determined by trial and error methods and the analysis may be completed.

Further small adjustments to the positions of the atoms were then made to improve the general agreement between calculated and observed values of *F*(*hkl*). This agreement is shown in Tables 1, 2, 3 and 4. It is far from perfect, and could no doubt be improved by further adjustment, but it did not appear to be profitable to continue the analysis as our trials indicated that the coordinates were correct to within a few hundredths of an Ångström unit.

The process of making final adjustments to the coordinates is somewhat tedious but quite straightforward. At high glancing angles the oxygen atoms contribute very little to *F*(*hkl*), and a knowledge of their approximate positions is sufficient to enable their contribution to be assessed with sufficient accuracy. These reflexions at high angles therefore provide suitable material for adjusting the positions of calcium, magnesium and silicon. It is always possible to find reflexions which are sensitive to slight movements of one atom and insensitive to movements of the other two. These are

330

used to fix more closely the coordinates of the important atom. When calcium, magnesium and silicon have been fixed, the reflexions at small glancing angles are used to determine the oxygen coordinates. It is more difficult to estimate these latter values. We estimate the probable error in coordinates to be 0,05 Å.

8. The atomic coordinates. Comparison of calculated and observed values of $F(hkl)$.

The coordinates assigned to the atoms are as follows, x, y, z being in Ångström units.

	θ_1	θ_2	θ_3	x	y	x
Calcium	0°	−110°	90°	0	−2,72	1,31
Magnesium	0	30	90	0	0,74	1,31
Silicon	76	148	85	2,05	3,66	1,24
Oxygen, O_1	136	145	50	3,67	3,58	0,73
Oxygen, O_2	51	90	115	1,38	2,22	1,68
Oxygen, O_3	56	173	0	1,51	4,27	0,00

An atom at $\theta_1\theta_2\theta_3$ is multiplied into a group of eight atoms in the unit cell by the symmetry elements. The contribution to $F(hkl)$ by these eight atoms is the product of the atomic F value by the factor

$$8 \cos(l\pi/2 - h\theta_1 - l\theta_3) \cos(l\pi/2 + k\theta_2)$$

where $h + k$ is even. It is zero when $h + k$ is odd.

In the case of the calcium and magnesium atoms, the factor becomes

$$4 \cos(l\pi/2 + k\theta_2).$$

Calculated and observed values of $F(hkl)$ are given in Tables 1, 2, 3, 4 in section 2. In general the agreement is satisfactory. There are a few cases where it is not good; for instance the calculated and observed values of $F(604)$ are in pronounced disagreement, as are also those of $F(330)$. We do not think these discrepancies cast doubt on the structure or on the experimental measurement but rather that the small errors in the coordinates produce errors in the computed atomic contributions which, in the rather exceptional case of their all having the same sign, and of the atoms being in a sensitive position may amount to a large value. In the case of $F(330)$ for instance, the calculated value is very sensitive to alteration in the θ_2

coordinates of magnesium, silicon, and two of the oxygen atoms. A movement by the silicon atom in the b direction of $b/100$ alters $F(330)$ by ± 10.

9. *Discussion of the structure*

The most interesting feature of the structure is the arrangement of silicon and oxygen atoms, referred to in the first section. This arrangement will be seen in Fig. 8. Four oxygen atoms, O_1 O_2, O_3, O'_3 are grouped around the silicon atom which is at height $0,41b$. They are nearly at the corners of a regular tetrahedron, their distances from the silicon atom being 1,57 Å, 1,62 Å, 1,60 Å, 1,64 Å, and their distances from each other 2,66 Å, 2,61 Å, 2,48 Å, 2,59 Å, 2,63 Å, 2,64 Å. Similarly a silicon atom at height $0,\overline{09}b$ is surrounded by four oxygen atoms at heights $0,\overline{10}b$, $0,\overline{25}b$, $0,\overline{02}b$, $0,02b$.

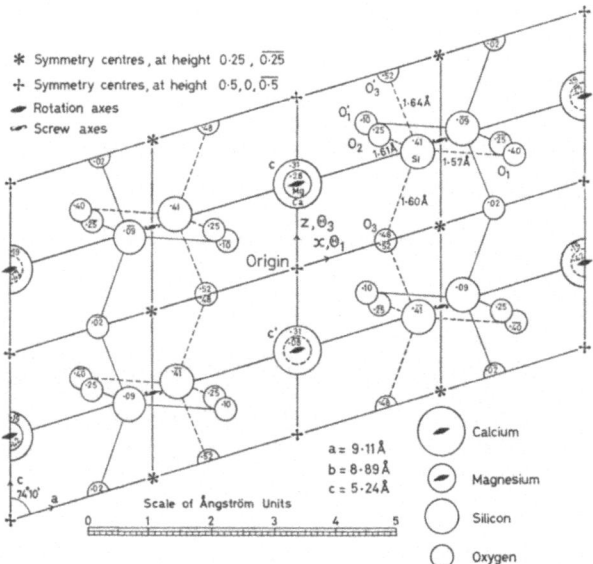

Fig. 8. Projection of diopside structure on plane (010).

This group is linked by the oxygen atom at $0,02b$, to another group centred around a silicon atom at $0,09b$, and the linking is continued indefinitely in the c direction. The lines in the figure joining the silicon and oxygen atoms make these chains clear. The continuous lines are wholly included in the unit cell. The broken lines show parts of similar chains, some atoms of which lie either above or below the unit cell; the atom marked O_3 is for convenience represented in the lower half of the figure by the corresponding atom a distance b below.

The cleavages of the crystal are (110), perfect; (100), (010) imperfect, all of which are parallel to the chains, suggesting that the binding is strong along the chain. The linking of tetrahedral groups of oxygen atoms around each silicon atom is a distinctive feature of the forms of silica SiO_2 examined by W. H. Bragg, Gibbs, and Wyckoff. In quartz, tridymite, and cristobalite every oxygen atom is shared between neighbouring groups, which are thus tightly bound in every direction. In diopside the linking only exists in a direction parallel to the shortest axis of the crystal, the c axis. The parallel chains so formed are bound to their neighbours on all sides by the calcium and magnesium atoms.

Each calcium atom is surrounded by eight oxygen atoms, and each magnesium atom by six oxygen atoms. This is shown in Fig. 9, a view

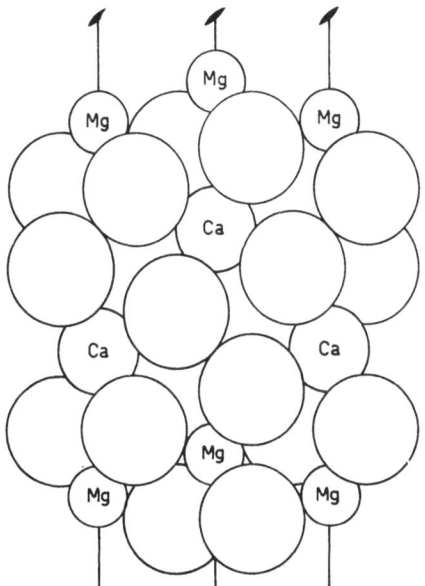

Fig. 9. Scheme of packing of the $CaMgO_6$ groups on alternate axes of the bc plane.

of the unit cell looking in the positive direction along the a axis. In order to make the arrangement clearer, the c axis has been lengthened from 5,24 Å to 6 Å, thus enabling the lower layers of atoms to be more easily seen.

We made this close study of the diopside structure in the hope that it would cast light on the structure of the whole pyroxene-amphibole series.

There is a close relationship between these two large groups of minerals, exemplified by the pyroxene diopside and the amphibole tremolite.

Diopside $CaMg(SiO_3)_2$

$$a : b : c = 1,0921 : 1 : 0,5893. \quad \beta = 74°10'$$

Cleavage   {110}, {100}, {010}.

Tremolite $CaMg_3(SiO_3)_4$

$$a : b : c = 0,551 : 1 : 0,294. \quad \beta = 73°58'.$$

Cleavage   {110}, {100}, {010}.

The tendency amongst the amphiboles to assume fibrous forms is well known, asbestos being an example. The possible connection between these forms and the silicon-oxygen chains is very interesting.

It is interesting to contrast the structure with Beryl $Be_3Al_2Si_6O_{18}$,[1] which also contains silicon and oxygen atoms in the ratio of one to three. In beryl similar tetrahedral silicon-oxygen groups are linked into closed rings of six by their corners, while in diopside they are linked in the same way but form extended chains.

In this analysis, since so little is known about the structure of meta-silicates, we did not think it justifiable to make any assumptions about the nature of the atomic arrangement. We were prepared to find groups SiO_3 or Si_2O_6, or closed or endless chains of tetrahedral groups, and were led to this structure by an exclusion of all other possibilities. The only stage at which a search for the parameters is shortened is that where the arrangement along the digonal axes is considered. Having proved the closeness of the oxygen atoms to the rotation axes, it seems justifiable to use the interatomic distances now so well established for compounds of this class to suggest approximate coordinates along these axes. The final values of these coordinates were determined entirely by X-ray measurement.

We wish to acknowledge our indebtedness to Mr. J. West, who made all the measurements on which the analysis is based.

175. The form in which the experimental results are expressed is convenient for the comparison of calculated and observed values. A lack of agreement amounting to a definite number of units in the value of $F(hkl)$ is as significant in the case of a strong reflexion as a weak one. This is so because a displacement of any atom is as likely to alter a small value of F as a large one by a given amount.

[1] *Roy. Soc. Proc.* A **111** (1926) 691.

This makes it possible when testing a proposed structure to take as an aim some figure for the maximum discrepancy between calculation and observation, and to attempt an adjustment of atomic positions until the errors throughout all the range are less than this value. The limit can then be reduced if closer analysis is desired. It is important to realize that the agreement is as good between a calculated value zero and observed value 20, as between a calculated value 80 and an observed value 100, although in the first case the proposed structure which is being tested gives no reflexion at all where one is actually observed, whereas in the second case there is obvious agreement. It is very helpful to have the numerical discrepancies equally significant throughout the scale.

176. Grundprinzip der Silikatstruktur ist, daß Si^{+4} stets vier O^{-2} (oder solche einfache oder zusammengesetzte Ionen F^{-1}, OH^{-1}, die das Ion O^{-2} isomorph vertreten können) koordinativ um sich zu gruppieren bestrebt ist. Diese O^{-2} (oder strukturell dem O^{-2} äquivalenten) Ionen sind, wie alle bisherigen Strukturuntersuchungen von Silicium-Sauerstoffverbindungen zeigen (man vergleiche die bedeutungsvolle Abhandlung von W. L. Bragg und J. West: The Structure of Certain Silicates, *Proc. Royal Soc.* A **114** (1927), 450–473)[1], tetraedrisch, bezw. verzerrt tetraedrisch um das Si-Ion angeordnet. Durch Tetraeder kann nun bekanntlich der Raum nicht lückenlos erfüllt werden, so zeigen z.B. die verschiedenen Modifikationen des SiO_2 im kristallisierten Zustande zwischen den O^{-2}-Tetraedern mit Si^{+4} als Kern, wobei jedes O^{-2} gleichzeitig zwei benachbarten Tetraedern angehört, Hohlräume.

Beim Olivin[2] und beim Monticellit[3] hat man z.B. SiO_4-Tetraeder, die keine mit den Nachbartetraedern gemeinsamen O-Schwerpunkte besitzen (Allgemeiner Typus der Orthosilikate, siehe unten!) und zwischen diesen finden sich die Ionen Mg^{+2} und Fe^{+2}, bezw. Mg^{+2} und Ca^{+2} eingelagert.

Weiters ist maßgebend für die Struktur der Silikatkristalle, daß das Ion Al^{+3} bezüglich seiner Koordinationszahl gegenüber O^{-2} infolge seiner Größe[4] im Verhältnis zu der von O^{-2} an der Grenze zwischen 4 und 6 steht und daher in den Silikaten gegenüber O^{-2} wohl mit beiden

---

[1] Die beiden Autoren weisen an mehreren Stellen auf die grundsätzlich analoge Anordnung zwischen Si- und O-Ionen in allen bisher untersuchten Strukturen von Si-O-Verbindungen hin.

[2] W. L. Bragg und G. B. Brown, *Zeitschr. f. Krist.* **63** (1926) 538 ff.

[3] G. B. Brown and J. West, *Zeitschr. f. Krist.* **66** (1927) 154 ff.

[4] Nach V. M. Goldschmidt (*Geoch. Vert. Ges.* 7 (1926) 28) beträgt der Radius von Al^{+3} rund 0,57 Å, nach L. Pauling (*Journ. Am. Chem. Soc.* **49** (1927) 771) 0,50 Å.

Koordinationszahlen auftreten kann. Dort, wo es die Koordinationszahl 4 besitzt, tritt es an Stelle von Si^{+4} in der Mitte von O^{-2}-Tetraedern auf, wodurch diese etwas ausgeweitet werden dürften. Von den übrigen silikatbildenden Elementen hat das Ion Be^{+2} gegenüber O^{-2} die Koordinationszahl 4, kann also in das Gitter ebenfalls für Si^{+4} eintreten[1]. Ähnliches gilt wohl auch von B^{+3}. Den restlichen silikatbildenden Elementen kommt gegenüber O^{-2} die Koordinationszahl 6 (oder mehr) zu. Die Koordinationszahlen gegenüber O^{-2} und selbstverständlich die Valenzzahlen sind maßgebend für den Bau der Silikate und dafür, ob gewisse Konfigurationen von SiO_2 mit anderen Oxyden im kristallisierten Zustande möglich sind oder nicht.

Bei den Feldspaten liegt nun ein Gitter vor, das in der Anordnung seiner Bausteine einem reinen SiO_2-Gitter grundsätzlich entsprechen muß. Bestimmte Si-Ionen sind entsprechend dem Gesagten und den nachfolgenden 'Formeln' durch Al^{+3} ersetzt. Die Zwischenräume zwischen den deformierten O^{-2}-Tetraedern mit Si^{+4}-, bezw. Al^{+3}-Kernen werden bei den natürlich vorkommenden Feldspaten im wesentlichen von den Ionen K^{+1}, Na^{+1}, Ca^{+2} und Ba^{+2} eingenommen. Die 'Formeln' der Feldspate wären somit wie folgt zu schreiben[2]:

$$SiO_2: (SiO_2)_4$$

$$\text{Kalifeldspat:} \quad \begin{bmatrix} (SiO_2)_3 \\ AlO_2 \end{bmatrix}^{-1} K^{+1}$$

$$\text{Natronfeldspat:} \quad \begin{bmatrix} (SiO_2)_3 \\ AlO_2 \end{bmatrix}^{-1} Na^{+1}$$

$$\text{Kalkfeldspat:} \quad \begin{bmatrix} (SiO_2)_2 \\ (AlO_2)_2 \end{bmatrix}^{-2} Ca^{+2}$$

$$\text{Baryumfeldspat:} \quad \begin{bmatrix} (SiO_2)_2 \\ (AlO_2)_2 \end{bmatrix}^{-2} Ba^{+2}$$

Die oben angewandte Schreibweise SiO_2, bezw. AlO_2 soll damit vereinbar sein, daß die Anordnung der O^{-2}-Tetraeder mit den Si^{+4}- und Al^{+3}-Kernen derartig ist, daß jedes O^{-2} zwei aneinanderstoßenden Tetraedern angehört, wie dies bei den Kristallen von SiO_2 selbst der Fall ist. Dadurch vermindert sich eben formelmäßig die Zahl der zu jedem Si^{+4}, bezw. Al^{+3} gehörigen O^{-2} auf 2. Dieser Haupttypus der Silikatstruktur sei vorläufig

---

[1] Auch für den Phenakit, für dessen Gitter die Positionen der Be-Ionen bisher noch nicht festgestellt werden konnten, nimmt W. L. Bragg (*Proc. Royal Soc.* A **113** (1927) 657) deren Lage im Zentrum von Tetraedergruppen als wahrscheinlich an.

[2] Typen der räumlichen Anordnung liefern die Strukturen von SiO_2, vom Beryll, usw.

als *Feldspattypus* bezeichnet. Hieher gehören die 'Open Structures' nach W. L. Bragg und J. West[1] (SiO_2, Beryll, Phenakit).

Analog wären die Metasilikate mit SiO_3, bezw. AlO_3 zu schreiben, da hier überall die O^{-2}-Tetraeder so gruppiert sein müssen (vermutlich kettenartig), daß nur je zwei O-Ionen eines jeden O^{-2}-Tetraeders gleichzeitig auch Nachbartetraedern zugehören. Als formelmäßige Darstellung wäre vielleicht vorzuschlagen:

$$\begin{bmatrix} \begin{array}{cccc} O & O & O & O \\ | & | & | & | \\ -Si-O-Si-O-Si-O-Si-O \\ | & | & | & | \\ O & O & O & O \end{array} \end{bmatrix}^{-8} R_4 \text{ (oder } R_2^{II} R_2^{I\ III})^2 \qquad (1)$$

Dieser Typus sei als *Metatypus* bezeichnet.

Für die Orthosilikate, bei denen die benachbarten SiO_4-Tetraeder keinen gemeinsamen O^{-2}-Schwerpunkt haben, wäre sinngemäß in der Formel, wie üblich, SiO_4 zu schreiben. Dieser Typus sei als *Orthotypus* bezeichnet.

Es sei aufmerksam gemacht, daß sich die Begriffe Ortho- und Metasilikat mit den Begriffen Ortho- und Metatypus nicht decken. Infolge der kristallographischen Vertretbarkeit des Si^{+4} durch Al^{+3} und Be^{+2} (B^{+3}?) reihen sich unter die genannten Haupttypen zahlreiche andere Silikate ein, die sonst nicht so eingereiht würden. W. L. Bragg und J. West[1] machen z.B. ausdrücklich auf den prinzipiellen Unterschied in der Struktur des Phenakites einerseits und der Struktur der übrigen, bisher untersuchten Orthosilikate andrerseits aufmerksam.

Für die Alumosilikate liegt, wie schon betont, der Schlüsselpunkt darin, ob Al^{+3} gegenüber O^{-2} mit der Koordinationszahl 4 oder 6 auftritt. Dementsprechend reihen sie sich in die obigen Haupttypen ein.

Es läßt sich noch nicht absehen, ob man mit den genannten drei Haupttypen das Auslangen finden wird. Manche Silikate werden vielleicht so lange, als noch keine Strukturuntersuchungen vorliegen, im Sinne der hier dargelegten Anschauung auf verschiedene Weise gedeutet werden können, und eine vollständige Klärung wird noch umfangreicher Arbeiten sowohl auf dem Gebiete der Strukturforschung als auch auf dem Gebiete der Silikatsynthese bedürfen.

Nun zu den Isomorphiebeziehungen und Mischkristallbildungen in der Gruppe der Feldspate:

[1] *Proc. Royal Soc.* A **114** (1927) 470. Ebenso W. L. Bragg, *Proc. Royal Soc.* A **113** (1927) 657.

Nach V. M. Goldschmidt betragen die Ionenradien[1]:

$$Na^{+1}, \; r = 0,98 \, \text{Å}; \quad Ca^{+2}, \; r = 1,06 \, \text{Å}$$

$$K^{+1}, \; r = 1,33 \, \text{Å}; \quad Ba^{+2}, \; r = 1,43 \, \text{Å}$$

Es ist nun ohne weiteres ersichtlich, daß in Anbetracht des durch obige 'Formeln' angedeuteten analogen Baues der Feldspate weitestgehende Mischbarkeit sowohl zwischen Albit und Anorthit, als auch zwischen Kalifeldspat und Baryumfeldspat vorkommen muß.

Es sei hervorgehoben, daß die hier in groben Strichen dargelegten Überlegungen durch das Studium der Abhandlungen V. M. Goldschmidt's über die Ionenradien und Isomorphiegesetze[2] angeregt wurden, und daß später ihre Entwicklung wesentlich durch gelegentliche Unterredungen mit Herrn Prof. Dr. V. M. Goldschmidt und Herrn W. Zachariasen gefördert wurde.

Mineralogisches Institut der Universität Oslo

177. With the aid of the general principles[3] governing the structures of complex ionic crystals I have formulated a structure for talc, pyrophillite, the micas, and the brittle micas which is substantiated by the x-ray examination of the minerals, explains their remarkable physical properties, and leads to a general chemical formula unifying the widely varying analyses reported for different specimens. This structure is described in the following paragraphs.

The monoclinic (pseudohexagonal) unit of structure of muscovite, $KSi_3Al_3O_{10}(OH, F)_2$, has been reported by Mauguin[4] to have $a = 5.17 \, \text{Å}$, $b = 8.94 \, \text{Å}$, $c = 20.01 \, \text{Å}$, and $\beta = 96°$. This unit contains 4 molecules of the above composition. We have verified this with oscillation and Laue photographs of fuchsite, a variety of muscovite; the unit found having $a = 5.19 \, \text{Å}$, $b = 8.99 \, \text{Å}$, $c = 20.14 \, \text{Å}$, and $\beta = 96°$.

The dimensions of the unit in the basal plane closely approximate those for the similarly pseudohexagonal crystal hydrargillite, $Al(OH)_3$, as well as of the hexagonal layers in two forms of silica, β-tridymite and β-cristobalite. The monoclinic (pseudohexagonal) unit of structure of hydrargillite has $a = 8.70 \, \text{Å}$, $b = 5.09 \, \text{Å}$, $c = 9.76 \, \text{Å}$, and $\beta = 85°29'$, and contains

---

[1] *Geochem. Vert. Ges.* **VII** (1926) 20, 26.

[2] Insbesondere: *Geochem. Vert. Ges.* I-VIII.

[3] Linus Pauling, 'Sommerfeld Festschrift', S. Hirzel, Leipzig, 1928, p. 11; *J. Am. Chem. Soc.* **51** (1929) 1010.

[4] Ch. Mauguin, *Comp. Rend.* **185** (1927) 288.

$8Al(OH)_3$. The crystal is composed of layers of octahedra, each consisting of $6OH^-$ ions grouped about an Al^{+3} ion, with each octahedron sharing three edges as shown in Figure 1. The electrostatic valence rule is satisfied; each OH^- ion is held by two bonds of strength $s = \frac{1}{2}$. These layers are superimposed, sharing no octahedral elements with one another.

In β-tridymite and β-cristobalite there are present the hexagonal layers of silicon tetrahedra shown in Figure 2. Their dimensions, $a = 5.03$ Å and $b = 8.71$ Å, agree closely with those of the hydrargillite layers and of mica. Another type of tetrahedral layer with the same dimensions can be obtained by pointing all the tetrahedra in the same direction (Figure 3). The oxygen ions forming the bases of the tetrahedra have their valences satisfied: the strength s of a silicon bond is 1, and each $O^=$ is held by two

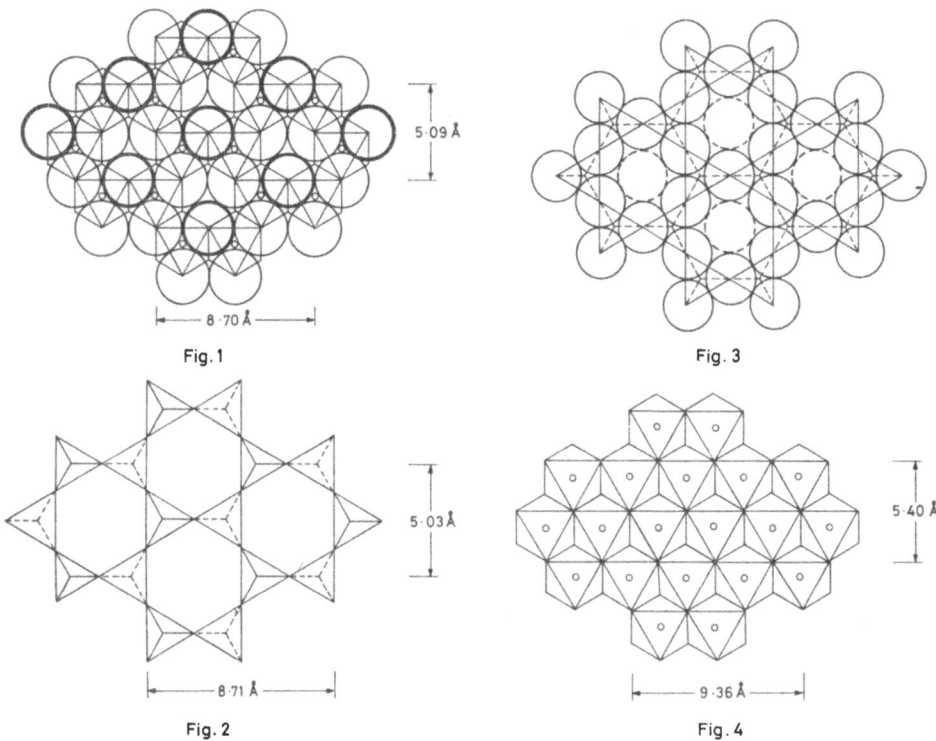

Fig. 1

Fig. 3

Fig. 2

Fig. 4

Fig. 1. A hydrargillite layer of octahedra. The light circles indicate oxygen ions, the heavier ones hydroxyl or fluorine ions in mica.

Fig. 2. A tetrahedral layer from β-cristobalite or β-tridymite. A silicon ion is located at the center of each tetrahedron, and an oxygen ion at each corner.

Fig. 3. A tetrahedral layer in which all the tetrahedra point in the same direction.

Fig. 4. A complete layer of octahedra (brucite layer).

339

such bonds, giving $\sum s = 2$. But the oxygen ions at the unshared tetrahedral corners have $\sum s = 1$ only; it is hence necessary that they be held by further bonds. Now the relative positions of these tetrahedron corners are such that the tetrahedral layer can be imposed on the hydrargillite layer with the tetrahedron corners coincident with two-thirds of the shared octahedron corners; namely, those indicated in Figure 1 by the large light circles. These positions are occupied by oxygen ions, which have $\sum s = 1 + \frac{1}{2} + \frac{1}{2} = 2$. The remaining positions, indicated by heavy circles, have $\sum s = \frac{1}{2} + \frac{1}{2} = 1$, and are occupied by hydroxide or fluoride ions. A similar tetrahedral layer is attached to the other side of the hydrargillite layer.

The resultant layer, with a total thickness of about 10 Å, is electrically neutral. A crystal built up by the superposition of such layers would have the composition $Si_4Al_2O_{10}(OH)_2$. It is very probable that the mineral pyrophillite, with this composition, has this structure.

Brucite, $Mg(OH)_2$, is built of complete octahedral layers such as that shown in Figure 4, with dimensions not much greater than those of a hydrargillite layer, and with the octahedron corners in the same positions. A neutral layer containing tetrahedra and octahedra closely similar to that described above can be made with this complete octahedral layer in place of the hydrargillite layer; the electrostatic valence rule will be again satisfied, $O^=$ having $\sum s = 1 + \frac{1}{3} + \frac{1}{3} + \frac{1}{3} = 2$ and OH^- having $\sum s = \frac{1}{3} + \frac{1}{3} + \frac{1}{3} = 1$. These layers probably occur in talc, $Si_4Mg_3O_{10}(OH)_2$, which is reported to be monoclinic, pseudohexagonal, and morphologically similar to mica.

By replacing one-fourth of the silicon ions in a pyrophillite layer by aluminum ions, which can have the coördination number 4 as well as 6, the layers become negatively charged. To regain neutrality further positive ions must be introduced, such as K^+ or Na^+. There is room for these ions between the layers, in the pockets formed by six oxygen ions on the top of one layer and six on the bottom of the layer above. The composition of such a crystal is $K \cdot Si_3Al \cdot Al_2O_{10}(OH, F)_2$, which is the formula generally assigned to muscovite. A similar treatment of talc gives $KSi_3AlMg_3O_{10}(OH, F)_2$, which is biotite.

Pyrophillite, talc, muscovite, and biotite have the following sequence of atom-planes along the pseudo-hexagonal axes:

Pyrophillite   Talc

$6O^=$ $6O^=$

$4Si^{+4}$ $4Si^{+4}$

$4O^= + 2OH^-$ $4O^= + 2OH^-$

$4Al^{+3}$ $6Mg^{++}$

$4O^= + 2OH^-$ $4O^= + 2OH^-$

$4Si^{+4}$ $4Si^{+4}$

$6O^=$ $6O^=$

Muscovite   Biotite

$z_2\ 6O^=$ $6O^=$

$u\ 3Si^{+4} + Al^{+3}$ $3Si^{+4} + Al^{+3}$

$z_1\ 4O^= + 2(OH^-, F^-)$ $4O^= + 2(OH^-, F^-)$

Origin $4Al^{+3}$ $6Mg^{++}$ } Charge -2

   } Charge -2

$4O^= + 2(OH^-, F^-)$ $4O^= + 2(OH^-, F^-)$

$3Si^{+4} + Al^{+3}$ $3Si^{+4} + Al^{+3}$

$6O^=$ $6O^=$

$2K^+$ } $+2$ $2K^+$ } $+2$

$6O^=$ } -2 $6O^=$ } -2

$3Si^{+4}$, etc. $3Si^{+4}$, etc.

The verification of the suggested structures by the comparison of the observed and calculated intensities of reflection of x-rays has been begun. So far the calculations have been carried out for 18 even orders of reflection from (001). The observed intensities given in Table 2 were obtained by the visual comparison of four photographs from fuchsite, identical except for varying exposure times of 15, 90, 300, and 960 minutes. The calculated intensities were obtained from the formula

$$I = \text{constant}.\,A^2,$$

$$\text{with} \quad A = \sum_n A_n e^{2\pi i(hx_n + ky_n + lz_n)}, \tag{1}$$

the summation being taken over all the atoms in the unit. Values of the atomic amplitude functions[1] used are given in Table 1. Assuming the group of atoms such as $4O^= + 2(OH^-, F^-)$ to depend on a single z-parameter, only three parameters are involved, z_1, u, and z_2. Giving the hydrargillite layer the thickness found for it in hydrargillite, and assuming

---

[1] The atomic amplitude functions take account of the atomic F-factor, the temperature factor, the Lorentz factor, and the polarization factor.

Table 1. Atomic A-values.

d/n	10	5	2.5	1.0	0.75	0.60 Å
A_O =	30	21	10	2.0	0.8	0.25
A_{Al} =	49	28	17	5.2	3.0	1.7
A_{Si} =	53	30	18	6.0	3.7	2.1
A_K =	72	45	24	7.8	5.0	3.2

Table 2. Observed and calculated intensities of reflection from (001) of fuchsite

Order of reflection	2	4	6	8	10	12	14	16	18
Observed intensity	40	40	120	20	150	3	10	30	0.5
Calculated intensity	40	47	121	18	124	7	12	30	1.4

Order of reflection	20	22	24	26	28	30	32	34	36
Observed intensity	6	9	2	2	0	0.8	0.1	0	1
Calculated intensity	4.3	4.5	4.8	5.2	0.0	1.6	0.2	0.6	1.7

the tetrahedra to be regular and 2.60 Å on an edge as in silicates in general, the parameter values $z_1 = 0.055$, $z_2 = 0.165$, and $u = 0.137$ were predicted.

The amplitude expression is

$$A = 2A_{Al} + (-1)^{1/2}A_K + (3A_{Si} + A_{Al}) \cos 2\pi l u$$
$$+ 6A_O(\cos 2\pi l z_1 + \cos 2\pi l z_2), \qquad (2)$$

with l having even values from 2 to 36. It was found that the values of A^2 were in general agreement with the observed intensities for the predicted parameter values. Slightly better agreement was obtained by reducing u to 0.135. The calculated intensities (the constant in equation 1 being given the arbitrary value 0.015) are included in Table 2, and represented graphically together with the observed intensities in Figure 5. The general agreement is striking, and lends strong support to the suggested structure.

The mica structure provides an interesting variation of the types of close-packing observed for large ions in crystals. The two central layers of $O^=$, OH^-, and F^- ions form close-packed planes with three spheres in a hexagonal unit of edge 5.2 Å, at positions 00, $\frac{1}{3}\frac{2}{3}$, and $\frac{2}{3}\frac{1}{3}$ relative to axes 120° apart. The outer layers, however, consist of oxygen ions occupying three of the four positions of a close-packed plane with four spheres in the same unit of edge 5.2 Å, at positions such as 00, 0$\frac{1}{2}$, and $\frac{1}{2}$0. The insertion of a fourth sphere at $\frac{1}{2}\frac{1}{2}$, indicated by the dotted circles in Figure 3,

342

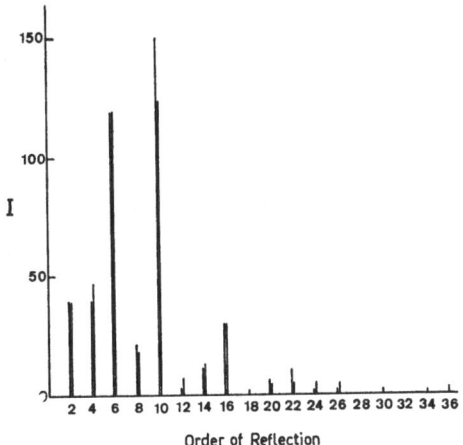

Order of Reflection

Fig. 5. Comparison of the observed and calculated intensities of reflection of x-rays from (001) of fuchsite, in the 2nd to 36th order. In each pair of vertical lines that on the left shows the observed intensity, that on the right the calculated intensity.

would complete this layer. Such a large variability in the effective dimensions of the large ions introduces some ambiguity in structure determinations involving the deduction of the type of close-packing from the size of the unit of structure.

The structure leads to a general formula for the micas: namely, $KX_nY_4O_{10}(OH, F)_2$, with $2 \leqq n \leqq 3$, in which X represents cations of coördination number 6 (Al^{+3}, Mg^{++}, Fe^{++}, Fe^{+3}, Mn^{++}, Mn^{+3}, Ti^{+4}, Li^+, etc.) and Y cations of coördination number 4 (Si^{+4}, Al^{+3}, etc.). The subscript n can have any value between 2 (hydrargillite layer) and 3 (complete octahedral layer). K^+ can be partially replaced by Na^+ and possibly to some extent by Ca^{++}. This formula represents satisfactorily the numerous recently published mica analyses almost without exception[1]. The distribution of the various ions X and Y must be such as to give general agreement with the electrostatic valence rule.

The clintonites or brittle micas have a similar structure, the layers having twice the electrical charge of those in mica, and being held together by calcium ions instead of potassium ions. The correspondingly stronger forces bring the layers closer together, the separation of adjacent layers being 9.5–9.6 Å in place of the value 9.9–10.1 Å for the micas. The general formula $CaX_nY_4O_{10}(OH, F)_2$, with $2 \leqq n \leqq 3$, holds for these minerals.

The physical properties of talc, pyrophillite, the micas, and the brittle micas are in agreement with the suggested structure. To tear apart one

---

[1] Mauguin showed that for various micas the unit of structure uniformly contains the number of atoms $O + F$ given by the above formula.

of the pseudohexagonal layers it is necessary to break the strong Si–O, Al–O, etc., bonds; as a consequence these individual layers are tough. But they can be easily separated from one another, giving rise to the pronounced basal cleavage shown by all these minerals. In talc and pyrophillite the layers are electrically neutral, and are held together only by stray electrical forces. These crystals are accordingly very soft, feeling soapy to the touch as do graphite crystals. To separate the layers in mica it is necessary to break the bonds of the univalent potassium ions, so that the micas are not so soft, thin plates being sufficiently elastic to straighten out after being bent. Separation of layers in the brittle micas involves breaking bonds of bivalent calcium ions; these minerals are hence harder, and brittle instead of elastic, but still show perfect basal cleavage. The sequence of hardness is significant; on the Mohs scale it is: talc and pyrophillite, 1–2; the micas, 2–3; the brittle micas, 3½–6.

I wish to acknowledge the assistance of Mr. J. Sherman in the preparation and analysis of the x-ray photographs.

Gates Chemical Laboratory, California Institute of Technology

178.   1. *Introduction*

A large proportion of the earth's crust consists of felspars and our knowledge of silicate structures cannot be considered satisfactory if we lack detailed information regarding the atomic arrangement in these important minerals. Machatschki[1] showed (in 1928) in a most interesting paper that it is very probable that the felspars are based on a framework of linked tetrahedra of oxygen atoms, each tetrahedron containing a silicon or aluminium atom, with the kations (K, Na, Ca, Ba) located in the interstices of the oxygen framework. Schiebold[2] published (in 1927) unit cell and space-group data for a number of felspars, and discussed the dimensional features which suggest that they may be pseudocubic or pseudotetragonal. In 1929 Schiebold[3] also published a more detailed account of his work, including a list of parameters determining the position of every atom in a pseudotetragonal unit cell which he says is the essential unit common to all felspar structures. The parameters listed are given as preliminary values, and more exact values are promised. Professor

[1] Machatschki, *Ctbl. Min.* 3 (1928) 97 [this Vol. paper **176**].
[2] Schiebold, *Fortschr. Min.* **12** (1927) 78; *Trans. Faraday Soc.* **25** (1929) 317.
[3] Schiebold, *Ctbl. Min.* (1929).

W. L. Bragg[1] pointed out in 1930 that Schiebold's structure appears to bring certain oxygen atoms much too close together, the interatomic distance being about one half the normal value. The work described in the present paper was undertaken at the suggestion of Professor Bragg, with a view to deciding whether Schiebold's felspar structure is essentially correct, or whether some quite different atomic arrangement may be suggested.

Sanidine was used for a first attempt because good crystals were available and because its monoclinic symmetry would simplify the calculations involved in a search for atomic parameters.

II. *The Experimental Data*

All the data used in the present investigation was obtained from two platy crystals of sanidine from a hornblende-sanidine block from Mt. Vesuvius, Italy. The density is given (by Mr. Bannister) as 2.57, and the refractive index as 1.52. A chemical analysis is not available at the moment but we have assumed (for the purpose of F-calculations) that one third of the potassium represented by the chemical formula $KAlSi_3O_8$ has been replaced by sodium: in any case the replacement of a portion of the potassium by sodium makes little difference to calculated F-values except for very small glancing angles.

A Laue photograph taken with the X-rays parallel to the face (010) shows a symmetry plane (010), as expected, thus indicating either that the crystal is truly monoclinic or that the departure from monoclinic symmetry is slight.

Rotation photographs about the crystallographic c axis [001] give the axial length as $c = 7.1$ Å, photographs about the b axis [010] give $b = 12.9$ Å, and all the reflections on both sets of photographs can be indexed by assuming for the third axial length $a = 8.4$ Å. These axial lengths are approximate (± 0.1 Å probably), and agree with those found by Schiebold. The angle of this monoclinic unit cell is $\beta = 116°$, and the cell contains 4 molecules of $KAlSi_3O_8$. All reflections observed have indices $\{hkl\}$ such that $(h + k)$ is even—i.e. the cell is centred on the (001) face—but no other halvings are observed so that the space group (assuming monoclinic holohedry) is $C_{2h}{}^3$ [$C2/m$, Ed.]. All this is in agreement with Schiebold's published results.

The intensities of reflection used for comparison with F-values calculated on the basis of an assumed structure, were obtained by visual estimation of the relative blacknesses of the spots on the rotation photographs.

[1] W. L. Bragg, *Z. Krist.* **74** (1930) 237.

The elements of symmetry in C_{2h}^3 are arranged as follows:

Symmetry centres I: 000, $\frac{1}{2}\frac{1}{2}0$; $\frac{1}{2}00$, $0\frac{1}{2}0$; $00\frac{1}{2}$, $\frac{1}{2}\frac{1}{2}\frac{1}{2}$; $\frac{1}{2}0\frac{1}{2}$, $0\frac{1}{2}\frac{1}{2}$;

           II: $\frac{1}{4}\frac{1}{4}0$, $\frac{3}{4}\frac{1}{4}0$, $\frac{1}{4}\frac{3}{4}0$, $\frac{3}{4}\frac{3}{4}0$; $\frac{1}{4}\frac{1}{4}\frac{1}{2}$, $\frac{3}{4}\frac{1}{4}\frac{1}{2}$, $\frac{1}{4}\frac{3}{4}\frac{1}{2}$, $\frac{3}{4}\frac{3}{4}\frac{1}{2}$;

Reflection planes: $(010)_0$ and $(010)_{\frac{1}{2}}$

Glide planes: $(010)_{\frac{1}{4}}$ and $(010)_{\frac{3}{4}}$ with glide $a/2$.

Digonal rotation axes: $[010]_{00}$ $[010]_{\frac{1}{2}0}$; $[010]_{0\frac{1}{2}}$ $[010]_{\frac{1}{2}\frac{1}{2}}$.

Digonal screw axes: $[010]_{\frac{1}{4}0}$ $[010]_{\frac{3}{4}0}$; $[010]_{\frac{1}{4}\frac{1}{2}}$ $[010]_{\frac{3}{4}\frac{1}{2}}$.

A point in a general position in the unit cell is converted by the operation of the symmetry elements into a group of eight equivalent points; four-fold points lie on reflection planes or rotation axes or at symmetry centres II, and two-fold points lie at symmetry centres I. Figure 1 shows the disposition of the symmetry elements in the unit cell of sanidine.

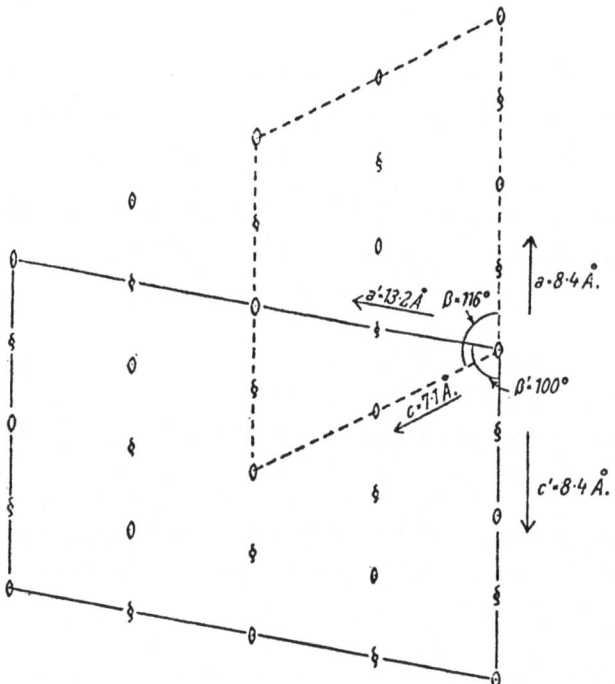

0 Rotation axis [010] with symmetry centres at heights 0, 6.45, 12.9 Å.
§ Screw axis [010] with symmetry centres at heights 3.22, 9.67 Å.
   Symmetry planes (010) at heights 0, 6.45, 12.9 Å.
   Glide planes (010) at heights 3.22, 9.67 Å.

Fig. 1. The unit cell and symmetry elements of sanidine. The unit cell in dotted line is the base-centred crystallographic cell abc containing 4 molecules of composition $KAlSi_3O_8$. The unit cell in full line is the pseudotetragonal cell $a'b'c'$ containing 8 molecules of composition $KAlSi_3O_8$.

III. *Determination of the Structure*

Reference to Figure 1 shows that if we consider the large monoclinic cell $a'b'c'$ with $\beta' = 100°$ we can consider it as divided (by reflection planes and pairs of rotation axes) into eight (nearly rectangular) boxes, each $6.5 \times 6.5 \times 4.2 \, \text{Å}^3$ approximately, and each containing $KAlSi_3O_8$—i.e. 8 molecules in the face-centred cell $a'b'c'$. If we assume that the felspar structure is based on a framework of linked tetrahedra, then each 'box' in the unit cell must contain four tetrahedra—i.e. $AlSi_3O_8$—and the dimensions of the 'box' suggest that the four tetrahedra are arranged in a ring, with two tetrahedron vertices pointing upwards and two downwards (thus accounting for the height of the 'box', 4.2 Å, which is 2×2.1 Å the height of a single tetrahedron of oxygen atoms with diameter 2.6 Å). Such a ring of four tetrahedra cannot fit squarely in a (nearly rectangular) box $6.5 \times 6.5 \, \text{Å}^2$ in size, and must be rotated so that it fits in the box with two oxygen atoms lying on the reflection planes and two lying on rotation axes at the sides of the box. Figure 2 represents such a ring of four tetrahedra fitted in the box. If now the space group symmetry elements operate

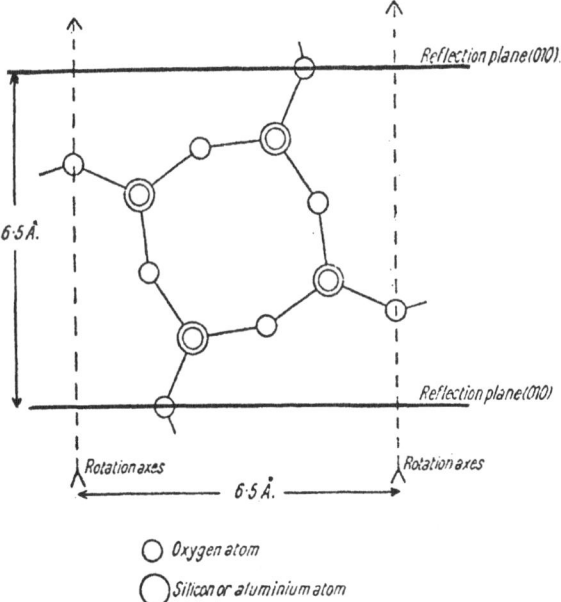

Fig. 2. Portion of the unit cell containing a ring of four tetrahedra. If the box is supposed rectangular, the rotation axes are at heights 0, 4.2 Å above the plane of the diagram, the bases of tetrahedra are in the plane of the diagram, two vertices are 2.1 Å above it, and two are 2.1 Å below it. This idealised diagram should be compared with Figure 4.

347

on this ring of tetrahedra, it is found that the four tetrahedra in the ring are linked everywhere to other tetrahedra, so forming a continuous tetrahedron framework throughout the crystal. In order to secure this continuous linkage it is not necessary to distort the tetrahedra seriously, though some of them are tilted slightly from the ideal position described above. So far as the author can discover, this is the only way of arranging linked tetrahedra in conformity with the symmetry elements of the space group $C_{2h}{}^3$. It should be noted that, as usual, Al and Si atoms inside tetrahedra are supposed indistinguishable.

Assuming this arrangement of tetrahedra it remains to assign positions to the potassium atoms. They number four per unit cell ($a = 8.4$ Å, $b = 12.9$ Å, $c = 7.1$ Å) and in accordance with the space group symmetry must therefore lie either on rotation axes or reflection planes, or at symmetry centres. (The possibility that the potassium ions might lie in positions forbidden by formal space group theory was not forgotten, but there is of course no reason to suppose that this is the case unless it is impossible to find a structure with strict space group symmetry which also explains the intensities of reflection.) Trials show that the intensities of reflections of low indices of types $\{0k0\}$, $\{00l\}$, and $\{0kl\}$ are explained if the potassium atoms are placed on the reflection planes, approximately midway between the points where adjacent rotation axes intersect the planes.

Table I. Coordinates of atoms in the unit cell of sanidine.

Atom	Number in cell	θ_1	θ_2	θ_3	x Å	y Å	z Å
O_{A_1}	4	0°	50°	0°	0	1.79	0
O_{A_2}	4	237	0	85	5.53	0	1.68
O_B	8	295	55	85	6.90	1.97	1.68
O_C	8	0	115	90	0	4.12	1.78
O_D	8	55	45	150	1.28	1.61	2.96
Si_1	8	0	67	78	0	2.40	1.54
Si_2	8	253	40	125	5.90	1.43	2.47
K	4	106	0	50	2.47	0	0.99

The coordinates are referred to the crystallographic unit cell with axes $a = 8.4$ Å, $b = 12.9$ Å, $c = 7.1$ Å, and $\beta = 116°$.

θ_1 θ_2 θ_3 are the angular coordinates used in calculating F-values and are given by the expressions $\theta_1 = 2\pi x/a$, $\theta_2 = 2\pi y/b$, $\theta_3 = 2\pi z/c$.

Silicon and aluminium atoms are not distinguished in the above Table; the groups $8 Si_1 + 8 Si_2$ must include 12 silicon atoms and 4 aluminium atoms.

For the purpose of F-calculations, one third of the potassium is supposed replaced by sodium.

From this point the determination of the structure consists of a laborious search for more accurate atomic parameters which will give calculated F-values in agreement with the observed relative intensities of a large number of reflections. Table I gives the atomic coordinates finally selected; Tables II, III, IV, and V [not reproduced] give calculated F-values and observed relative intensities of reflection for a large number of planes. The agreement between observation and calculation is almost perfect for reflections $\{0kl\}$, and good for general reflections $\{hkl\}$, and it seems likely that the structure is essentially correct. In considering these comparisons of observed and calculated intensities, it is perhaps worth while to recall that the real symmetry of sanidine cannot be monoclinic holohedry; for the Al : Si ratio is always 1 : 3, so that the 16(Al + Si) atoms which in this structure are divided into two groups $8Si_1$ and $8Si_2$ without attempting to distinguish Si from Al, must really be divided into four groups of which three contain 4Si each and one contains 4Al. This is possible either if the symmetry is triclinic or if it is monoclinic but not holohedral; the latter suggestion seems reasonable because adularia is reported[1] to be pyroelectric. In any case, the tetrahedra of oxygen atoms around aluminium atoms will probably be rather larger than those around silicon atoms, and the fact that this difference is not taken into account may explain some of the discrepancies found. It is also to be remembered that if some potassium atoms are replaced by smaller sodium atoms, there will certainly be corresponding alterations in their oxygen atom environment.

It is not claimed that the above structure is uniquely determined to the exclusion of all other possible structures: this cannot be the case when quantitative data alone is used. But the very close agreement between observation and calculation for the many reflections considered is strong evidence in its favour.

IV. *Important Features of the Structure*

On account of the low symmetry it is unfortunately almost impossible to show all the features of the structure in a single diagram, but it is hoped that the more important may be described with the aid of special sectional diagrams.

In the structure proposed there are two kinds of tetrahedral groups of oxygen atoms, one around silicon (or aluminium) atoms of type Si_1, the other around atoms of type Si_2. In the first case the silicon-oxygen distances

---

[1] Hankel, *Wied. Ann.* **1** (1877) 280.

are 1.55, 1.7, 1.7, 1.75 Å, and the oxygen-oxygen distances 2.55, 2.6, 2.7, 2.7, 2.8, 2.95 Å, in the second case they are 1.6, 1.65, 1.7, 1.7 Å and 2.55, 2.55, 2.7, 2.8, 2.85, 2.9 Å. These interatomic distances are close to the values found in other silicates, and both tetrahedral groups are nearly regular. The potassium atom has as nearest neighbours six oxygen atoms at about 2.85 Å distance, with four more at about 3.1 Å. This is the type of environment to be expected for this large atom. Every oxygen atom in the structure is held in common by two tetrahedra, and is also linked to either one or two potassium atoms (taking into account bonds of length 3.1 Å as well as those of length 2.85 Å). In other structures based on linked tetrahedron frameworks the bonds of a Si–O–Si or Si–O–Al linkage are

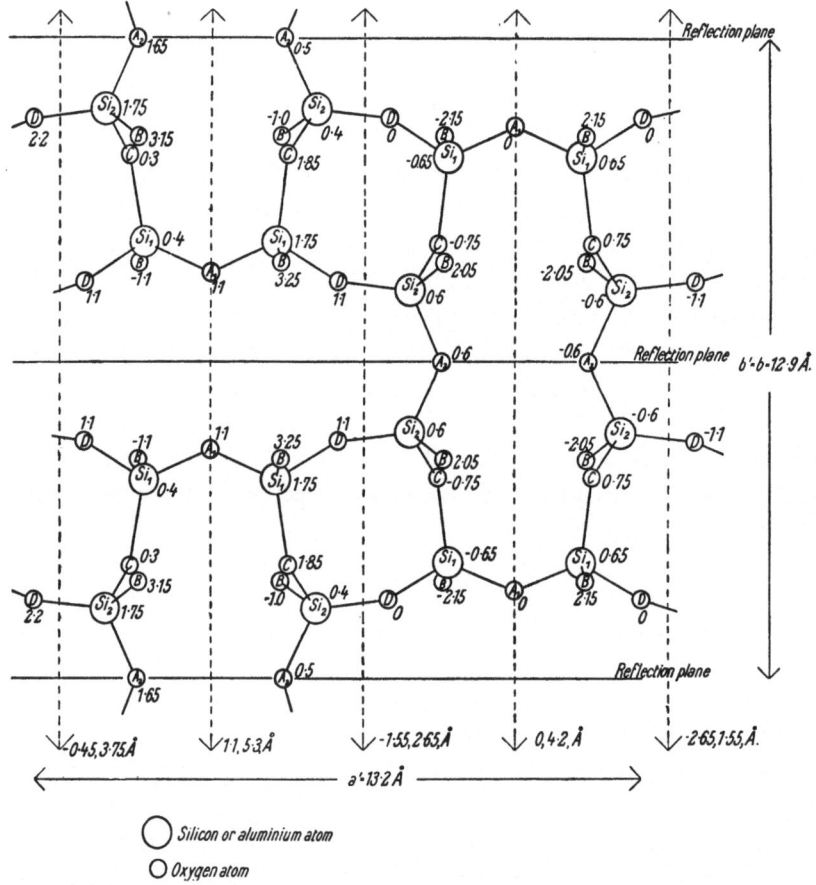

Fig. 4. Sheets of tetrahedra in the sanidine structure. The axis a' slopes up from right to left of the diagram. The lettering of the atoms corresponds to that in Table I, and their heights (above the plane of the diagram) are given in Å. Reflection planes are shown by full lines, rotation axes by dotted lines (heights in Å above the plane of the diagram).

usually inclined at an angle of 120° to 160°; in the sanidine structure the bond angles are 120°, 130°, 140°, 150°, 160° approximately.

The nature of the tetrahedron framework may best be described with reference to the large all-face centred monoclinic cell $a'b'c'$ (outlined in Figure 1) with $a' = 13.2$ Å, $b' = 12.9$ Å, $c' = 8.4$ Å, $\beta' = 100°$. If this cell is viewed along the positive direction of the c'-axis (i.e. along the negative direction of the crystallographic a-axis) then the base of the cell (the plane (001) of the cell $a'b'c'$) consists of a sheet of tetrahedra as shown in Figure 4. In this diagram the c'-axis is normally down into the plane of the diagram, b' is in the plane, and a' slopes up from right to left of the diagram. The sheet of tetrahedra consists of rings of four

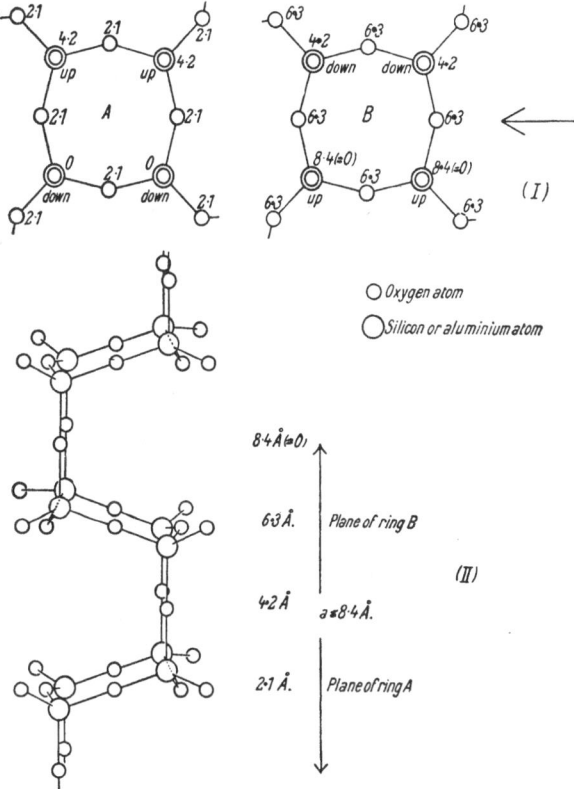

Fig. 5. Idealised diagram of continuous chains of tetrahedron rings parallel to the a-axis. In (I) the rings are viewed along the a-axis, the numbers indicating the heights of oxygen atoms above the base plane. Ring B is to lie above Ring A in such a way that the oxygen atoms at height 4.2 Å above the base plane are common to both rings. The four tetrahedra thus linked together (2 from A, 2 from B) form a new ring of 4 tetrahedra. In (II) the linked rings are viewed in the direction indicated by the arrow in the upper diagram.

351

W. H. TAYLOR

tetrahedra linked together so as to form also rings of eight, and the whole sheet, when viewed in projection as in Figure 4, is pseudo-tetragonal; when the heights of the atoms are taken into account, the sheet is no longer pseudo-tetragonal.

There are two simple ways of describing the linkage of successive sheets to form a three-dimensional framework. First we may consider the sheet as our unit of structure: then the next sheet above that shown in Figure 4 is produced by the operation of either the rotation axes (which are perpendicular to the reflection planes) or the glide planes (midway between reflection

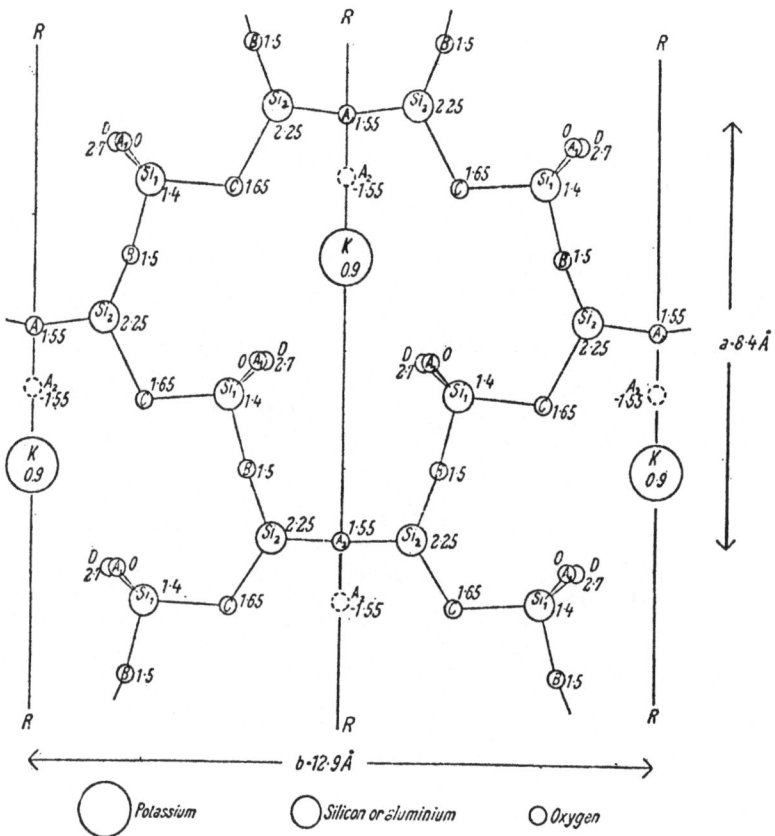

Fig. 6. A portion of the unit cell of sanidine viewed along the normal to the face (001). The lines *RRR* represent reflection planes, the lettering corresponds to that of Figure 4 and of Table I, and the numbers give the distances of the atoms in Å below the plane of the diagram (except for 4 atoms A_2 drawn in dotted line which are above the plane of the diagram). The diagram illustrates (1) the nature of the environment of the potassium atoms, (2) the extent to which the chains of tetrahedron rings are tilted from the ideal positions represented in figure 5. One chain lies between each pair of reflection planes, and the diagram shows only portions of two tetrahedra from each ring of four.

planes, parallel to them, with glide $a/2$ (crystallographic axes) or $c'/2$ (referred to the large cell)). This new sheet is linked to the original sheet by oxygen atoms 'B' each of which is common to one tetrahedron from each sheet. Thus successive sheets are linked together throughout the whole crystal.

Alternatively, we may consider the ring of four tetrahedra as the unit of structure. Such rings occur at intervals $c'/2$ perpendicular to the plane of Figure 4, one ring being formed from the next by the operation of rotation axes or glide planes as above. In each four-ring two adjacent tetrahedron vertices point up, the other two point down; the two which point up are linked to the two downward vertices of the next four-ring above, the two which point down are linked to the upward vertices of the next four-ring below. Thus every four-ring of the sheet shown in Figure 4 is linked to similar four-rings above and below, in such a way that pairs of tetrahedra from successive four-rings also form new rings of four. This is illustrated in Figure 5, which is idealised for the sake of simplicity, and shows how continuous chains of four-rings are formed parallel to the c'-axis (crystallographic a-axis). The actual tilt of the four-rings is indicated in Figure 6.

Figure 6 shows the nature of the environment of the potassium atom, which is about 3.6 Å from the six nearest silicon (or aluminium) atoms.

In view of the fact that this determination of the sanidine structure is to be regarded as preliminary to a full investigation of other felspars, it is preferable to defer a detailed discussion of the relation between physical properties and structure until the work has been completed.

The author is indebted to Professor W. L. Bragg, F.R.S., who suggested this investigation.

Manchester University

179. The Crystal Structure of Bixbyite, by L. PAULING and M. D.
SHAPPELL (1930)

It is shown that the unit of structure of bixbyite contains $16(Mn, Fe)_2O_3$.
The lattice is the body-centered cubic one and the space group is T_h^7 [$Ia3$].
Two possible arrangements alone of the metal atoms are found to be
compatible with the X-ray data (oxygen atoms being neglected), the first
with $8(Mn, Fe)$ in $8e$, $24(Mn, Fe)$ in $24e$ with $u = 0.030$ and the second
the same except with $u = -0.030$ [$24e$: centre in $000 + (000, \frac{1}{2}\frac{1}{2}\frac{1}{2}) +$
$+ \{u0\frac{1}{4}\circlearrowright; u\frac{1}{2}\frac{3}{4}\circlearrowright\}$; $8e$: centre $= (000, \frac{1}{2}\frac{1}{2}\frac{1}{2}) + \{\frac{1}{4}\frac{1}{4}\frac{1}{4}, \frac{1}{4}\frac{3}{4}\frac{3}{4}\}\circlearrowright]$. These two
physically distinct arrangements give the same intensities of X-ray reflections.
For let us consider the position $24e$. Its structure factor is:

$$S_{hkl} = 8\overline{M}[\cos 2\pi(hu + l/4) + \cos 2\pi(ku + h/4) + \cos 2\pi(lu + k/4)]$$
$$\text{for } h, k, l \text{ all even};$$
$$= 8\overline{M}\cos 2\pi(hu + l/4) \text{ for } h \text{ even}, k \text{ odd}, l \text{ odd};$$
$$= 8\overline{M}\cos 2\pi(ku + h/4) \text{ for } h \text{ odd}, k \text{ even}, l \text{ odd};$$
$$= 8\overline{M}\cos 2\pi(lu + k/4) \text{ for } h \text{ odd}, k \text{ odd}, l \text{ even}.$$
$$= 0 \text{ otherwise.}$$

It is seen that the value of the structure factor is the same for a given
positive as for the same negative value of u, except for a difference in sign
in some cases. But the positive and the negative parameter values correspond
to structures which are not identical, but are distinctly different, as can be
seen when the attempt to bring them into coincidence is made. This is a
case where *two distinct structures give the same intensity of X-ray reflections
from all planes.*

The presence of atoms in $8e$ (or $8i$) does not change this result, so that
an unambiguous structure determination for a crystal containing only atoms
in $24e$ (or in $24e$, $8e$, $8i$) could not be made with X-ray methods alone, despite
the dependence on only one parameter.

In the case of bixbyite a consideration of the positions of the oxygen
atoms enables the decision between the two alternatives to be made for
the $(24 + 8)$ metal atoms.

Gates Chemical Laboratory, California Institute of Technology, Pasadena

F. AN APPLICATION OF ISOMORPHOUS
SUBSTITUTION IN THE TRIAL METHOD

180. Die Struktur der Tuttonschen Salze, by W. Hofmann (1931)

▬▬

Verfasser dieser Arbeit begann seine röntgenographischen Untersuchungen mit dem Salz $(NH_4)_2Mg(SO_4)_2 \cdot 6H_2O$; doch zeigte sich bald, daß die Bestimmung der Struktur dieses Salzes nur bei gleichzeitiger Berücksichtigung anderer Glieder der Reihe durchzuführen sei. So erstreckten sich die Untersuchungen schließlich auf neun Salze, nämlich die Sulfate von $(NH_4)Mg$, $(NH_4)Zn$, $(NH_4)Cd$, $(NH_4)Fe$, KMg, TlMg, TlZn und die Selenate von $(NH_4)Mg$, $(NH_4)Zn$.

Zahl der Moleküle in der Elementarzelle 2. Raumgruppe C_{2h}^5. Zweiwertiger Metall in der zweizähligen Lage 000, $\frac{1}{2}\frac{1}{2}0$.

Man erhält für die allgemeine Punktlage folgende zusammengehörigen Koordinatenwerte: (m, n, p); $(\bar{m} + \frac{1}{2}, n + \frac{1}{2}, \bar{p})$; $(m + \frac{1}{2}, \bar{n} + \frac{1}{2}, p)$; $(\bar{m}, \bar{n}, \bar{p})$.

Als Strukturamplitude der allgemeinen Punktlage berechnet man für

Indizessumme $h + k$ gerade: $\quad S = 4\psi \cos 2\pi nk \cos 2\pi(mh + pl)$,

Indizessumme $h + k$ ungerade: $S = -4\psi \sin 2\pi nk \sin 2\pi(mh + pl)$.

VI. BESTIMMUNG DER LAGE DES EINWERTIGEN METALLES UND DES SCHWEFELS

a) *Allgemeines*

Bei der großen Zahl von variabeln Parametern, die in den Strukturfaktor der einzelnen Ebenen nach obigem eingehen, mußte eine Methode gefunden werden, um verschiedene dieser Parameter unabhängig von den übrigen zu bestimmen. Ich verwendete zu diesem Zweck die Isomorphiebeziehung zwischen den verschiedenen Gliedern der Reihe und bestimmte auf diesem Wege die sechs Parameter des einwertigen Metalles und des Schwefels.

Die drei ersten Parameter wurden bestimmt, indem Salze mit verschiedenem einwertigen Metall untersucht wurden, nämlich neben $(NH_4)Mg$-sulfat KMg-sulfat und TlMg-sulfat; die drei Parameter des Schwefels ergaben sich in analoger Weise durch den Übergang von $(NH_4)Mg$-sulfat zu dem entsprechenden Selenat. Die Schwierigkeit bestand darin, daß man die Diagramme der einzelnen Salze wegen ihres verschiedenen Absorptionsvermögens nicht direkt miteinander vergleichen darf. Ein quantitativer Vergleich zwischen verschiedenen Gliedern wäre nur möglich bei absoluten Intensitätsmessungen[1]. Diese Schwierigkeit konnte, wie im folgenden gezeigt wird, bis zu einem gewissen Grad dadurch beseitigt werden, daß auch Verbindungen mit verschiedenem zweiwertigen Metall (dessen Lage bekannt ist), nämlich Zn und Cd an Stelle von Mg, untersucht wurden.

Die Grundlage der Arbeitsmethode bildet eine formale Auffassung, die das Gitter jeden Gliedes der Reihe in eine einfache Beziehung setzt zu dem Gitter von (NH_4)Mg-sulfat als der untersuchten Substanz mit den leichtesten Atomen. (NH_4)Zn-sulfat z.B. wird aufgefaßt als (NH_4)Mg-sulfat mit einem Zusatzgebilde vom Streuvermögen $\Delta\psi = (\psi_{Zn} - \psi_{Mg})$ in den Punktlagen des Magnesiums, das Gitter von (NH_4)Mg-selenat als (NH_4)Mg-sulfatgitter mit einem Zusatzgebilde $(\Delta\psi = (\psi_{Se} - \psi_S))$ an der Stelle $(u'v'w')$ des Schwefels usw.

Die Strukturamplituden eines der oben erwähnten Salze ergeben sich dann einfach als Strukturamplitude von (NH_4)Mg-sulfat + Zusatzamplitude ΔS. Letztere hat im Falle des (NH_4)Zn-sulfates die Form: $\Delta S = = 2(\psi_{Zn} - \psi_{Mg})$ {wenn $h + k$ gerade}, bzw. $\Delta S = 0$ {wenn $h + k$ ungerade}, im Falle des (NH_4)Mg-selenates (und entsprechend bei Änderung des einwertigen Metalles):

$$\Delta S = 4(\psi_{Se} - \psi_S)\cos 2\pi v'k \cos 2\pi(u'h + w'l) \qquad \{h + k \text{ gerade}\},$$

$$\Delta S = -4(\psi_{Se} - \psi_S)\sin 2\pi v'k \sin 2\pi(u'h + w'l) \qquad \{h + k \text{ ungerade}\}.$$

Im Falle der Änderung des zweiwertigen Metalles sind die Zusatzamplituden also von vornherein bekannt, was umgekehrt dazu dient, das Vorzeichen—im Bezug auf den gewählten Koordinatenursprung—und den ungefähren Betrag der Amplituden von (NH_4)Mg-sulfat festzustellen.

In die betreffenden Ausdrücke gehen bei Ersetzung des einwertigen Metalles oder des Schwefels höchstens drei Unbekannte ein, für $(h00)$, $(0k0)$, $(00l)$ ist es sogar nur je eine einzige. Die unbekannten Größen müssen so bestimmt werden, daß sich aus ihnen die relativen Intensitäten bei den betreffenden Gliedern der Reihe (nur unter sich verglichen) richtig berechnen lassen.

Bei relativen Intensitätsmessungen können die konstanten Faktoren weggelassen werden, so daß man zur Berechnung der Intensitäten folgenden Ausdruck erhält:

---

[1] W. L. Bragg und J. West, *Z. Krist.* **69** (1928) 118.

$$\rho \sim I \text{ (gesamte Energie der Sekundärstrahlung)} \sim S^2 \frac{1 + \cos^2 2\alpha}{\mu \sin 2\alpha},$$

wobei μ = linearer Absorptionskoeffizient, S = Strukturamplitude.

Die Streufunktionen für Mg^{2+}, K^{1+}, O^{2-} wurden der gleichen Arbeit von Bragg entnommen, die Funktionen für Zn, Cd, Tl, Se, S (letztere beide zur Berechnung von $\psi_{Se} - \psi_S$) nach den dort angegebenen, von L. H. Thomas abgeleiteten Formeln berechnet, die Werte für S^{6+} entstammen einer Arbeit von R. W. James und W. A. Wood[1], diejenigen für $(NH_4)^{1+}$ einer experimentellen Untersuchung von R. W. G. Wyckoff[2]. Die Wirkung von μ fällt bei Vergleichen innerhalb eines Diagrammes weg und wurde daher bei den Intensitätsberechnungen nicht berücksichtigt; doch ist es zweckmäßig, wenigstens die relativen Werte bei den verschiedenen Salzen zu kennen, um in gewissen Fällen auch Vergleiche zwischen verschiedenen Diagrammen anstellen zu können.

b) *Intensitätsänderungen bei Ersatz des zweiwertigen Metalles*

$h + k$ ungerade

Die Strukturfaktoren müßten, da $\Delta S = 0$, bei (NH_4)Zn- und (NH_4)Cd-sulfat die gleichen sein wie bei (NH_4)Mg-sulfat; die Intensität sollte entsprechend der Zunahme von μ abnehmen, bzw. beim Zn- und Cd-Salz ungefähr die gleiche sein. Daß die wirklichen Verhältnisse, wenigstens qualitativ, dem entsprechen, geht aus den Tabellen V–VII [not reproduced, Ed.] hervor.

$h + k$ gerade

Für den Übergang von (NH_4)Mg-sulfat zu (NH_4)Zn-sulfat gilt: $\Delta S = = 2(\psi_{Zn} - \psi_{Mg})$. Wenn trotz erhöhter Absorption die Intensitäten beim Zn-Salz gleich groß oder größer sind wie beim Mg-Salz, muß dies auf einer Vergrößerung von S beruhen, d.h. die Amplitude von (NH_4)Mg-sulfat hat positives Vorzeichen. In manchen Fällen kann man die Zunahme der Strukturamplitude sozusagen unmittelbar sehen, ohne Rücksicht auf Absorption und Aufnahmebedingungen, wenn nämlich dem Reflex von (NH_4)Mg-sulfat ein ungefähr gleich starker benachbart ist, für den $h + k$ ungerade. Dieser Reflex, dessen Amplitude man trotz veränderter Intensität bei (NH_4)Zn-sulfat als praktisch unverändert ansehen muß, bildet die Vergleichsmarke, an der man die Veränderung der Amplituden benachbarter Linien unmittelbar beobachten kann.

[1] R. W. James und W. A. Wood, *Proc. Roy. Soc.* (A) **109** (1925) 598.
[2] R. W. G. Wyckoff und A. H. Armstrong, *Z. Krist.* **72** (1929) 319.

Wenn die Amplituden von (NH_4)Mg-sulfat negativ und sehr klein sind, kann in seltenen Fällen bei ganz kleinem Glanzwinkel schon durch die Zusatzamplitude von (NH_4)Zn-sulfat eine Überkompensation erfolgen, so daß der Reflex in fast gleicher Stärke wiederkehrt.

Im allgemeinen werden jedoch Reflexe mit negativer Amplitude bei (NH_4)Zn-sulfat sehr stark abgeschwächt oder völlig ausgelöscht, um dann eventuell bei (NH_4)Cd-sulfat mit umgekehrtem Vorzeichen wiederzukehren. Die Vorzeichendiskussion für (NH_4)Mg-sulfat sei an einigen Beispielen erläutert [partly reproduced, Ed.].

(200) (NH_4)Mg-sulfat: (200) $\leqq$ (210),
        (NH_4)Zn-sulfat: (200) > (210),
        (NH_4)Cd-sulfat: (200) $\gggtr$ (210): Positives Vorzeichen.

(400)   Auslöschung bei (NH_4)Zn-sulfat, daher negatives Vorzeichen bei (NH_4)Mg-sulfat.

(110) (NH_4)Mg-sulfat: (110) > (220), (330),
        (NH_4)Fe-sulfat: (110) $\lll$ (220), (330),
        (NH_4)Zn-sulfat: (110) < (220), (330),
        (NH_4)Cd-sulfat: (110) $\sim$ (330): Negatives Vorzeichen.

Die Auslöschung von (400), (060), (002) bei (NH_4)Zn-sulfat ermöglicht es, die Amplituden von $(h00)$, $(0k0)$, $(00l)$ bei (NH_4)Mg-sulfat auch ihrem ungefähren Betrag nach abzuschätzen.

Zu diesem Zweck wurden die Hauptspektren bei (NH_4)Mg-, (NH_4)Zn-, (NH_4)Cd-sulfat in einem selbstregistrierenden Instrument der Firma Zeiss zusammen mit einer geeichten Schwärzungsskala photometriert. Durch Dividieren der Intensitäten mit dem Faktor $(1 + \cos^2 2\alpha)/\sin 2\alpha$ wurden Größen gewonnen, die proportional S^2 gesetzt wurden; aus ihnen wurde die Wurzel gezogen, der Proportionalität mit $|S|$ zugeschrieben wurde. Der Proportionalitätsfaktor für je ein Hauptspektrum wurde nun dadurch festgelegt, daß die Amplitude der bei (NH_4)Zn-sulfat ausgelöschten Ordnung jeweils = negative Zusatzamplitude gesetzt wurde, wodurch nun S auch für die übrigen Ordnungen festgelegt war.

c) *Intensitätsänderungen bei Ersatz des einwertigen Metalles und des Schwefels*

Die Verhältnisse sind in beiden Fällen analog, so daß diese gemeinsam diskutiert werden. Die Zusatzamplituden—für $(h00)$, $(0k0)$, $(00l)$ cos-Funktionen mit einer Variabeln—wurden für die verschiedenen Ordnungen

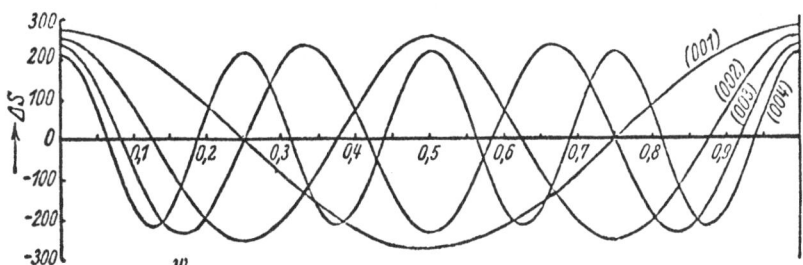

Fig. 5. Zusatzamplituden von TlMg-sulfat für Interferenzen (00*l*).

einer Fläche graphisch aufgetragen (Fig. 5). Durch eine bloß qualitative systematische Betrachtung dieser Kurven konnte schon eine erste, ziemlich enge Beschränkung der Parameterwerte erzielt werden. Für die noch bleibenden Möglichkeiten wurden die Intensitäten der verschiedenen Ordnungen berechnet, indem die dritte Stelle des gesuchten Parameters von 5 zu 5 Einheiten abgeändert und als Amplitude $S = S_{(NH_4)Mg-s.} + \Delta S$ verwendet wurde. Dadurch konnten die sechs gesuchten Parameter auf einen genügend kleinen Bereich eingeengt werden. Zur letzten Einschränkung, bzw. zur Kontrolle, wurden verschiedentlich hohe Ordnungen (aus Weißenberg-Aufnahmen) herangezogen; ihre Intensität ist im wesentlichen durch die Lage der schwersten Atome (Tl, Se, S) bedingt, so daß man deren Koordinatenwerte, wenn sie ungefähr bekannt sind, noch genauer festlegen kann. Das Ergebnis der Parameterbestimmung war:

Parameterfestlegung für das einwertige Metall.

$$u^* = 0{,}120 \quad (\bar{u}^* + \tfrac{1}{2},\ u^* + \tfrac{1}{2},\ \bar{u}^*).$$

$$v^* = 0{,}143 \quad (\bar{v}^* + \tfrac{1}{2},\ v^* + \tfrac{1}{2},\ \bar{v}^*).$$

$$w^* = 0{,}345 \quad (\bar{w}^*).$$

Parameterfestlegung für den Schwefel.

$$u^{*'} = 0{,}090 \quad (\bar{u}^{*'} + \tfrac{1}{2},\ u^{*'} + \tfrac{1}{2},\ \bar{u}^{*'}).$$

$$v^{*'} = 0{,}135 \quad (\bar{v}^{*'} + \tfrac{1}{2},\ v^{*'} + \tfrac{1}{2},\ \bar{v}^{*'}).$$

$$w^{*'} = 0{,}260 \quad (\bar{w}^{*'}).$$

e) *Eindeutige Festlegung eines Koordinatentripels für den Schwefel und das einwertige Metall.*

Das Ergebnis der Parameterbestimmung für das einwertige Metall sind vier Werte für u: $(u^*, \bar{u}^* + \frac{1}{2}, u^* + \frac{1}{2}, \bar{u}^*)$, vier Werte für v: $(v^*, \bar{v}^* + \frac{1}{2}, v^* + \frac{1}{2}, \bar{v}^*)$, zwei Werte für w: $(w^*, \bar{w}^*)$. Aus diesen Werten können durch Kombination $4.4.2 = 32$ Koordinatentripel gebildet werden. Da immer vier von diesen strukturell gleichwertig sind, ergeben sich also acht Möglichkeiten für die Lage des einwertigen Metalles und ebenso viele für die des Schwefels. Ich konnte auf direktem Wege eine eindeutige Auswahl unter diesen Kombinationen treffen. Zu diesem Zweck mußten die Zusatzamplituden allgemeiner Flächen von KMg-sulfat, TlMg-sulfat, $(NH_4)Mg$-selenat gegenüber $(NH_4)Mg$-sulfat berechent werden. Sie besitzen für die einzelnen Kombinationen verschiedene Werte oder Vorzeichen, so daß die beobachteten Intensitätsänderungen gegenüber $(NH_4)Mg$-sulfat es ermöglichen, gewisse Kombinationen auszuschließen.

Die Parameter wurden bestimmt zu:

$$N: \quad u = 0{,}120, \quad v = 0{,}357, \quad w = 0{,}345,$$

$$S: \quad u' = 0{,}090, \quad v' = 0{,}635, \quad w' = 0{,}260,$$

Der Berechnung der Lage der Sauerstoffatome wurden 'Wasseroktaeder' und 'SO$_4$-Tetraeder' zugrunde gelegt. Für die Orientierung der letzteren ergab sich ein enger Bereich möglicher Winkelwerte, der zu einer befriedigenden Übereinstimmung mit dem Experiment führte. [Values of oxygen parameters not reproduced. The metal and sulphur parameters agree roughly with a more recent redetermination (1962). Ed.]

Vor allem danke ich meinem verehrten Lehrer, Herrn Prof. Dr. E. Schiebold, für die ständige Förderung der Arbeit durch wertvolle Ratschläge und Anregungen.

Leipzig, Institut für Mineralogie und Petrographie der Universität

G. THE FIRST INVESTIGATION OF A GLOBULAR PROTEIN

181a. X-ray Photographs of Crystalline Pepsin, by J. D. Bernal and Dorothy Crowfoot (1934)

■

181b. X-ray Single Crystal Photographs of Insulin, by Dorothy Crowfoot (1935)

■

181a. Four weeks ago, Dr. G. Millikan brought us some crystals of pepsin prepared by Dr. Philpot in the laboratory of Prof. The Svedberg, Uppsala. They are in the form of perfect hexagonal bipyramids up to 2 mm in length, of axial ratio $c/a = 2.3 \pm 0.1$. When examined in their mother liquor, they appear moderately birefringent and positively uniaxial, showing a good interference figure. On exposure to air, however, the birefringence rapidly diminishes. X-ray photographs taken of the crystals in the usual way showed nothing but a vague blackening. This indicates complete alteration of the crystal and explains why previous workers have obtained negative results with proteins, so far as crystalline pattern is concerned[1]. W. T. Astbury has, however, shown that the altered pepsin is a protein of the chain type like myosin or keratin giving an amorphous or fibre pattern.

It was clearly necessary to avoid alteration of the crystals, and this was effected by drawing them with their mother liquor and without exposure to air into thin capillary tubes of Lindemann glass. The first photograph taken in this way showed that we were dealing with an unaltered crystal. From oscillation photographs with copper $K\alpha$-radiation, the dimensions of the unit cell were found to be $a = 67$ A, $c = 154$ A, correct to about 5 per cent. This is a minimum value as the spots on the c row lines are too close for accurate measurement and the c axial length is derived from the axial ratio. The dimensions of the cell may still be multiples of this. Using the density measured on fresh material[2] as 1.32 (our measurements gave 1.28), the cell molecular weight is 478,000, which is twelve times 40,000, almost exactly Svedberg's value arrived at by sedimentation in the ultra-centrifuge. This agreement may however be quite fortuitous as we have found that the crystals contain about 50 per cent of water removable at

---

[1] G. L. Clark and K. E. Korrigan (*Phys. Rev.* (II) **40** (1932) 639) describe long spacings found from crystalline insulin, but no details have been published.

[2] J. H. Northrop, *J. Gen. Physiol.* **18** (1930) 739.

room temperature. But this would still lead to a large molecular weight, with possibly fewer molecules in the unit cell.

Not only do these measurements confirm such large molecular weights but they also give considerable information as to the nature of the protein molecules and will certainly give much more when the analysis is pushed further. From the intensity of the spots near the centre, we can infer that the protein molecules are relatively dense globular bodies, perhaps joined together by valency bridges, but in any event separated by relatively large spaces which contain water. From the intensity of the more distant spots, it can be inferred that the arrangement of atoms inside the protein molecule is also of a perfectly definite kind, although without the periodicities characterising the fibrous proteins. The observations are compatible with oblate spheroidal molecules of diameters about 25 A and 35 A, arranged in hexagonal nets, which are related to each other by a hexagonal screw-axis. With this model we may imagine degeneration to take place by the linking up of amino acid residues in such molecules to form chains as in the ring-chain polymerisation of polyoxymethylenes. Peptide chains in the ordinary sense may exist only in the more highly condensed or fibrous proteins, while the molecules of the primary soluble proteins may have their constituent parts grouped more symmetrically around a prosthetic nucleus.

At this stage, such ideas are merely speculative, but now that a crystalline protein has been made to give X-ray photographs, it is clear that we have the means of checking them and, by examining the structure of all crystalline proteins, arriving at far more detailed conclusions about protein structure than previous physical or chemical methods have been able to give.

Department of Mineralogy and Petrology, Cambridge

181b. Since insulin was first prepared crystalline[1] in 1926, several efforts have been made to obtain X-ray photographs of the crystals. The first attempts of W. H. George[2] by the powder method failed to show any pattern indicative of a crystal structure, and though later long spacings were reported by G. L. Clark and K. E. Korrigan[3], it was impossible to base any unambiguous interpretation on their results. The fact that pepsin could be made to give a single crystal X-ray diffraction pattern[4] suggested

[1] J. Abel, *Proc. Nat. Acad. Sci.* **12** (1926) 132.
[2] *Proc. Leeds Phil. Lit. Soc.* **1** (1929) 412.
[3] *Phys. Rev.* **40** (1932) 639.
[4] J. D. Bernal and D. Crowfoot, *Nature* **133** (1934) 794 [this Vol. preceding paper].

that the problem of insulin, which is in many respects a more stable crystalline species, could be attacked in the same way if large enough crystals could be grown. This was made possible by D. A. Scott's study of the crystallisation of insulin in the presence of salts of zinc and of other metals[1].

The crystallisation was therefore carried out by a modification of Scott's method from a phosphate buffer solution containing a little acetone and some zinc chloride at a pH of 6.2–6.5. The solution was cooled very slowly from 50° to room temperature over a period of three days, at the end of which time sufficiently large crystals had grown.

The crystals have the form of very flat rhombohedra which often grow in pairs united at the ends of their trigonal axes. The larger ones present the appearance of six lobed stars and are as much as 0.2 mm across and 0.05 mm thick. These show a positive uniaxial figure when viewed along the trigonal axis. The crystals prove to be perfectly stable in air (unlike pepsin) with unchanged birefringence and reflecting power, and it was accordingly possible to examine them dry by X-ray methods.

X-ray photographs have been taken on three crystals, one rotating about the trigonal axis and the others about the normals to $(10\bar{1}0)$ and $(11\bar{2}0)$. Examples are shown in Fig. 1 [Fig. unsuited for reproduction, Ed.]. Copper $K\alpha$-radiation was used and exposure times of about 15 hours for a single 5° oscillation photograph with a plate distance of 6 cm. The crystals have so far proved unaltered by exposure for more than 100 hours to X-radiation. The photographs taken indicate a simple rhombohedral cell of $a = 44.3$ A and $\alpha = 115°$ correct to about 2 per cent. This referred to hexagonal axes corresponds to a cell three times as large with $a = 74.7$ A, $c = 30.6$ A, which shows no halvings but those required by the rhombohedral lattice. The structure may also be described in terms of a pseudocubic body-centred cell twice the size of the primitive cell with $a = 47.7$ A, $\alpha = 103°6'$. No planes of symmetry are present and the space group is therefore $R3$. The cell molecular weight calculated for the primitive rhombohedral cell and the density 1.315 measured by Dr. Eyer[2] is $39,300 \pm 800$. (Density measurements on the acutal crystals used gave 1.306 ± 0.003.) As Abel has measured the water lost by heating the crystals at 104° in a vacuum at 5.35 per cent of the air-dried weight[3], the weight of insulin in the cell (cell molecular weight —water of crystallisation) may be deduced as 37,200, which is very close

[1] *Biochem. J.* (1934) 1596.

[2] K. Freudenberg, *Z. physiol. Chem.* **204** (1932) 233.

[3] J. Abel, E. M. K. Geiling, C. A. Roudler, F. K. Bell and O. Wintersteiner, *J. Pharm. Exp. Ther.* **31** (1927) 65.

to the weight of one molecule of insulin reported by The Svedberg[1].

It therefore appears that the crystal unit cell contains only one molecule of insulin, although a cell containing $3n$ sub-molecules is not excluded by the X-ray data. The laws of crystal symmetry rigorously applied would require this molecule to have trigonal symmetry, but it is possible also that the crystal attains apparent trigonal symmetry by a statistical regularity of arrangement of molecules about the lattice points, or that the X-ray effects so far observed are first approximation mass effects to which further work may add a fine structure, due to the arrangement of the atoms within the molecules, which our methods are as yet insufficiently delicate to detect. The measurements obtained do, however, fix quite definitely the arrangement of the molecules with respect to one another and their approximate size and shape, since this follows directly from the crystal lattice, while the variation of the intensities of the spectra strongly suggests that the arrangement of atoms within the molecules is also of a perfectly definite kind.

The crystal structure of insulin is of an eight co-ordination type, as the possible reference to the pseudo-cubic body-centred cell of twice the size most clearly indicates. Each insulin molecule is surrounded by eight others, two at the short distance of 30 A above and below along the trigonal axis, and siz at the longer distance of 44 A along the edges of the primitive rhombohedron. The shape of the molecule therefore appears approximately as an oblate spheroid of diameters 44 A and 30 A. D. A. Scott gives the atomic percentage of zinc in insulin crystals as 0.00795, or 3 atoms of zinc per molecule of insulin[2]. It therefore seems reasonable to suggest that these atoms are required to bind the molecules into nets parallel to the c plane, one between each pair of insulin molecules along the six points of contact, the closer linkage along the c axes being due to other causes. This would provide some explanation of the rôle played by zinc and other bivalent metals in promoting the crystallisation of insulin.

It is of particular interest to compare this crystal structure of insulin with that which may be deduced for pepsin from the X-ray measurements previously reported. It has been found that the true cell of pepsin has a c dimension three times as long as that at first suggested, namely, 461 A, and that the structure should probably be referred to rhombohedral axes[3] $a = 162$ A, $\alpha = 23°50'$. The first order of the reflections from the c plane to occur is, however, the 45th, while the strongest order is the 48th, which

[1] *Nature* **127** (1931) 438; B. Sjögren and T. Svedberg, *J. Amer. Chem. Soc.* **53** (1931) 2657.
[2] Private communication to J. D. Bernal.
[3] Unpublished observations of J. D. Bernal.

indicates that the most marked periodicity along c is one of only 9.6–10.2 A, very much the same as the distance—10 A—between layers of atoms along the c axis of insulin. It seems significant, further, that the length of a in pepsin—67 A—referred to the original hexagonal axes, is so similar to that of insulin—74.7 A—when given hexagonal axes. These two dimensions define a crumpled layer structure in which the molecules are arranged in networks of six-sided rings of the non-planar cyclohexane type which occurs, for example, in diamond and wurtzite. From the side of such a ring projected on to (0001), $a/\sqrt{3}$, or 38.7 A in pepsin, and the thickness of the order of 10 A, a length for one radius of the pepsin molecule may be calculated $= 20$ A. In insulin, the layers are so arranged that atoms in one fall as nearly as possible into the spaces of the one below, which makes a very compact structure. In pepsin we may imagine that the layers of rings are slid relatively to one another, to bring atoms of the lower ring directly beneath those of the upper ring in such a way that each is approximately tetrahedrally co-ordinated. The effective depth of a single layer is then equal to the thickness of the ring system plus the diameter of a single molecule, and may be calculated to be 51.2 A, with a spheroidal pepsin molecule of diameter 41.2 A in this direction. A combination of the crystallographically possible ways of sliding the ring systems is able to give the required length of c, nine times that of the depth of one layer and fifteen times the c dimension of insulin.

This kind of structure proposed for pepsin is of a very much looser type than that of insulin. Each molecule is surrounded by only four others and there are large channels through which free movement of water and dissolved substances may occur within the crystal structure. On drying, such a structure would collapse, in agreement with the fact that, in contradistinction to insulin, the crystals of pepsin lose their birefringence on exposure to air and only show crystalline X-ray diffraction effects when immersed in the mother liquor. Various observations[1] suggest that loose 4 co-ordination structure of this kind may be general among certain classes of protein crystals which belong either to a hexagonal type with an axial ratio about 2.3 similar to that of pepsin, or to a cubic type which shows diamond cleavages. Wherever the attraction between the adjacent protein molecules is of the same order of magnitude as that between the protein molecules and the medium, a low co-ordination structure type may be expected. In insulin, on the other hand, where the molecules can be strongly attracted

---

[1] A. F. W. Schimper, *Z. Krist.* **5** (1881) 131.

together with the assistance of metal atoms, the structure is very much more condensed and shows a high co-ordination number.

I have to thank Prof. Pyman and Messrs. Boots Pure Drug Co., Ltd., for a gift of the insulin used in this research.

Department of Mineralogy, Oxford

CHAPTER XI

Fourier Method

Beim Anblick solcher [Laue] *Photogramme tritt natürlich die Frage an uns heran: Können wir mit Hilfe der Röntgenstrahlinterferenzen die Struktur eines Kristalles vollständig und eindeutig bestimmen? Damit wäre eins der wichtigsten Probleme der Kristallographie seiner Lösung nahe gebracht. Aus einer Interferenzfigur läßt sich diese Kenntnis nicht schöpfen. Wir stehen den Kristallen dabei ähnlich gegen über, als sollten wir ein optisches Gitter ohne Mikroskop, allein aus seinen Spektren heraus untersuchen. Die Gesamtheit dieser Spektren enthält freilich, wie ja besonders Abbe betont hat, alle Elemente, aus denen sich das mikroskopische Bild zusammensetzt. Aber zur Konstruktion dieses Bildes genügt die Kenntnis der Lage und Intensität der Spektren nicht. Es kommt noch wesentlich auf die Phasen an, mit welchen diese gegeneinander schwingen. Um die Kristallstruktur zu 'mikroskopieren', müßte man mindestens auch noch die Phasendifferenzen zwischen den verschiedenen Interferenzpunkten eines Photogrammes feststellen, und diese Aufgabe dürfte wohl eine recht schwierige sein.*

M. LAUE,
Physik. Zeitschr. **14** (1913) p. 1174

see further Vol. I paper **25**:
W. H. BRAGG, *Bakerian Lecture* (1915)

A. THEORETICAL AND APPLICATION TO ROCK-SALT

182. The Transfer in Quanta of Radiation Momentum to Matter, by W. DUANE (1923)

■

183. The Quantum Integral and Diffraction by a Crystal, by A. H. COMPTON (1923)

■

184. The Quantum Theory of the Fraunhofer Diffraction, by P. S. EPSTEIN and P. EHRENFEST (1924)

■

185. The Calculation of the X-ray Diffraction Power at Points in a Crystal, by W. DUANE (1925)

■

186. The Distribution of Diffracting Power in Sodium Chloride, by R. J. HAVIGHURST (1925)

■

187. A Technique for the X-ray Examination of Crystal Structures with many Parameters, by W. L. BRAGG and J. WEST (1928): § 9 Is a Direct Method of Analysis possible?

■

188. An Optical Method of Representing the Results of X-ray Analysis, by W. L. BRAGG (1929)

■

189. A Note on the Representation of Crystal Structure by Fourier Series, by W. L. BRAGG and J. WEST (1930)

■

182. This note describes an attempt to formulate a theory of the reflection of X-rays by crystals, based on quantum ideas without reference to interference laws.

2. The fundamental hypothesis of the theory now presented is that the *momentum* of radiation is transferred to and from matter in quanta, and further, that the laws of the conservation of energy and of momentum apply to these transfers.

3. In order to illustrate the meaning of this hypothesis, let us take a particular example, namely, that of the reflection of an X-ray by a crystal. For the sake of simplicity, let us assume that the axes of the crystal lie at right angles to each other and take the particular case in which the X-ray strikes the crystal in a direction parallel to one set of principal planes. The problem thus becomes a two-dimensional one. Suppose

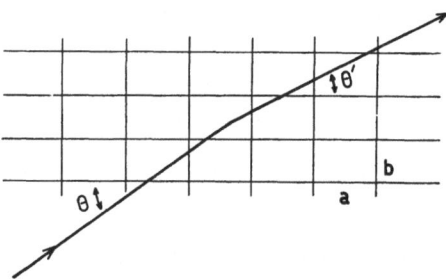

that the distances between parallel planes are a and b, respectively, and let an X-ray of energy $h\nu$ and momentum $h\nu/c$ pass through the crystal at an angle θ with the planes, called x, as represented in the figure. At some point in its path the X-ray may be deflected so as to travel in a direction, making an angle θ' with the x planes. At this point of deflection the X-ray transfers some momentum and energy to the crystal. The mass of the crystal being so large, however, the velocity acquired by it must be very small and, consequently, the energy transferred from the radiation to the crystal may be neglected. Neglecting the amount of energy transferred to the crystal, we may put $h\nu' = h\nu$ and we may consider that the X-ray has the same energy and total momentum after the transfer as before.

The momenta transferred to the crystal in the directions x and y, respectively, are expressed by the following equations:

$$M_x = h\nu/c.(\cos \theta - \cos \theta'), \quad M_y = h\nu/c.(\sin \theta - \sin \theta'). \tag{1}$$

According to the fundamental hypothesis the momentum M_x transferred to the crystal is to be delivered in quanta. We may write, therefore, M_x proportional to the constant h. Now, let us apply to the problem dimensional reasoning. The momentum has the dimensions ml/t and the constant h has the dimensions ml^2/t. In order, therefore, that both sides of our equation may have the same dimensions, we must divide h by a length. Since this is a crystalline phenomenon the only quantity of importance having the dimensions of a length in the direction of the axis of X is the distance a between the crystal planes. Since any whole number, τ, of quanta may be transferred, we will multiply τ into the quotient of h by a and we reach the conclusion that the momentum M_x must be equal to

$\tau h/a$ multiplied by some numerical constant. The equation will take its simplest form if we put this constant factor of proportionality equal to unity. We thus get the equation: $M_x = \tau h/a$. Similarly, $M_y = \tau' h/b$. Another way of looking at these equations is to regard them as peculiar applications to the transfers of momenta of the quantum equation, $\int p dq = nh$, expressed in generalized coördinates*. Substituting for M_x and M_y in equations (1), we get immediately:

$$h\nu/c.(\cos\theta - \cos\theta') = \tau h/a, \quad h\nu/c.(\sin\theta - \sin\theta') = \tau' h/b. \text{**} \quad (2)$$

4. In the above equations τ and τ' represent whole numbers. If $\tau = \tau' = 0$, $\theta' = \theta$, and the incident radiation quantum passes through the crystal without deflection. It will keep on traveling through the crystal until it reaches a point where either τ or τ' differs from 0. Suppose that τ' differs from 0 at some point, and that τ remains 0, then, from the first of equations (2) it follows that $\cos\theta' = \cos\theta$. θ', however, cannot equal θ, for in this case τ' does not equal 0. Consequently, θ' must equal $-\theta$. This means that the X-ray at the point is deflected downward in a direction making an angle with the x-planes equal to the angle of incidence. In other words, the angle of reflection by the x-planes equals the angle of incidence on them.

Substituting $\theta' = -\theta$ in the second of equations (2) we get:

$$2h\nu/c.\sin\theta = \tau' h/b, \quad (3)$$

which represents the relation that must exist between ν and θ in order that τ' may differ from zero. Putting $\nu/c = 1/\lambda$, equation (3) reduces to:

$$\tau'\lambda = 2b\sin\theta.$$

This is Braggs' law of reflection of an X-ray by a crystal.

By similar reasoning, if $\tau' = 0$ and τ differs from 0, we deduce the equation:

$$\tau\lambda = 2a\cos\theta,$$

which represents reflection according to Bragg's law from the y-planes. If both τ and τ' differ from 0, the equation obtained reduces to Bragg's equation representing the reflection from a set of planes other than the principal planes. In the case where the axes of the crystal are not at right angles to each other, we apply the law of the transfer of momenta in quanta to the total component of the radiation momentum in the direction of each axis and we equate this component to $\tau h/a$ where a is the parameter

---

* [See the next paper. Ed.]
** [These are the Laue equations: $a(\cos\theta - \cos\theta') = \tau\lambda$, $b(\sin\theta - \sin\theta') = \tau'\lambda$. Ed.]

of the crystal along the axis. This gives us equations which reduce to Bragg's equation for the reflection from each set of planes, as in the orthogonal problem.

Department of Physics, Harvard University

183. Duane has recently pointed out[1] that if the momentum of a crystal grating perpendicular to the crystal face is nh/a, where n is an integer, h is Planck's constant and a is the distance between successive atomic layers, and if the momentum of the incident radiation quantum is $h\nu/c$, then Bragg's diffraction formula $n\lambda = 2a \sin \theta$ is a necessary consequence. It is worth while to point out that the general statement of the quantum postulate, $\int p \, dq = nh$, leads directly to the result that the momentum of the crystal changes by integral multiples of h/a as Duane assumes.

Let us express our quantum postulate in the form

$$\bar{p} \equiv \frac{\int p \, dq}{\int dq} = \frac{nh}{q_1}, \tag{1}$$

where $\bar{p}$ is the displacement average of the momentum, $q_1 = \int dq$ is the displacement necessary to bring the system back to its original condition. Applying this expression to the case of a beam of infinite plane waves of wave-length λ, it is clear that after the beam has propagated itself through a complete wave-length it is again in its original condition. Thus $q_1 = \lambda$. The momentum of the beam in the direction of propagation is therefore

$$p = nh/\lambda.$$

This corresponds, according to the relativity theory, to an energy

$$\varepsilon = pc = nhc/\lambda = nh\nu.$$

If $n = 1$, we thus have $\varepsilon = h\nu$, which is in accord with the results of photoelectric experiments. Thus the momentum of the light ray is

$$p = h/\lambda. \tag{2}$$

Let us consider in a similar manner the motion of an infinite three dimensional grating, such as a crystal. If a_x is the distance between the layers of atoms in the X-direction, the condition of the grating after it has moved a distance a_x is indistinguishable from its original condition, since layers of atoms again occupy the original positions. Thus in equation (1), $q_1 = a_x$, and hence

$$\bar{p}_x = n_x h/a_x = p_x, \tag{3}$$

[1] W. Duane, *Proc. Nat. Acad. Sci.* (May, 1923) [this Vol. preceding paper].

since $p_x = \bar{p}_x$ for uniform motion. Just as in the case of the light ray, where p was the momentum of the whole ray, so here p represents the momentum of the whole crystal. This equation states that the momentum of the crystal along the X-axis changes by integral multiples of h/a_x, i.e., that

$$\delta p_x = \frac{h}{a_x} \delta n_x, \tag{4a}$$

where δp_x is the change in the X component of the momentum, and δn_x is an integer. Similarly, if a_y and a_z are the distances between the layers of atoms along the Y and Z axes, respectively,

$$\delta p_y = \frac{h}{a_y} \delta n_y \quad \text{and} \quad \delta p_z = \frac{h}{a_z} \delta n_z. \tag{4b, c}$$

These expressions (4) ascribe a momentum to the crystal which is quantized in precisely the manner assumed by Duane. He had shown that the dimensions of the equations demanded a length in the denominator of the right-hand members, and the lengths a were the only ones which appeared suitable; but the constant of proportionality, unity, remained arbitrary. [See, however, the sentence with the * mark, this Vol. p. 371. Ed.] We now see that this quantized momentum of the crystal is a direct consequence of the fundamental quantum postulate.

Attention may well be called to the fact that the present quantum conception of diffraction is far from being in conflict with the wave theory. In fact we were able to quantize the incident radiation only in view of the fact that it repeats itself at regular space intervals. Thus even from the quantum viewpoint electromagnetic radiation is seen to consist of waves.

Ryerson Physical Laboratory, University of Chicago

184.   1. *Introductory*

In an important paper published on the pages of these *Proceedings*, W. Duane[1] makes a successful 'attempt to formulate a theory of the reflection of X-rays by crystals, based on quantum ideas without reference to interference laws.' A. H. Compton[2], enlarging upon a hint contained in Duane's paper, has recently pointed out that the latter's hypothesis

[1] W. Duane. *Proc. Nat. Acad. Sci. Washington* **9** (1923) 159 [this Vol. paper **187**].
[2] A. H. Compton. *Ibid.* **9** (1923) 359 [this Vol. paper **183**].

can be justified by the application of the general rules of the theory of quanta to the translatory motions of a crystal lattice.

3. Infinite Linear Grating. Let us begin our considerations with the single case of a one dimensional grating the elements of which are arranged in a straight line. In this line (the distance in which from a fixed point we denote by x) the material points with which the light quanta may collide are distributed with a certain density ρ. We shall call ρ for short the 'electronic density', and the notion of a grating involves that this electronic density is a periodic function of x with a period a called 'the spacing of the grating'.

If the grating is moving at a constant velocity in the direction x relatively to a resting point in this line, the density in that point will change as a periodic function of the time and return to its original value after the grating has moved through the distance a or a multiple of it. That is the reason why Compton regards the spacing a as the region over which the quantum integral $\int p\,dq = nh$ must be extended, giving the relation

$$p = hn/a \qquad (4)$$

If the distribution of electronic density was a sinusoidal one represented by the formula

$$\rho = A \sin (2\pi x/a + \delta) \qquad (5)$$

the change of density in a fixed point, due to the motion of the grating, would be a simple harmonic oscillation. By means of the Fourier theorem any distribution of electronic density can be built up of sinusoidal terms, in other words, any grating, infinite or finite can be represented as a super-position of infinite sinusoidal gratings of the type (5). This case deserves, therefore, a particularly close study. The principle of correspondence tells[1] us that to every harmonic term in the expression of ρ there corresponds a quantum change of motion accompanied by a change of momentum p given by our equation (4) if we substitute in it for n/a the coefficient of $2\pi x$ in the argument of the sine.

If, therefore, the Duane-Compton relation[1] tells us that a grating can only pick up momentum in multiples of the quantity h/a, the principle of correspondence permits us to go farther and to say that a *sinusoidal grating of the constitution (5) will experience only changes of momentum in amounts $\pm h/a$ and not in multiples of it*[2].

[1] Strictly speaking the principle of correspondence must be applied to the totality of the grating and the incident light wave, because without an exciting wave the uniform motion of a grating does not produce any radiation.

[2] We have to include the sign minus because (5) can be written also $\rho = -A \sin (- 2\pi x/a - \delta)$.

The general expression for ρ in an infinite grating is

$$\rho = \sum_{n=0}^{\infty} A_n \sin(2\pi nx/a + \delta) \qquad (6)$$

To a term of this series with the coefficient n/a of the argument $2\pi x$ there corresponds a change of momentum given by the same value of the coefficient of h in equation (4). In a lattice of such a constitution momentum can be, therefore, picked up in a large variety of ways. Moreover, the principle of correspondence gives us additional information with respect to the relative frequency of the different possible changes of momentum: the probability of the grating's picking up the momentum nh/a is proportional to the square of the coefficient of the corresponding term of our series, that is to A_n^2.

The change of momentum experienced by a grating in its collision with a light quantum determines the direction of emergence of the light quantum according to the equation

$$\alpha - \alpha_0 = \lambda n/a. \qquad (7)$$

The above statement on the probabilities of momenta means, therefore, that the intensity of the spectrum of the nth order will be proportional to A_n^2.

In order to show that these conditions are in agreement with the interference theory of gratings we have only to prove that a sinusoidal grating produces the same effect from the point of view of the latter theory, that is that a distribution of electronic density represented by the formula

$$\rho = A_m \sin 2\pi mx/a \qquad (8)$$

will give two absolutely sharp reflected beams at angles following from (7) by putting $n = \pm m$ with intensities proportional to A_m^2. The proof is easily given: An element dx of the grating gives a contribution to the amplitude of light, emitted in a direction α, which is proportional to the modulus of the expression

$$\rho e^{i(2\pi/\lambda)x(\alpha-\alpha_0)} \, dx \qquad (9)$$

The total amplitude in that direction is, therefore

$$S = C . A_m \int \sin 2\pi m \frac{x}{a} . e^{i(2\pi/\lambda)x(\alpha-\alpha_0)} \qquad (10)$$

We shall evaluate this expression first for a finite grating taking as limits of integration $\pm Na$ and then go over to $N = \infty$.

375

$$S = \frac{iCA_m}{2\pi} \left\{ \frac{\sin\left(\dfrac{m}{a} - \dfrac{\alpha - \alpha_0}{\lambda}\right)Na}{\dfrac{m}{a} - \dfrac{\alpha - \alpha_0}{\lambda}} - \frac{\sin\left(\dfrac{m}{a} + \dfrac{\alpha - \alpha_0}{\lambda}\right)Na}{\dfrac{m}{a} + \dfrac{\alpha - \alpha_0}{\lambda}} \right\}.$$

We see that the amplitude is proportional to A_m and that it has two maxima in the two directions $\alpha - \alpha_0 = \pm m\lambda/a$. Moreover, the maximum amplitude is proportional to N, and when N becomes infinite the intensity in the maximum completely dominates so that the whole energy is thrown into the directions of the maxima and the latter become absolutely sharp. There is, therefore, a complete identity in the Fraunhofer diffraction produced by an infinite sinusoidal grating from the point of view of the classical theory and from the point of view of the theory of light quanta sketched above. As any linear diffracting system can be built up by infinite sinusoidal gratings this identity will hold for the totality of all phenomena of Fraunhofer reflection. *The considerations of the first half of this section contain, therefore, the complete translation of the theory of Fraunhofer diffraction into the language of the quantum theory.*

5. The Space Lattice

The generalization for the three-dimensional case does not involve any new ideas. (...)

Norman Bridge Laboratory of Physics, Pasadena

185. Epstein and Ehrenfest[1] have shown, that, insofar as Fraunhofer diffraction is concerned, the proposed quantum theory and the wave theory reach identical conclusions, so that the Fourier series analysis may be regarded as deducible from the wave theory.

If we reverse the line of thought, and attempt to deduce the density, $\rho(x, y, z)$, of the diffracting power (or the density of the electron distribution) in a crystal from the measured intensities of the various reflected beams by adding together the corresponding terms in the Fourier series, we find that these intensities do not determine the phase angles, δ. In other words, an indefinitely large number of distributions of diffracting power will produce beams of rays of precisely the same intensities in the same di-

[1] *Proc. Nat. Acad. Sci.* (April, 1924) 133 [this Vol. preceding paper].

rections. It becomes necessary, therefore, to make further fundamental assumptions.

As one of these assumptions, we may suppose that the distribution of diffracting power conforms to the symmetry of the crystal. This symmetry fixes the values of many, sometimes of all, of the δ's as being either $\pi/2$ or zero. For example, if the crystal has three mutually perpendicular planes of symmetry and if we take the intersections of these planes as axes of coördinates, the terms in the series can contain cosines only, for they must have the same values when we reverse the algebraic sign of either x, y, or z. In this case, therefore, the δ's must be odd multiples of $\pi/2$. What the multiples of $\pi/2$ really are is immaterial, since the coefficient, A, is the square-root of a measured quantity and its algebraic sign is not determined by the diffraction data.

As a second example, suppose that the crystal has three mutually perpendicular, two-fold axes of symmetry. In this case, if we take these axes as the axes of coördinates, each term in the Fourier series may contain the product of three cosines, but, unless the coördinate planes are also planes of symmetry, the series must contain terms with trigonometric sines. These sines, however, must occur in pairs. A term cannot contain the product of one sine and two cosines, nor can it be the product of three sines; for the term must have the same value when we change the algebraic signs of any two of the three coördinates x, y, z. In this case, therefore, the symmetry of the crystal fixes the values of the δ's in each term as being either all three $\pi/2$, or one of them $\pi/2$ and the other two zero.

The symmetry conditions often determine, also, the values of certain constants A, as being equal to each other. If the crystal possesses such complete symmetry as that of sodium chloride, all the A's having the same values of n_1, n_2 and n_3, but interchanged in any manner, must be equal to each other.

The permissible values of the δ's and the coefficients, A, which must equal each other can be deduced without difficulty from the symmetry conditions, where the symmetry is less complete than in the above mentioned examples.

Neither the intensities of the diffracted beams of rays nor the symmetry conditions determine the algebraic signs of the coefficients, A. It follows, therefore, that an indefinitely large number of distributions of diffracting power not only will produce diffracted beams of the same relative intensities but will also conform to any given symmetry conditions. Sometimes such considerations as the exact position of the point chosen as origin of coordinates, the fact that $\rho(x, y, z)$ should not change its sign, the amount of diffracting material near a given point as compared with the number of electrons in an atom, etc., may enable us to decide which algebraic signs should be used for the more important terms. In general, if the crystal possesses three mutually perpendicular planes of symmetry, the intersections of these planes must be a point at which the density has either a maximum or a minimum value, and, if there is an atom at this point, it is natural to assume that the value of $\rho(x, y, z)$ is a *maximum* there. There may, however, be other points in the crystal at which $\rho(x, y, z)$ has maximum values. For instance, in sodium chloride, we may suppose that the value of $\rho(x, y, z)$ at the center of a chlorine atom is greater than that at the center of a sodium atom. Further, if in a case like sodium chloride, we take the origin of coördinates at the center of a chlorine atom and, therefore, at a point where $\rho(x, y, z)$ has its *greatest* maximum value, it is natural to suppose that the terms in the Fourier series are all positive at that point. Taking this point as origin of coördinates, this means that all the coefficients, A, are positive.

If we analyze a crystal such as sodium chloride in accordance with the theory here proposed, we do not make any assumptions to the effect that the crystal contains atoms or molecules. We simply assume that the fundamental principles of the theory represent the facts and that the distribution of the diffracting power conforms to the crystal symmetry. In addition, we assume that if we take the origin of coördinates at the center of the heaviest atom, all of the coefficients in the Fourier series have positive values. It will be shown by Dr. Havighurst in subsequent notes that the analysis based on these fundamental assumptions leads to the conclusion that the diffracting power groups itself around points corresponding to the positions of the atoms as determined by other methods of X-ray analysis.

In deducing the intensities of the beams of X-rays from the experimental observations, a number of corrections must be applied as follows: (*a*) A correction for the absorption of the radiation by the crystal. This includes not only the ordinary absorption but also the selective absorption, both primary and secondary, that occurs when the analyzing crystal lies in such a position as to reflect the incident beam of rays. (*b*) Corrections due to the methods employed in making the measurements. These depend upon whether large, single crystals are used or the well-known powder method. They also depend upon the exact way in which the intensity is estimated, i.e., whether by the photographic or the ionization method and in the latter case, whether or not the crystal or ionization chamber is turned at a constant velocity through a given angle. (*c*) Corrections due to the polarization factor. (*d*) Corrections due to the fact that the diffraction of X-rays by a crystal, although very approximately Fraunhofer diffraction, is not strictly speaking exactly so. The quantum theory, as developed above, applies only to Fraunhofer diffraction, that is to diffraction in which the rays in the incident beam are parallel to each other and, also, those in the observed deflected beam are parallel to each other. We know that the equations deduced from the classical wave theory do not in general represent the experimental facts observed in the scattering of radiation. This seems to be true in all cases in which there are transformations of energy to or from radiant energy. As a first approximation, however, we may perhaps use the classical wave theory to deduce the corrections to be applied for the very slight lack of parallelism in the beams of X-rays. The above corrections have been so thoroughly discussed from the point of view of the classical wave theory in the literature on the subject[1] that it is not necessary to enter into details here.

Strictly speaking, the theory developed here represents the distribution throughout the space occupied by the crystal of what we may call the time average of the density of the diffracting power. The passage from this to the space distribution of the time average of electron density involves an additional assumption. If we assume that the two densities are proportional to each other, the Fourier series represents the distribution in space of the time average of the electron density. It is interesting to compare the results obtained by this Fourier series analysis, as represented in the calculations made by Dr. Havighurst, with the results obtained by Compton, Debye and Scherrer, Bragg, James and Bosanquet and Bijvoet and Karssen (l.c.).

Except in such cases as those above mentioned, in which the electron distribution is calculated, the ordinary analysis of crystals by means of X-rays determines the positions of certain points, which are supposed to coincide with the mean positions of the centers of the atoms. Undoubtedly the points do coincide with the centers in simple structures. In the more complicated structures, however, in which the analysis fixes the values of certain parameters, the points may or may not coincide exactly with the centers. The values of the parameters are calculated from the relative intensities of the diffracted beams of rays, and in this calculation it is assumed, tacitly or otherwise, that the diffracting power has certain space

---

[1] Debye, *Ann. Physik* **43** (1914) 49 [Vol. I paper **44**]; Darwin, *Phil. Mag.* **27** (1914) 315 and 675 [Vol. I papers **46** and **53**]; W. H. Bragg, *Phil. Mag.* **27** (1914) 881 [Vol. I papers **24** and **43**]; *Phil. Trans.* A **215** (1915) 253 [Vol. I paper **25**]; A. H. Compton, *Physic. Rev.* **9**, Jan. (1917) 29 [Vol. I paper **51**]; Debye and Scherrer, *Phys. Zeitschr.* **19** (1918) 474 [Vol. I paper **52**]; W. L. Bragg, James and Bosanquet, *Phil. Mag.* **41** (1921) 309, *ibid.* **42** (1921) 1 [Vol. I papers **60** and **61**]; Darwin, *Phil. Mag.* **43** (1922) 800; Bijvoet and Karssen, *Rec. Trav. Chim.* **42** (1923) 859; James, *Phil. Mag.* **49** (1925) 585.

distributions about the points. Undoubtedly, the values of the parameters of the points can be determined with considerable accuracy, for a slight change in them makes a great change in the relative intensities of reflections of higher orders. The relative intensities of these reflections of higher orders, however, depend upon the distribution of diffracting power about the points and a slight change in this distribution also corresponds to a very great change in the intensities. Hence the closeness with which the points determined by the analysis lie to the actual mean positions of the atoms depends to a considerable extent upon the precision with which the assumed distribution of diffracting power agrees with its real distribution. In the method of analysis described above the positions of the points corresponding to maximum densities of diffracting power can be determined with considerable accuracy.

Department of Physics, Harvard University

186. Duane[1] has applied Epstein and Ehrenfest's[2] quantum treatment of the problem of Fraunhofer diffraction to the determination of the distribution of diffracting power in a crystal. The use of his ideas leads to the following expression for the 'electron density' at a point (xyz) in the unit cell of a crystal of NaCl:

$$\rho_{(xyz)} = \sum_{n_1} \sum_{n_2} \sum_{n_3} A_{n_1 n_2 n_3} \cos 2\pi n_1 x/a \cos 2\pi n_2 y/a \cos 2\pi n_3 z/a \qquad (1)$$

where n_1, n_2, n_3 are the Miller indices of the different crystal planes multiplied by the number representing the order of reflection; a is the length of side of the unit cell; $A_{n_1 n_2 n_3}$ is the square root of the intensity of the reflection from the plane $(n_1 n_2 n_3)$, and has the same value for all combinations (plus and minus) of these numbers. The expression 'electron density' will be used for the purpose of brevity, but it must be remembered that what is obtained from (1) is not necessarily the actual distribution of electrons, but rather the distribution of diffracting power.

In order to determine the electron density at a point in a crystal of NaCl, it is necessary to have experimental values for the A's out to fairly high values of $n_1 n_2 n_3$. The best measurements of these quantities have been made by Bragg, James and Bosanquet[3]. Their theoretical expression for the intensity of reflection of X-rays at an angle θ from a single crystal face is:

[1] Duane, *Proc. Nat. Acad. Sci.* **11** (1925) 489 [this Vol. paper **185**].
[2] Epstein and Ehrenfest, *Proc. Nat. Acad. Sci.* **10** (1924) 133 [this Vol. paper **184**].
[3] W. L. Bragg, James and Bosanquet, *Phil. Mag.* **42** (1921) 1 [this Vol. paper **61**].

$$I \propto \frac{N^2 f^2 \lambda^3 (1 + \cos^2 2\theta) \exp(-b \sin^2 \theta/\lambda^2)}{F(\mu) \sin 2\theta} \tag{2}$$

The procedure adopted in this paper is to solve (2) for values proportional to $f^2 \exp(-b \sin^2 \theta/\lambda^2)$, using the data of B., J., B. The result is a function of the space distribution of the time average electron density of the crystal. Because the work of James[1] has made the evaluation of the constant b so uncertain, it was thought best to make no attempt to correct for the Debye factor, which would result in a determination of the distribution of electron density at the absolute zero. B., J., B. split each value of f into two parts, one due to chlorine and the other to sodium, and determined the electron distributions in the separate atoms. In the theory we are using, there is no *a priori* assumption as to the existence of atoms or ions in the crystal. The average electron density will be found to have certain maxima in the unit cell which must correspond to the atoms, and the fact that the maxima have different magnitudes points to the fact of the existence in the crystal of atoms of different kinds.

The square root of $f^2 \exp(-b \sin^2 \theta/\lambda^2)$ is proportional to the A in (1). B., J., B. did not measure the intensity of reflection of every crystal plane, but they measured the intensities of reflection from 18 planes, going as far as $n_1 n_2 n_3 = 10, 0, 0$. The experimental values of the A's have been plotted against $\sin \theta$, smooth curves drawn through the points, and values of A for all the values of $n_1 n_2 n_3$ included on the curves have been read off. It was necessary to extrapolate the curve for planes which alternate with Cl and Na, as the observations did not extend over quite as great a range as those on the other class of planes. Values of $A_{n_1 n_2 n_3}$ were thus obtained for the first 46 sets of planes. These values are shown in Table 1 [not reproduced]. They are to be substituted in expression (1).

In Figure 1, the electron density has been plotted against the distance from the origin, which, except for the lowest curve, is the center of a chlorine atom. The first three curves, from the top down, represent the variation of the electron density along the cube edge, body diagonal and face diagonal. If they were all plotted to the same scale of abscissae, the corresponding peaks on the curves would be seen to be practically identical. Along the cube edge and body diagonal, there is evidently an alternation of two kinds of atoms. But along the face diagonal, all atoms are alike. The bottom curve shows the distribution along a line parallel to that represented by the curve just above it, with the origin shifted to the point $(\frac{1}{2}, 0, 0)$. Again there is a succession of similar atoms but the peaks are not so high

---

[1] James, *Phil. Mag.* **49** (1925) 585.

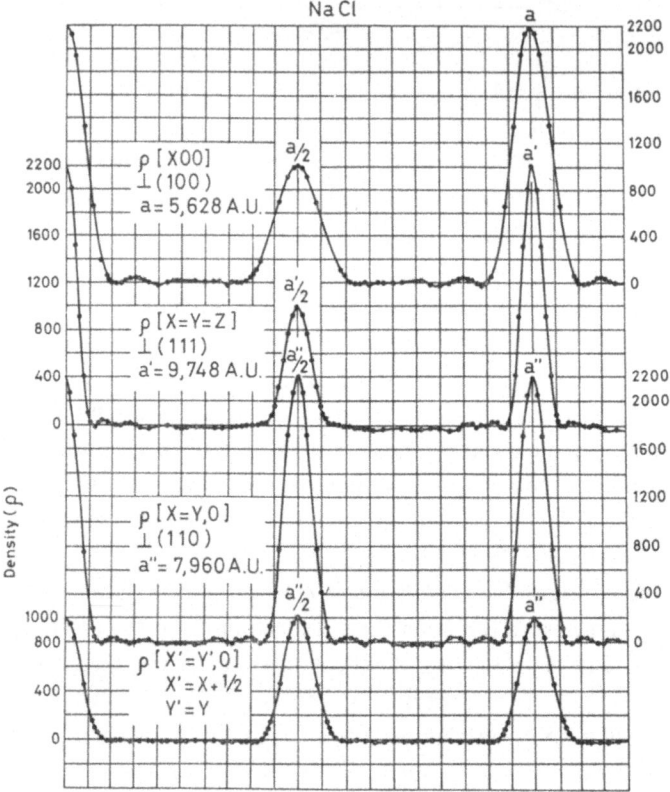

Fig. 1.

as those of the other curve. They very clearly represent sodium atoms. The three upper curves give density distributions in 26 directions from the origin.

There is a small but persistent difference in the character of the peaks representing the two kinds of atoms. It will be noticed that the density in the peak representing chlorine drops to zero and then goes up again to a small positive value, finally dropping down to a haphazard variation about zero. This might perhaps be attributed to the fact that all the terms of the Fourier's series have not been used, but it should hardly be found in the same place on all three curves if that were its cause. Furthermore, if the curves from which the A's were obtained are extrapolated, and the values of the next half-dozen coefficients in the series thus determined, it is found that the 'humps' still persist and become slightly more pronounced. The implication of this little hump is that there is an outer shell of electrons in the chlorine atom, in which some of the electrons remain for a time, but no similar shell is to be found in the sodium atom.

Jefferson Laboratory, Harvard University

187.   *9. Is a direct method of analysis possible?*

The usual method of analysis is clearly one of 'trial and error'. Analysis proceeds by ruling out configurations which are clearly incompatible with experimental results, until only one is left which explains the observations.

The interesting question arises as to whether this method of indirect approach to the solution is necessary, or whether on the other hand there exists some set of formulae in which experimental measurements can be substituted, and which then automatically provides the required solution. We can contrast two methods of analysing the crystal. The first method is in principle one of trial and error, although the range of possible co-ordinates for the atom is sometimes so narrowed by a knowledge of absolute values that these coordinates may almost be said to be found directly. The other method of Fourier analysis, has claims to be called direct because observed values of $F(hkl)$ are substituted in a series which when evaluated gives a faithful image of the distribution of scattering matter in the crystal.

As was emphasized by Duane, the Fourier analysis makes no assumption as to the existence of discrete atoms in the crystal. On the contrary it yields a continuous distribution of scattering matter in which we may recognize the atomic positions by maxima occurring here and there.

The method brings out clearly the optical analogy between the examination of the crystal by X-ray diffraction, and the analysis of the form of a small body by viewing it through a microscope when illuminated with light (Abbe's diffraction theory of microscopic vision). Waves are scattered by the unit of the crystalline cell, or by the small body, with different efficiency in different directions: the message as to the form of the body from which they originate, finally interpreted by the eye in visual observation or by computation in the case of X-ray analysis, exists at an intermediate stage impressed upon the scattered group of waves. In the first case, the waves recombine to form an image on the retina or photographic plate; in X-ray analysis this intensity can only be measured in certain directions permitted by the crystal lattice, and the image must be built up by calculation. This difference is not, however, a fundamental one. The important distinction between the two processes is that in visual observation the phases of the waves scattered in various directions play an essential part in the final image formation, whereas in X-ray analysis the information conveyed by the phase is lost, because we only measure the intensities of the scattered waves in one direction at a time. This is the fundamental difference which makes it impossible to build up by calculation an image of the crystalline cell seen under X-ray illumination analogous to the image of a small body seen under the microscope.

This difficulty appears in the Fourier analysis as an ignorance of the phases of the members in the series. The difficulty has been discussed by Duane, who shows that in very simple crystals of high symmetry it can be evaded. In NaCl, for instance, it may be safely assumed that all components of a Fourier series composed only of cosine terms have the same phase if we take as origin the centre of the heavier Cl atom. If a crystal has centres of symmetry, and one of these is taken as origin, we know that the phase angle is $0°$ or π, or alternatively that $F(hkl)$ is positive or negative. There are other special cases where the phase can be found (see for example Cork's work[1] on the alum crystals).

In general, in the analysis of a complex crystal, when this uncertainty of phase in a member of the Fourier series is taken into account, the distinction between the directness of the Fourier method and the indirect approach of the 'trial and error' method largely disappears. The values of $F(hkl)$ may be interpreted analytically in an infinite number of ways all of which explain equally well the experimental results. Some of these ways may be inadmissible as leading to negative values of the density of scattering matter. With a limited number of values of $F(hkl)$ available however, this criterion is not enough to settle the phases, or in the simpler case of a centro-symmetrical crystal, the signs of the coefficients.

We are forced to adopt as a criterion that the Fourier analysis must lead to an image of the crystal which contains masses of scattering matter approximating in extent and quantity to the atoms which we know exist in the crystal. But this is precisely what has been assumed in any method of using atomic F curves. We have calculated the scattering efficiency of each atom as a whole, and built up the calculated value of $F(hkl)$ for a finite number of atomic contributions instead of from a continuous distribution throughout the cell. In fact, a process which will fix the sign (or phase) of the members of the Fourier series by the criterion that the series leads to recognisable atoms, is identical with the process we call 'analysis of the crystal structure'. *It would appear that the crystal structure must first be analysed before the Fourier series can be used.*

To sum up, the experimental results do not uniquely determine the image we are trying to find, which is the pattern of scattering matter in

[1] Cork, *Phil. Mag.* **4** (1927) 688 [this Vol. paper **192**].

the crystal cell. An infinite number of images correspond to any given series of experimental results. In order to determine the structure uniquely, we apply the criterion that the correct image out of all those possible must resolve into masses which correspond in form, extent and number to the atoms in the unit cell. This introduces the 'trial and error' principle into the Fourier analysis. On the other hand by using atomic F curves it is ensured from the outset that the calculations will lead to a reasonable image of the structure with a correct number of atoms of the correct form. The simple cases where the phases of the members of the Fourier series can be determined by special considerations are also those in which analysis in the usual way is quite easy and direct. There would appear to be no way of avoiding trial and error methods in the general case; it is forced on us by the impotence of the X-ray measurements to determine phase relationship between the scattered waves.*

It will be realized that none of the foregoing considerations detract from the value of the Fourier method of analysis. If the structure of a crystal with centres of symmetry has been approximately determined, it is then possible to use the Fourier series to suggest final adjustments of the parameters instead of the much more laborious method of trial and error. In addition the series enables a graphical representation of the distribution of scattering matter to be made in a very striking and direct way. The two methods supplement each other.

Crystal analysis would become a matter of routine if a general formula could be found, which when the experimental measurements where substituted, yielded by automatic computation the required answer of the crystal structure. The difficulty of finding such a formula is very fundamental and has not yet been overcome. It is necessary to make assumptions about the structure and to use methods of trial and error, and this in its turn leads to ingenious devices for shortening the search and to the treatment of each crystal as a special case.

188.   2. *The analogy with image formation in the microscope*

This treatment of X-ray diffraction depends on an optical principle developed fifty years ago by Abbe in discussing the resolving power of the microscope. Abbe's principle was given mathematical form by A. B. Porter[1], and

---

* [Paper **195** shows in how short a period hereafter the phase problem, against all expectations, would fully be mastered for the case of centrosymmetrical crystals containing a heavy atom. Ed.]

[1] A. B. Porter, *Phil. Mag.* **2** (1906) 154.

W. H. Bragg based his work on this paper of Porters'. The principle is illustrated in Fig. 1.

In Fig. 1, $O_1O_2O_3$ represent lines of a grating whose rulings are perpendicular to the plane of the diagram. A train of monochromatic plane waves falls perpendicularly on this grating. The waves passing through any line, such as O_1, spread in all directions and are received by the lens of a microscope objective represented by a simple line in the figure. After passing through the lens they converge an form an image I_1, of the line O_1.

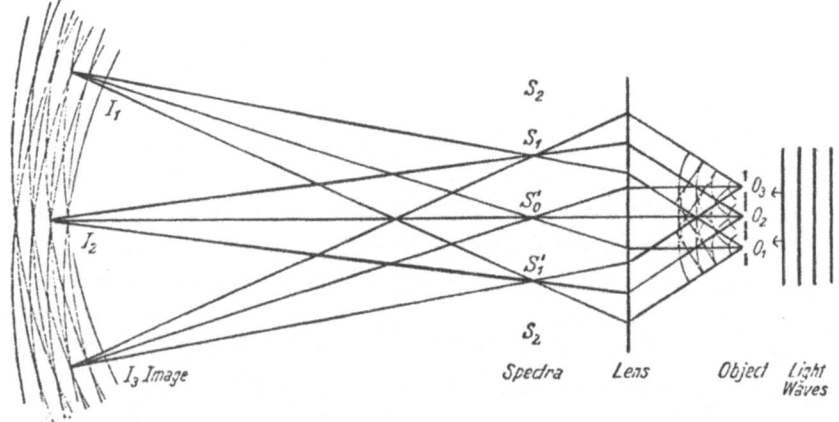

Fig. 1. Abbe's treatment of image formation in the microscope.

The formation of this image may be considered as taking place in two stages. In the first stage, the waves scattered by all the lines $O_1O_2O_3$ etc. recombine to form parallel trains, the 'spectra' due to the grating. The lens focusses these spectra at points $S_2S_1S_0S_1'S_2'$ etc. These may now be considered as sources of monochromatic wave trains, which spread out and form interference fringes in the image space as shown in the figure. These fringes compose the image of the grating.

The figure shows the interference pattern due to the spectrum of zero order S_0 and the two spectra of first order S_1 and S_1' alone. For simplicity it may be taken that the grating is symmetrical about the point O_2. The amplitudes of the waves in the image space coming from the points $S_1S_0S_1'$ will be taken to be $s_1s_0s_1'$ and in this case $s_1 = s_1'$. The amplitude of the light waves in the image space will be clear from the diagram. If a is the distance between I_2 and I_1, and x the distance from I_2 of the point considered, the amplitude then will be

$$s_0 + 2s_1 \cos(2\pi x/a).$$

If all spectra are taken into consideration, the amplitude will be

$$s = s_0 + 2s_1 \cos(2\pi x/a) + 2s_2 \cos(4\pi x/a) + \ldots$$
$$= \sum_{-\infty}^{\infty} s_k \cos(2\pi k x/a).$$

In this case the amplitude of the light in the image space will map out a perfect representation of the amplitude of the light immediately after passing through the grating. This can easily be seen by reversing the light from $S_1 S_0 S_1{}'$ etc., and making it pass back through the objective, when it builds up the amplitude everywhere of the light which passes through the grating. The whole diagram is in fact symmetrical about the plane of the spectra, the only distinction between object and image being one of scale. Abbe's treatment of resolving power is based upon the consideration of the extent to which the image will reproduce the object, when a limited number only of the spectra are transmitted by the objective. The greater the number of spectra, the more clearly can we see in the microscope the precise form of the lines of the grating.

Suppose the series of lines $O_1 O_2 O_3$ to be given a small movement in a downward direction; the spectra $S_1 S_0 S_1{}'$ remain fixed in position. There must be, however, a movement upwards of the images I_1, I_2, I_3 and the way in which this is effected is shown by the diagram. Although S_1 remains in the same position, the waves coming from it will be retarded in phase owing to the movement of the grating, and those coming from $S_1{}'$ will be advanced. Hence the maxima of the interference fringes all move upwards. The form of the image depends both on the *phase and amplitude of the waves* coming from the spectra.

Let the simple grating now be replaced by a two dimensional grating (such as the pattern of markings on the surface of a diatom). The spectra now form a doubly infinite series of points, in rows and columns. Each pair of points sends out wave trains which interfere in the image space. The interference fringes are at right angles to the line joining the two spectra, and their spacing is inversely proportional to the distance between the spectra. Each pair of spectra must now be labelled by two indices h, k and the total amplitude of vibration in the image space is given by the formula

$$s = \sum_{-\infty}^{\infty} \sum_{-\infty}^{\infty} 2s(h, k) . \cos 2\pi(hx/a + ky/b).$$

It has again been assumed for the sake of simplicity that there is a centre of symmetry of the pattern on the axis of the instrument. $s(h, k)$ is the amplitude of the wave in the image space due to the spectrum in the column h and row k, and $s(h, k) = s(\bar{h}, \bar{k})$.

As A. B. Porter expresses it (*Phil. Mag.* **2** (1906) p. 156), a double process of Fourier analysis takes place when light falls on a grating. A simple harmonic grating, if such could be realized, would only give spectra of the first order. If a complex light train falls on it, the light is resolved into a series of harmonic vibrations spread out into a spectrum, so that the grating gives a Fourier analysis of the light. On the other hand, if a plane monochromatic wave train falls on a complex grating a series of spectra are formed, and the amplitude of the light waves in these spectra are proportional to the coefficients of a Fourier series which, when summed, reproduces the form of the grating. In this case the light performs a Fourier analysis of the grating.

In a precisely similar way we use a single periodicity in a crystal to analyse a complex beam of X-rays (analysis of X-ray spectra) and a single X-ray wave length to analyse a complex periodicity in the crystal (crystal analysis).

The above discussion merely summarizes the optical principle due to Abbe, and the application of Fourier series to X-ray analysis by W. H. Bragg, Ewald, Duane and Compton, and it introduces no new points. In the present paper, I wish to draw a somewhat closer analogy between the examination of a two-dimensional pattern under the microscope, and the examination of all faces round a zone in X-ray analysis of crystals. In Fig. 1, the light falling on the object is first converted into a series of inter-ference maxima, the spectra $S_1 S_0 S_1'$ etc. These act as new light sources and send out waves which interfere in the image space and produce the image. If there is a large number of spectra, the image will be a faithful representation of the object. All data necessary to build up the image must be available if we know the amplitudes and phases of the waves coming from the spectra, for these are their only characteristics. We may therefore consider the complex system of spectra in the S space as a sort of code message, containing complete information about the form of the object: it is automatically decoded into an image by optical interference in the microscope.

We have a very similar case in X-ray analysis, for we have a table of spectra obtained by experimental measurement. The main difference, as has been often stressed, is that we can only measure intensities and so all information about phase is lost. Further, since we are dealing with a three-dimensional grating, there is no way of forming experimentally the complete series of spectra simultaneously, as can be done with light and a two-dimensional grating. It can be done formally, and in fact this is represented by Ewald's 'Reciprocal Lattice'. In the reciprocal lattice there is a three-dimensional array of points. Each is defined by three indices h, k, l and corresponds to the diffracted beam characterized by the same indices. The line joining the point to the origin of the reciprocal lattice is perpendicular to the plane (hkl) of the crystal, and its length is inversely proportional to the spacing of planes parallel to (hkl). There is thus a precise analogy between the points of Ewald's reciprocal lattice and the series of spectra S in Fig. 1. Each spectrum $S(h, k)$ is at right angles to the line (hk) in the two-dimensional grating, and its distance from the origin is inversely proportional to the spacing of the lines parallel to (hk). If there were some experimental way of forming simultaneously all possible crystal spectra by using a fourth dimension, they would be arranged on the reciprocal lattice. Their positions, amplitudes, and phases would again contain all information necessary to reform a perfect image of the crystal structure. We may associate with each point of the reciprocal lattice a corresponding amplitude $F(hkl)$. At the origin the amplitude $F(000)$ is equal to Z, the total number of electrons in the unit cell. This is the spectrum of 'zero order', the amount of radiation scattered in the direction of the incident rays.

Now the case of X-ray diffraction can be reduced to one of two dimensions by considering only the diffraction of planes around one given zone. We may, for convenience, call the zone the a axis of the crystal, and choose suitable axes b and c to outline a unit crystal cell. The amplitudes and phases of all spectra $F(0kl)$ are supposed to be known. If the crystal is centro-symmetrical $F(0kl)$ will be real, but may be positive or negative. The scattering matter in the crystal is now supposed to be projected on the bc plane, when it forms a two-dimensional grating with periodicities b and c along these axes. The density of scattering matter per unit area of the projection, when a single layer of unit cells is projected on the face (100), is given by the formula

$$\rho(y, z) = \frac{1}{A} \left\{ \sum_{-\infty}^{\infty} \sum_{-\infty}^{\infty} F(0kl) \cos(2\pi ky/b + 2\pi lz/c) \right\}^1$$

corresponding to the formula for the amplitude in the image in the microscope

$$s = \sum_{-\infty}^{\infty} \sum_{-\infty}^{\infty} s(kl) \cos(2\pi ky/b + 2\pi lz/c).$$

3. The optical representation of the projected crystal cell

Each equal pair of spectra $F(0kl)$ and $F(0\bar{k}\bar{l})$ produces a periodic variation of density in the image. Its amplitude is equal to $2F(0kl)$ and its crests are on the successive $(0kl)$ planes. It is therefore possible to build up an image of the crystal structure if some means can be devised of adding together

---

[1] A is the area of the face (100) of the unit cell.

389

periodic variations crossing the unit cell in all directions, whose positions and amplitudes correspond to the interference fringes produced by the spectra. The experimental arrangement employed to effect this was as follows. A photographic plate was prepared with a series of light and dark bands on it, the amount of light passing through the plate having as closely as possible a harmonic alternation between the transparent and opaque bands. Actually this was achieved by photographing a row of opaque rods, placed at a distance apart of twice their diameters, illuminated from behind and thrown out of focus. By trial we succeeded in getting a photograph which showed the correct alternation of light and shade (Fig. 2).

This plate was placed in a projection camera and illuminated from behind, and the shadows of the bands thrown on a screen. The distance of the camera from the screen could be varied and the shadows thrown in any position. A large sheet of photographic paper was pinned on the screen. A hinged flap shielded it from the light when necessary. On this flap an outline of the face bc of the crystal cell was drawn to scale.

The bands due to each pair of spectra (k, l) $(\bar{k}, l)$ were now thrown in turn on the same sheet of paper. The camera was adjusted until these bands crossed the unit cell drawn on the flap with correct orientation and

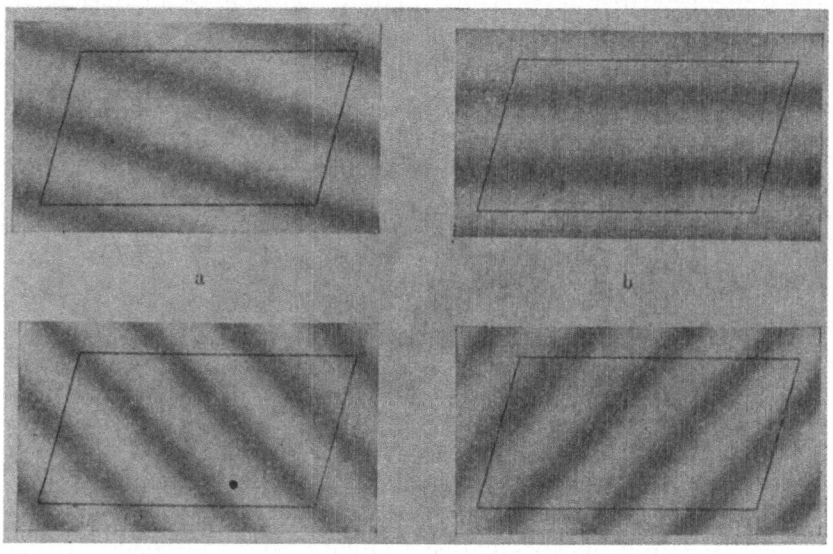

Fig. 2. Sinusoidal alternations of light and shade. The bands in the figure represent the contributions to the image due to the following spectra
a) $F(102)$, phase negative. b) $F(002)$, phase positive.
c) $F(302)$, phase negative. d) $F(30\bar{1})$, phase positive.

position (see Fig. 2). The shutter of the camera was closed, the flap lifted, and then the shutter opened for a time proportional to $F(0kl)$. The flap was then replaced, and the process repeated for the next pair of spectra. In the examples given here, about forty such sets of bands were thrown on each sheet of photographic paper. The paper was then developed, and the result is a projection of the unit cell with all its atoms in their correct positions.

Manchester University

189.   1. *Introduction*

In the present paper we wish to discuss the extent to which the representation is a faithful image of the actual crystal structure. It will be seen that the term 'image' is appropriate, for it suffers from defects precisely analogous to those caused by diffraction in an optical image. It is interesting to see how close the analogy is and how constants of X-ray analysis can be defined which have exactly the same significance as the 'resolving power' and 'numerical aperture' of a microscope.

2. *The Fourier Representation*

If the series is calculated for points distributed throughout the unit cell the atoms appear as concentrations of greater density. Though they sometimes overlap each other in the projection, certain of them often stand out clear of their neighbours. In such cases it is possible to sum up the total amount of scattering matter, and so to obtain a direct count of the number of electrons in the atom.

There are many problems of crystal structure which it would be interesting to solve in this way. Havighurst and Compton have used Fourier series with a single variable to test such questions as the ionization of sodium and chlorine in sodium chloride. The Fourier series with two variables is on the face of it a more powerful method, because it separates the atoms more from each other. Further, the series is just as easy to manipulate for large unit cells containing many atoms as for small cells of a simple type such as that of sodium chloride. A cell of large dimensions yields measurements of scattering at small glancing angles, and these are essential in finding out the positions of the outer electrons in the atoms. An instance of a problem on which light may be cast by the Fourier representation is the distribution of electrons between silicon and oxygen in silicates.

The difficulty of such investigations arises partly from errors in the experimental measurement of diffraction (values of F), and partly from the termination of the Fourier series before its coefficients have become very small. The series is terminated because it is impossible to measure more than a certain number of reflexions with a given X-ray wave-length, and its effect is precisely like that which limits the resolving power of an optical instrument. In the present paper we wish to examine the question of resolving power alone, and will take an ideal case where we suppose no experimental errors to have been made.

3. Diffraction Effects in the Image

It has been shown by James, Waller and Hartree[1] that the observed X-ray diffraction by sodium chloride corresponds very closely with that to be expected from the Hartree model of atomic structure. We have therefore tested the Fourier representation by taking an ideal crystal of sodium chloride composed of such Hartree atom-models, calculating its diffraction as if observations were being made experimentally for a limited range of crystal planes, expressing these results in the form of a double Fourier series, and then seeing with what faithfulness the projection represents the original model[2].

The most convenient projection which separates the sodium and chlorine atoms is that upon the plane (110); the unit cell then has the dimensions a by $a/\sqrt{2}$, and contains two atoms of each kind. It is only necessary to represent one quarter of such a cell in order to see the pattern.

Planes $(0kl)$ in our adopted unit cell correspond to planes (kkl) in the *true* unit cell of rock-salt.

We have supposed the measurements to be made up to a glancing angle θ_0 equal to $\pi/6$ ($\sin\theta_0 = 0.5$), and for a wave-length of 0.615 Å. (RhK$_\alpha$ radiation). The series has been calculated for values of F at room-temperature (290° absolute) and for an ideal crystal of atom-models unaffected by any

---

[1] R. W. James, I. Waller, and D. R. Hartree, *Proc. Roy. Soc.* A **118** (1928) 334.

[2] Havighurst (*Phys. Rev.* **29** (1927) 1) has made a precisely similar examination with a single Fourier series and a model of the sodium atom, and comes to the conclusion that 'the value of the retention of the temperature factor in the experimental F-curves as a means of securing a rapidly convergent series and therefore of obtaining a more trust-worthy Fourier analysis is clearly demonstrated'. As will be seen, we arrive at the same conclusion about a temperature factor for the double Fourier series. It is interesting to examine this series in detail, however, because it is so widely applicable to complex crystals, where the single Fourier series is helpless, and, further, because the optical analogies are so striking.

movement or by the os-called zero-point energy. The F values adopted for each atom in the two cases are taken or deduced from those given by James, Waller, and Hartree. In the case of the crystal at room-temperature the calculated electron density throughout the unit cell is shown in Fig. 1. The numbers represent the summation of the Fourier series

$$\sum_{-\infty}^{\infty} \sum_{-\infty}^{\infty} F(0kl) \cos 2\pi(ky/b + lz/c),$$

and, after being doubled (see note above), must be divided by the area of the cell (22.40 Å) in order to get the actual density. It is convenient to use the Ångström as a unit of length in such a projection. Although the coefficients of the series are still appreciable when the series terminates, they are small compared with the initial coefficients. The corresponding

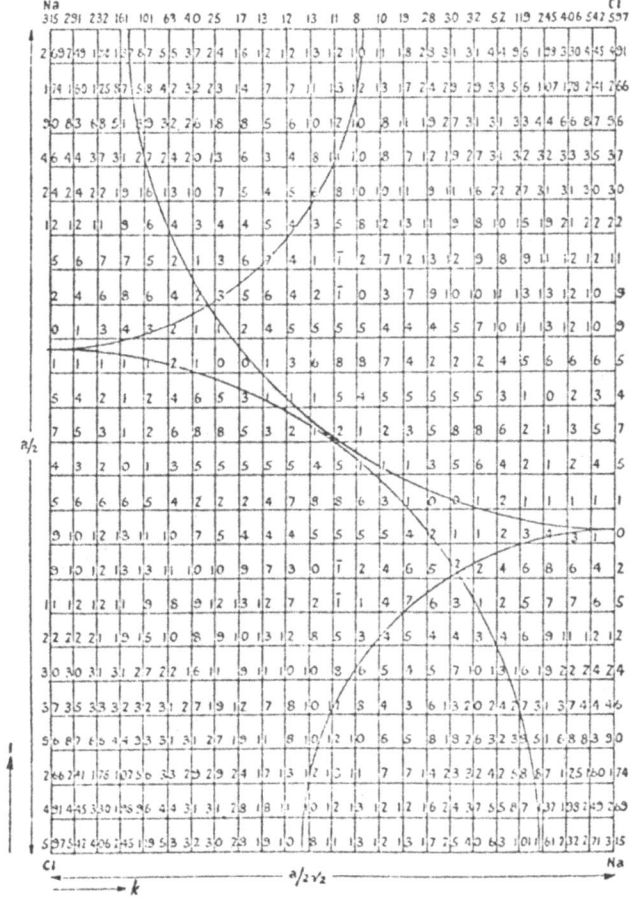

Fig. 1. Projection on (100) (i.e., (110) true cell of side a = 5.628 Å).
Crystal at room-temperature.

projection for the ideal crystal composed of atoms unaffected by movement is shown in Fig. 2.

4. *Discussion of Fig.* 1

In regions widely removed from the centres of the atoms the density is very small, and it nowhere has an appreciable negative value. Although sodium and chlorine overlap in the direction of the short axis (perpendicular to the (110) plane in the normal unit cell) they are fairly well separated elsewhere. The image reproduces correctly the spherical symmetry of the atom-models, and an attempt can be made to separate the one atom from the other, apportioning the density between the two atoms in the part

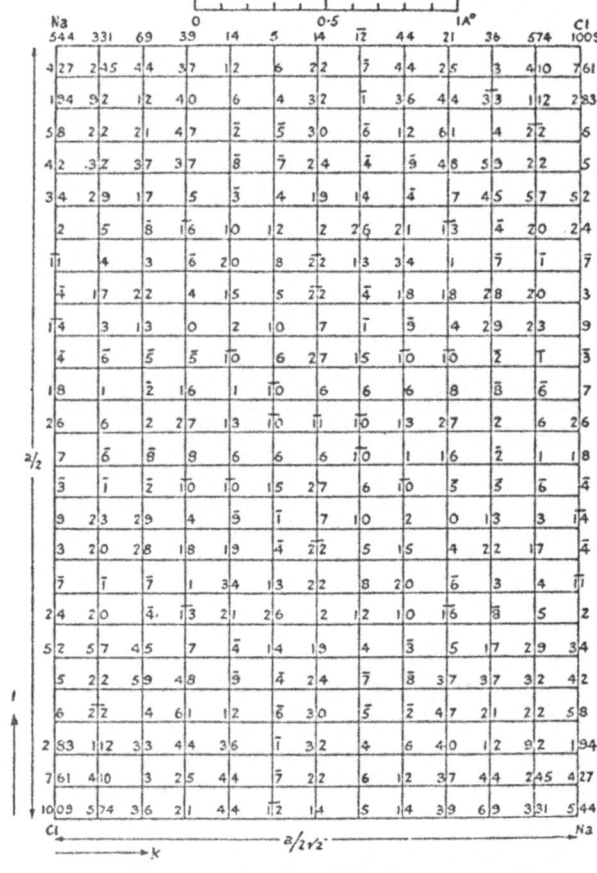

Fig. 2. Projection on (100). Crystal with atoms at rest.

where they overlap, so that the components are in accord with the values found elsewhere at corresponding distances. The result of an electron count gave 17.5 for chlorine and 10.5 for sodium, and is close to the values 18 and 10 for the ionized atoms.

The count does not, however, decide definitely the question as to whether the atoms are ionized or not, although the figures on which the projection is based were appropriate to ionized atoms. At first sight this may seem to be due to an error in drawing the boundary between the two atoms. Such an error could arise from the presence of irregularities in the projection —discussed below as 'diffraction effects'—caused by terminating the Fourier series whilst the coefficients are still appreciable. These irregularities as we shall see later, can seriously affect an electron count, and although in Fig. 1, by making use of a temperature factor, they are by no means so marked as in Fig. 2, they are still present.

A consideration of Table IV will, however, show that the defect is more fundamental, and that the count would still be incorrect even if the irregularities just mentioned were removed by the use of a higher temperature

Table IV.

Cl^-		Na^+	
Radius of circle	Number of electrons within circle	Radius of circle	Number of electrons within circle
3.19 Å	17.96	1.06 Å	9.96
2.40	17.8	0.575	9
1.55	17	0.446	8
1.14	16	0.367	7
0.936	15	0.298	6
0.772	14	0.239	5
0.692	13	0.186	4
0.522	12	0.128	3
0.440	11	0.074	2
0.314	10	0.027	1
0.256	9	0	0
0.213	8		
0.181	7		
0.149	6		
0.117	5		
0.096	4		
0.064	3		
0.043	2		
0.027	1		
0	0		

coefficient. The boundaries chosen for the two atoms are indicated by the arcs in Fig. 1, which have radii 1.73 Å for Cl⁻ and 1.08 Å for Na⁺. On the other hand the Hartree model atoms have been projected on a diametral plane and the figures in Table IV give the radii of circles within which successive integral numbers of electrons are contained in the projection. Table IV shows that whereas the radius 1.08 Å for Na⁺ includes in effect all the electrons of Na⁺, the radius 1.73 Å for Cl⁻ only includes about 17.4 electrons of Cl⁻. The remainder falls within the boundary of the sodium atom, and being automatically included in the count for Na⁺ in the projection, is wrongly ascribed to it. This seems to suggest that a correct electron count is not possible. However, we have to remember that the original F curves used were calculated for free ions, and since presumably F values measured experimentally will correspond to ions whose electrons are compressed into a smaller region by the mutual approach of the atoms, it would seem possible that a correct count might be made in an actual experiment provided the difficulties of obtaining accurate F values are successfully overcome. At the same time it is clear that considerable caution should be exercised in interpreting the results and that rigorous conditions must be satisfied before they can be relied upon.

5. *Discussion of Fig. 2*

In this case the numbers show oscillations between positive and negative values all over the unit cell. The centres of atoms are marked by higher peaks than in Fig. 1 but they are surrounded by concentric rings of positive and negative density which, by overlapping, produce the complicated aspect of the projection. *These irregularities have no counterpart in the atomic structure.* They are due to the termination of the Fourier series whilst the coefficients are still large, and are precisely analogous to the diffraction effects produced by an optical instrument. The atomic model has a large concentration of scattering matter at its centre, and an effect is produced like the diffraction rings around the image of a luminous point. In the present case it is *amplitude* and not *intensity* which is plotted, so that the diffraction rings give both positive and negative contributions and are more important as compared with the central maximum. A negative value for the density has no physical significance in this case, and its existence shows that the picture is artificial.

Electron counting is much more difficult here than in Fig. 1. If, for instance, we count merely the number of electrons in the central region of positive density for chlorine we obtain 11.3, whereas we should have 18.

Clearly the scattering matter represented by the outer rings must be included in the count, but it is difficult to know where to draw the boundary between the atoms, since the oscillations overlap in all directions. The precise point at which counting stops influences considerably the total number obtained, whereas in Fig. 1 it matters little where the boundary is placed, since the region between the atoms has a vanishing density.

6. *The Optical Analogy*

The origin of these circular bands of positive and negative density is clear if we treat the projection as analogous to an image formed by an optical instrument with a circular stop.

Consider the image I of a luminous point P (Fig. 6). The rays from P which traverse the instrument are limited by a circular stop at some point in the system which is equivalent in its effect to the 'entrance pupil' B. After traversing the system a wave coming originally from P converges to build up the image at I. The angular aperture is measured by $2u_0$ and the angle of projection by $2u_0'$.

The customary treatment of diffraction divides a wave front such as S into elements, the secondary waves from which combine to form the image

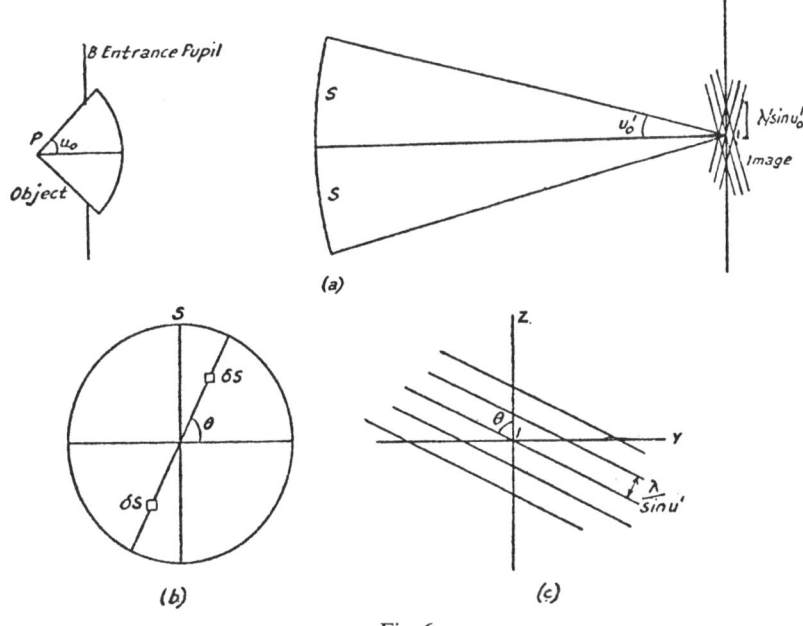

Fig. 6.

at I. A pair of elements of area δS symmetrically placed at opposite sides of the wave centre contributes an amplitude proportional to

$$\delta S \cos(2\pi t/T) \cos 2\pi(\cos \theta \sin u' . y/\lambda + \sin \theta \sin u' . z/\lambda),$$

where $2u'$ is the angle the elements subtend at I. The elements and the wave system they produce are shown in Fig. 6 (*b* and *c*). The resultant amplitude of vibration for a given point *yz* in the image plane is got by integrating this expression over the whole wave surface. The pattern has circular symmetry, the amplitude of vibration at a distance *r* from the centre of the image being given by*

$$\text{constant} . J_1(2m)/m$$

$$= \text{constant}\left(1 - \frac{1}{2}\left(\frac{m}{1}\right)^2 + \frac{1}{3}\left(\frac{m^2}{2}\right)^2 - \frac{1}{4}\left(\frac{m^3}{3.2}\right)^2 + \cdots\right),$$

where $m/\pi = r(\sin u_0')/\lambda$.

The square of this expression yields the familiar diffraction pattern for a circular aperture, but we are here concerned with amplitude and not intensity.

The curve for the amplitude $J_1(2m)/m$ is plotted in Fig. 7. Its maxima, minima, and points where it crosses the axis occur as follows:

	Magnitude	m/π
Central maximum	1.0	0
Zero value	−	0.61
Minimum (first diffraction ring)	0.132	0.81
Zero value	−	1.116
Second maximum (second diffraction ring)	0.064	1.333
Zero value	−	1.619

In the same figure the amplitude for a slit whose width is equal to the diameter of the stop is shown for comparison, the expression in this case being

$$(\sin 2m)/2m.$$

Each pair of equal and opposite elements in the wave converging to form the image builds up in the image plane a contribution to the amplitude like that of a set of standing waves. Elements close to the centre give standing waves of long wave-length, whilst those farther apart give waves of short wave-length because the wavelets from those elements cross at a greater angle. The shortest wave-lengths which go to compose the image have a length $\lambda/\sin u_0'$, as shown in the figure. This corresponds to a distance in

* [see, e.g., W. C. Elmore and M. A. Heald, *Physics of Waves* Chapters 10.5 and 10.4 (Mc Graw-Hill Book Comp. (1969). Ed.]

the object of

$$\lambda/n \sin u_0 = \lambda/a,$$

where a is the numerical aperture of the instrument, n being the refractive index of the medium in which the object is placed.

To sum up, the diffraction image at I is built up by integrating these standing waves, which cross each other in all directions and have wave-lengths varying from infinitely great to a minimum value corresponding to a distance λ/a in the object. These waves are all in phase at the centre of the image where the effect is greatest.

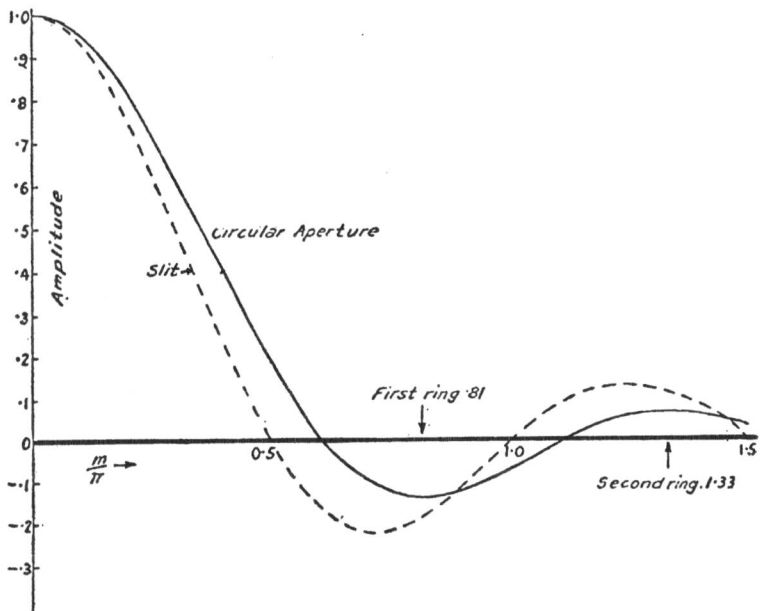

Fig. 7. Optical diffraction rings for circular aperture and slit.

Suppose now that the projection of a crystal structure is being found which for simplicity we take to consist of a simple network of single electrons. The image of such a structure is given by the Fourier series

$$\rho(y, z) = (1/A) \sum_{-\infty}^{\infty} \sum_{-\infty}^{\infty} F(0kl) \cos 2\pi(ky/b + lz/c),$$

in which all values of F are unity and A is the area of the cell. All spectra are measured for planes whose spacings are not less than a given amount d_0, thus imposing limitations on possible values of k and l. The smaller d_0, and the larger the dimensions of the unit cell, the greater will be the number of spectra. It is easy by considering the reciprocal net to show that the number of spectra approximates to

$$\pi A / d_0{}^2.$$

The image of each electron is formed by sets of cosine waves crossing in all directions, as in the optical case, only now their number is finite, and we are dealing with a Fourier series instead of the integral over the wave surface in the optical case. Nevertheless, when the number of spectra is very large, the form of the image will be like that due to optical diffraction. The electron density at each point in the image at a distance r from corner of the net is given by

$$\rho(r) = \frac{1}{A} \cdot \frac{\pi A}{d_0{}^2} \sum_{-\infty}^{\infty} \sum_{-\infty}^{\infty} \cos 2\pi(ky/b + lz/c)$$

$$\approx \frac{\pi}{d_0{}^2} \cdot J_1(2m)/m,$$

where $m/\pi = r/d_0$.

The size and shape of the unit cell of the network do not affect this last expression. When the number of spectra is very large, the image of each net point approximates to what it would be if it were the only point in the field of view; expressed otherwise, the diffraction patterns are only slightly affected by overlapping. It is as if we were looking through the microscope at a regular two-dimensional array of luminous points producing cross spectra in the region S (Fig. 6). Waves from these spectra recombine in the image plane to reproduce an array of images, each of which is surrounded by diffraction rings, much as it would be if it were the only point in the field.

Now, the shortest wave which goes to build up the image of our network corresponds to the shortest distance d_0 between reflecting planes. If all spectra are measured up to a glancing angle θ_0, then

$$d_0 = \lambda/2 \sin \theta_0.$$

In the optical image the corresponding distance is λ/a. Thus, *the numerical aperture in X-ray analysis must be defined as $2 \sin \theta_0$*. In the microscope points cannot be resolved unless their distance apart is greater than $0.6\lambda/a$. Similarly in X-ray analysis detail cannot be distinguished unless it is on a coarser scale than $0.6\,d_0$ or $0.6\,\lambda/2 \sin \theta_0$.

7. Diffraction Effects in the Fourier Projections

In the Hartree atom-models there is a strong concentration of scattering matter at the centre of the atom. This gives rise to diffraction rings on the projection like those due to a bright point at the atom centre. The first diffraction ring has its greatest negative value at $0.81\, d_0$, the next its greatest positive value at $1.33\, d_0$, and so forth. In the present case the maximum value of θ is $\pi/6$, so that the 'angular aperture', $2\sin\theta_0$, is unity, and d_0 is equal to λ, i.e., $0.615\,\text{Å}$.

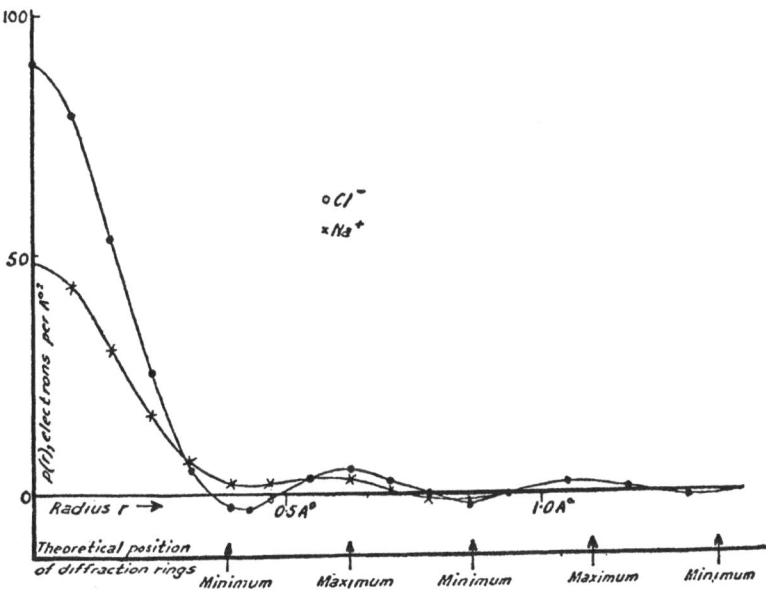

Fig. 8. Diffraction rings of Cl⁻ and Na⁺ compared with optical rings.

In Fig. 8, $\rho(r)$ is plotted against r for sodium and chlorine, using the data of Fig. 2 and disentangling as far as possible the overlapping system of rings. The corresponding positions of the maxima and minima due to the diffraction rings of a luminous point are marked in the diagram. It is obvious that the rings of positive and negative density in the projection correspond precisely to optical diffraction rings. They have no counterpart in the atomic structure, and are merely due to the termination of the Fourier series whilst the coefficients are still large. The image of either atom consists of a faithful representation of the more diffused outer parts, together with a system of diffraction rings, due to the large concentration at the centre, superimposed upon it.

8. Conclusions

The use of a Fourier series which is incomplete, in that the coefficients are still large when it is terminated, leads to an image of the crystal structure which has defects like those produced by optical diffraction. Such defects are strongly marked in Fig. 2, whereas in Fig. 1 they are much less apparent. The coefficients fall to a much smaller value in this latter case, because a 'temperature factor' $e^{-B \sin^2 \theta}$ has been added to allow for thermal agitation. Although this agitation will increase the overlapping of the atoms to a certain extent, the effect is more than compensated for by the disappearance of the diffraction effects, and the line of demarcation between the atoms is actually much more evident. An electron count in Fig. 1 can be made with considerable accuracy, whereas in Fig. 2 it is impossible.

This suggests that it would be desirable in all cases to multiply the coefficients by an arbitrary factor $e^{-B \sin^2 \theta}$, choosing a value of B which makes the final coefficients very small. If, for instance, the maximum value of $\sin \theta_0$ is 0.5, a value of 2.3 for B gives the following factors:

$\sin \theta$	$e^{-B \sin^2 \theta}$
0.1	0.912
0.2	0.692
0.3	0.437
0.4	0.229
0.5	0.100

It is somewhat surprising to see to what a small value of the average amplitude of thermal agitation this value of B corresponds. The factor in terms of the average mean square of the amplitude, $\overline{u_x^2}$, is

$$e^{-(8\pi^2 \sin^2 \theta / \lambda^2) \cdot \overline{u_x^2}}.$$

For $\lambda = 0.615$ Å, and $\dfrac{8\pi^2 \overline{u_x^2}}{\lambda^2} = 2.3$ as above, we have

$$\overline{u_x^2} = (0.21 \text{ Å})^2.$$

Such an amplitude hardly affects the extent to which the atoms overlap. The Fourier series will give a very faithful picture of the actual crystal structure with this assumed thermal agitation. Electron counting should therefore be more reliable, and one can be sure that no false detail due to diffraction will appear.

Compton ('X-rays and Electrons', p. 164) develops a formula for the amount of scattering matter $U_r dr$ between radii r and $r + dr$ in the atom

$$U_r dr = \frac{8\pi r}{D^2} \sum_1^\infty n F_n \sin\left(\frac{2\pi n r}{D}\right) dr,$$

where each F value is measured or deduced for an order of reflexion n from a simple crystal of spacing D composed of the one kind of atom. This series converges very slowly. Using experimental determinations of F, Compton obtained a curve for chlorine, for instance, with a series of 'humps'. By comparing the curve with the Hartree model (with which we know the experimental measurements to be in accord) it is clear that the humps are similar to the diffraction effects treated above, and do not correspond to the actual distribution in the chlorine atom. Havighurst[1] used the same formula and pointed out the importance of a temperature coefficient in making the series converge. Even in Havighurst's results for chlorine, however, humps appear which have no counterpart in the Hartree model, and, indeed, Havighurst expresses doubts of their reality. It seems clear that we can only trust details of structure indicated by the Fourier analysis when the actual or applied temperature coefficient causes the series to converge so fast that the last terms are vanishingly small. Unless this is the case, many real-looking features appearing in the representation must be distrusted.

By making the final coefficients of the Fourier series vanishingly small in this artificial way we are sacrificing a part of the information about the crystal structure which the experimental determinations have provided. A comparison of Figs. 1 and 2 shows what is being lost. The higher terms define more sharply the positions of the atomic centres by making the peaks in the density higher and more concentrated. They give, in fact, more exact information about the positions of the atomic centres in more complicated crystals where these are not fixed by symmetry considerations. The electron count, on the other hand, is mainly determined by the earlier terms in the series. The higher terms 'trim' the crude picture of the structure given by the earlier terms, and in order to make this trimming effective and to cut out false detail it is quite justifiable to multiply the coefficients by an arbitary temperature factor. The introduction in an optical instrument of a screen with a graded absorption increasing towards the edge would cut out false detail due to diffraction in just the same way.

[1] R. J. Havighurst, *Proc. Nat. Acad. Sci.* **11** (1925) 502 [this Vol. paper **186**]

B. FOURIER SYNTHESES

B1. Signs from Previous Trial Structure

190. The Determination of Parameters in Crystal Structures by means
of Fourier Series, by W. L. Bragg (1929)

▬

191. The Crystalline Structure of Anthracene. A Quantitative X-ray
Investigation, by J. M. Robertson (1933)

▬

190. *Summary.* The representation of the scattering matter in a crystal by
Fourier series, first used by W. H. Bragg and later developed by Duane,
Havighurst and Compton, is applied here to the determination of the
parameters in a complex crystal.

A series is used which gives the projection of the scattering matter in
the unit cell on each of its faces in turn. For instance, when projection is
made on the face (100) of the cell, the formula for the density $\rho(y, z)$ of the
scattering matter at a point y, z is as follows:

$$\rho(y, z) = (1/bc \sin \alpha) \sum_{-\infty}^{+\infty} \sum_{-\infty}^{+\infty} F(0kl) \cos 2\pi(ky/b + lz/c).$$

In this formula, $F(0kl)$ is the value of the structure factor for the reflexion
$(0kl)$ measured in absolute units. The formula applies to a cell of any shape,
provided that it has a centre of symmetry. $F(000)$ is taken to be the number
of electrons in the unit cell.

This series is evaluated for the crystal diopside $CaMg(SiO_3)_2$. The signs
of the coefficients $F(0kl)$ had been fixed by a previous analysis of the
crystal[1]. The projections are shown as contoured diagrams in Figs. 1A,
2A and 3A. The positions of the atoms agree very closely with those given
by the previous analysis, Figs. 1B, 2B and 3B, made by finding values
for the parameters which gave agreement between calculated and observed
values of F. A comparison of the two sets of 14 parameters is shown in
Table IX.

It is possible to count the numbers of atomic electrons in the projections.
They are approximately as follows: Ca 16.5, Mg 12.5, Si 11.5, O 8.5–9.
It is interesting to note that the oxygen does not appear to be an ion O^{-2}
with 10 electrons.

---

[1] B. Warren and W. L. Bragg, *Z. Krist.* **69** (1928) 168 [this Vol. paper **174**].

The groups of F values used for any projection may be conveniently described as the weights attached to a network of points on a central section of Ewald's reciprocal lattice.

Table IX.

	Parameters determined by Fourier analysis			Values given in previous paper on diopside		
	u	v	w	u	v	w
Ca	0	−0.299	0.25	0	−0.306	0.25
Mg	0	0.089	0.25	0	0.083	0.25
Si	0.211	0.407	0.236	0.211	0.411	0.236
O_1	0.375	0.419	(0.14)	0.378	0.402	0.139
O_2	0.142	0.253	(0.32)	0.142	0.250	0.320
O_3	0.145	(0.50)	0.00	0.155	0.480	0.00

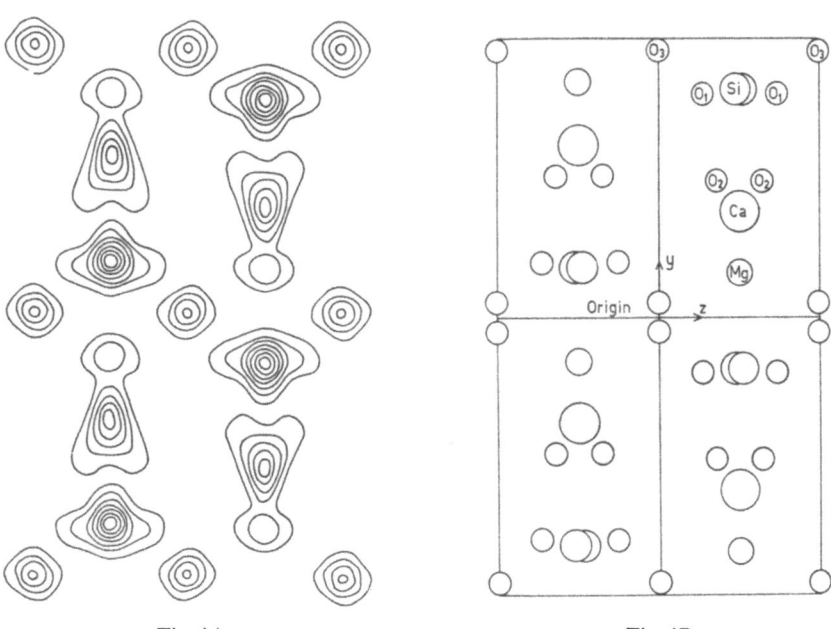

Fig. 1A.   Fig. 1B.

Fig. 1A. Summation of Fourier series for projection of diopside on face (100). The distribution of scattering matter is indicated by contour lines drawn through points of equal density in the projection.

Fig. 1B. Atomic positions.

405

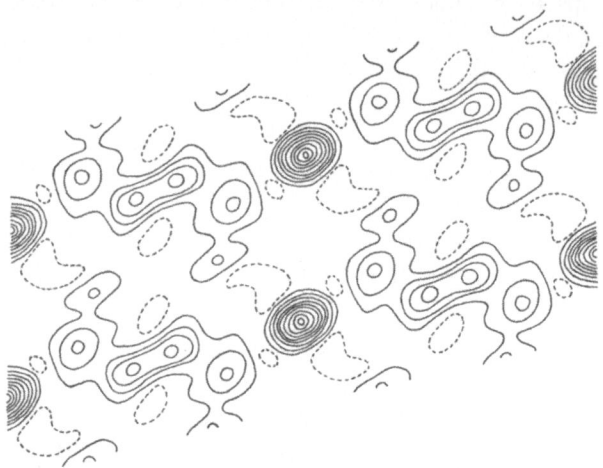

Fig. 2A.

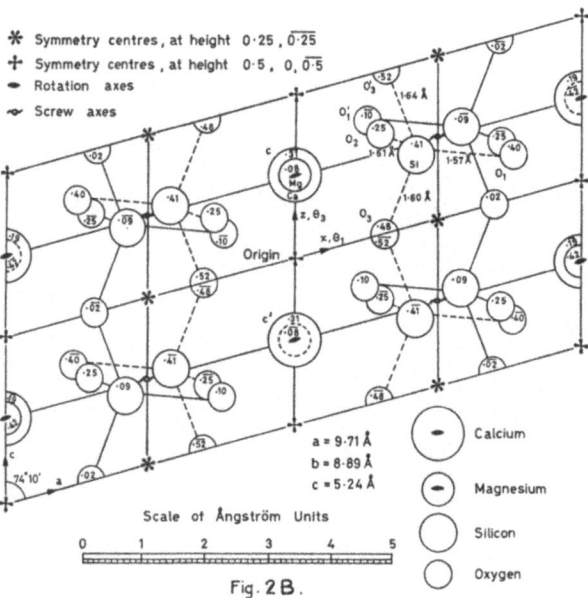

＊ Symmetry centres, at height 0·25, 0·25
＋ Symmetry centres, at height 0·5, 0, 0·5
● Rotation axes
～ Screw axes

a = 9·71 Å
b = 8·89 Å
c = 5·24 Å

Scale of Ångström Units

0   1   2   3   4   5

Calcium

Magnesium

Silicon

Oxygen

Fig. 2B.

Fig. 2. Projection on face (010).

The employment of Fourier series in analysing complex crystals is discussed, and it is concluded that it may be used in conjunction with an analysis of the usual type made by assigning parameters to the atoms, and may considerably shorten the labour of analysis. The series is particularly of value in discovering the positions of the lighter atoms and in leading directly to precise values of the parameters.

406

Fig. 3A.

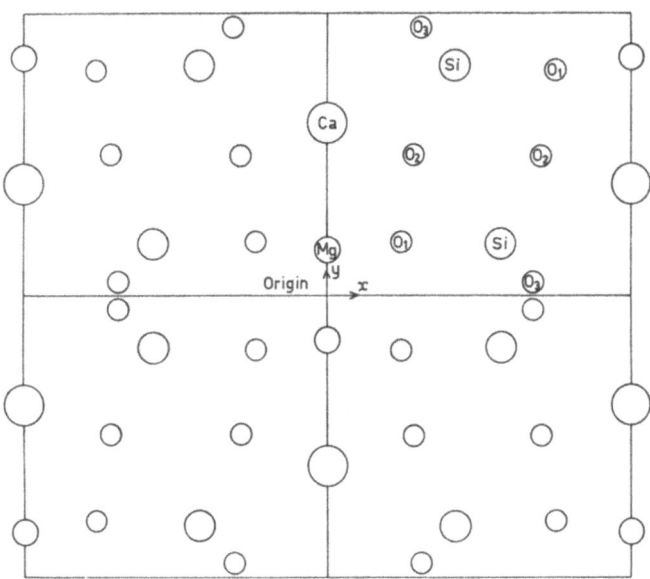

Fig. 3B.

Fig. 3. Projection on face (001).

407

In a recent paper by Mr. West and the author in the Zeitschrift für Krystallographie, entitled 'A Technique for the X-ray Examination of Crystal Structures with many Parameters'[1], examples were given to show that the use of absolute measurements of X-ray diffraction enabled these complex crystal structures to be solved with directness and accuracy. The use of the same measurements in the method of Fourier series affords further evidence in support of the effectiveness of such absolute measurements.

It is with great pleasure that I acknowledge my indebtedness to my father, Sir William Bragg, for suggestions which materially contributed to the work described in this paper. At the time when I was following up the connection between our usual methods of analysis and the analysis by Fourier series, a connection briefly treated in the paper by Mr. West and myself, my father showed me some results which he had obtained by using relative values of the first few terms of two- and three-dimensional Fourier series to indicate the general distribution of scattering matter in certain organic compounds. It was largely as a result of his suggestions that I was encouraged to make all the computations for these two-dimensional series, using the extensive absolute measurements which we had made on certain crystals.

191.

The results obtained illustrate the great power and beauty of the X-ray method of investigating problems of molecular structure in the solid state. Perhaps the most striking feature of the results is the amazing verification which they afford of the stereochemical conceptions of organic chemistry. Of course it may be argued that these fundamental formulas did not stand in any need of verification—they were firmly established by chemical methods long before the diffraction of x-rays was discovered, indeed before the discovery of x-rays at all. But the experimental methods employed in building up the molecular maps are so remote from the reactions and syntheses which establish the chemical structural formulas, that the verification afforded is of considerable philosophical interest. The structural formulas are now endowed with a new degree of reality—not necessarily more profound, but certainly quite a different reality.

The greatest difference lies in the exact metrical representation of the structures which has now been achieved. The interatomic distances appear as constants which are definitely characteristic of certain types of binding between the atoms.

---

[1] Z. Kristallogr. **69** (1929) 168.

> *These constants can be related to the heats of formation and other physical properties, and furnish data for theoretical investigation.*
>
> J. M. ROBERTSON,
> *Chem. Rev.* **16** (1935) p. 433

The aromatic hydrocarbons naphthalene and anthracene were among the first organic compounds to be investigated by the X-ray method. The early work[1] on these crystals afforded evidence that the long axes of the molecules lay along the *c* direction in the crystal, but less definite indication was obtained regarding the lateral disposition of the molecules. An attempt was made to settle this point by a study of the intensities of the principal X-ray reflections by visual estimate, and this led to a distorted 'tetrahedral' structure being advanced.[2] Somewhat later it was shown by Sir William Bragg that this structure was untenable, the discrepancies between the calculated and the measured values of certain reflections being too great[3]. In this work absolute measurements of intensity were obtained for the more important planes, and the evaluation of the atomic positions carried out by Fourier analysis. This led to a structure consisting of molecules with flat or slightly distorted carbon rings, whose planes made an angle of about 25° with the *bc* plane, and with the long axis of the molecule tilted about 6° away from the direction of the *c* axis towards a more upright position. Almost at the same time Banerjee[4] independently arrived at a structure, based upon the measurement of some intensities, together with the results of optical and magnetic measurements, which agreed very closely with Bragg's structure.

At the commencement of the present work, then, it could be taken that the structure of anthracene was approximately known. It was very important, however, to confirm these results by more accurate and more extensive intensity measurements, and also to attempt to work out the finer details of the structure. For example, it was not known if the carbon rings were quite flat like the graphite structure, or slightly puckered, perhaps intermediate between the diamond and graphite type: the evidence seemed to point to the latter possibility. The carbon to carbon distance was not accurately known. The only estimate obtained was 1.48 Å, which is again intermediate between the diamond (1.54) and the graphite (1.42) distance.

[1] W. H. Bragg, *Proc. Phys. Soc.* **34** (1921) 33 [this Vol. paper **170**]; **35** (1923) 167; 'X-rays and Crystal Structure' (1925) 229; *Z. Kristallog.* **66** (1927) 22; *Nature* **121** (1928) 327.
[2] J. M. Robertson, *Proc. Roy. Soc.* A **125** (1929) 542.
[3] *Nature* **125** (1930) 456 [this Vol. paper **172**].
[4] *Indian J. Phys.* **4** (1930) 557; *Nature* **125** (1930) 456 [this Vol. paper **171**].

As the anthracene structure involves 21 parameters without considering the hydrogen atoms, its complete determination is a very complicated task. With the results already obtained it was obviously desirable to apply the Fourier analysis, and this method has been extensively used in the present investigation. Briefly, the procedure has been as follows. First of all the parameters were determined as far as possible by trial and error, and then a preliminary Fourier analysis was carried out for two zones of reflections (about the b and c crystallographic axes). From the results of this analysis the values of the co-ordinates of the atoms were refined, and it was now found that a much better agreement was obtained between the observed and calculated values of the reflections. This was chiefly due to the presence of a hitherto unsuspected tilt in the axis of the molecule away from the ac plane. With the revised values of the co-ordinates it was now possible to make sure of the phase constants of some of the weaker reflections which had previously been doubtful. The intensity measurements were now checked and made as accurate as possible and a final double Fourier analysis carried out for the zones about the three crystallographic axes. The results of this final analysis are set out and discussed in this paper.

Crystal Data

Anthracene. Melting point 217°C. Monoclinic prismatic. $a = 8.58$, $b = 6.02$, $c = 11.18$ A. $\beta = 125°$. Space group C_{2h}^5 (P2$_1$/a). 2 molecules of $C_{14}H_{10}$ per unit cell. Total number of electrons per unit cell $= F(000) = 188$.

Measurement of Intensities

Like most organic compounds, anthracene forms crystals which are usually small and friable, unsuitable for cutting or grinding into sections. The best method for measuring the integrated reflections from such crystals therefore appears to be that of completely immersing a small single crystal in the X-ray beam. As this method, although tested[1], has not been much used in quantitative work, Robinson in this laboratory has conducted a parallel investigation into the validity of the whole method, with special reference to anthracene, and has also determined the absolute values of some of the anthracene reflections. For the purposes of this paper, then, the writer

[1] Bragg, *Proc. Phys. Soc.* **33** (1921) 304.

has only required to measure the relative values of the integrated reflections from anthracene. These measurements were then calibrated with Robinson's absolute values, using one of his measured crystals for the purpose.

Two methods of measuring the intensities have been used, the ionization spectrometer and the integrating photometer devised by Robinson. The photographic method seems to be the most suitable for obtaining a complete survey of a large number of reflections, such as those required for making the two dimensional Fourier analyses. There is less chance of missing certain reflections, which may be comparatively weak, yet valuable for completing the detail in the analyses. But in general the more important reflections have been measured both by the ionization and photographic methods, and the following list is typical of the agreement obtained by the two methods. The figures give the integrated intensity in arbitrary units.

hkl	Ionization spectrometer	Robinson's photometer
200	1000	982
20$\bar{1}$	652	645
20$\bar{2}$	253	258
001	711	667
002	192	194
003	52	59
004	81	72
005	32	31

The effect of extinction on the strong reflections is discussed by Robinson, but we believe that by using sufficiently small crystals, weighing about one-tenth of a milligram, the effect is largely eliminated. In making the relative measurements the smallest possible crystals were used in estimating the strong reflections, although slightly larger ones were occasionally used in measuring the weaker reflections, which do not appear to be affected by extinction.

Most of the measurements were made with copper radiation ($\lambda = 1.54$). With this wave-length the absorption of the beam in the crystal is quite considerable, but again, with very small crystals, the effect is small. It did not seem practicable to apply a separate correction to each of the relative measurements to allow for the shape of the crystal, and the largest errors will probably be due to the varying thickness of crystal presented to the beam at each reflection. Several crystals of different shapes were usually measured, with good agreement except in extreme cases. In general, crystals

Table I. Measured and Calculated Values of the Structure Factor.

hkl	sin θ Cu Kα	F calc	F measured
200	0.219	+60.5	59
400	0.438	− 6.5	3
600	0.657	− 8.5	7
020	0.256	−19.5	24.5
040	0.511	− 2	< 3
001	0.084	+34	30
002	0.168	−20.5	22
003	0.252	+12	14.5
004	0.336	−20	22
005	0.420	−12	16
006	0.504	+ 5	< 4
007	0.588	− 1	< 4
008	0.672	− 4.5	< 4
009	0.756	− 7	3

with a nearly square section perpendicular to the rotation axis were sought, those in the form of thin flakes being avoided.

Table I [reproduced for the $h00$, $0k0$ and $00l$ reflections only, Ed.] gives the value of the structure factor for each reflection in absolute units, and it is compared with a value calculated from the co-ordinates found for the atoms (which are given in table VIII), and from an F-curve for carbon based on the measured reflections from graphite[1]. The value of the structure factor is obtained from the measured reflection by the usual formulae for the 'imperfect' crystal.

Fourier Analysis of Experimental Results

From the preceding tables it will be seen that there is a fairly good average agreement between the measured and calculated values of F. The atomic f-curve derived from graphite, upon which the calculated values are based, cannot be expected to fit the anthracene results exactly. No allowance has been made for the hydrogen atoms; and it will be seen from the following analysis that slightly differing curves ought to be applied to the different atoms in the molecule. But the agreements obtained seem to be sufficient to determine the phase constant of each member beyond any reasonable doubt; that is, in this case, whether the sign of the structure factor is positive

[1] Bernal, *Proc. Roy. Soc.* A **106** (1924) 749; Lonsdale, *ibid.* **123** (1929) 499.

or negative. With this information it is then possible to proceed with the Fourier analysis, which has the great advantage of presenting the structure directly from the experimental measurements, only the sign of the terms being taken from the calculated results.

The method of the double Fourier series was first developed by W. L. Bragg[1] and the theory is described in his paper. The arrangement of the results adopted here is also similar to that employed by him. By this method the distribution of scattering matter in the unit cell as projected along any zone axis can be calculated. The density of scattering matter per unit area, $\rho(y, z)$, for the projection along the a axis is

$$\rho(y, z) = \frac{1}{A} \sum_{-\infty}^{+\infty} \sum_{-\infty}^{+\infty} F(0kl) \cos 2\pi(ky/b + lz/c),$$

A being the area of the plane upon which the projection is made. Corresponding formulae apply for the projections along the other zone axes, the coefficients in the Fourier series being the structure factors for the planes in the zone. For convenient reference these coefficients with their signs are collected below in Tables II, III and IV [not reproduced, Ed.] for the projections along the three crystallographic axes.

It will be seen that the values of the coefficients which terminate the series are mostly fairly small compared with the initial values. In a comparatively soft organic compound like anthracene (melting point 217°C) the temperature factor is probably fairly large. This has the advantage of making the series fairly rapidly convergent without the necessity of applying an artificial temperature factor as has been suggested for certain inorganic compounds[2].

Tables V, VI and VII [not reproduced. Ed.] give the summations for the projections along the three crystallographic axes, the figures corresponding to the series

$$S(y, z) = \sum \sum F(0kl) \cos 2\pi(ky/b + lz/c)$$
$$S(x, z) = \sum \sum F(h0l) \cos 2\pi(hx/a + lz/c)$$
$$S(x, y) = \sum \sum F(hk0) \cos 2\pi(hx/a + ky/b).$$

The distribution of scattering matter in these projections is shown in Figs. 1a to 4a. In each case the projection has been made on a plane perpendicular to the axis. The contour lines are drawn through points of equal density at intervals of 100 in the values of S. Alongside each diagram is a drawing on the same scale showing the positions of the atoms

[1] Proc. Roy. Soc. A **123** (1929) 537 [this Vol. preceding paper].
[2] Bragg and West, Phil. Mag. **10** (1930) 839 [this Vol. paper **189**].

J. M. ROBERTSON

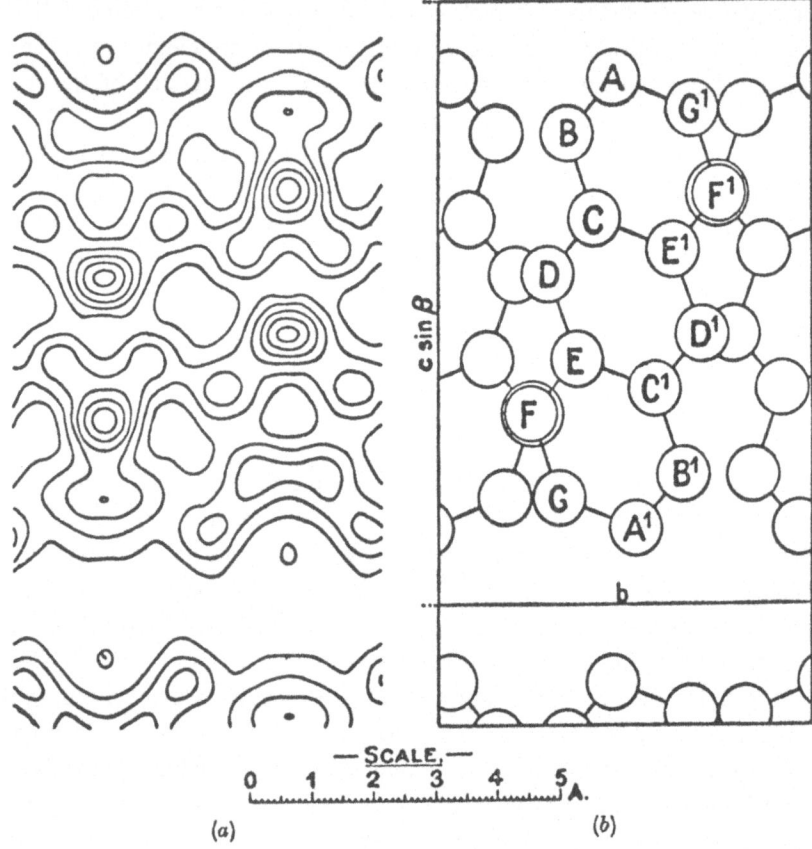

(a)   (b)

Fig. 1. Projection along *a* axis.

and how they are linked together to form the anthracene molecules, Figs. 1*b* to 4*b*.

The projection along the *a* axis, Fig. 1, shows one complete unit cell and a small part of the next cell, which is added to show the extent of the gap between the ends of the molecules. In this projection the different molecules do not separate, because the centre one, which lies half a translation along the *a* axis in front of the others, overlaps them on either side. Some of the individual atoms, however, are quite clearly separated.

The projection along the *b* axis is the most striking, in that the individual molecules as well as many of the atoms are clearly separated. Fig. 2 shows a single molecule on a large scale, occupying one-half of the unit cell. Fig. 3 shows on a smaller scale the mutual relations of six molecules. The centre ones, which are dotted, are identical with the others in this projection, but are actually removed half a translation along the *b* axis (perpendicular to the paper).

414

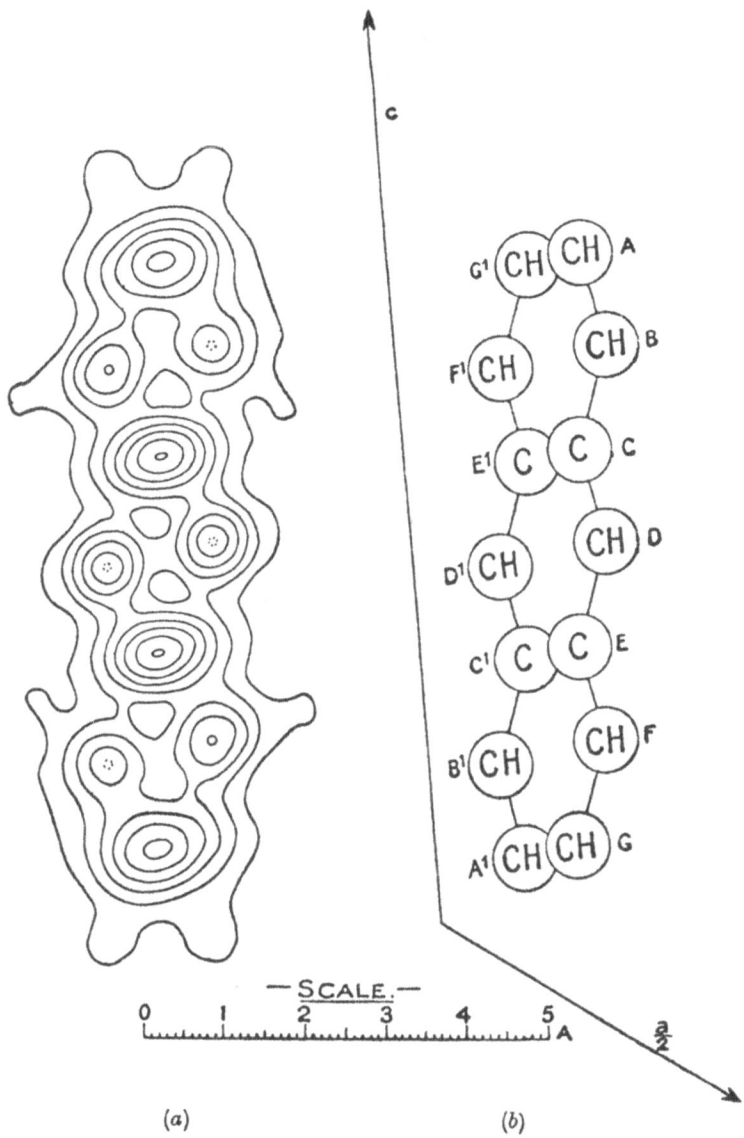

Fig. 2. Projection along *b* axis. The dotted centres at atoms D and B mark densities of 560 and 460 respectively.

In the projection along the *c* axis, Fig. 4, the molecules, seen end on, are very clearly separated, but not the individual atoms. The diagram shows the mutual relation of five molecules, the dotted line being the boundary of the unit cell.

415

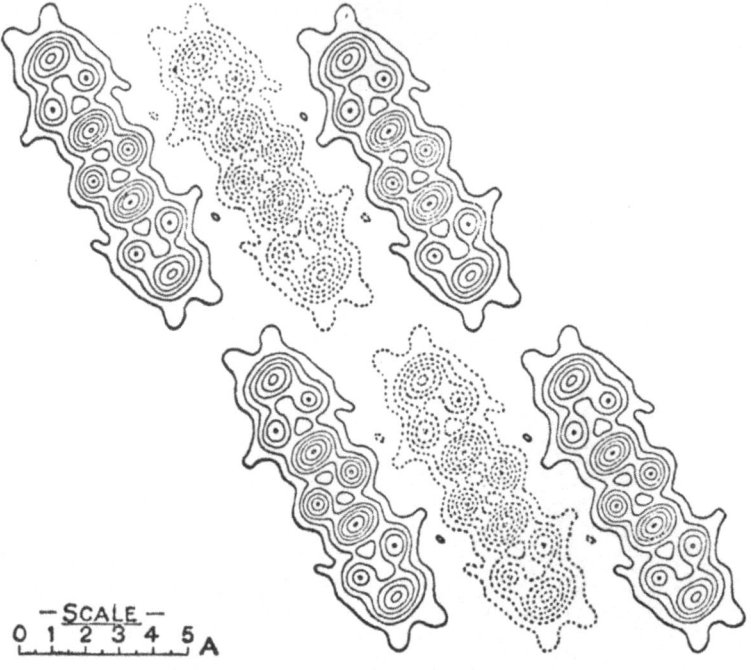

Fig. 3. Projection along b axis showing mutual relation of molecules.

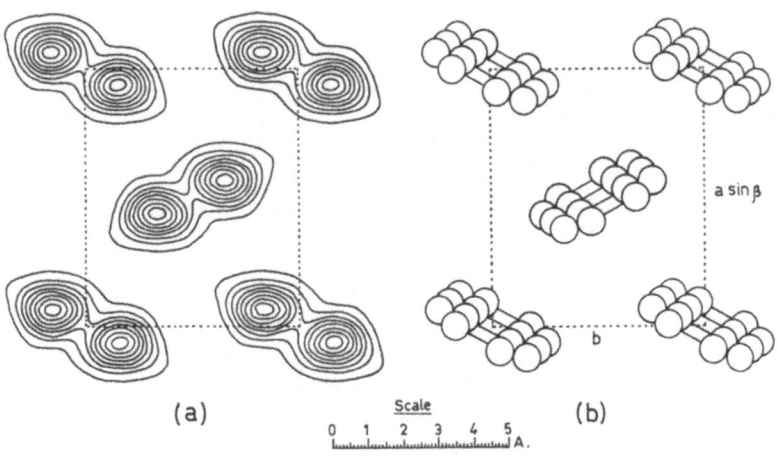

(a)   (b)

Fig. 4. Projection along the c axis.

The Structure Deduced from the Fourier Analysis

Several encouraging features are immediately evident from the contour diagrams. The carbon atoms which are sufficiently removed from their neighbours to separate clearly show a fairly high degree of spherical symmetry. Without such symmetry, of course, the analysis of organic compounds by atomic *f*-curves, which is a necessary preliminary to any Fourier analysis, would be very difficult. Again, in the projection along the *b* axis, the outlying atomic centres (B, D and F) are seen to lie quite accurately on a straight line, and this line is parallel to the line joining the peaks of the unresolved centres, AG', CE', etc. This regularity in the molecule is most striking, and is in harmony with our ideas of the chemical structure.

It will be noticed that in the projection along the *b* axis the outlying atoms, B, D and F, stand out clearly as isolated peaks, while in the *a* axis projection the inner atoms, A, C and E are quite clearly separated. Thus in one or other of the projections we get a clear picture of every atom in the molecule, with the exception of G, which is always somewhat obscured by its neighbours.

The molecule is inclined at varying angles to all the crystallographic axes; nevertheless, it is easy to apply a direct test of the form of the carbon rings. If these rings are regular plane hexagons, then the short cross lines AG', CE', EC' and GA' will be parallel to and one-half as long as the long cross lines BF', DD' and FB' in every projection. The test is rendered a little difficult because of the unresolved centres, but when the most reasonable positions are assigned to these centres it is found to hold to quite a high degree of accuracy.

It is now easy to obtain the actual orientation of the molecule in space. As the crystal is monoclinic it is convenient to use an axis, *c'*, perpendicular to the *a* and *b* axes. The apparent angle which the long axis of the molecule makes with this *c'* (vertical) direction in the projection along the *a* axis is 8.0°, and in the projection along the *b* axis, 30.1°. From these figures a simple calculation gives the actual angles, χ, ψ and ω which the long axis of the molecule makes with the *a*, *b* and *c'* axes as

$$\chi = 119.7° \qquad \cos \chi = -0.495$$
$$\psi = 96.9° \qquad \cos \psi = -0.121$$
$$\omega = 30.7° \qquad \cos \omega = +0.860$$

(The angle with the *c* crystallographic axis is 8.5°.)

In the same way for the cross lines CE', DD', etc., the average apparent angle with the *c'* (vertical) direction measured on the *a* axis projection is

417

69.6° and on the b axis projection 46.9°. For the actual angles χ', ψ' and ω' with the a, b and c' axes these figures give

$$\chi' = 69.6° \qquad \cos \chi' = +0.349$$
$$\psi' = 28.6° \qquad \cos \psi' = +0.878$$
$$\omega' = 70.9° \qquad \cos \omega' = +0.327$$

The real angle between the long axis of the molecule and the cross lines DD', etc., is arc cos(cos χ cos χ' + cos ψ cos ψ' + cos ω cos ω') and with the above figures this works out at 89.9°. This is a very satisfactory result, because it shows in a more precise way that the molecule is built from regular hexagon rings.

The radius of these hexagons, which is equal to the carbon to carbon distance, can readily be measured on the projections. In the projection along the b axis it is one-half of the distance between the outlying atoms BF', DD' and FB'. This is quite accurately constant, and is equal to 0.676 A in this projection. Combined with the direction cosines given above, this gives a value of 1.41 A for the real radius.

In the projection along the a axis the average distance between the inner atoms, CE', EC', etc., is 1.324 A, which combined with the direction cosines again gives a value of 1.41 A for this interatomic distance.

The periodicity along the molecule, that is, the distance between the lines joining the centres AG', BF', CE', DD', etc., measured along the axis of the molecule, is quite reasonably constant. In the a projection it averages 1.07 A, and in the b projection 1.20 A. Combined with the direction cosines these figures give 1.23 and 1.21 A for the real periodicities. The mean value, 1.22 A, is equal to $(1.41 \times \sqrt{3/2})$ A, the value required by a regular hexagon structure.

Table VIII. Co-ordinates. Centre of Symmetry as Origin monoclinic axes.

Atom, cf. fig. 2	x A	$2\pi x/a$	y A	$2\pi y/b$	From a projection z A	From b projection z A	$2\pi z/c$ (mean)
A	0.81	33.8	0.19	11.4	4.17	4.10	133.0
B	1.06	44.6	0.94	56.5	3.14	3.11	100.5
C	0.54	22.5	0.49	29.5	1.57	1.55	50.3
D	0.81	34.1	1.24	74.4	0.55	0.57	18.0
E	0.28	11.9	0.78	46.8	—1.0	—0.98	—31.9
F	0.56	23.5	1.53	91.4	—2.02	—1.98	—64.4
G	0.02	0.6	1.06	63.6	—3.60	—3.54	—115.0

All the above information is, of course, contained in a statement of the co-ordinates of the atoms. But it has been given in the preceeding form because it is sometimes easier to measure the angle between certain lines in these diagrams than to locate the atomic centres accurately. The parameters as measured directly from the diagrams are now given in Table VIII. The figures in bold type refer to atoms which separate clearly as more or less circular masses in one or other of the diagrams. The remaining atoms are too near their neighbours and appear as unresolved ovals, but the probable position of the centres can be estimated fairly well. It will be seen that only 14 of the 21 parameters can be directly measured from the diagrams. The z co-ordinates are estimated separately from the b and c axis projections, and the agreements obtained may be considered satisfactory, remembering that the two projections are based on different sets of experimental measurements.

Intermolecular Distances

The distances between atoms on neighbouring molecules are in marked contrast to the distance of 1.41 A which has been found for the linked carbon atoms within the molecule. The molecule in the centre of the (001) face is derived from the corner molecule by a reflection in the (010) plane, or by a rotation about the b axis, and a shift of $\frac{1}{2}a$, $\frac{1}{2}b$. Each molecule consists of two chains of seven carbon atoms, one chain being the inversion of the other. The co-ordinates of one such chain, forming an asymmetric unit, are given in Table VIII. The whole unit cell can be built up from these co-ordinates (x, y, z) in the following way:

$$x, \quad y, \quad z \ \dots \ (1) \text{ Standard unit}$$
$$-x, \quad -y, \ -z \ \dots \ (2) \text{ Inversion of (1)}$$
Standard molecule.

$$4.29+x, \ 3.01-y, \quad z \ \dots \ (3) \text{ Reflection of (1)}$$
$$4.29-x, \ 3.01+y, \ -z \ \dots \ (4) \text{ Inversion of (3)}$$
Reflected molecule.

It is then easy to calculate the distances between the centres of atoms in adjacent molecules. The distance between the standard and the reflected molecule varies from atom to atom owing to the inclination of the molecules, but the shortest distance found is about 3.77 A. Between the molecules at the ends of the b axis, the closest distance of approach is 3.80 A. Between those at the ends of the c axis the gap is somewhat greater, reaching a minimum of 4.06 A. Owing to the molecules being inclined at about 30° to the vertical position, the distance between the ends of a standard molecule and a reflected molecule one translation along the c axis is less, reaching a minimum of about 3.67 A. This is the closest distance of approach which has been found between the centres of atoms in adjacent molecules.

Comparison with Other Structures

It is interesting to compare these distances with those determined in other structures involving the carbon atom. The carbon to carbon distance within the molecule of 1.41 A compares with a corresponding distance of 1.42 A in graphite[1] and hexamethylbenzene[2]. In the graphite structure there are two atoms on a vertical axis 3.41 A apart (the distance between the layers) and six symmetrically distributed about it at 3.70 A. The (001) spacing in hexamethylbenzene is 3.70 A. For the long chain hydrocarbon $C_{29}H_{60}$ Müller[3] found the distance of nearest approach between two centres on two neighbouring molecules placed end to end to be approximately 4.0 A, and for the same hydrocarbon the distance of nearest approach between two centres on two adjacent molecules placed sideways to lie between 3.6 and 3.9 A.

The Electron Distribution

The Fourier projections which have been made reveal some particularly interesting features, which are most noticeable in the projection along the b axis, Fig. 2. On the central carbon atoms, D and D′, which are the 'meso-' positions in anthracene, the electron density reaches a height of 560 in the value of S. These atoms are very circular and rise to moderately sharp peaks. The atoms in the 'benz-' positions, however, are somewhat flatter and more spread out, the electron density rising to a value of 500 on F, and only about 460 on B. The height of the peak is thus seen to fall away as we move farther out from the centre of the molecule. The same effect is noticeable for the unresolved centres, CE′ being 700 and AG′ only a little over 600 units at the maximum. Further, the form of the first contour line shows quite pronounced bulges around the atoms in the end benzene rings, which are absent from the atoms of the central ring.

Considerable caution, however, is necessary in accepting finer detail of this kind in the structure. The Fourier series is necessarily incomplete, the

[1] Bernal, loc. cit.
[2] Lonsdale, loc. cit.
[3] Müller, Proc. Roy. Soc. A **120** (1928) 437.

measurements being limited by the wave-length and the experimental conditions. This limitation is very liable to introduce false detail to the picture, similar to diffraction effects in the case of an optical instrument[1]. In the present analysis the concluding coefficients are generally fairly small, but it is quite possible that if still more terms could be added they would have the effect of smoothing out some of the detail.

At the same time, the fall in the peak values of the electron density as we pass out from the centre of the molecule, combined as it is with a slight broadening of the structure, are rather pronounced experimental facts which it seems somewhat difficult to explain in this way. Are they perhaps connected with the well-known chemical difference between the 'meso-' and the 'benz-' positions in anthracene? In most anthracene reactions the meso-positions are first attacked, and substituents only enter the end benzene rings at a later stage. It would seem that these chemical facts must ultimately be explained by some difference in the electron configuration.

Again, the effect might be explained by thermal agitation if we imagine the molecule oscillating slightly about its centre. The more outlying atoms would then have the largest amplitudes, with the consequent effects of broadening the picture and lowering the peak values of the electron density. In opposition to this view it may be considered that oscillations about any axis with such a high moment of inertia are improbable. It must be remembered, however, that the crystal molecule is far from being a free body in space. It is surrounded by other molecules, and in fact it appears to be somewhat more closely bound to its neighbours on either side than to those at either end of the c axis. Between the ends of the molecules, in the (001) plane, the electron density falls to very low values. It is only in this region that occasional small patches of negative density occur. The (001) plane, moreover, is the cleavage plane, and by far the most important natural face on the crystal. Hence the bonds between the ends of the molecules must be very weak. Between the sides of the molecules, on the other hand, the (100) and the (010) planes never appear as faces on the crystal. The (110) face is developed, but never to anything like the extent of the (001) face. It may be possible, then, that the relatively looser binding between the ends of the molecules will permit an oscillation about their centres capable of explaining the observed facts. The point could perhaps be decided by an investigation of the structure at different temperatures, especially near the melting point.

The form of the scattering centres given by the Fourier analysis is conveniently studied by making a relief model of the projections. When this is

[1] *Cf.* Bragg and West, *loc. cit.*

done for the projection along the *b* axis, the difference between the meso- and the benz-carbon atoms is very noticeable. On calculating the electron distribution in the carbon atom from the measured reflections of graphite, it is found that the graphite carbon atom is similar to the atoms in the end rings of the anthracene structure, but that in the anthracene middle rings the atoms are more sharply defined.

It is difficult to make an accurate electron count because even the centres which stand out most clearly actually overlap to a considerable extent in all the projections. An estimate has been made, however, of the electron distribution along the molecule in the following way. The atoms appearing in the projection along the *b* axis, Fig. 2, were divided into groups of two by drawing parallel lines between the pairs AG', BF', CE', DD', etc., at a constant distance apart, equal to the mean distance between these groups. The procedure will be clear from Fig. 5. The number of electrons in each section was then obtained by adding the electron density of the appropriate points in the projection along the *b*-axis. Fig. 5 gives the result of this electron count. The figures in brackets are the number of electrons which we should expect to be associated with the centres from the chemical structural formula.

It is interesting to observe that the tertiary carbon atoms give the lowest

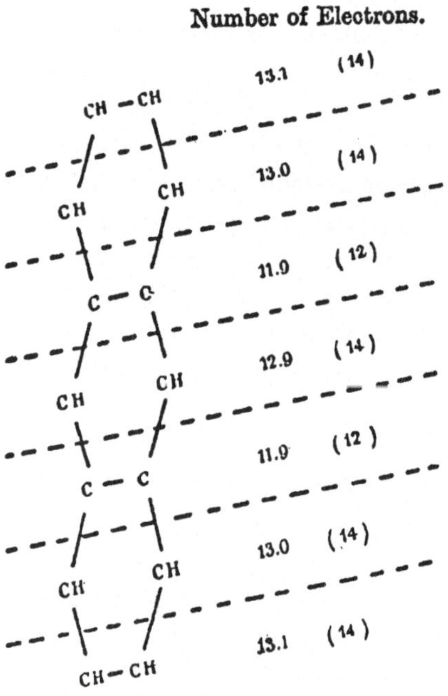

Fig. 5. Electron count in anthracene.

count of 11.9; at the same time it can hardly be said that the effect of the hydrogen atoms is fully evident. The bulges seen on the first contour line at the ends of the molecule are not included in the estimation. They account for a residue of about 2 electrons.

In conclusion, I wish to thank my wife for the help received in dealing with the large amount of numerical work involved in this analysis.

An investigation of this kind, which represents the X-ray method at its full power, is only possible after much gradual development of methods on the part of many people, and I am particularly indebted to Sir William Bragg, F.R.S., not only for his continual interest in the work, but for placing at my disposal many of his earlier, and often unpublished, results.

To the managers of the Royal Institution I am grateful for the facilities provided at the Davy Faraday Laboratory.

B2. Signs from Heavy Atoms

(and, occasionally, their isomorphous substitution)

HEAVY ATOMS PARAMETER-FREE

192. The Crystal Structure of the Alums, by J. M. CORK (1927) *with an insertion* **192a** *from*: Die Kristallstruktur der Alaune, by R. W. G. WYCKOFF (1923)

▬

193. The Crystal Structure of the Alums, by H. LIPSON and C. A. BEEVERS (1935)

▬

HEAVY ATOM CONFIGURATION DETERMINED BY TRIAL AND ERROR

194. The Crystal Structure of Sodium Chloride by W. H. ZACHARIASEN (1929)

▬

HEAVY ATOM CONFIGURATION DETERMINED BY PATTERSON SYNTHESIS

195. The Crystal Structure of Nickelsulphate heptahydrate $NiSO_4.7H_2O$, by C. A. BEEVERS and C. M. SCHWARTZ (1935)

▬

HEAVY ATOMS PARAMETER-FREE

192.   *Introduction*

The alums make an interesting group of crystals for X-ray analysis. Although the molecule is rather complicated, there are so many isomorphous members in the group, differing from one another simply by the replacement of a single atom, that the solution of the structure is not hopeless, especially regarding the positions of the heavier atoms.

Crystallographic considerations assign them to the pyritohedral class of the cubic system.

Vegard and Schjelderup[1] on the basis of spectrometer measurements assigned positions to the atoms in the unit cell so as to make observed and calculated intensities agree. This assignment was made without regard to the space group and led to a complicated and improbable arrangement. P. Niggli[2], using the data of Vegard and Schjelderup, assigned the alums to the space group $T_h{}^2$ and suggested possible arrangements. R. W. G. Wyckoff[3], by means of Laue and rotation photographs, concluded that the correct assignment of the space group was $T_h{}^6$ rather than $T_h{}^2$.

192a. Da die Raumgruppe als $T_h{}^6$ [Pa3] festgestellt worden ist, und 4 Moleküle der Zusammensetzung $R'R'''(SO_4)_2.12H_2O$ im Elementarwürfel vorhanden sind, müssen die Atome dieser Alaune folgende Lagen haben[2]:

$$K \text{ oder } N \text{ Atome } 4b: \quad 000; \tfrac{1}{2}\tfrac{1}{2}0; \tfrac{1}{2}0\tfrac{1}{2}; 0\tfrac{1}{2}\tfrac{1}{2},$$

$$\text{Al Atome} \quad 4c: \quad \tfrac{1}{2}\tfrac{1}{2}\tfrac{1}{2}; \tfrac{1}{2}00; 0\tfrac{1}{2}0; 00\tfrac{1}{2},$$

$$\text{S Atome} \quad 8h: \quad uuu; u+\tfrac{1}{2}, \tfrac{1}{2}-u, \bar{u}; \bar{u}, u+\tfrac{1}{2}, \tfrac{1}{2}-u;$$
$$\tfrac{1}{2}-u, \bar{u}, u+\tfrac{1}{2}; \bar{u}\bar{u}\bar{u}; \tfrac{1}{2}-u, u+\tfrac{1}{2}, u;$$
$$u, \tfrac{1}{2}-u, u+\tfrac{1}{2}; u+\tfrac{1}{2}, u, \tfrac{1}{2}-u.$$

Sauerstoffatome der Sulfatreste: Da diese Raumgruppe keine 32-zähligen Punktlagen besitzt, können alle vier Sauerstoffatome der Sulfatreste nicht genau gleich sein. Eine annehmbare Struktur wird nicht erzielt, wenn alle vier als ungleichwertig angenommen werden, und auf den Körperdiagonalen des Elementarwürfels plaziert werden. Infolgedessen muß für acht dieser Sauerstoffatome die Anordnung 8h gewählt werden mit anderem u-Wert als für Schwefel, während 24 gleichwertige Atome allgemeine Lagen des $T_h{}^6$ besetzen. Es ist ersichtlich, daß in diesem Kristall drei der Sauerstoffatome der Sulfatgruppe gleichwertig sind und vom vierten abweichen (wie in dem von Niggli vorgeschlagenen Strukturbild).

Wassermoleküle: In ähnlicher Weise müssen die 48 Sauerstoffatomen des Wassers in zwei Gruppen geteilt werden, die je eine allgemeine Lage des $T_h{}^6$ in ihren gleichwertigen Stellen besetzen.

Da die Definition der Lage der Schwefel- und Sauerstoffatome auf 11 variabelen Parametern beruht, konnte bei unserem jetzigen Mangel an Kenntnissen über Streuung nichts unternommen werden, die Lage dieser Atome im Elementarwürfel näher zu präzisieren.

[1] *Annalen d. Physik* **54** (1917) 146.
[2] *Phys. Zeits.* **19** (1918) 225.
[3] *Amer. Journ. of Sci.* **5** (1923) 209 [this Vol. paper **192a**].

192. In the present paper new intensity measurements have been made by the X-ray spectrometer upon all orders of the three principal plane sets, (111), (110), and (100), for ammonium, potassium, rubidium, caesium, and thallium-aluminium alums and potassium chrome alum. The dimensions of the unit cell have been accurately determined in each case. In addition, rotating photographs have been taken to verify the space group assignment. Using the method of Fourier analysis, as simplified by Duane, and used so successfully in the case of many relatively simple structures, an attempt has been made to locate the position of some of the heavier atoms in this more complicated structure. In addition, use is made of the atomic structure factor curves to check the agreement between calculated and experimental F values, particularly in the case of high order reflexions, where the effect of the lighter atoms may be neglected. This, together with consideration of the atomic domain, or dimensions required for the particular atoms or groups of atoms in other structures, and the symmetry of space group, limits very markedly the number of possible arrangements.

Spectrometric Intensity Results

Intensity measurements were made using a Bragg ionization spectrometer.

For the case of the ideal mosaic crystal Darwin[1] has developed a relationship involving the integrated reflexion $E\omega/I$, or ρ, and the structure factor F, which may be approximated as follows:

$$\rho = F^2 \cdot \frac{n^2 \lambda^3 e^4}{4m^2 c^4} \cdot \frac{1 + \cos^2 2\theta}{\sin 2\theta} \cdot \frac{1}{\mu + \alpha\rho} \cdot e^{-B \sin^2 \theta}.$$

In this formula n represents the number of unit cells per cubic centimetre, λ and c the wave-length and velocity of X-rays, e and m the charge and mass of the electron respectively, θ the glancing angle of incidence for X-ray reflexion, α, a constant for a particular crystal specimen, being an indication of the perfection of the crystal, so that $\alpha\rho$ represents the effective increment to the absorption coefficient due to extinction, μ is the linear absorption coefficient, and $e^{-B \sin^2 \theta}$ a temperature factor due to the heat motion of the atoms.

If, as an approximation, α be assumed zero and the temperature factor be taken as unity, then for each value of ρ, a value of the structure factor F may be obtained. The value of the absorption coefficient for ammonium alum was measured experimentally using several thicknesses of the absorber,

[1] *Phil. Mag.* 27 (1915) 315, 675 [Vol. I papers 46 and 53]; *ibid* 43 (1922) 800.

and that of the other alums found by computation from this, making use of the data of Wingårdt[1] on atomic absorption coefficients.

The abnormal spacing characteristic of the space group T_h^6 is, that for planes of indices $\{hk0\}$ the spacing is halved if h is odd, for $\{0kl\}$ if k is odd, for $\{h0l\}$ if l is odd. No reflexions were found contrary to this requirement, thus confirming Wyckoff's assignment of space group and co-ordinates.

There are thus 24 generally equivalent positions. The eight sulphur atoms in each cell must be distributed, one in each small cell, on the trigonal axis. They therefore introduce only one parameter. Eight of the oxygen atoms would be arranged similarly to the sulphur atoms, while the remaining 24 would occupy the general positions, introducing in all four more parameters. If the sulphate group, however, maintains in the alums the form it appears to possess in other structures, these 24 atoms would give 3 to each small cell, arranged about the 3-fold axis, forming with the oxygen atom on the axis an equilateral tetrahedron whose base is perpendicular to the 3-fold axis. The oxygen atoms having a diameter of about 2.7 ångströms are thus in close packing array having the sulphur at the centre of mass of the tetrahedron, the sulphur atom being so small that it can well be accommodated by the space before the four oxygens. Thus, if the position of the relatively heavy sulphur atom can be obtained, and the orientation of the tetrahedron along the diagonal, then only the angular position of the tetrahedron thought of as rotating about the 3-fold axis is left undetermined. Having located the SO_4 tetrahedron and considering the dimensions of the monovalent and trivalent metal atoms, only certain available space remains for the other oxygen planes normal to the trigonal axes. The 96 hydrogen atoms making 4 sets of 24, occupying generally equivalent positions, can scarcely be considered, as their scattering power is so small. The space occupied by them must be small. However, in the hexagonal close-packed assemblage of oxygen atoms only 13.94 cubic centimetres per atom is required, whereas in the alum structure, treating the monovalent metal atom as an oxygen atom, about 21.7 cubic centimetres per atom is allowed, so that considerable space is left for the hydrogen atoms, if necessary.

Fourier Analysis of Electron Distribution

In the form developed by A. H. Compton the sheet electron density P_z for the unit cell at a distance z along a direction perpendicular to the plane

[1] *Zeit. f. Physik* **8** (1922) 363.

set giving the information is given by the series

$$P_z = \frac{Z}{d} + \frac{2}{d}\left(F_1 \cos\frac{2\pi z}{d} + F_2 \cos\frac{4\pi z}{d} + \dots + F_n \cos\frac{2\pi n z}{d}\right),$$

where Z denotes the number of electrons in the unit cell and F_n is the structure factor for the nth-order reflexion from the plane set being considered, and d is the spacing of the planes in the direction considered. By dividing by the number of molecules in the unit cell, the electron distribution within the molecule is given.

The value of F given in Table II [not reproduced], being the square root of a quantity, may be either positive or negative in sign. In this series of alums, however, the correct sign is readily observed, as the origin is taken as the centre of a monovalent metal atom, and when this is replaced by a heavier atom F should increase algebraically. If F decreases numerically when the heavier metal atom is substituted, then it must be negative in sign. In order to make the series more rapidly convergent, the F values were multiplied by a factor of the form of the Debye temperature factor.

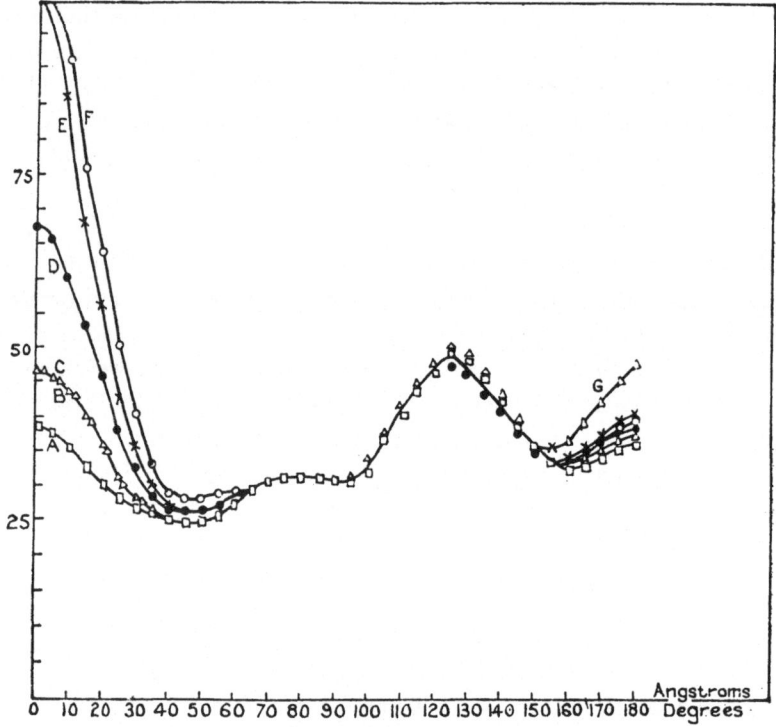

Fig. 2. Electron Distribution in Planes parallel to the (111) set.
A: NH₄Al; B: KAl; C: KCr;
D: RbAl; E: CsAl; F: ThAl.

The values of P_z for planes normal to the (111) set for the different alums, computed at intervals of 0.1 ångströms, are shown in Fig. 2. This plane set gives alternately layers of monovalent and trivalent metals. Thus if the distance from one monovalent metal layer to the next adjacent similar layer be taken as 360° or 7.06 ångströms for the alums, then the trivalent metal layer will lie at 180° or 3.53 ångströms, and if the sulphur atoms were each at the centres of the small cubes they would lie at 90° or 1.76 ångströms. If, however, the sulphur is displaced along the diagonal of the small cube any number of degrees, then there will appear sulphur planes of two types. That is, if we imagine a shift of the sulphur of ϕ degrees toward the aluminium, then for a single molecule of alum one plane of $\frac{3}{4}$ atom of sulphur will occur at $90° + \phi$ and another plane of $\frac{1}{4}$ atom at $90° - 3\phi$, and symmetrical positions.

The remarkable similarity of the curves for the different alums indicates clearly the effect of changing either the monovalent or trivalent metal atom. At 0° occurs the electron density on the plane through the centres of monovalent metal atoms. Since number of electrons is represented in the figure by areas, the difference between any two of these curves should give the difference in the number of electrons in the corresponding atoms. Thus the potassium-aluminium curve subtracted from the caesium-aluminium curve gives a difference of 37.4 instead of 36. Similarly, good agreement is obtained in each case except for thallium where the discrepancy is larger. The fact that these curves should agree so well, considering that they represent independent measurements and calculations employing absorption coefficients varying from NH_4 to Tl by a factor of 20, is most striking. The difference between potassium-aluminium and potassium-chromium alum is evident only in the region of 180°, as would be expected. The curves do not at any place drop to zero ordinate values, since as there are so many atoms their fields must overlap. The important question is the interpretation of the peak at about 124°. If this is the $\frac{3}{4}S$ sulphur peak it would represent a displacement of the sulphurs along the diagonal from the centre of the small cube of 34° or 1.98 ångströms toward the trivalent metal atom. If this were the case, the position of the corresponding peak could be predicted using the (110) and (100) plane sets. These curves were drawn giving families similar to those shown in Fig. 2. The central peak in these curves, however, indicated a smaller shift of the sulphur from the centre of the small cubes than indicated by the (111) set above. This might well be the case if the peaks were not due to the sulphur alone but rather to the region of maximum density of the sulphur and adjacent oxygen layers.

To test this conclusion various positions might be assigned the sulphur

429

atom and the structure factor computed from the atomic structure factor curves of Hartree[1]. Since the effect of the oxygen falls off rapidly as the diffracting angle increases compared to sulphur, for high orders of reflexion the oxygen may be neglected as an approximation. The mean of these calculations seemed to assign a position to the sulphur shifted about 26/180 or 1.55 ångströms from the small cube centre toward the aluminium. The difference between this value and the peak indicated in Fig. 2 is readily accountable by the effect of oxygen atoms if the tetrahedron is properly fixed in its angular position on the cube diagonal, and might be of assistance in deciding between possible arrangements.

This position of the sulphur leads to at least two possible configurations for the remaining atoms, both compatible with dimensional requirements and the crystal symmetry, but perhaps not equally reasonable. In one of these configurations the apex of the SO_4 tetrahedron approaches the aluminium. If the following atomic radii[2] are assumed: oxygen, 2.7 ångströms; sulphur, 0.6 ångströms; aluminium, 1.1 ångströms and monovalent metal, 2.7 ångströms, then the distance between hexagonally close packed layers of oxygen or monovalent metal becomes 2.2 ångströms and there is room for sulphur and also approximately aluminium in the interstices between the oxygen. In the above configuration, however, there would be just sufficient room for the oxygen at the apex of the tetrahedron to touch the aluminium. Toward the monovalent metal there would be just room for three groups of three oxygen atoms: the first, that of the SO_4 group, and the other two the six water molecules ascribed to each cell. Since the SO_4 tetrahedron can take only one angular position to account for the apparent larger shift from the (111) curves, and since space limitations are just satisfied by a hexagonal close packing of the three oxygen groups, the position of every oxygen atom would follow.

There are two objections to this arrangement. The aluminium atom would be left in line between two oxygen atoms; and the water of crystallization would be associated entirely with the monovalent metal, whereas it is perhaps more natural to associate water of crystallization with the smaller atom.

The following alternative arrangement possesses neither of these objections and satisfies the space requirements almost equally well. In this arrangement the SO_4 tetrahedron shifts with its base towards the aluminium. Between the three oxygens of the base of the tetrahedron and the aluminium is a closely packed group of three water oxygens, the other three water oxygens

[1] *Phil. Mag.* **50** (1925) 289.

[2] W. L. Bragg, *Phil. Mag.* **2** (1926) 258 [this Vol. paper **117**].

lie between the apex of the tetrahedron and the monovalent metal. Six water molecules are thus associated with each metal atom. Since the angular position of the SO_4 tetrahedron on the trigonal axis is limited to one position to give the increased apparent shift for the (111) plane set, so also would be the position of the six oxygens about the trivalent metal. This would not be the case, however, with the six oxygens about the monovalent metal. It would now be possible to write the general equation for the complete structure factor and assigning different parameters to the oxygens in this last general position to compare calculated and experimental F values for low-order reflexions. However, since the extinction factor plays such an important part in those calculations at small reflecting angles, and since its effect in the alums is uncertain, such calculations are deemed hardly worth while.

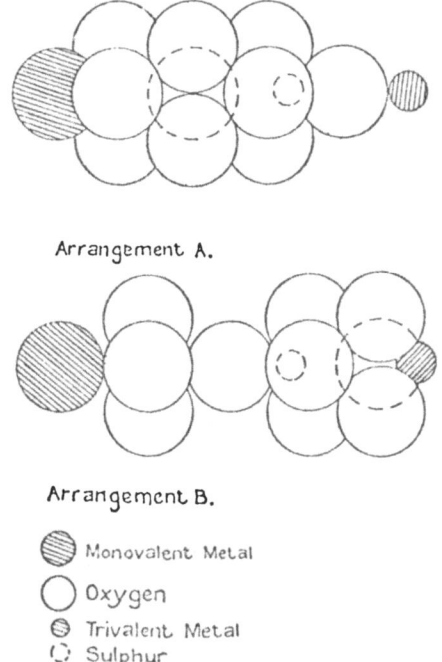

Arrangement A.

Arrangement B.

◉ Monovalent Metal
◯ Oxygen
◉ Trivalent Metal
⬡ Sulphur

Fig. 3. Possible Arrangement of Atoms along the Trigonal Axis.

The two configurations mentioned above are shown in Fig. 3. The positions of the various atoms for the second more probable arrangement (B) are as follows:

Monovalent metal:

{000}   and three other related positions.

Trivalent metal:

{.5, 0, 0}   „   „   „   „   „

Sulphur:

{.322, .322, .322} „ seven „   „   „

SO$_4$ oxygen—Vertex:

{.245, .245, .245} „   „   „   „   „

SO$_4$ oxygen—Base:

{.304, .304, .407} „ twenty- „   „   „
                       three

Water oxygen,

    trivalent metal:
    {.483, .483, .392} „   „   „   „   „

    monovalent metal:
    {x, y, z}   „   „   „   „   „

In order to confirm the arrangement suggested here more members of the alum group should be examined, particularly those in which the sulphur is replaced by a heavier element such as selenium.

This opportunity is taken to express appreciation to Mr. W. H. Taylor for taking the rotation photographs, and to Prof. W. L. Bragg for providing facilities for the work as well as kind advice during the investigation.

193.   1. *Introduction*

The literature on the structure of the alums is fairly extensive, for though the formula is complex, the problem is greatly simplified by the high symmetry of the crystal. Nevertheless, none of the proposed structures has been fully supported by X-ray measurements.

The first published work was that of Vegard and Schjelderup[1], but, though they arrived at the correct unit cell, their structure involved an improbable arrangement of atoms, in which the identity of even the SO$_4$ group was lost. On this ground it was strongly criticized by Schaefer and

[1] *Ann. Physik* **54** (1917) 146.

Schubert[1], and Niggli[2] showed that it was also incompatible with space-group theory. From Vegard and Schjelderup's measurements he assigned the alums to the space-group T_h^2 (Pn 3), but Wyckoff[3], by means of Laue and rotation photographs, showed that this was incorrect, and that the space-group was T_h^6 (Pa 3).

This was verified by Cork[4], who further made absolute measurements of reflections from various alums, and attempted to deduce therefrom the positions of the heavier atoms. He partly succeeded, and suggested possible arrangements of atoms in the complete structure. One of his proposals was supported by Vegard and Esp[5], from a study of powder photographs, but they did not adduce any conclusive X-ray evidence in its favour.

We have attempted from a study of various hydrated crystals[6] to deduce some general laws concerning the nature of water of crystallization, and it appeared to us that none of the proposed structures substantiated our theories. It was thus considered desirable to make another attempt to solve the problem.

3. Cork's Structure

Cork (loc. cit.) assumed that the potassium was at the origin and then attempted to find the sulphur parameter by determining the electron distribution projected on to

(a) the cube edge,
(b) the face diagonal,
(c) the cube diagonal.

The latter only can fix the sulphur uniquely, for in (a) and (b) the projections of K and Al coincide. Thus Cork published only the latter Fourier synthesis, obtaining the signs of the terms by comparison of the isomorphous crystals measured by him. This showed a large peak at $\theta = 124°$[7] which he naturally attributed mainly to the sulphur, but he observed that the parameter so obtained did not agree at all well with that derived from the other two Fourier syntheses.

[1] Ann. Physik **55** (1918) 397; **59** (1919) 583; also **58** (1919) 291.

[2] Phys. Z. **19** (1918) 225.

[3] Amer. J. Sci. **5** (1923) 209.

[4] Phil. Mag. **4** (1927) 688 [this Vol. preceding paper].

[5] Ann. Physik **85** (1928) 1152.

[6] Z. Kristallog. **82** (1932) 297; **83** (1932) 123; and Proc. Roy. Soc. A **146** (1934) 570.

[7] $u = 0.35$ in Fig. 4.

433

He adduced from this evidence the two possible orientations of the sulphate group which would explain the apparent shift of the sulphur peak along the (111) diagonal, and thence built up possible structures by grouping the remaining atoms as closely as possible about the triad axes.

These structures involved the contacts of one water to two oxygens of the same SO_4 group. Now we believe that a water molecule cannot do this, as it means a valence angle $O–H_2O–O$ of about 60°, and there is good evidence that this angle is about 120°.[1]

4. Determination of the Structure*

We first made an estimate of the sulphur parameter by using a large number of reflections from $KAl(SO_4)_2 . 12H_2O$. These reflections, however, did not include any of the h odd, k odd, l odd type, and it follows, therefore, that no distinction can be drawn between a parameter value of u and of $\frac{1}{2} - u$. It was assumed that the effect of the oxygens and waters would in general be small, and thus the sulphur parameter which determines intensities was easily found as 0.31 (or 0.19). This compares with Cork's estimate of 0.322. The discrepancy may be attributed to the few reflections used by Cork, for the effect of the oxygens and waters is bound in certain cases to be large.

The F's of all the (hk0) planes were then estimated from our spot intensities. Some of these spots (namely the h00's and the hh0's) had been measured by Cork, and therefore could be used to relate the value of ΘF^2 to the spot intensity. A family of curves was constructed relating ΘF^2 to θ for the strong spots, the medium spots, and so on. From these curves the ΘF^2 for any spot could be read off. We were fortunate in obtaining some (unpublished) absolute measurements by Parker and White-house in Manchester, and found that these agreed well with our estimations.

The signs of the F's were found by the comparison of reflections from three different alums. Since the potassium and aluminium atoms have no contribution to the reflections with k odd, the signs of the F's of these planes cannot be found by comparison of crystals in which these atoms only are substituted. Recourse was therefore had to selenium alum, $KAl(SeO_4)_2 . 12H_2O$. Since the difference between the atomic numbers of S (16) and Se (34) is considerable, the signs of most of the F's were quite definitely indicated. If the F value was increased by the change S to Se then its sign was taken to be that of the S contribution, which could be

---

* [For positions of cations, *see* this Vol. p. 425]
[1] Beevers and Lipson, *Proc. Roy. Soc.* A **146** (1934) 570.

calculated from the S parameter (0.31 and 0.19 give the same sign here since the h odd, k odd, l odd planes are not involved). In some cases, however, the contribution of the sulphur was too small for the change to show. Photographs were therefore taken with chrome alum, $KCr(SO_4)_2 \cdot 12H_2O$, which gave the signs of those F's with k even, and it was considered then that the signs known would be sufficient to give an approximate structure from which the remaining signs could be found. Using the method described by us[1] for a two-dimensional Fourier synthesis we proceeded on these lines.

The signs of all these F's proved to be the same as the signs of the contributions from K, Al and S.

The resulting contour map of electron density is shown in Fig. 1.

In estimating atomic positions form such a projection certain possibilities of error must be borne in mind. These may be divided into four types:

(*a*) Errors in the magnitudes of the F's will produce irregularities which may affect the positions of the peaks. Probably errors of this type are particularly large in this case, since many of the F's were estimated only from rotation photographs, which will give the smaller F's with fair accuracy, but the larger ones with appreciable errors.

(*b*) The omission of F's outside a certain range of θ will produce diffraction effects[2] which, however, do not seem to be noticeable in this particular case.

(*c*) The omission of F's below a certain value will produce errors which may be quite considerable, but in those cases in which some of the heavy atoms lie on special positions these errors may be allowed for. Alum may be considered a typical example. The K and Al atoms have no contribution to certain reflections, and these will in general be weaker than those which include K and Al. Therefore, more of these reflections will be omitted as being below a certain strength. The K and Al positions have a higher symmetry than the structure as a whole, and the resulting Fourier projection will tend to have this higher symmetry. For alum each peak will have faint reflections in the lines $x = 0$, 0.25, etc. The sulphurs, for example, being the heaviest atoms not on the centres of symmetry, have quite definite 'ghosts'. One can be seen in Fig. 1 at (0.19, 0.19). A similar ghost can be seen in the Fourier synthesis of copper sulphate pentahydrate[3].

(*d*) In projecting through as large a distance as 12A the projections of many atoms are bound to be irresolvable and the parameters thus not accurately measurable.

[1] *Phil. Mag.* **17** (1934) 855.
[2] Bragg and West, *Phil. Mag.* **10** (1930) 823 [this Vol. paper **189**].
[3] Beevers and Lipson, *Proc. Roy. Soc.* A **146** (1934) 570.

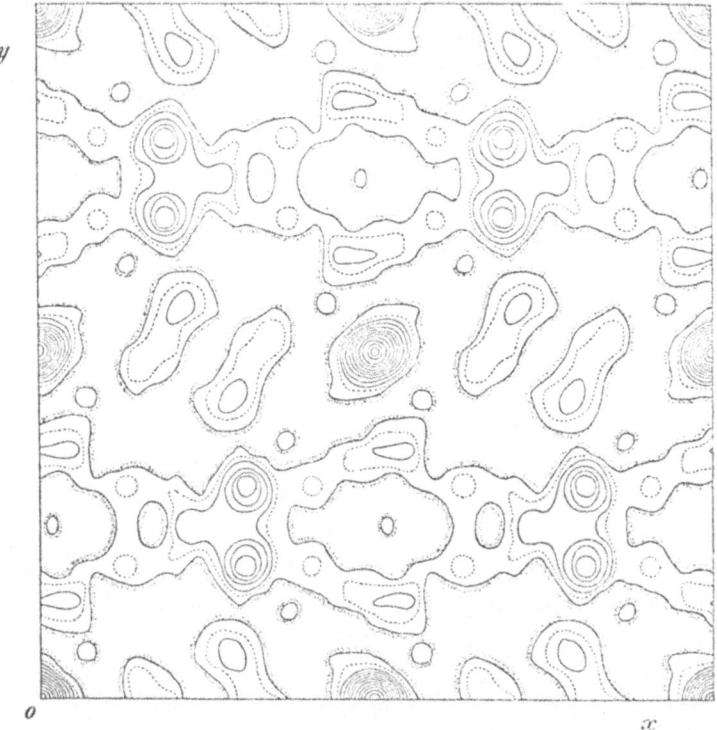

Fig. 1. Contour map of electron density projected on to a cube face. Contours are drawn at 100 (lower side shaded), at 150 (dotted), and 200, and then at intervals of 100.

Bearing these considerations in mind the following interpretation was made of Fig. 1.

The existence of a large peak at (0.31, 0.31) and at other related positions, verified the value of the sulphur parameter.

The oxygens in special and general positions are easily identified by the known shape and size of the SO_4 group.

The remaining peaks in the projection are in two sets, corresponding with the two sets of general waters. Considering them as grouped about (000) and $(00\frac{1}{2})$, i.e., (00) in the diagram, one set of peaks must have parameters $(x_1 y_1)$, $(y_1 z_1)$, $(z_1 x_1)$, and the other $(x_2 y_2)$, $(y_2 z_2)$, $(z_2 \bar{x}_2)$. This is because the triad axis through (000) is the zone axis [111] while the triad axis through $(00\frac{1}{2})$ is the zone axis $[1\bar{1}\bar{1}]$, and it enables a distinction to be made between the two sets.

In the final determination of parameters some assistance was gained from interatomic distances. In particular, the general oxygens of the SO_4 group were somewhat close together and parameters had to be slightly modified to give the usual distances. These modifications (which never

436

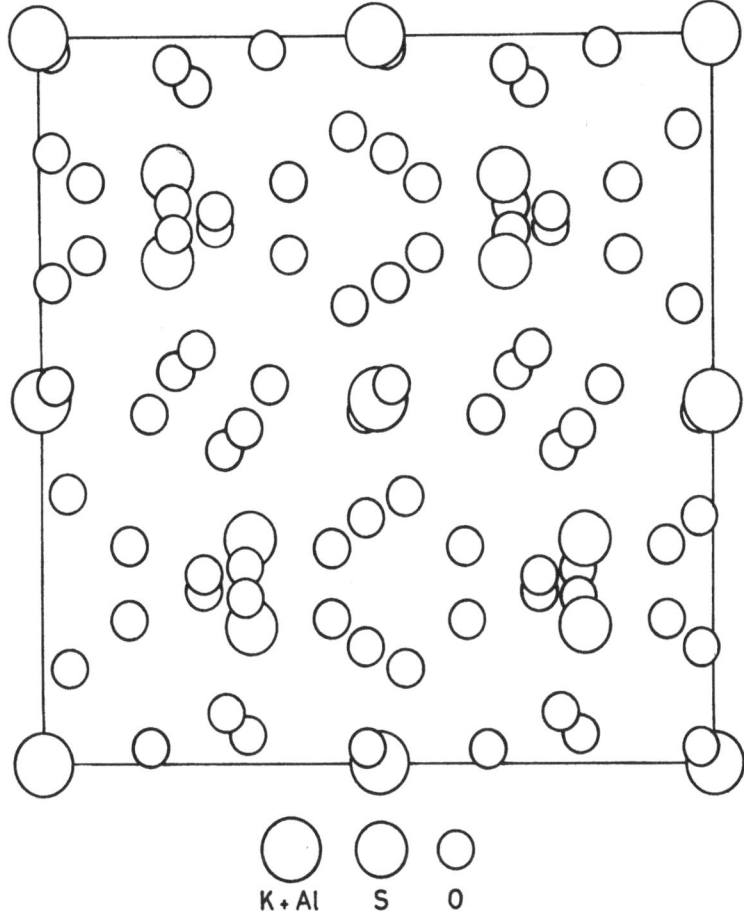

Fig. 2. Atomic positions corresponding with Fig. 1.

exceeded 0.02) gave better agreement of intensities also. Putting the atoms exactly on the Fourier peaks ought, of course, to give perfect agreement of the $hk0$ intensities. Actually the agreement was very good, but three reflections out of ninety were definitely wrong. Modifying the parameters to improve distances made the three erring intensities very much better and brought the agreement as a whole to within experimental error.

Fig. 2 shows the positions adopted.

In this method of arriving at the structure no immediate distinction could be drawn between K and Al, since both these atoms coincide in the projection. However, the atoms comprising the group round the origin were found to be at about 2A from the origin, whereas those round $(\frac{1}{2}\frac{1}{2}\frac{1}{2})$ were at a distance of 3A from that point. Since the usual radius of Al is

437

0.65A and of K, 1.35A, this strongly suggests that it is the aluminium atom which is at (000) and the potassium which is at $(\frac{1}{2}\frac{1}{2}\frac{1}{2})$, in contradiction to Cork's result.

We may, of course, continue to regard the K as being at the origin but then the S parameter must be taken as 0.19 instead of 0.31. Adopting this point of view for the time being we carried out a Fourier synthesis along the cube diagonal of selenium alum, in order definitely to prove the S position. The F's of the *hhh* planes of selenium alum were measured

Table III. F's of Orders of (111).

Order	Se alum	S alum	Order	Se alum	S alum
1	$\overline{12}$	$\overline{9}$	5	29	12
2	$\overline{13}$	4	6	25	21
3	16	31	7	$\overline{4}$	$\overline{8}$
4	0	1	8	0	4

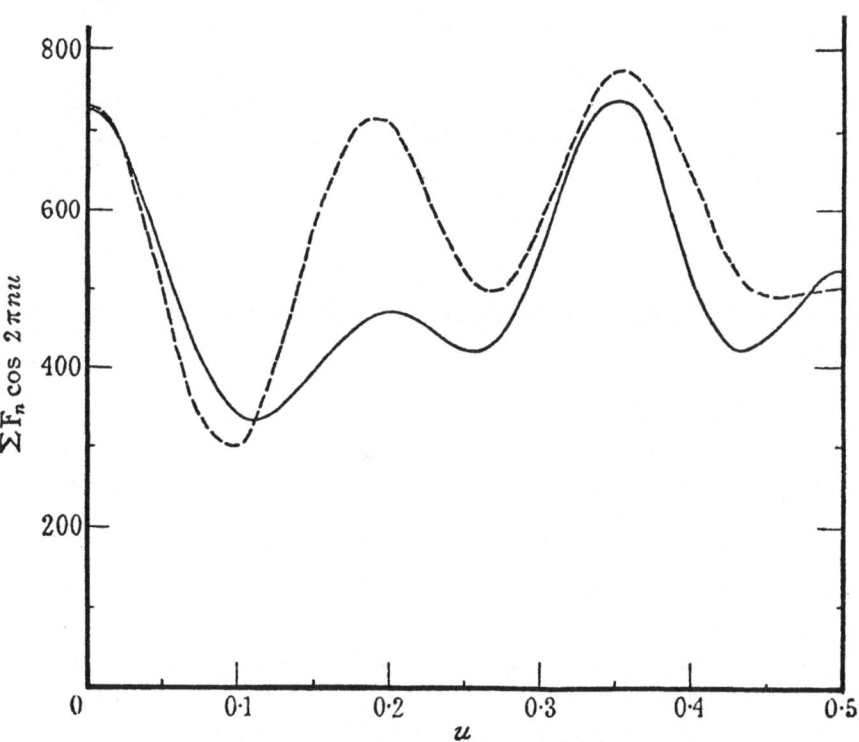

Fig. 3. Electron density, projected on to the cube diagonal of $KAl(SO_4)_2.12H_2O$ (full line) and $KAl(SeO_4)_2. 12H_2O$ (dotted line).

by us at Manchester and are shown in Table III, in comparison with Cork's results for the sulphur alum.

The signs can be deduced with certainty from comparison with Cork's results.

The resulting Fourier curve is shown in Fig. 3, together with the corresponding curve for $KAl(SO_4)_2.12H_2O$, derived directly from Cork's measurements. It will be seen that the two curves are almost idenitcal except for the region around $u = 0.19$. This is quite definite proof that if the sulphur parameter be taken as 0.31 then it is the aluminium atom which is at the origin.

The complete structure is expressed by:

4 Al on (000), $(0\frac{1}{2}\frac{1}{2})$ $(\frac{1}{2}0\frac{1}{2})$ $(\frac{1}{2}\frac{1}{2}0)$;

4 K on $(\frac{1}{2}\frac{1}{2}\frac{1}{2})$ $(\frac{1}{2}00)$ $(0\frac{1}{2}0)$ $(00\frac{1}{2})$;

8 S on (0.31 0.31 0.31) and 7 associated points;

8 O on (0.24 0.24 0.24) and 7 associated points;

24 O on (0.30 0.27 0.43) and 23 associated points;

24 H_2O on (0.02 $\overline{0.02}$ 0.16) and 23 associated points;

24 H_2O on (0.04 0.13 0.30) and 23 associated points.

The comparison of intensities calculated on this structure with observed intensities is shown in Table IV [not reproduced].

5. Reconciliation with Cork's Work

We have seen in the previous paragraph that the structure requires the sulphur atom to be nearer to the potassium than to the aluminium on the same triad axis, and it remains to show that the structure accounts for Cork's determination of the projection parallel to (111), which at first sight seems to require the sulphur to be nearer to the aluminium.

We have repeated Cork's three Fourier syntheses using his values of the F's but without multiplying them by a factor of the form $e^{-B \sin^2 \theta}$ as Cork did. Those of our curves projected on to the cube edge and the face diagonal have peaks which, of course, agree with a sulphur parameter of 0.31 or 0.19, but the projection on to the cube diagonal has a large peak with a parameter of 0.35. This latter curve is shown in Fig. 4, in comparison with Cork's curve (i.e., the curve obtained by using the 'reduced F's').

It will be seen that although Cork's curve has a suggestion of a peak at 0.19, ours has a much more pronounced one. This peak is quite big enough to be the S peak. In Fig. 4 the projections of all the atoms of the

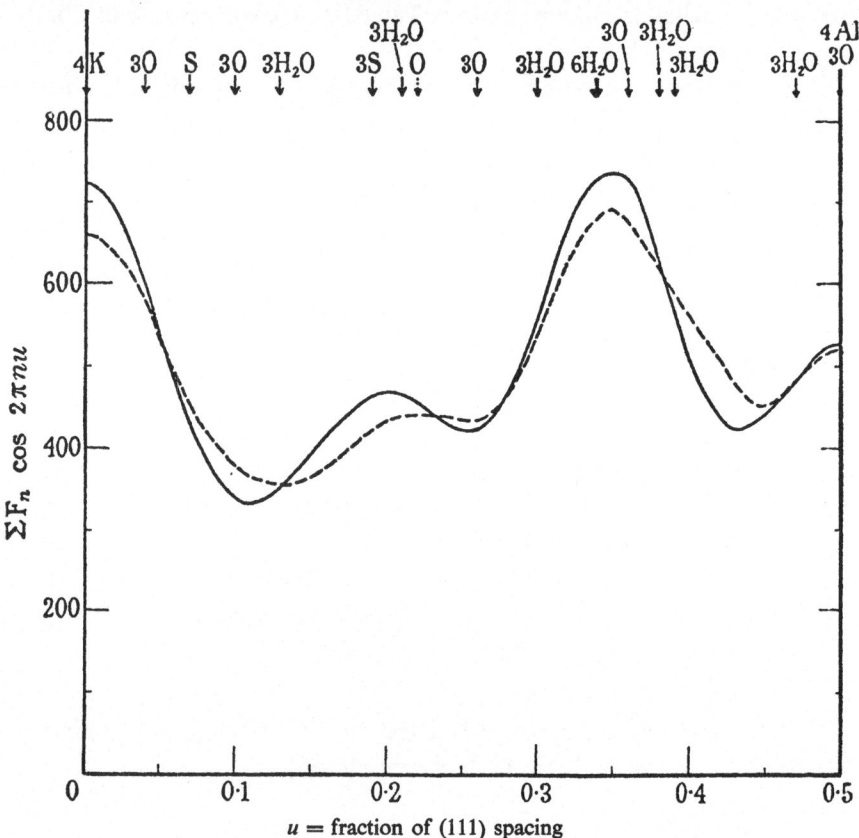

Fig. 4. Electron density, projected on to the cube diagonal, of $KAl(SO_4)_2.12H_2O$, estimated by us (full line) and by Cork (dotted line), both from Cork's measurements. The projections of the various atoms are shown by the small arrows above.

unit cell are shown. The peak at 0.19 is a 3S one, while that at 0.50 is a 4Al one. Thus the relative heights of the two peaks are about right. It will be seen that a large number of oxygens and waters combine to give the very large peak at 0.35.

6. *Discussion of the Structure*

We shall describe first of all the nature of the groupings around S, Al, and K, and then the manner in which the groups fit together.

Around S are four O's, one special and three general, forming a tetrahedron with the following distances:

440

	A
S—special O	1.47
S—general O	1.54
Special O—general O	2.45
General O—general O	2.53

Having regard to the limits of accuracy we may therefore say that the sulphate groups are regular tetrahedra of side 2.5A.

The waters around Al are six in number with the parameters $(0.02\ \overline{0.02}\ 0.16)$, etc. The distance Al–$H_2O$ is therefore 1.97A and the grouping is a very nearly regular octahedron, the two H_2O–H_2O distances involved being 2.81 and 2.77A. Such a group is very similar to the groups of 6O around Al in the silicates.

The second set of waters having the parameters $(0.04\ 0.13\ 0.30)$ form much more open groups of six waters around K. The distance H_2O–K is 2.94A, while the smallest H_2O–H_2O distance in the group is 3.23A. This latter is too large to be regarded as a contact so that the arrangement of the H_2O's around K is governed solely by their external contacts.

The nature of the contacts between the groups may be best understood by consideration of that potassium at the centre of the unit cell, Plate Fig. 5. This has six Al's at a distance of 6.07A, these six being related by the trigonal axis through K and by the centre of symmetry. There are eight sulphate groups in the unit cell, six at 5.80A and two at 4.03 from K. The latter are on the same trigonal axis, and although the nearest, have no direct contracts to the K waters. The K waters thus make contact with six Al.6H_2O groups, and six SO_4 groups.

The exact way in which these contacts are made may be seen from a stereoscopic view of Fig. 5. Each K water touches one Al water, one special O of the SO_4 group, and one general O. Each Al water, on the other hand, touches beside the K water, only one (general) O. It follows that each general O touches one K water and one Al water, and each special O touches three K waters. Diagrams of the arrangements of these contacts or bonds for the two different kinds of water molecules are shown in Fig. 6. The K water has a tetrahedral disposition of bonds while the Al water has its three bonds almost exactly coplanar. A 'bond structure' which gives the strengths of these bonds can easily be devised, and is shown in Fig. 7. The function of the K waters seems to be to average out, as it were, the trivalency of Al and monovalency of K, and so present to the SO_4 groups surfaces more or less uniformly charged.

The behaviour of the water of crystallization in alum fits in very beautifully indeed with the ideas gained from the study of other hydrated

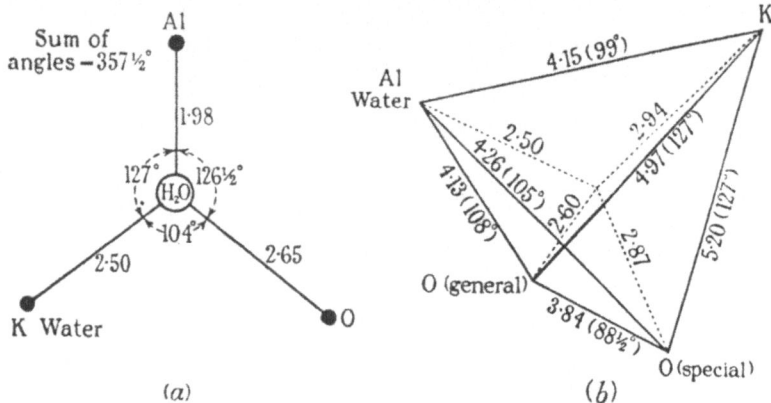

Fig. 6. Disposition of the bonds from (*a*) each Al water, (*b*) each K water. (Not to scale).

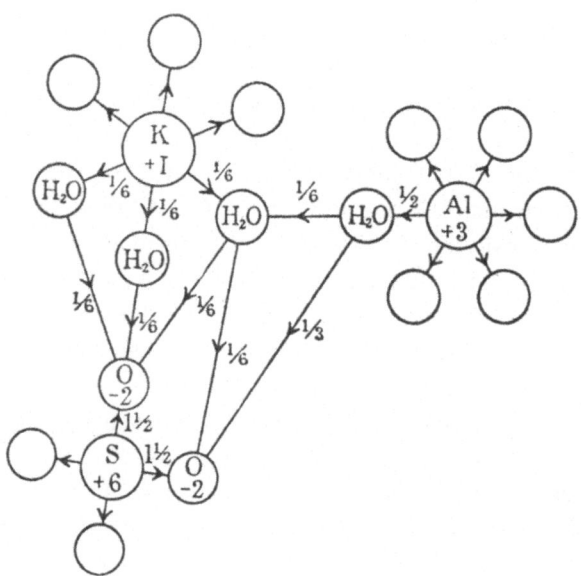

Fig. 7. The electrostatic bond structure.

crystals[1], and with the theoretical model of Bernal and Fowler[2]. It is necessary to postulate that water is a spherical molecule with an arrangement of two positive bonds and two negative bonds which is tetrahedral but is subject to considerable modification under certain conditions, especially as regards the negative bonds. Thus the aluminium ion is able to attract

[1] *Proc. Roy. Soc.* A **146** (1934) 570.
[2] *J. Chem. Phys.* **1** (1933) 515 [this Vol. paper **124a**].

both the negative bonds of six waters and so forms a co-ordination group. The potassium, on the other hand, can satisfy only one of the negative bonds and also the positive ones must be satisfied externally, and the spatial conditions required to do this obtain for the alums, but cannot obtain for many other structures. Thus the association of six waters with a potassium atom is rare.

 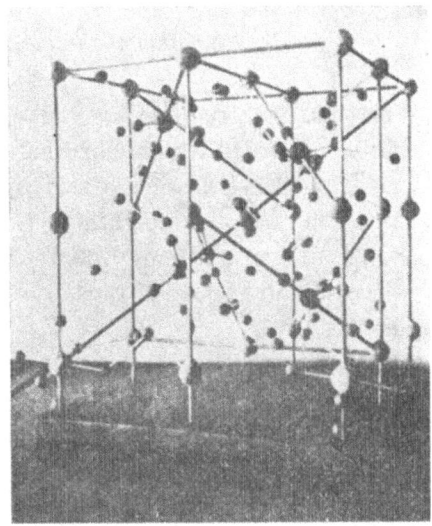

Fig. 5. Model of the structure of alum. (The two pictures should be viewed sterescopically.) The potassiums and aluminiums are arranged on a rock-salt lattice (an aluminium being at the origin). The octahedral groups around the two front lower aluminiums are inserted in the model, but apart from these only the atoms actually within the unit cell are included. The sulphate groups may be seen on the triad axes. A more complete description is given in the text.

George Holt Physics Laboratory,
The University of Liverpool

HEAVY ATOM CONFIGURATION DETERMINED BY TRIAL AND ERROR

194.   *1. Introduction*

Professor W. L. Bragg proposed me to carry out an accurate determination of the structure of Sodium Chlorate. This compound crystallizes isometric tetartohedral. Several attempts to determine the atomic arrangement have been made[1], but they have lead to very different results. It was therefore thought advisable to carry out the investigation right from the beginning without depending on previous results. The results finally obtained agree, however, very closely with those of Dickinson and Goodhue.

5. *Determination of the Parameters for Sodium and Chlorine*

For reflexions occurring at large glancing angles the influence of oxygen can be neglected. An examination of reflexions with large glancing angles leads for the 4Na and 4Cl in Space Group T^4 [P2$_1$3] to $u_{Na} = 23°$, $v_{Cl} = 150°$, the accuracy being $\pm 2°$. In the present case it is possible to fix the oxygen positions with the greatest accuracy quite uniquely and, what is more valuable, directly, by using the two-dimensional Fourier analysis.

6. *Two-dimensional Fourier Analysis*

It is generally said that a Fourier analysis of the electron density distribution can be carried out only if the structure has a centre of symmetry. That, however, is true only for a three-dimensional analysis. For two-dimensional analysis it is sufficient that the projection of the structure on some plane or other has a symmetry centre. Sodium chlorate is such a case. Allthough the structure has no centre of symmetry Fourier analysis is made possible by the fact that twofold axes are present in the projection of the structure on the cube face.

The calculation depends in the first place on the numerical values of the structure amplitudes, which we obtain from the absolute intensity measurements, and in the second place on the signs of the structure amplitudes which we in general can obtain only if we know the structure, at least approximately. In the present case two parameters have been determined in the usual 'trial and error' method, so we know the structure as far as sodium and chlorine are concerned. In the following I will show that this knowledge enables us to tell the sign of all observed amplitudes save three

---

[1] Kolkmeijer, Bijvoet, Karssen, *Pr. Roy. Acad. Sci. Amsterdam* **23** (1921) 644. Dickinson, Goodhue, *J. Am. Chem. Soc.* **43** (1921) 2045. Vegard, *Z. Phys.* **12** (1922) 289; *Videnskapsselsk. Skr.* Nr. 16 (1922). Wulff, *Z. Krist.* **57** (1922) 190. Kiby, *Z. Phys.* **17** (1923) 213.

or four. And the remaining amplitudes where the sign cannot definitely be given are all small, so they can only introduce errors in the calculation which are less than the experimental error in the observed figures.

If we calculate the contribution from sodium and chlorine for every amplitude and compare these values with the maximum contribution of oxygen, we find that all amplitudes save those for 110, 310, 130, 320, 230 and 410 necessarily must have the same sign as the contribution from chlorine and sodium, which we know. Now the expression for the contribution from oxygen has such a form that it is very difficult, with any values of *xyz*, to obtain so big effect from oxygen that beyond the neutralizing of the Na and Cl contribution, an amplitude of considerable height is produced with sign opposite to that of the Na, Cl contribution. It is therefore not unlikely that the signs of the remaining six amplitudes also are the same as those for the cation contribution, but it is at the same time not impossible that the signs are opposite for some of the amplitudes. A trial showed that this might be the case for at most 4 of them. We will, however, assume that the signs remain unaltered for these 6 amplitudes also, when taking the oxygen effect into account; the error introduced is negligible. Finally it is to be noticed that several reflexions are found absent for which the contribution from Na and Cl is considerable. This, of course, means that the oxygen effect for these reflexions is taken into account in our calculation.

The distribution of electron density calculated in the manner outlined in the paper of W. L. Bragg[1] and on the basis of the considerations given above, showed maxima of density in the following points:

1. Symmetrical 'peak' of maximum height 1420 in position corresponding to $x = +150° \ y = +150°$. Is due to a chlorine atom. [Origin at $(-\frac{1}{4}, \frac{1}{4})$ from left upper corner of figures. Ed.]
2. Oblong 'peak' of maximum height 970 in position corresponding to $x = 27° \ y = 17°$. The oblongness indicates the presence of 2 atoms, one of which must be the sodium atom at $x = 23° \ y = 23°$. The other atom, thus partially hidden under the Na atom, must be one of the 3 oxygen atoms.
3. Symmetrical 'peak' of maximum height 410 in position corresponding to $x = 109° \ y = 213°$. This can only be accounted for as due to an oxygen atom.
4. Unsymmetrical 'peak' (the unsymmetry obviously due to partial overlapping of the neighbouring sodium and chlorine peaks) of maximum

---

[1] *Proc. Roy. Soc.* **123** (1929) 537.

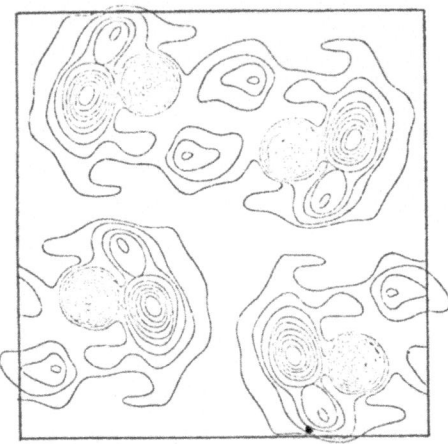

Fig. 1. Fourier diagram of the NaClO₃-structure projected on a cube-face.

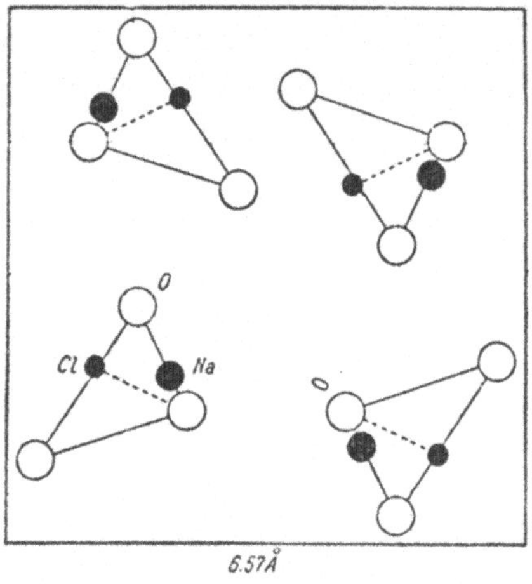

Fig. 2. Atomic positions of the NaClO₃-structure in the projection on a cube-face.

height 520 in position corresponding to $x = 180°$ $y = 112°$. It must be ascribed to the presence of an oxygen atom.

No other maxima are present, so we have consequently obtained the correct number of atoms. The positions of the oxygen maxima will be seen to correspond to the following values of the parameters xyz: $x = 110°$ $y = 215°$ $z = 180°$. We can now calculate the oxygen contribution and test the doubtful signs of the six amplitudes 110, 310, 130, 320, 230 and 410. We then find that we have got the correct signs for all of them save 130, and we can thus correct in our results for this fault.

The distribution of electron density so obtained is given in Fig. 1. The curves connect points with the same value of $a^2\rho$ (namely 100, 200 and so on). The positions of the different atoms are easily seen by comparison with Fig. 2. The contour lines often seem to assume curious forms; but it will be noticed that the unsymmetrical distribution is due to the presence of neighbouring atoms, i.e. to the overlapping of the electron density of the next atom in the projection. That the resulting picture of the distribution is very accurate is indicated by the fact that we do not get negative values for the density. There are as a matter of fact some spots with negative values, but their sum corresponds to a negative value of lesst han, 6 electron, which, of course, is quite negligible as compared with the positive values corresponding to 208 electrons.

From the diagram (Fig. 1) we derive the following set of values for the 5 parameters:

$$\text{Na} \quad u = +\ 23° : ,064$$
$$\text{Cl} \quad v = +150° : ,417$$
$$\text{O} \begin{cases} x = +109° : ,303 \\ y = +213° : ,592 \\ z = +180° : ,500 \end{cases}$$

F-values calculated with these figures are given in table 4. [not reproduced].

I wish to express Professor W. L. Bragg F.R.S. my warmest thanks for his kind help and advice throughout the investigation and for his hospitality and kindness during the time I have been working in his laboratories.

Manchester

HEAVY ATOM CONFIGURATION DETERMINED BY PATTERSON SYNTHESIS

195.   1. *Introduction*

Nickel sulphate heptahydrate is a member of an iso-dimorphous group of crystals.

The crystal class is orthorhombic bisphenoidal, and the space-group $P2_12_12_1$ (V_4). There are four molecules to the unit cell, and all atoms must lie in general positions. No previous attempt to find the 39 parameters involved has been published.

For the purposes of this paper we shall choose an origin of coordinates which is shifted by $\frac{1}{4}$ in the x direction from the origins of Wyckoff and Astbury and Yardley, so that the equivalent points of the space-group become:

$$(xyz) \quad (\bar{x}\bar{y}\tfrac{1}{2}+z) \quad (\tfrac{1}{2}+x\ \tfrac{1}{2}-y\ \bar{z}) \quad (\tfrac{1}{2}-x\ \tfrac{1}{2}+y\ \tfrac{1}{2}-z).$$

This brings the origin to a position on one of the screw-axes parallel to c, which means that a projection of the structure on to the (001) plane has a centre of symmetry.

2. Determination of x and y parameters of Ni and S

The first work done was an attempt to find the Ni and S positions in the x, y plane by the method of A. L. Patterson[1]. This method involves a double Fourier series using the values of F^2 as the coefficients, and is thus independent of the signs of the F's. It gives peaks whose positions with respect to the origin correspond to vectors between atoms in the crystal. There are, of course, a very large number of such vectors in a crystal containing many atoms in general positions. However the analysis of the resulting projection is not as difficult as would at first appear, for the Ni-Ni vectors should be most prominent (Ni being the heaviest atom present) and should be related to each other. Thus if the Ni coordinates are (x_1, y_1) in the structure we have to expect three Ni-Ni peaks at $(\frac{1}{2} - 2x_1, \frac{1}{2})$, $(2x_1, 2y_1)$, $(\frac{1}{2}, \frac{1}{2} - 2y_1)$ in the projection.

The values of F^2 used were estimated from the $(hk0)$ spot intensities. To place these estimates on an absolute scale a number of intensities were measured on the ionization spectrometer, using the slip method described

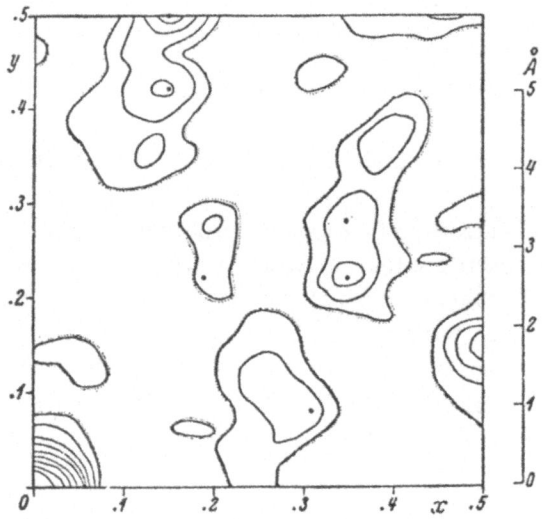

Fig. 1. Patterson synthesis of the $(hk0)$ intensities. Contours are drawn at levels of 110 and every 10 up to 150, and then every 20. The origin peak goes up to 247.

[1] *Physic. Rev.* **46** (1934) 372. [See the equivalent paper, this Vol. paper **196**.]

by Bragg and West[1]. The calculation of the Fourier series was carried out by the method described by Beevers and Lipson[2].

The resulting summation is shown in Figure 1. It is only necessary to consider one quarter of the full cell, and to imagine vertical reflection planes at the edges. It will be seen that along the line $y = \frac{1}{2}$ there is indeed a large peak, with x coordinate .15, suggesting that $x_1 = .175$. We next look out for a peak at $x = 2x_1 = .35$, and find two peaks, with $y = .22$ and $y = .28$, respectively, suggesting that $y_1 = .11$ or .14. The third peak should therefore be at $(\frac{1}{2}, .28)$ or $(\frac{1}{2}, .22)$. No such large peak is to be found, however, although the background of the summation is everywhere high, (of the order of 100).

We shall see later that the $(\frac{1}{2}, \frac{1}{2}-2y_1)$ Ni-Ni peak is actually at $(\frac{1}{2}, .28)$ but does not stand out well from the background.

The x and y parameters of Ni may be found with certainty by a determination of the difference of the ($h00$) and ($0k0$) F's observed from crystals of MgSO$_4$.7H$_2$O and NiSO$_4$.7H$_2$O. For these sets of planes the structure-factors are $\cos 2\pi h x$ and $\cos 2\pi k y$, respectively*. The intensities were measured with the X-ray spectrometer using the crystal-slip method.

Table II shows the results for the ($h00$) planes. Columns 1 and 2 give the F values of the various planes for the two crystals, column 3 the difference in the atomic scattering factors of Ni and Mg (this is the greatest possible change in the F's) and column 4 is the change of F (i.e. column

Table II.

Plane	F values		Change of f value	Ratio	cos 2πhx		
	MgSO$_4$	NiSO$_4$	Ni-Mg			x	
					.06	.08	.10
200	5$\frac{1}{2}$	0	16	.34	.73	.54	.31
400	6	0	13	.46	.06	$\overline{.43}$	$\overline{.81}$
600	<1	12	11	1.00	$\overline{.64}$	$\overline{.99}$	$\overline{.81}$
800	<1$\frac{1}{2}$	6	9	.50–.78	.99	.64	.31
1000	<1$\frac{1}{2}$	0	8	< .19	$\overline{.81}$	.31	1.00
1200	<2	7$\frac{1}{2}$	8	.69–1.10	$\overline{.19}$	.97	.31
1400	<2	7	7	.71–1.28	.54	.73	$\overline{.81}$
	(1)	(2)	(3)	(4)	(5)		

[1] Bragg and West, *Z. Kristallogr.* **69** (1928) 120.
[2] *Philos. Mag.* **17** (1934) 855.
* [In fact a quarter of the structure factor is considered. Ed.]

1 minus column 2) divided in each case by column 3. Column 4 should therefore be equal to the structure factor ($\cos 2\pi hx$) and the remaining columns show the value of $\cos 2\pi hx$ for $x = .06, .08, .10$. Comparison of columns 4 and 5 shows that the Ni x parameter is .08 or the associated value .17, which gives the same value of $\cos 2\pi hx$. This confirms very well the value .17 deduced from the Patterson synthesis. This value of x was therefore adopted.

An identical procedure may be carried out for the ($0k0$) planes with the results of Table III. The best agreement here is obtained with $y = .10$ (or the associated value .15).

Table III.

Plane	F values		Change of f value	Ratio	cos 2πky		
	MgSO₄	NiSO₄	Ni-Mg		.08	y .10	.12
0 2 0	8½	8	16	.03	.54	.31	.06
0 4 0	9½	<3	13	.50–.96	$\overline{.43}$	$\overline{.81}$	$\overline{.99}$
0 6 0	<1	7	11	.54–.73	$\overline{.99}$	$\overline{.81}$	$\overline{.19}$
0 8 0	13	10	9	.33	$\overline{.64}$	.31	.97
0 10 0	<3½	8	8	.56–1.44	.31	1.00	.31
0 12 0	<2	<2	8	0–.50	.97	.31	$\overline{.93}$
0 14 0	8	13½	7	.79	.73	$\overline{.81}$	$\overline{.43}$
	(1)	(2)	(3)	(4)	(5)		

The Patterson synthesis had given roughly the same two alternative values of y and choice between them was made from a consideration of the change in intensity (when Mg is replaced by Ni) for the more general ($hk0$) reflections. Spot intensities from rotation photographs were sufficient for this purpose, and there was good agreement between change of intensity

Table IV.

Plane	Observed Intensities		Structure Factor of Ni			
	MgSO₄.7H₂O	NiSO₄.7H₂O	x (.17	y .15)	x (.17	y .10)
2 3 0	0	m, w–m	.26		.80	
3 5 0	0	w–m	.00		1.00	
9 7 0	0	0	.93		.30	

and the value of the Ni structure factor calculated for the position (.17, .10). The other possibility (.17, .15) gave bad agreement here. Three examples are given in Table IV. The Ni position was therefore taken to be close to (.17, .10).

The Patterson synthesis can now be used to derive the S position. After the Ni-Ni peaks the next largest peaks to be expected in Figure 1 are those corresponding to Ni-S vectors. If the Ni is at $(x_1\ y_1)$, and equivalent points, and the S at $(x_2\ y_2)$ then there will be four such peaks at positions $(x_1 - x_2\ y_1 - y_2)\ (x_1 + x_2\ y_1 + y_2)\ (\frac{1}{2} + x_1 - x_2\ \frac{1}{2} - y_1 - y_2)\ (\frac{1}{2} - x_1 - x_2\ \frac{1}{2} + y_1 - y_2)$. If we write $x = x_1 + x_2,\ x' = x_1 - x_2,\ y = y_1 + y_2,\ y' = y_1 - y_2$ the expected peaks are at $(x'y')\ (xy)\ (\frac{1}{2} + x'\ \frac{1}{2} - y)\ (\frac{1}{2} - x\ \frac{1}{2} + y')$. Now three of the remaining large peaks in the F^2 synthesis together with a smaller peak at (.19 .22) do satisfy these conditions. The four sets of parameters are (.19 $.\overline{22}$) (.15 .42) (.31 .08) (.35 .28), from which we may derive

$$x_1 = .17,\ y_1 = .10,\ x_2 = .\overline{02},\ y_2 = .32.$$

These values of x_1 and y_1 confirm those found previously. It is interesting to note, therefore, that both Ni and S parameters were found independently from the Patterson synthesis, but the Ni y value derived from a consideration of the Ni-S peaks was not wholly supported by the Ni–Ni peaks, a fact which did not give us enough confidence to use the Ni and S parameters derived from the Patterson synthesis alone for further work. However the Ni parameter values being confirmed by the ($h00$) and ($0k0$) intensity changes described above showed that the discrepancy was in the F^2-synthesis, and these parameters were accepted as good approximations.

3. Determination of remaining parameters

The oxygen and water x and y parameters were found from a double Fourier synthesis using the ($hk0$) planes. The signs of the F's were obtained from two considerations, a) the signs of the Ni and S contributions to the structure factors, b) the change of intensity when Ni is replaced by Mg. The signs obtained by the two methods agreed in general. The first synthesis was carried out using only those planes whose signs were beyond doubt. The projection obtained indicated roughly the positions of the oxygens and waters, but the peaks were badly shaped. However the approximate structure thus obtained permitted about twenty more planes to be brought into the series and a second synthesis gave the result of Figure 2 (upper portion). Comparison of observed and calculated intensities is given in Table VI [not reproduced, Ed.].

451

The (hk0) projection suggested that six of the waters are around Ni in an approximate octahedron and that the SO_4 group is of the usual shape and size. Making assumptions as to interatomic distances, it should therefore have been possible to deduce the whole structure. However it was not found easy to obtain in this way values of the z's which gave very good agreement of the (h0l) intensities. A double Fourier synthesis on the (010) face was therefore carried out. The F's were again obtained from rotation photographs, and the signs were calculated from the approximate structure. The resulting projection is shown in Figure 2 (lower portion).

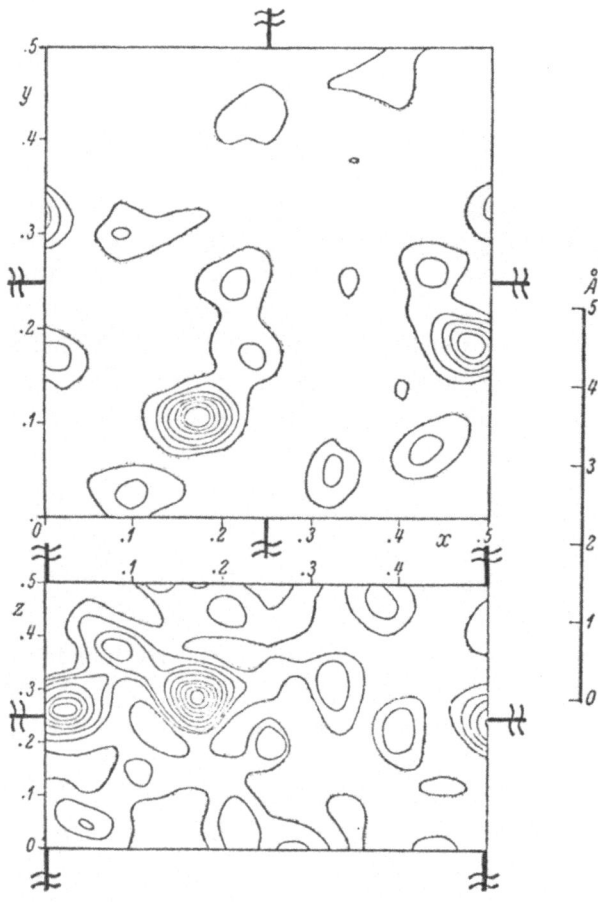

Fig. 2. Double Fourier syntheses projected on to (001) and (010). Contours are drawn at intervals of 50 in $\Sigma\Sigma F \cos 2\pi(hx + ky)$, and at equivalent contours in the (h0l) projection. The zero contour is omitted and the 50 contour is dotted on its lower side to avoid confusion.

[In the Fig. the range of z should read $-.25$ to $+.25$ instead of 0 to .5. Ed.]

The values of the x's obtained from the two projections agreed very well, and the parameters chosen are shown below. The atoms were moved from the positions obtained directly from the projections, to the positions given below to improve inter-atomic distances. The projections are not considered to be accurate to within about .1 Å, on account of errors in the F's and the finite termination of the series.

	x	y	z		x	y	z
Ni	.170	.110	.04	5 (H_2O)	.01	.17	.01
S	.475	.185	.49	6 (H_2O)	.21	.25	.19
1 (O)	.44	.08	.37	7 (H_2O)	.21	.18	$\overline{.21}$
2 (O)	.61	.19	.48	8 (H_2O)	.32	.04	.05
3 (O)	.44	.19	.69	9 (H_2O)	.11	$\overline{.04}$	$\overline{.09}$
4 (O)	.43	.28	.37	10 (H_2O)	.11	.03	.29
				11 (H_2O)	.23	.44	$\overline{.07}$

[§ 4, not reproduced, gives a detailed discussion of the structure. Ed.]

5. Acknowledgements

We have to thank Mr. Lipson, of Liverpool, for his collaboration in the analysis of the rotation photographs, and Prof. W. L. Bragg for his advice and constant interest.

Summary

The x and y parameters of Ni and S in the crystal $NiSO_4.7H_2O$ are derived by means of a) replacement of Ni by Mg, b) a Patterson synthesis. These are used for two-dimensional Fourier syntheses which give all the parameters to a fair approximation.

The structure has SO_4 groups and octahedral groups of six H_2O around Ni, and an extra water molecule. The water molecules are of two types, and the coordination group contains both types exactly as in the lower hydrate $NiSO_4.6H_2O$.

Manchester

CHAPTER XII

The Patterson-synthesis

$$G_h = \Sigma m_s \exp. 2\pi i (h.r_s) \qquad (16'')$$

$$|G_h|^2 = \Sigma_{ss'}\, m_s m_{s'} \cos 2\pi(h.(r_s - r_{s'})) \qquad (31)$$

Wir sehen daß die Beträge $|G_h|$ nur die Differenzen *der Basiskoördinaten, $r_s - r_{s'}$, enthalten.*

P. P. EWALD,
Z. Krist. **56** (1921) p. 129

Now there has been developed a procedure for the straightforward interpretation of experimental data which provides much definite information about the structure of the crystal, leading in some cases to the immediate determination of the structure and in others to the limitation to a restricted number of possibilities. This method, developed by Patterson after preliminary work by Warren, involves the calculation of Fourier series showing the magnitudes and orientations of vectors drawn between atoms in the crystal or their projections on crystallographic planes or lines.

L. PAULING,
Current Sci. (1937) p. 21

196. A DIRECT METHOD FOR THE DETERMINATION OF THE COMPONENTS OF INTERATOMIC DISTANCES IN CRYSTALS

by

A. L. Patterson

In a recent publication[1] the author has given a brief account of a new method for the direct determination of the components of interatomic distances in crystals. The method is based on the application of the results of the theory of Fourier series. It is the object of the present paper to give a more detailed account of the underlying theory and to describe the application of the method to specific examples. In addition, two new results are presented which promise to increase the usefulness of the method for the investigation of complex structures.

The paper is divided into two parts. In Part I, which contains the theory of the method, a Fourier Series whose coefficients are the observed values of F^2 (*hkl*) is derived. This series is shown to be a weighted average distribution of electrons about any point in the crystal. It is then shown that the peaks of this distribution occur at points which correspond very closely in direction and magnitude with the interatomic distances in the crystal. A second series is then set up, with coefficients derived from those of the first, which corresponds closely to a weighted average distribution of atomic centres. This gives the interatomic distances with greater resolution than the first. Additional results which are of value in practical applications are also derived. In Part II, several known structures are discussed in terms of the new method.

PART I. THEORETICAL

1. *The Available Data*

After the determination of the unit cell and space group of a crystal, the next step in the analysis of a complex structure is the determination of the intensity of X-ray reflections from a number of the planes of the crystal.

[1] Patterson, A. L., *Physic. Rev.* **46** (1934) 372.

From these intensities, a set of quantities F^2 (*hkl*) can be derived[1] which are a measure of the actual reflecting power of the various planes after correction for polarization and Lorentz factors, absorption, etc. The quantities F (*hkl*) are in general complex numbers[2] which can be calculated for the known atomic *f*-facors and any set of assumed structure factors for the crystal. The usual method of trial and error consists in calculating these F's for various values of the atomic coordinates and comparing their absolute values with observation.

There is, however, another approach. It has been shown[3] that the electron density in a crystal $\rho(xyz)$ can be expressed as a three dimensional Fourier series with coefficients F (*hkl*) in the form

$$\rho(xyz) = \sum_{hkl=-\infty}^{\infty} \sum \sum F(hkl) \exp 2\pi i(hx/a + ky/b + lz/c). \tag{1}$$

We see that if we knew the values of the quantities F (*hkl*), crystal analysis would merely be a matter of substituting in the formula (1). But we only know F^2 (*hkl*) and in general we can only determine the appropriate phases by the trial and error process mentioned above.

When the Fourier series point of view was first developed it seemed a hopeful method of attack, but so far its chief contribution to X-ray crystallography has been to provide an elegant method of depicting the results of analyses carried out by other methods. Several attempts have been made to develop workable methods from this point of view[4] but the results have in general been disappointing.

The problem of the determination of the unknown phases by direct experiment is hopeless at the present state of our knowledge. It remains then to make as efficient use of the available data as possible. The available data consist (i) in the knowledge that the crystal is built up of a known number of atoms whose individual atomic numbers and *f*-factors are known at least approximately; and that these atoms are arranged in a known space group; and (ii) in a knowledge of the relative or absolute values of as many of the quantities $F^2(hkl)$ as we may require.

The properties of a three dimensional series such as (1) depend entirely on the properties of one dimensional series. Since one dimensional series

[1] See for example Bragg, W. L., and West, J., *Z. Kristallogr.* **69** (1928) 118.
[2] We shall write $F^2(hkl)$ for the squared absolute value of $F(hkl)$ whether it be real or complex. In the latter case, its numerical value will of course be $|F(hkl)|^2$.
[3] See Bragg, W. L., *Proc. Roy. Soc. London* A **123** (1929) 537, for details and earlier references.
[4] See Bragg, W. L., *loc. cit.* also West, J., *Z. Kristallogr.* **74** (1930) 306; and Patterson, A. L., *Z. Kristallogr.* **76** (1930) 177, 187.

are much easier to handle we shall confine our discussion to them for the moment, to obtain several important results which can then be extended immediately to multiple series.

2. *Derivation of the 'F²-series'*

Let us consider the one dimensional Fourier series[1]

$$\rho(x) = \sum_{h=-\infty}^{\infty} F(h) \exp 2\pi i h x/a. \tag{2}$$

We shall assume that $\rho(x)$ is real and that therefore $F(\bar{h}) = \tilde{F}(h)$, where $\tilde{F}(h)$ is the conjugate complex of $F(h)$. Let any such distribution be represented by the curve of Fig. 1.

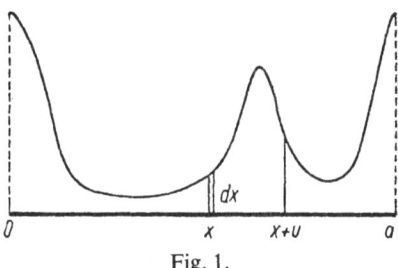

Fig. 1.

The density distribution about any point x is given as a function of a parameter u by the expression $\rho(x+u)$. It now seems reasonable to allot a weight to the distribution about x of amount $\rho(x)dx$, the total amount of material in the interval between x and $x + dx$. The weighted distribution about the point x is then of the form

$$\rho(x)\rho(x + u)dx. \tag{3}$$

Let us now average quantities of the type (3) for all values of x within the period. We then obtain

$$A(u) = 1/a . \int_0^a \rho(x)\rho(x + u)dx \tag{4}$$

where $A(u)$ can be described as the weighted average distribution of density about any point. If now we substitute the series (2) in (4) and make use of the orthogonal properties of the complex exponential

$$1/a . \int_0^a \exp 2\pi i(n - m)x/a . dx = \begin{cases} 0; n \neq m \\ 1; n = m \end{cases} n, m, \text{ integers}; \tag{5}$$

---

[1] Such a series might represent the average density of electrons in planes parallel to (100) for an orthorhombic crystal, in which case of course we would have $F(h) = F(h00)$.

we immediately obtain the result

$$A(u) = \sum_{h=-\infty}^{\infty} F^2(h) \exp 2\pi i h u / a. \tag{6}$$

We have called series of the type (6) simply 'F^2-series'[1]. Since the F^2-series involve nothing but observed data it will obviously be of great importance to crystallography if we can give them a physical meaning. This can readily be done by making use of the integral form (4). If we consider the quantity (3) we notice that it can only be large when both $\rho(x)$ and $\rho(x + u)$ are large; if either of these functions is small, the contribution to the integral (4) is small. We can therefore expect large values of $A(u)$ if there are peaks in $\rho(x)$ at distance u apart*. If the density $\rho(x)$ is an electron density in a crystal, it will have a number of peaks, each corresponding to an atom. These peaks will then produce peaks in the F^2-series which correspond in position to the interatomic distances in the distribution $\rho(x)$.

Let us illustrate this point by a simple example (Fig. 2a). The density function $\rho(x)$ has a peak (atom) of area 3 at the origin, and two peaks each of area 2 units at the points $3d/8$ and $5d/8$. The whole configuration is repeated with period d. The peaks in this simple example are all error curves with the same width at half value ($d/12$). The coefficients of the F-series are in this case computed from the known shape and position of the peaks. These F values are squared, and the F^2-series is synthesized

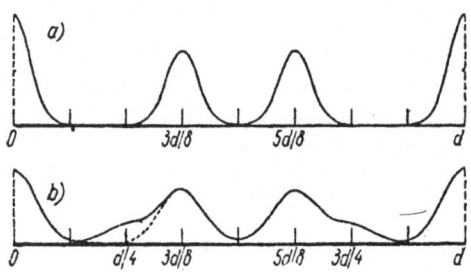

Fig. 2. Example of F^2 series: (a) F-series, (b) F^2-series.

---

[1] The equivalent expressions (4) and (6) are known to the theory of Fourier series as the 'Faltung' of $\rho(x)$. In discussing crystal problems, it has seemed convenient to use the term 'F^2-series' for series of the type (6) and 'F-series' for the type represented by (1) and (2).

* [The following note is taken from Patterson's paper cited sub[1] this Vol. p. 457: 'The result obtained here is an extension of the application to crystals of the theory of scattering of X-rays in liquids reported by Gingrich and Warren at the Washington meeting of the American Physical Society (B. E. Warren and N. S. Gingrich, *Phys. Rev.* **46** (1934) 368) and arose in a discussion of that work.' Ed.]

(Fig. 2b). We see that the peaks due to the distances between the atoms 3 and 2 appear clearly at the points $3d/8$ and $5d/8$, while on the sides of these peaks, the minor peaks due to interatomic distance between the two atoms of area 2 are clearly visible. Their exact position can only be found by making use of our knowledge that the peaks must be symmetrical. When this is done as indicated in the figure the distance is found to be $d/4$ as we should expect.

Several properties of the F^2-series are illustrated by this simple example. The peaks of the F^2-series are flatter than those of the F-series. This is a general property of all series which are built up from separate peaks. In the present case the half value width is $\sqrt{2}d/12$ for the F^2-series. We notice too that some of the peaks which correspond to interatomic distances may be distorted or they may be completely hidden by other peaks. These two properties suggest possible improvements to the method. It may be possible to sharpen up the peaks of the F^2-series, and it may be possible to remove some of the peaks which do not give us any new information. These two possibilities will be realized in the following sections.

3. Some Properties of Atomic Series

We shall give the name 'atomic series' to Fourier series which are built up of a number of functions, each of which consists of only one peak within the unit cell. The series of Fig. 2a is obviously an atomic series, and of course all series, which we shall meet in the Fourier analysis of crystal structures will also be atomic series. We shall now investigate some of the properties of such series. We shall define the symbol $[\rho_1 : \rho_2]$ by the relation

$$[\rho_1 : \rho_2] = 1/a . \int_0^a \rho_1(x)\rho_2(x + u)dx. \tag{7}$$

We see that the F^2-series can then be written in the form $[\rho : \rho]$. Let us assume that the functions $\rho_1(x)$ and $\rho_2(x)$ are two of the atomic functions which form part of an atomic series. The series $\rho_1(x)$ has one peak at x_1 whose f-factor[1] is f_1, and similarly $\rho_2(x)$ has one peak at x_2 whose f-factor is f_2. These series can then be written in the form

$$\rho_1(x) = \sum_{-\infty}^{\infty} f_1 \exp 2\pi i (h/a)(x - x_1) \tag{8a}$$

$$\rho_2(x) = \sum_{-\infty}^{\infty} f_2 \exp 2\pi i (h/a)(x - x_2). \tag{8b}$$

[1] These f-factors correspond directly to the atomic f-factors of X-ray diffraction theory. The atoms of the present series are assumed to be at rest. The f-factors are of course functions of h, being the coefficients of a distribution of period d having an atom at the origin.

If we now substitute the two series (8) in (7) and make use of (5) we get

$$[\rho_1 : \rho_2] = \sum_{-\infty}^{\infty} f_1 f_2 \exp 2\pi i(h/a)[u - (x_2 - x_1)]. \tag{9}$$

This series represents a single peak at the point $u = (x_2 - x_1)$, whose f-factor is $f_1 f_2$.

We are now in a position to derive rigorously the result of the previous section for atomic series.

Let $\rho(x)$ be built up from a number N of atomic functions $\rho_r(x)$ each of which consists of an atom (f_r) at the point x_r. We can then write

$$\rho(x) = \sum_{r=1}^{N} \rho_r(x) = \sum_{r=1}^{N} \left\{ \sum_{h=-\infty}^{\infty} f_r \exp 2\pi i(h/a)(x - x_r) \right\} \tag{10}$$

and

$$F(\bar{h}) = \sum_{r=1}^{N} f_r \exp 2\pi i(hx_r/a). \tag{11}$$

We now form the F^2-series by means of the integral (4), making use of the notation (7). We obtain the relation

$$[\rho : \rho] = \sum_{r=1}^{N} [\rho_r : \rho_r] + \sum_{\substack{r,s=1}}^{N} {}' [\rho_r : \rho_s] \tag{12}$$

where the prime indicates that the values $r = s$ are to be excluded from the double summation. We see that the F^2-series for an atomic function having N atoms is also an atomic series having N^2 atoms. These N^2 atoms can be divided into two types; there are N of the type $[\rho_r : \rho_r]$ and $N(N-1)$ of the type $[\rho_r : \rho_s]$. For the first group we have

$$[\rho_r : \rho_r] = \sum_{-\infty}^{\infty} f_r^2 \exp 2\pi i(hu/a) \tag{13}$$

an atom function having a single peak at the origin. For the second we have

$$[\rho_r : \rho_s] = \sum_{-\infty}^{\infty} f_r f_s \exp 2\pi i(h/a)[u - (x_s - x_r)] \tag{14}$$

an atom function having a single peak at the point $u = x_s - x_r$. In the first case the f-factor is f_r^2 and in the second, $f_r f_s$.

We now see in detail how the result of section 2 is obtained for the F^2-series of an atomic series. A large peak at the origin is produced by the terms of type (13), while the terms of type (14) each give rise to one peak at a value of u given by the distance between the two atomic peaks in the F-series.

This process can now be followed in the example of Fig. 2 without further explanation.

We notice that the first term of the expression (12) which consists of terms of the type (13) gives us no information, merely expressing the obvious fact that all atoms are at zero distance from themselves. We could therefore remove the peak at the origin and still retain all the information as to interatomic distances. To remove the peak at the origin, we use a new set of F^2 values which we shall call F_0^2. These are given by

$$F_0^2 = F^2 - \sum_1^N f_r^2. \tag{15}$$

If we compute the F_0^2-series we shall find that it is identical with the original F^2-series except for the peak at the origin. Actually in the example of Fig. 2 we gain nothing by removing the peak at the origin as the F^2-series has no structure in that neighbourhood. There are cases, however, in which this procedure may be useful, as will be shown below.

4. *Introduction of a Temperature Effect*

In the discussion of the previous section the atoms have been considered at rest. In the present section we shall introduce a temperature motion of the atoms such as is always present in real crystals. We shall find that the presence of this temperature effect makes possible a transformation which is of considerable value in the interpretation of the F^2-series.

According to the theory developed by Debye, Waller, and others, the effect of temperature motion on X-ray diffraction can be allowed for in the one dimensional case by multiplying the atomic f-factor by a quantity $\exp(-\frac{1}{2}\gamma h^2 T)$; where T is the absolute temperature; and γ is a constant which can be calculated for a few very simple crystals, but which is in general unknown. We shall define the quantity t_r, the temperature factor for the rth atom by the expression

$$t_r = \exp(-\tfrac{1}{2}\gamma_r h^2 T). \tag{16}$$

We can now write

$$\rho(x) = \sum_{-\infty}^{\infty} F(h) \exp 2\pi i(hx/a) \tag{17}$$

where

$$F(\bar{h}) = \sum_1^N t_r f_r \exp 2\pi i(hx_r/a) \tag{17a}$$

$\rho(x)$ now represents the time average electron density for a crystal having the given set of F values.

Let us now consider an entirely new distribution $\rho'(x)$ given by

$$\rho'(x) = \sum_{-\infty}^{\infty} T(h) \exp 2\pi i(hx/a) \tag{18}$$

where

$$T(\bar{h}) = \sum_1^N Z_r t_r \exp 2\pi i(hx_r/a) \tag{18a}$$

463

and Z_r is the atomic number of the rth atom. We see that the series (18) is an atomic series, having peaks at x_r, the f-factors for these peaks being $Z_r t_r$. We can see then that the distribution represented by (18) consists of N point charges of charge Z_r electrons each moving about the points x_r with the heat motion originally possessed by the rth atom. As we are now dealing with point charges instead of charges distributed in peaks defined by the factors f_r, the peaks which compose the series (18) will be very much sharper than those of (17). The new series contains, however, all the information about interatomic distances that was contained by (17). Its F^2-series would have much sharper peaks than those of (17) and would be consequently much more suitable for the determination of interatomic distances. We shall now see if it is possible to set up series of the type (18) in a practical case.

We notice that (18a) is derived from (17a) by dividing each term of the sum by the appropriate quantity f_r/Z_r; this might be termed the f-factor for the average electron of the rth atom. If all the atoms in the crystal are of the same kind, the derivation of the series (18) is simple. We write immediately

$$T(h) = ZF(h)/f; \tag{19}$$

and obviously, if we know the F^2 values, we can immediately derive the T^2-series and use it to obtain interatomic distances. Unfortunately the matter is not quite so simple when the crystal contains atoms of several different kinds. There is, however, an approximation which we can make which is adequate for experimental purposes. If we examine any table of atomic f-factors[1] and compute f/Z for each, we find that it varies in a similar manner for all atoms, though it is by no means the same. The general similarity in type of these functions leads us to make use of a compromise. We set up an average f-factor per electron, i.e.,

$$\bar{f} = \sum_1^N f_r / \sum_1^N Z_r \tag{20}$$

the average now being taken over all the electrons in the crystal.

If we now form the series $F/\bar{f}$ from (17) we find that

$$F(\bar{h})/\bar{f} = \sum_1^N (f_r t_r/\bar{f}) \exp 2\pi i(hx_r/a) \tag{21}$$

and we can see immediately that while their shape may be changed considerably, the positions and areas of the peaks are unaltered.

We can thus make use of the factor $\bar{f}$ to sharpen up our F^2 series. We merely compute the value of $\bar{f}$ for the atoms known to be in the crystal

---

[1] e.g. James, R. W., and Brindley, G. W., Z. Kristallogr. **78** (1931) 470.

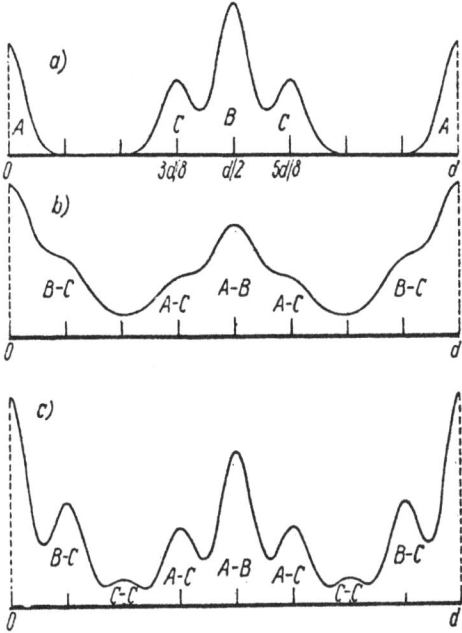

Fig. 3. Sharpening of peaks. (a) F-series, (b) F^2-series, (c) $F^2/\bar{f}^2$ series.

and divide our F^2 values by the appropriate $\bar{f}^2$ and compute the $F^2/\bar{f}^2$ series and use it for the determination of interatomic distances.

The effectiveness of this method in increasing the resolving power is well illustrated by another simple example. The distribution of Fig.3a consists of an atom A of atomic number 3 at the origin, two atoms C of atomic number 2 at the points $3d/8$ and $5d/8$, and an atom B of atomic number 4 at the point $d/2$. For simplicity the atoms are all of the same shape, and the temperature factor for all atoms is the same as $\bar{f}$. The F^2-series is shown in Fig. 3b, and the $F^2/\bar{f}^2$ series in Fig. 3c. In the latter curve, the B–C distance at $d/8$, the C–C distance at $d/4$, the A–C distance at $3d/8$, and the A–B distance at $d/2$ are all clearly shown. The presence of all these peaks with the exception of C–C is indicated by the F^2-series, but the A–B peak is the only one whose position could be determined with any accuracy.

5. Removal of Peaks due to Atoms in Special Positions

In many crystals, certain of the atoms are fixed by the space group symmetry. The distances between such fixed atoms are therefore known. In the F^2-series the peaks due to these distances may possibly be hiding peaks

465

due to unknown structure, and it would therefore be of interest to obtain a method for the removal of such peaks. We now consider a distribution $\rho(x)$ of the type (17) which involves a temperature motion, and form its F^2-series as in (12). The peaks of the first type can now be written

$$[\rho_r : \rho_r] = \sum_{-\infty}^{\infty} f_r^2 t_r^2 \exp 2\pi i (hu/a) \tag{22}$$

while those of the second type are of the form

$$[\rho_r : \rho_s] = \sum_{-\infty}^{\infty} f_r f_s \, t_r t_s \exp 2\pi i (h/a)[u - (x_s - x_r)]. \tag{23}$$

Suppose now that the two atoms 1 and 2 are in the special positions, x_1 and x_2. We can then immediately calculate the contribution of these two atoms to the F^2-series and remove it at the same time that we remove the peak due to the origin. Thus the contribution of the peaks 1 and 2 above is

$$[\rho_1 : \rho_2] + [\rho_2 : \rho_1] = 2\sum_{-\infty}^{\infty} f_1 f_2 t_1 t_2 \cos 2\pi i (h/a)(x_1 - x_2) \exp 2\pi i (hu/a). \tag{24}$$

Thus if we remove the peak at the origin, and the contribution of the fixed atoms, we can set up a new series whose coefficients F_1^2 are given by

$$F_1^2 = F^2 - \sum_1^N f_r^2 t_r^2 - 2f_1 f_2 t_1 t_2 \cos 2\pi (h/a)(x_1 - x_2). \tag{25}$$

It will be noted that those portions of the F^2-series are removed which are due only to the atoms in the fixed positions. The peaks which are produced by the distances between the fixed and the unknown atoms are of course allowed to remain.

This process is well illustrated on the example of Fig. 3. Fig. 4a shows the F^2-series of Fig. 3b with the peak at the origin removed, while in Fig. 4b the atoms A and B are assumed to be fixed by symmetry at 0 and $d/2$, respectively, and their contribution is removed. The removal of the peak at the origin brings up the B–C peak at $d/8$, and the removal of the A–B peak brings up the A–C peak at $3d/8$. No amount of manipulation of this type can bring up the C–C peak, as it is concealed by the two unknown

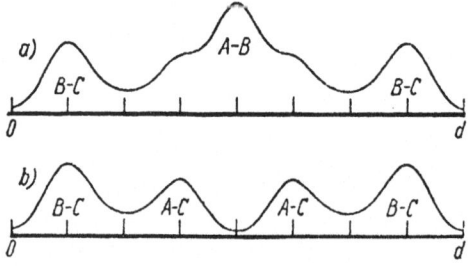

Fig. 4. Removal of peaks (a) Series of Fig. 2 with peak at origin removed. (b) Same with known peak A-B removed.

peaks at $d/8$ and $3d/8$. Of course the knowledge of the A–B, A–C, and B–C distances would be quite sufficient in the present simple case to determine the original distribution completely.

We note particularly that this process of removal of portions of the F^2-curve requires a knowledge of the temperature factor. Fortunately there are cases when this factor can be estimated with sufficient accuracy to make the method practical. Our simple example indicates, however, that the method described in section 3 is much more powerful, and so far this has proved to be the case in practice.

6. *Extension to Two and Three Dimensions*

We can make the statement without reservation that every result which we have derived above has its direct analogue in two or three dimensions. Many of our results hold without change when we have given a more general definition to the symbols involved. In the present section we shall discuss only those extensions which we shall use most, and leave the details of the rest to the reader.

We consider first a periodic distribution $\rho(xyz)$ such as that given by (1). The distribution about the point (xyz) is given as a function of three variables (uvw) by the function $\rho(x + u, y + v, z + w)$. We weight this by the total amount of material in the element of volume $dxdydz$, ie. $\rho(xyz)dxdydz$; and average this weighted distribution over all points in the elementary cell. We get

$$A(uvw) = 1/V . \int\int\int_0^{abc} \rho(xyz)\rho(x + u, y + v, z + w)dxdydz, \qquad (26)$$

where V is the volume of the unit cell.

We see that the function $A(uvw)$ will show peaks only when $\rho(xyz)$ has peaks at both (xyz) and $(x + u, y + v, z + w)$. We then have the fundamental result that a peak at (uvw) for the function $A(uvw)$ will correspond to an interatomic distance in the crystal given in direction and magnitude by the vector whose components are (uvw).

If we substitute the series (1) in the integral (26) and use the condition (5) we immediately derive the F^2-series for three dimensions:

$$A(uvw) = \sum\sum\sum_{hkl = -\infty}^{\infty} F^2(hkl) \exp 2\pi i(hu/a + kv/b + lw/c). \qquad (27)$$

As in the case of one dimension the discussion of interatomic distances becomes exact for atomic series. These of course have an obvious significance in three dimensions. We define our atomic series as before by the relation

$$\rho(xyz) = \sum_1^N \rho_r(xyz) \tag{28}$$

where

$$\rho_r(xyz) = \sum_{hkl=-\infty}^{\infty}\sum\sum f_r t_r \exp 2\pi i \cdot$$

$$\cdot \left[\frac{h(x - x_r)}{a} + \frac{k(y - y_r)}{b} + \frac{l(z - z_r)}{c} \right]. \tag{29}$$

The distribution thus consists of N atomic distributions, each with a maximum density at the point $x_r y_r z_r$, with f-factor f_r and temperature factor t_r.[1]

We now define the symbol $[\rho_r : \rho_s]$ in three dimensions by the integral

$$[\rho_r : \rho_s] = 1/V \cdot \int\int\int_0^{abc} \rho_r(xyz)\rho_s(x + u, y + v, z + w)dxdydz \tag{30}$$

and it follows immediately that the $F^2(hkl)$ series is given by an expression identical with (12).

$$A(uvw) = [\rho : \rho] = \sum_1^N [\rho_r : \rho_r] + \sum_{rs=1}^N \sum' [\rho_r : \rho_s]. \tag{12'}$$

The contribution of the terms $[\rho_r : \rho_r]$ is again a large peak at the origin with f-factor equal to $\sum_1^N f_r^2$, while the term $[\rho_r : \rho_s]$ contributes a peak at the point $(uvw) = (x_s - x_r, y_s - y_r, z_s - z_r)$ whose size and shape is defined by the factor $f_r f_s t_r t_s$.

We now have all the results necessary to carry out an analysis of the three dimensional problem exactly as we did for one dimension. We could compute the F^2-series and use it for determining interatomic distances, we could remove the peak at the origin, or peaks at special positions by using the results of (15) or (25) or, we could compute $F^2/\bar{f}^2$ and use it to obtain interatomic distances. However, the computation of three dimensional

---

[1] The functions f_r and t_r are complicated functions of the indices (hkl) and of the lattice constants. In practice they are tabulated as functions of $\sin \vartheta/\lambda$, where ϑ is the Bragg angle for the plane in question when a wavelength λ is used. By the Bragg equation, $\sin \vartheta/\lambda = 1/2d$. The spacing d can be calculated from the indices and the lattice constants and thus the appropriate f for the plane can be determined. This property of the f's which is familiar to those who work in the field will not be expressed in the formulae in the text. Whenever an f-value or a t-value or any quantity derived from these is mentioned in connection with a Fourier coefficient of indices (hkl), the appropriate value derived as sketched above is the value to be used. This process is of course best carried out graphically by reciprocal lattice methods. See for example Bernal, J. D., *Proc. Roy. Soc. London* B **113** (1926 117.

series is not at all easy, and we very seldom have the necessary values of $F^2(hkl)$ for the large number of general planes which would be required. We are therefore forced to confine ourselves to two dimensional series. Luckily it is possible to obtain all the information which we require from such series.

The particular two dimensional series which interests us most is the projection of the density $\rho(xyz)$ on any one of the axial planes; for this projected density we write e.g. for the ab plane

$$\rho(xy) = \sum_{hk=-\infty}^{\infty} F(hk0) \exp 2\pi i(hx/a + ky/b). \tag{31}$$

If we now define the square bracket symbol in two dimensions by

$$[\rho_r : \rho_s] = 1/S \int\int_0^{ab} \rho_r(xy)\rho_s(x + u, y + v)dxdy \tag{32}$$

where S is the area of the elementary parallelogram, we can take over into two dimensions all the results which we have obtained for one and three dimensions. We shall not write any of these here. We shall require many of them in the practical examples of the second part and we shall refer to them there as they are needed.

7. *Plane Groups*

As most of the examples which we wish to present in Part II are two-dimensional series, we must make a brief digression here to refer to the theory of plane groups[1]. There are seventeen plane groups, and any plane projection of a distribution which belongs to one of the 230 space groups can be allotted to one of these plane groups.

For the interpretation of the F^2-series corresponding to these projections, it will be necessary to have available detailed information as to the relations which exist between the equivalent points of the plane group and the distances between them which locate the peaks of the F^2-series. As this information and the details of its derivation are given for all seventeen plane groups in the following paper[2] we shall confine ourselves here to the discussion of one simple example.

Let us take the case of a crystal belonging to the space group C_i^1. There is one molecule per cell, and that molecule contains one atom of number

[1] Polya, G., *Z. Kristallogr.* **60** (1924) 278 and also Niggli, P., *idem.* p. 283.
[2] Patterson, A. L., *Z. Kristallogr.* **90** (1935) 543.

Z_0 and two each of numbers Z_1 and Z_2. We suppose that the single atom occupies the special position (000) and that the pairs of atoms are in pairs of general positions. Let us now consider the projection of this structure on the axial plane ab. This projection will belong to the plane group C_2^1 and will have one atom Z_0 at (00), one atom Z_1 at each of $(x_1, y_1)(\bar{x}_1, \bar{y}_1)$ and one Z_2 at each of $(x_2, y_2)(\bar{x}_2, \bar{y}_2)$. We can then classify the peaks of the F^2-series according to their relative volumes as follows:

Table I. Interatomic Distances.

Vol.	ΣZ_r^2	Z_1^2	Z_2^2	$2Z_0Z_1$	$2Z_0Z_2$	$2Z_1Z_2$
Peak	00	$2x_1 2y_1$	$2x_2 2y_2$	$x_1 y_1$	$x_2 y_2$	$x_1-x_2, y_1-y_2;$ $x_1+x_2, y_1+y_2.$
		$-2x_1 2y_1$	$-2x_2 2y_2$	$-x_1 y_1$	$-x_2 y_2$	$-x_1+x_2, -y_1+y_2;$ $-x_1-x_2, -y_1-y_2.$

We notice two properties of these results which are useful in identifying peaks in practical cases. The relative areas are of great importance, and in addition we see that the coordinates of the peaks in the F^2-series are not independent of one another. This is particularly well illustrated by the discussion of copper sulphate which belongs to this group.

PART II. APPLICATIONS

We shall now apply the method which we have developed to the discussion of a few known structures. There are several difficulties which arise in practice which have not as yet been given adequate theoretical discussion. It has therefore been thought best to give the method a very thorough trial on these known structures to see how closely the results obtained by the present method agree with those obtained by trial and error. We can thus obtain an estimate of the accuracy which is to be expected from the present method.

In the next section we shall mention briefly the difficulties which arise in practical application, and the possible errors which they may introduce. We shall then discuss in some detail the application of the present method to the study of the structures of potassium dihydrogen phosphate, hexachlorobenzene, and copper sulphate pentahydrate[1] [only last example

[1] Preliminary results of the application of the new method to two first mentioned substances have already been published (l.c.). Here the method which has been used to derive these results will be emphasized. References to the original structure determinations

inserted. Ed.]. It is found that the present method gives results which are in very close agreement with those obtained by other methods. The new method, however, gives these results in a direct and simple manner.

8. *Possible Sources of Error*

The accuracy of any method of structure determination by means of X-rays, depends entirely on the accuracy with which the quantities $F^2(hkl)$ can be measured. That this accuracy is sufficient for the application of the ordinary methods of trial and error seems well proven by the successful determination of many complicated structures. Intensities measured by the methods which we have available also seem to give satisfactory electron densities when used in the Fourier synthesis of crystal structures. If the measured values give a reasonable picture of the F-series for a crystal there is no reason to suppose that they would be any less satisfactory for the F^2-series. The examples given below support this view.

It seems then that we are justified in assuming that the measured intensities give an adequate picture of the F^2-series for a crystal. Whether or not we can interpret this F^2-series and obtain from it accurate values of interatomic distances is another question. In the earlier sections of the paper, we have seen how the peaks of the F^2-series can be distorted, and can in some cases disappear completely from view, owing to the presence of large neighboring peaks. This difficulty is inherent if we confine ourselves solely to the use of the F^2-series; but we shall see below that in practice there is no difficulty in locating the principal peaks of the F^2-series with considerable accuracy; although many minor peaks may be obscured.

We have suggested (Part I) two methods for avoiding the difficulty discussed in the previous paragraph. Each of them introduces new difficulties peculiar to itself. We have already mentioned the difficulty met with in removing the peak at the origin and the peaks due to atoms in special positions. To do this we require a knowledge of the temperature factor, which is generally unknown. We are therefore forced to make use of estimates of this factor. In some cases this is quite satisfactory, but in general we shall not be able to put a great deal of confidence in structure revealed by the removal of peaks. This structure may be real, or it may be due entirely to

will be given below. It might be noted here that no new data have been obtained for any of these crystals, and that the results which we obtain are in close agreement with the published structures. This is not surprising as the substances were chosen as well established structures which would illustrate the various aspects of the new method.

errors made in estimating the temperature factor. Of course such suggestions of structure may be very valuable, as they can always be tested by direct computation of the structure factors.

In applying the method of Section 4, a very important difficulty must be recognized. In using this method we compute the series $F^2/\bar{f}^2$. The quantities $\bar{f}^2$ decrease rapidly for large values of the indices (hkl). The quantities $F^2(hkl)$ also decrease in a similar manner. Experimentally, however, it is impossible to follow their decrease beyond a fixed limit, at which point the peaks become indistinguishable from the background of general scattering. When we divide the measured F^2-values by the rapidly decreasing $\bar{f}^2$-values, it is obvious that the importance of the small intensities from planes with high indices is increased. It is obvious, too, that there may be F^2-values which were too small to be measured, which could become of considerable size and importance when divided by their appropriate $\bar{f}^2$. We can express this argument in another way by stating simply that if we had a crystal built of point atoms we would obtain many more high index reflections than we do from a similar one built of the diffuse atoms which Nature provides. Thus in the case of the F^2-series, we ignored only those coefficients which were too small to measure, a procedure which is quite logical; while in the case of the $F^2/\bar{f}^2$-series, we are forced quite arbitrarily to cut off at a point which has no relation to the properties of the series itself. Fortunately, however, theory indicates that the positions of the peaks are only slightly sensitive to the ignoring of coefficients above an arbitrary set of indices. The size and shape of the peaks are more sensitive, but we are interested in them in a qualitative manner only.

It seems that the possible sources of error which we have discussed qualitatively in this section are all amenable to accurate theoretical treatment. This is, however, far beyond the scope of the present paper, and while we hope to be able to provide this treatment in a later paper, we must be content for the present to justify our confidence in the method, solely on the accuracy of the results which it gives for known structures. We feel that this is sufficient to warrant the application of the method to the investigation of new structures provided that the limitations are well recognized; and provided that the final criterion for a correct structure is always taken to be the comparison of the observed and calculated F-values.

We have assumed throughout this paper that the observed F^2-values are corrected for secondary extinction. Attempts have been made to discuss theoretically the error which would be introduced by neglecting this effect, but formal difficulties have so far prevented their completion. For the present we must therefore assume that extinction has been corrected for, or that it has been shown to be sufficiently small to be neglected in the ordinary trial and error method. In this latter case it is probable that it is also small enough to be neglected in our present discussion.

9. *Note on the Synthesis of Two Dimensional Fourier Series*

In most of the examples which follow, use is made of two dimensional Fourier series. It seems, therefore, worth while here to discuss briefly the methods which are used in their computation. We have used throughout the method of Beevers and Lipson[1]. The first series were computed by a method which is identical with theirs. For the later series their method was modified for use on a computing machine provided with facilities for automatic multiplication.

Beevers and Lipson in the scheme which they publish compute 50×50 points. We have found it more convenient to use 48×48 points within the unit cell, as this introduces the cosine values 1, 0.5, and 0 with considerable frequency, a fact which is very convenient for computation.

12. *Copper Sulphate Pentahydrate*

The structure of copper sulphate has recently been investigated by Beevers and Lipson[2]. It crystallizes in the space group C_i with two molecules CuSO$_4$, 5H$_2$O in the unit cell. Complete intensity data were available for the $(hk0)$ zone.

The series $F^2(hk0)$ was computed[3] as shown in Fig. 8 .The intense peaks at $\frac{1}{2}\frac{1}{2}$ show that some of the atoms occupy a face centered lattice. Obviously such peaks could not be produced by the oxygen atoms. If the Cu atoms were centered by the S atoms we would expect to find larger peaks elsewhere in the plot which would correspond to Cu–Cu distances. As this is not the case it is necessary to assume that the Cu atoms occupy the special positions of the plane lattice at 00 and $\frac{1}{2}\frac{1}{2}$. We would then expect the Cu–S distances to be given by the peaks next in size. The S parameters[4]

---

[1] Beevers, C. A., and Lipson, H., *Philos. Mag.* **17** (1934) 855. I am indebted to Prof. W. L. Bragg for the details of this method before its publication.

[2] Beevers, C. A., and Lipson, H., *Proc. Roy. Soc. London* A **146** (1934) 570. I am very much indebted to these authors for a manuscript of their paper before publication. I am also indebted to Prof. W. L. Bragg, who told me of their work, and suggested this substance as an example for trial by the new method.

[3] I wish to thank Miss. E. L. Knight for her assistance in this computation.

[4] Expressed as fractions of the cell translation.

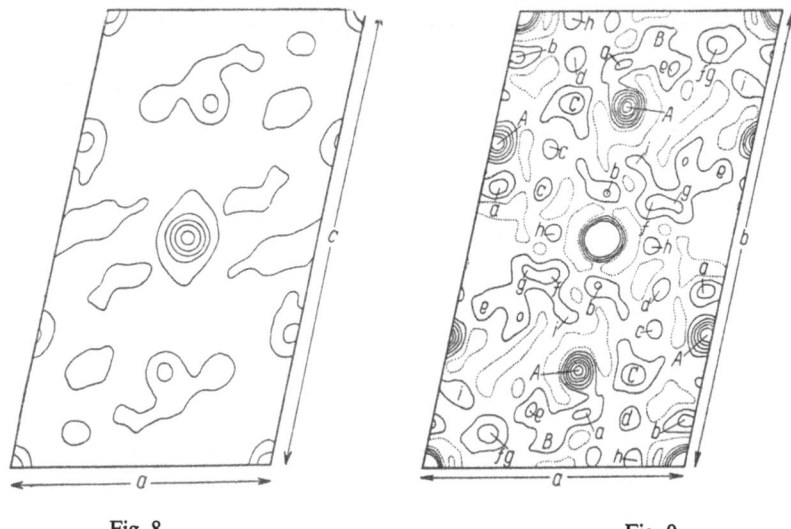

Fig. 8.   Fig. 9.

Fig. 8. $F^2(hk0)$ for CuSO$_4$, 5H$_2$O.

Fig. 9. $F^2(hk0)/\bar{f}^2$ for CuSO$_4$, 5H$_2$O: Letters correspond to Tables V and VI. Dotted contours are at height zero, and enclose negative areas. Contours at intervals of 2 units.

are thus determined immediately to be $X = 0.02$, $Y = 0.29$. From the remaining peaks of Fig. 8 it was possible to locate several of the oxygen atoms by the method which is used below. Those numbered 2, 3, 7, and 9 could be located approximately, but in general there was considerable doubt as to the exact positions of these points, and no conclusions could be drawn as to the positions of the other O atoms.

It was then decided to compute the series $F^2(hk0)/\bar{f}^2$. The result of this computation is plotted in Fig. 9. By comparison with Fig. 8, we see that the principal peaks are unchanged in position. There are, however, many new peaks which must be identified. For the purpose of our argument, we can treat all oxygen atoms including the water of crystallization as identical. We can then write down the following atomic positions:

$$\text{Cu: } 00 \; \tfrac{1}{2}\tfrac{1}{2}$$

$$\text{S: } X Y \bar{X} \bar{Y}$$

$$O_n: x_n y_n \; \bar{x}_n \bar{y}_n; \; n = 1, 2, \ldots 9.$$

We have already determined the Cu and S positions, so that the next most intense peaks are provided by the Cu–O distances. We note that all such positions occur in pairs, i.e., $x_n y_n$ and $\tfrac{1}{2} - x_n$, $\tfrac{1}{2} - y_n$ together with their equivalents which will take care of themselves.

Table V. Coordinates of Peaks. CuSO$_4$. 5H$_2$O; $F^2(h0l)$.

(Cell translation = 48 units.)

Height	u	v	$\tfrac{1}{2}-u$	$\tfrac{1}{2}-v$		Height	u	v	$\tfrac{1}{2}-u$	$\tfrac{1}{2}-v$	
27	0	0				3	14.5	23.4	9.5	0.6	h
17	24.0	24.0				3	12.0	18.9	12.0	5.1	d
12	0.8	13.8	23.2	10.2	A^1	3	13.2	9.5	10.8	14.5	c
11	23.0	10.2	1.0	13.8	A	3	12.9	5.3	11.1	18.7	d
5	10.9	14.3	13.1	9.7	c	3	19.1	$\overline{6.9}$	4.9	$\overline{17.1}$	e
5	21.0	5.5	3.0	18.5	a	3	20.0	$\overline{16.4}$	4.0	$\overline{7.6}$	i
5	13.2	$\overline{20.4}$	10.8	$\overline{3.6}$	fg	3	1.8	$\overline{21.1}$	22.2	$\overline{2.9}$	B
4	22.6	19.2	1.4	4.8	b	2	6.8	23.9	17.2	0.1	
4	3.0	18.4	21.0	5.6	a	2	8.0	11.3	16.0	12.7	
4	0.9	4.8	23.1	19.2	b	2	6.3	7.4	17.7	16.6	
4	8.0	$\overline{3.7}$	16.0	$\overline{20.3}$	f	2	24.0	5.7	0	18.3	
4	12.4	$\overline{3.4}$	11.6	$\overline{20.6}$	g	2	9.2	0.7	14.8	23.3	h
4	12.2	$\overline{8.1}$	11.8	$\overline{15.9}$		2	3.5	$\overline{7.9}$	20.5	$\overline{16.1}$	i
4	6.4	$\overline{17.9}$	17.6	$\overline{6.1}$	e	2	24.0	15.2	0.0	9.8	

We now construct a table (Table V) of the coordinates (uv) of the peaks in the plot, in order of their heights[2]. We then calculate $\tfrac{1}{2}-u$, $\tfrac{1}{2}-v$ for each point and group together the pairs of peaks which have the same coordinate values. This correlation is shown in Table V by the letters a, b, c, etc. We are thus led to a set of pairs of possible oxygen parameters. For each pair there are two possible values of $x_n y_n$, either u, v or $\tfrac{1}{2}-u$, $\tfrac{1}{2}-v$. The choice between these alternatives must obviously be made on packing or bonding considerations, the final criterion being of course intensity calculations or a reasonable Fourier synthesis. We cannot discuss this further without determining the z parameters. As this process can be carried out by normal methods of no interest to the present subject, we shall confine ourselves to the tabulation (Table VI) of the alternative parameter values as compared with those obtained by Beevers and Lipson, who have of course supplied that part of the argument which we have omitted. It is apparent that the numerical agreement is extremely good. We are however led to one parameter value (labelled h) which has no reality as a Cu–O distance. This is undoubtedly a disadvantage to the method, but such a false value can very readily be eliminated, particularly when the

---

[1] The letter A indicates the CuS distance, and B the S–S distance. The small letters correspond to Cu-O distances.

[2] The weight of a peak is of course strictly proportional to its area. The heights are, however, much easier to obtain, and we are justified in using this classification as long as too much importance is not attached to it.

Table VI. Possible Atom Parameters (xy): $CuSO_4, 5H_2O$.

(Expressed as fractions of cell translation.)

		Alternatives				B and L	
		x	y	$\frac{1}{2}-x$	$\frac{1}{2}-y$	x	y
—	Cu	0	0			0	0
—	Cu	0.50	0.50			0.50	0.50
AB	S	0.02	0.29			0.01	0.29
i	O_1	0.92	0.16	0.58	0.34	0.89	0.15
c	O_2	0.21	0.30	0.29	0.20	0.24	0.31
e	O_3	0.87	0.37	0.63	0.13	0.86	0.38
—	O_4	—	–	–	–	0.02	0.30
f	$(H_2O)_5$	0.83	0.08	0.67	0.42	0.83	0.08
d	$(H_2O)_6$	0.26	0.11	0.24	0.39	0.29	0.11
b	$(H_2O)_7$	0.48	0.40	0.02	0.10	0.48	0.41
g	$(H_2O)_8$	0.76	0.43	0.74	0.07	0.76	0.42
a	$(H_2O)_9$	0.44	0.12	0.06	0.38	0.43	0.12
h	—	0.19	0.01	0.31	0.49	–	–

numerical accuracy is so good. Actually, one member of this false pair (h) is due to an S–O distance, while the other is due either to a combination of a large number of O–O distances or is a ghost of the peak at $\frac{1}{2}\frac{1}{2}$ due to the premature cutting off of the series, a difficulty discussed in section 8 which is inherent in the method[1]. We see then that the present method fixes the xy parameters of the copper and sulphur atoms without ambiguity. It provides in addition 9 pairs of possible locations for the 9 oxygen atoms. The choice between these pairs is all that remains for trial and error methods to solve. Such methods, together with the analysis in the z direction, will show that one of the pairs is illusory and that the extra oxygen atom has the same xy coordinates as an S atom, and consequently could not be observed in this projection. The complete argument has not been carried out here, but it is perfectly safe to predict its result in the present case. Beevers and Lipson were able to choose the appropriate parameters when all values were possible. It is obviously much easier to do this when the choice is limited as it is by the methods of this paper.

13. Conclusion

In this paper an account has been given in some detail of a method for the direct determination of the components of interatomic distances in

---

[1] That such effects are present is shown by the large negative areas in the plot. These cannot be real, and must be due to the premature cutting off of the series.

crystals; and the application of the method has been illustrated on several examples. It has been shown that the numerical values of the distances obtained are in very good agreement with those obtained by ordinary methods. The only problem which remains for trial and error methods to solve arises from the ambiguity which is caused by the fact that we cannot always decide to which pair of atoms any one interatomic distance should be allotted. In addition, there is the possibility in some cases, particularly with the series $F^2/\bar{f}^2$, that extra peaks will appear owing to the incompleteness of the series. Such peaks must also be excluded by trial and error. It must therefore be emphasized that this method makes no claim to eliminate the method of trial and error from crystal analysis, since the final test must always be the calculation of intensities from atoms in positions defined by variable parameters. It does, however, seem that this method enables one very rapidly to narrow down the choice of these parameters to such an extent that it permits the determination of structures which by the usual method would take so long that their investigation would not be possible.

In conclusion it is a pleasure to express my indebtedness to Professor Norbert Wiener, whose wide knowledge of the Fourier series and the Fourier integral has been a source of information and inspiration in the development of this method. I also wish to thank Professor J. C. Slater for the continued hospitality of this laboratory; and Professor B. E. Warren, whose advice and discussion have been invaluable.

Eastman Physical Laboratory, Massachusetts Institute of Technology,
Cambridge, Massachusetts

Index of Names

(Numbers in ordinary print refer to pages, fat numbers to papers—the paper-number is found in the head-line of each left page after the number of the chapter).

Index of Subjects

(Numbers in ordinary print refer to pages, fat numbers to papers—the paper-number is found in the head-line of each left page after the number of the chapter).

Abbe's diffraction theory, 383, **188**

abnormal spacings
see absences

absences
for body or face centering, 9, 17
for glide planes, 10
for screw axis, 17
for space-groups V_h^{25}–V_h^{28}, 9, 10, 14, 20

alloys
Cu–Zn system, **162**
Hume–Rothery compounds, **164–167**
structure of γ-Brass, **163**

ambiguity in intensity/structure relation, **179**

atomic radius
see radius atomic

atomic scattering
see scattering atomic

Bond
covalent, **121**
octahedral, 64
square, 64
tetrahedral, 64
trigonal prism, 64
distances
in AsJ_3, 75
in $Bi_4Si_3O_{12}$, 75
$CuSbS_2$, 74
MnS_2, 73
PdO, 73
see also radius covalent
orbitals, 61ff
strength, 63
hydrogen–, 98
hydroxyl–, **124**
ionic–
see radius ionic
metallic–, **166, 167**

bragg's equation
correspondence principle and–– for sinusoïdal grating, **184**
derivation for reflection of light quanta, **182–184**

Classification of crystal structures, **122**
cleavage, 6

close-packing
as fabric with embroidered pattern, 317
cubic and hexagonal, 315

coordination number
radius ratio and––, 83
distance correction for––, 49, 85

coordination structure hypes
and building principles, **122, 123**
for compounds AX, 79
for compounds AX_2, 79ff.
principle of electrostatic valence, 95
rule of Parsimony, 96, 98

crystal
class, 9, 23
molecule, 6
structure
see structure determination,
random structures
of chondrodite series, 317
of chisthene, 316
of hydrargillite, 100
of ice, **124a**
systems, 4

Diffraction rings
optical, 397
in the X-ray image, 396, 401

dicarboxylic acids
see long chain compounds

disordered structures
see random structures

disordered layer sequence
in chlorite crystals, **155**
in cadmium bromide crystals, **156**
in cristobalite crystals, **157**

Electron-compounds, **164–167**
electron count in anthracene, 422

energy
potential- of ionic crystal, 46, 49, 85

equipoint arrangement
Ag_2HgJ_4, 229
lithium ferrite, 220
magnesium ferrite, **151**

extinction, correction for–, 324

extinctions
see absences

481